AF335588

FLUID SEALING

This volume consists of papers presented at the 14th International Conference on Fluid Sealing, held on 6-8 April 1994, in Firenze, Italy, organised by BHR Group Limited.

Technical Advisory Committee

Mr B D Halligan (Chairman)	James Walker & Co Limited
Mr P Dolan	B P Engineering
Mr M S Dosanjh	British Gas plc
Mr B Dühring	Shamban Europa A/S, Denmark
Mr A Gabelli	SKF Engineering and Research Centre BV, Netherlands
Mr C Gee	Gee Graphite Limited
Mr H F Ibbott	Esso Engineering (Europe) Ltd
Dr B S Nau	BHR Group Limited
Dr N A Peppiatt	Hallite Seals International Ltd
Mr J Plumridge	EG & G Sealol
Prof M T Thew	University of Southampton
Mr N Wallace	Flexibox Limited

Overseas Corresponding Members

Mr D Buchdahl	EDF, France
Prof Y Gu	The University of Petroleum, China
Dr F Hirano	Kyushu University, Japan
Mr D Janßen	Martin Merkel GmbH & Co KG, Germany
Dr D E Johnston	Carl Freudenberg, Germany
Mr C X Latty	Latty International s a, France
Dr A O Lebeck	Mechanical Seal Technology Inc, USA
Prof L Marchand	École Polytechnique Montreal, Canada
Dr R Metcalfe	Atomic Energy of Canada Ltd, Canada
Prof H K Müller	Universität Stuttgart, Germany
M. G Pierron	PMB & Associés, France
Prof A Strozzi	University of Udine, Italy
Prof R F Salant	Georgia Institute of Technology, USA

Organised and sponsored by:

BHR Group Limited, Cranfield, Bedford, MK43 OAJ, UK
Tel: 0234 750422   Fax: 0234 750074

Co-sponsored by the Institution of Mechanical Engineers

14TH INTERNATIONAL CONFERENCE ON
Fluid Sealing

Edited by Mr B Halligan

BHR Group Conference Series
Publication No. 9

Papers presented at the *14th International Conference on Fluid Sealing*, organized and sponsored by BHR Group Limited, and held in Firenze, Italy, on 6–8 April 1994

Mechanical Engineering Publications Limited
LONDON

ISBN   0 85298 920 2

A CIP catalogue record for this book is available from the British Library.

Other titles in the BHR Group Conference Series

1   Offshore Loss Prevention - A Systematic Approach
         Edited by C P A Thompson

2   Fluid Power - The Future for Hydraulics
         Edited by N Way

3   Effective Membrane Processes - New Perspectives
         Edited by R Paterson

4   Multi Phase Production
         Edited by A Wilson

5   Bioreactor and Bioprocess Fluid Dynamics
         Edited by A W Nienow

6   Slurry Handling and Pipeline Transport - Hydrotransport 12
         Edited by C A Shook

7   Pipe Protection
         Edited by A Wilson

8   Advances in Water and Effluent Treatment
         Edited by M J D White

Produced by Technical Communications (Publishing) Ltd.
Printed by Information Press Ltd, Oxford England.

CONTENTS

FOREWORD

Brian Halligan, James Walker & Co Ltd, UK
Chairman, Technical Advisory Committee

STATIC SEALS AND VALVES

The service envelope for biaxially orientated reinforced PTFE sheet sealing materials — 3
J R Hoyes, S Woolfenden, A D Stancliffe, TBA Sealing Materials Ltd, UK

Thermal transient sealing concepts - the "carrier ring" spiral wound gasket — 37
G Briggs, S P Noble, Flexitallic Ltd, UK

Experimental investigation of leakage in static seals subjected to reciprocating vibration — 57
F Rosengen, Chalmers University of Technology, Sweden

Emission control packing for high temperature/high pressure valves and related devices — 71
T Ueda, T Shiomi, K Ootawa, Nippon Pillar Packing Company Ltd, Japan

Stresses and deformation of compressed elastomeric o-ring seals — 83
I Green, C English, Georgia Instutute of Technology, USA

Finite element analysis of bolted flange connections — 97
V Wright, T & N Technology Ltd; J Hoyes, TBA Sealing Materials Ltd; G Briggs, Flexitallic Ltd, UK

Up-date on high temperature leakage behaviour of compressed fibre materials and an alternative approach for new joint design rules — 107
J Latte, C Rossi, ISTAG AG, Switzerland

Elastomeric-seal life prediction — 123
E Ho, B S Nau, BHR Group Limited, UK

Fibre combinations for use in asbestos-free 'it' calendered sealing materials — 135
D A Thomas, J R Hoyes, TBA Sealing Materials Ltd, UK

A review of corrosive processes in sealed joints — 151
A Hirschvogel, SIGRI Great Lakes Carbon GmbH, Germany

ROTARY SEALS

Impact tests of rubber materials for seals — 161
F Hirano, H Miyagawa, K Imado, Oita University; Y Kawahara, NOK Corporation, Japan

Sealing liquid with air — 193
N Stanger, W Haas, H K Müller, Universität Stuttgart, Germany

Finite element-analysis of PTFE-shaft-seal — 205
G Wüstenhagen, H K Müller, K-D Meck, Universität Stuttgart, Germany

MECHANICAL SEALS I - practical

Trends in mechanical seal performance at three process plants in the oil industry   219
P A Connor, Conoco; M T Thew, University of Southampton, UK

Seal check systems check which double seal is leaking   243
J A M ten Houte de Lange, Flexibox BV, Netherlands

Knife edge mechanical seal for low emission of high viscosity fluids   259
Y Goto, K Ohba, Nippon Pillar Packings Co Ltd, Japan

New developments in bellow seals for improved performance and reliability   273
J G Evans, Flexibox Ltd, UK

A shaft sealing system with gas barrier   291
A I Golubev, N V Degterev, Gidromash, Russia

Eccentric seals for nuclear pumps   297
R Metcalfe, T A Graham, W C Wong, AECL Research, Canada

Sealing of volatile liquids   323
J A M ten Houte de Lange, Flexibox B.V., Netherlands

MECHANICAL SEALS II - analysis/modelling

The accuracy of analytical solutions for the temperature distribution in mechanical face seals   341
I Etsion, M Groper, Technion, Israel

Optimisation and performance prediction of grooved face seals for gases and liquids   351
B Tournerie, J Huitric, D Bonneau, J Frene, Universite de Poitiers, France

Thermohydrodynamic calculation of end face seals   367
G Knoll, H Peeken, H-W Höft, Institut für Maschinenelemente Aachen, Germany

Analysis of a hydrostatic gas seal with a compliant face   385
R F Salant, Georgia Institute of Technology, USA

Modelling of plain face gas seal dynamics   397
S E Leefe, BHR Group Limited, UK

MECHANCIAL SEALS III - techniques and experiments

Long-term-tests of mechanical seals for hot water application   427
H-J Franke, R Lachmayer, Technische Universität Braunschweig; J Nosowicz, Feodor Burgmann Dichtungswerke GmbH & Co, Germany

Optimisation of the pumping ring in a mechanical seal with an integrated cooler for feed-water pumps   441
D Buchdahl, EDF/DER; R Martin, EDF/SEPTEN; G Gueret, Latty International; M Blanc, GEC Alsthom Bergeron Rateau, France

RECIPROCATING SEALS

Sealing techniques for subsea control applications 461
M C Theobald, FSSL Ltd, UK

Seal failure in fluid power applications - causes and appearances 479
H Weiss, Busak & Shamban GmbH & Co, Germany

Hydraulic rod seals with laser-structured back-surface 493
U Frenzel, H K Müller, Universität Stuttgart, Germany

Effects of surface finish on reciprocating seal performance 505
R K Flitney, B S Nau, BHR Group Limited, UK

The effects of multistage contact pressure distribution on sealing and frictional characteristics 519
in reciprocating seals
Y Kanzaki, Y Kawahara, NOK Corporation; M Kaneta, Kyushu Institute of Technology, Japan

Study of the behaviour of an hydraulic accumulator seal 533
G Morin, S Delattre, Principia Recherche Developpement S.A., France

FOREWORD

FOREWORD

It is pleasing to see that the most severe recession to have affected many countries in the industrialised world has not stunted the pace of development in the fluid sealing sector - witness the scope of papers in these Proceedings.

Although viewed by some as an inventory consumable, there has been a growing awareness - particularly in the last decade - of the vital role performed by dynamic and static seals in any plant operation where a measure is being taken of the real cost of maintenance downtime, product leakage and process interruption.

A real understanding is developing of the benefits of moving seal specification one notch higher so that products are not working at the limit of claimed performance as a standard routine to the detriment of reliability. Raw cost is seldom the correct arbiter for selection and certainly not when a view is taken of real cost-effectiveness.

More enlightened attitudes in this arena will lead to more beneficial partnerships between seal supplier and customer - whether OEM or end user. In turn, the seal producer is likely to be more forthcoming with reliable design information against which selection decisions and performance envelopes may be set. Or, there would be a reasonable expectation by the user that the seal industry should generate the data needed where it does not exist.

There is a danger that seal purchasers treat large product segments as commodities and, indeed, this has happened in certain sectors where one-stop-shop factoring has proliferated in 0-ring supply, other hydraulic seal utilities and cut gaskets. Many of these are perfectly reasonable operations based upon reputable suppliers. However, as an increasing level of legislation is imposed on the seal manufacturer in terms of the processes used, the implications of product liability and the measures required to support a proper defence and, not least, the prospect of increasingly penal fugitive emission standards, the real cost of making seals becomes a different equation.

Unfortunately, it is European and North American producers who bear the brunt of such cost escalations and there is a danger of a real impact on competitiveness in favour of emerging suppliers who are not similarly constrained. It will serve no user interests if the established names in international seal supply become trading as opposed to manufacturing entities.

During the last International Sealing Conference in Brugge there was much discussion of such topics which saw more than the usual informal debate off Conference time - which has long been such a benefit of the series. The intervening period has seen the establishment of the European Sealing Association ESA e.V. which now has some forty companies representing pan-European interests in the field of gaskets, pump and valve packings and latterly mechanical seals.

The Association comprises seal manufacturers, raw material suppliers and an increasing number of major users. It has working groups examining standardisation and fugitive emissions with a determination to influence impending legislation. Exchanges of information are regular with the Fluid Sealing Association in the USA to extend the forum for matters of global concern.

There is a strong message from the ESA to Brussels and other legislatory authorities that our industry is active in pushing sealing technology forward in a responsible manner to the real benefit of plant reliability, the environment and the safety of the user workforce.

These Conferences organised by BHR Group Limited are a key element of this message of technological awareness in the field of fluid sealing and they continue to offer an international bridge between the excellence of many technical cultures beyond just Europe.

Brian Halligan
James Walker & Co Ltd, UK
February 1994

STATIC SEALS AND VALVES

14th International Conference on Fluid Sealing, Firenze, Italy,
6-8 April 1994. Organised by BHR Group Limited, Cranfield,
Bedford, MK43 0AJ, UK; Tel: 0234 750422

The Service Envelope for Biaxially Orientated Reinforced PTFE Sheet
Sealing Materials

By: J.R. Hoyes
    S. Woolfenden
    A.D. Stancliffe

Of: TBA Sealing Materials Ltd
    PO Box 21
    Rochdale
    OL12 7EU
    England

Introduction

In a previous paper, Reference 1, the route taken by TBA Sealing
Materials during the development of a non-asbestos sheet sealing
material for extreme acid and alkali service was detailed.

The outcome of this work was a pair of sheet sealing materials that
are now well established. These are biaxially orientated reinforced
PTFE sheet materials with a mineral filler component of silica and
barium sulphate respectively.

This paper discusses the continuation of this project in three areas.

a)   initial work on the definition of the operational pressure -
      temperature envelope for these products;

b)   forms of the materials having further levels of reinforcement
      giving additionally boosted performance capability;

c)   development of a material for low stress sealing and use with
      deformed and/or brittle flanges.

Overview of the Previous Development Work

Three possible materials were initially considered to be contenders
for products to replace 'it' calendered, asbestos reinforced sheet
bound by either polyvinyl chloride and polyisobutylene, such as
CAF* 393 which complies with the specification UKAEA 70142C, or
chlorosulphonated polyethylene such as CAF* 399 which complies with
DIN 3754 It S.

* Note CAF is the registered trade mark of TBA Sealing Materials Ltd

4

These routes were:

1)   Non-asbestos form of CAF* 393 or CAF* 399.

2)   Exfoliated graphite.

3)   PTFE product either

    a)   Pure PTFE
    b)   Skived or moulded
    c)   Biaxially orientated.

The 'it' route was rejected because the low temperature service limits for the original asbestos materials were unlikely to be exceeded and because the easier material to produce, having chlorosulphonated polyethylene as the binder, was susceptible to oxidising fluids.

Similarly, the exfoliated graphite route was rejected due to the lack of toughness of this material, the ease with which the surface was damaged and the susceptibility to oxidising fluids. However, the superb stress retention was noted.

Concerns about the inherent stability of exfoliated graphite have led to tables such as that in Figure 1 being published.

Figure 1
Resistance to sulphuric acid 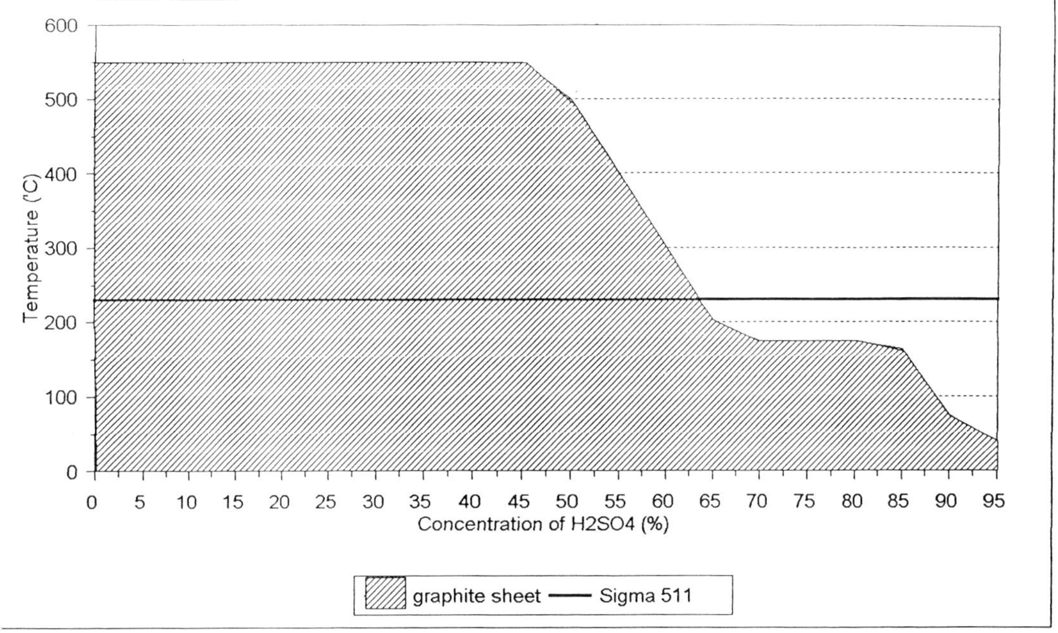
Temperature (°C)
600
500
400
300
200
100
0
0 5 10 15 20 25 30 35 40 45 50 55 60 65 70 75 80 85 90 95
Concentration of H2SO4 (%)
graphite sheet
Sigma 511

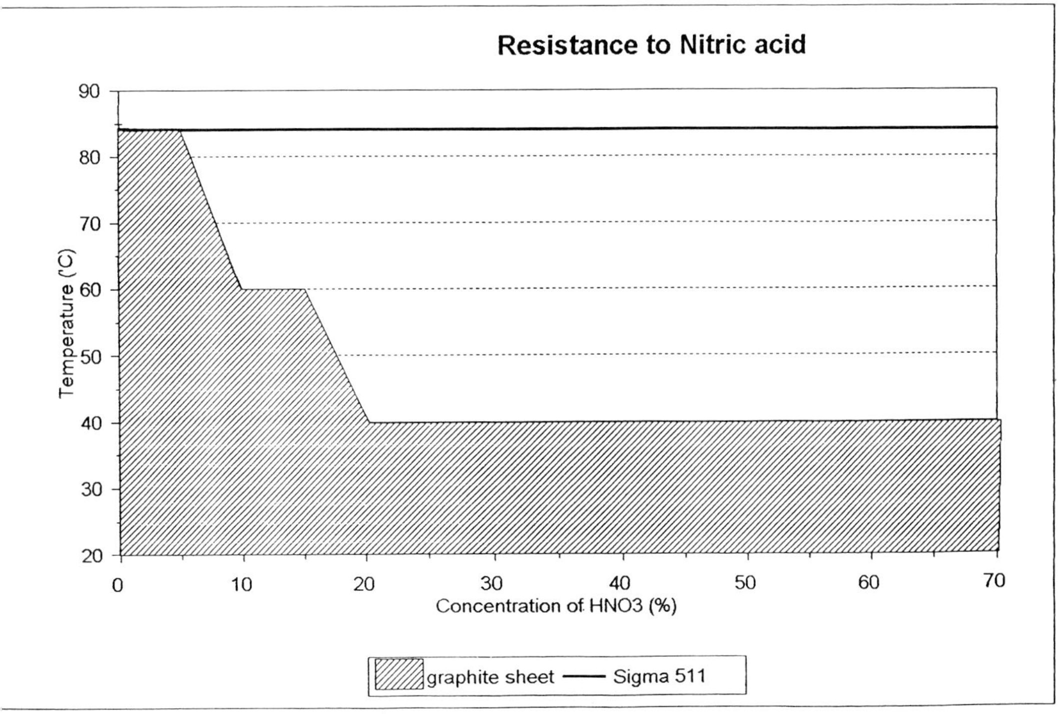

Resistance to Nitric acid
Temperature (°C)
90
80
70
60
50
40
30
20
0 10 20 30 40 50 60 70
Concentration of HNO3 (%)
graphite sheet
Sigma 511

Traditional PTFE products, pure, skived or moulded, suffer from creep but offer excellent chemical resistance combined with superb toughness, ease of use and a useful temperature capability.

Having been involved in the use of PTFE for sealing applications since 1947 and having various forms of PTFE products in our portfolio, it is not surprising that it proved to be within our capability to produce biaxially orientated PTFE sheet materials having the following advantages:

> Chemical resistance
> Robustness
> Enhanced creep resistance
> Excellent sealing characteristics
> Good load bearing capability
> Uniform tensile properties
> Temperature capability of at least 260°C
> Structure based on web of interlocked PTFE fibrils
> All components meet requirements of FDA
> Can achieve WRC approval.

The extreme chemical resistance of the established TBA Sealing Materials products is demonstrated by the results given below in Table 1.

<u>Table 1</u>

<u>Effect of Strong Acids on Biaxially Orientated PTFE Sheet</u>
(Immersion Period of 1 Hour)

PTFE + SILICA	As Received	98% H_2SO_4 230°C	95% Nitric Acid Boiling 84°C
Tensile Strength(MPa)	14.0	14.9	14.5
Thickness Change(%)	-	+0.28	-0.13
Weight Change(%)	-	+0.27	+0.67
Compression(%)	5.3	6.2	5.9
Recovery(%)	40.9	40.4	42.9
PTFE + BARYTES			
Tensile Strength(MPa)	13.5	13.8	12.7
Thickness Change(%)	-	+1.4	+0.2
Weight Change(%)	-	-1.5	-0.4
Compression(%)	4.8	5.9	5.4
Recovery(%)	48.1	46.5	52.8

One method of determining the upper temperature capability of a
material is by determining its weight loss at various temperatures
over time. For both PTFE and graphite some work of interest has been
done at TBA Sealing Materials and is shown in Tables 2 and 3.

<u>Table 2</u>

<u>Long Duration Weight Loss for PTFE and Graphite in Standard Atmospshere</u>

Test Temperature	Test Duration	Grade 1	Grade 2	Exfoliated Graphite
350°C	10 days	0.22	0.155	0.51
	17 days	0.17	0.32	0.62
	45 days	-	-	-
400°C	10 days	3.03%	1.58%	0.65%
	17 days	4.68%	2.52%	0.98%
	45 days	12.60%	6.25%	2.67%
450°C	10 days	99.9%	99.4%	5.81%
	17 days	-	-	11.53%
	45 days	-	-	31.8%

Using a tube furnace some further weight loss work has been conducted in an oxygen atmosphere. The results obtained are shown in Table 3.

<u>Table 3</u>

<u>Weight Change Figure in Oxygen for PTFE and Graphite</u>

Test Temperature	Test Duration	PTFE/Silica	Exfoliated Graphite
250°C	7 days	-0.05%	-1.28%*
300°C	Further 7 days	+0.34%	+0.12

* No change after 41 hours at 250°C.

A conclusion of the work undertaken up to this stage was that PTFE had a good claim to be suitable as a universal sealing material for use in the chemical and other industries up to the maximum service temperature that was appropriate. As a result a two stage objective was set.

1. Develop forms of biaxially orientated sheet PTFE with the widest possible areas of applicability by further boosting the performance.

2. Demonstrate the operating envelope over which such products could be used.

The existing pair of products are giving good service worldwide. They are increasingly being specified by major users and are available in sheets of 60" x 60" (1.524m x 1.524m) or rectangles of a width of 60" (1.524m) with a length of up to 120" (3.048m).

<u>Performance Envelope for Biaxially Orientated PTFE Sheet</u>

It is inevitable that at some stage these materials, if used as a universal sheet material on a site, would be used to seal steam. In anticipation, steam service trials at 180°C have been carried out at TBA Sealing Materials under cycling conditions with both forms previously discussed. When used in BS 10 Table E and Table H 2" flanges, raised and full face, there have been no leaks over test periods of many weeks. In all cases 8.8 steel bolts have been used with an assembly torque of 160 Nm (118 lbft).

The performance limits quoted for biaxially orientated PTFE sheet materials are 260°C (500°F) or more and a pressure of 85 bar (1250 lbf/in^2) or more.

Work on the Shell Thermal Cycle test rig at Flexitallic indicated superb performance of 1.5mm (1/16") biaxially orientated PTFE including silica at 200°C and a nitrogen gas pressure of 45.6 bar (670 lbf/in^2). This pressure being the maximum allowed at that temperature under ANSI 300 lb class conditions.

Work has been undertaken at Flexitallic for TBA Sealing Materials under ANSI 400 lb class conditions which have allowed the maximum simultaneous performance bounds to be explored further. The same material at 1.5mm (1/16") was tested at 250°C, 275°C and 300°C at the maximum pressures allowed for the flange at these temperatures. The Shell Thermal Cycle test procedure was used and the pressure drops after three cycles of heating from room temperature to test temperature were recorded. The gasket stress on assembly was 65 MPa (9427 lbf/in^2). The results are given below in Table 4.

Table 4

Shell Thermal Cycle Test Results

Test Temperature		Test Pressure		Pressure Drop	
°C	°F	Bar	lbf/in^2	Bar	lbf/in^2
250	482	57.1	840	0.68	10
275	527	55.8	820	0.82	12
300	572	54.4	800	1.29	19

Whilst this work is not complete, the provisional operating envelope can now be identified. Figure 2 shows the provisonal envelope for biaxially orientated PTFE including silica in relation to the class limits of BS 1560:1989 for carbon steel flanges.

Figure 2

Performance envelope of PTFE sheet material

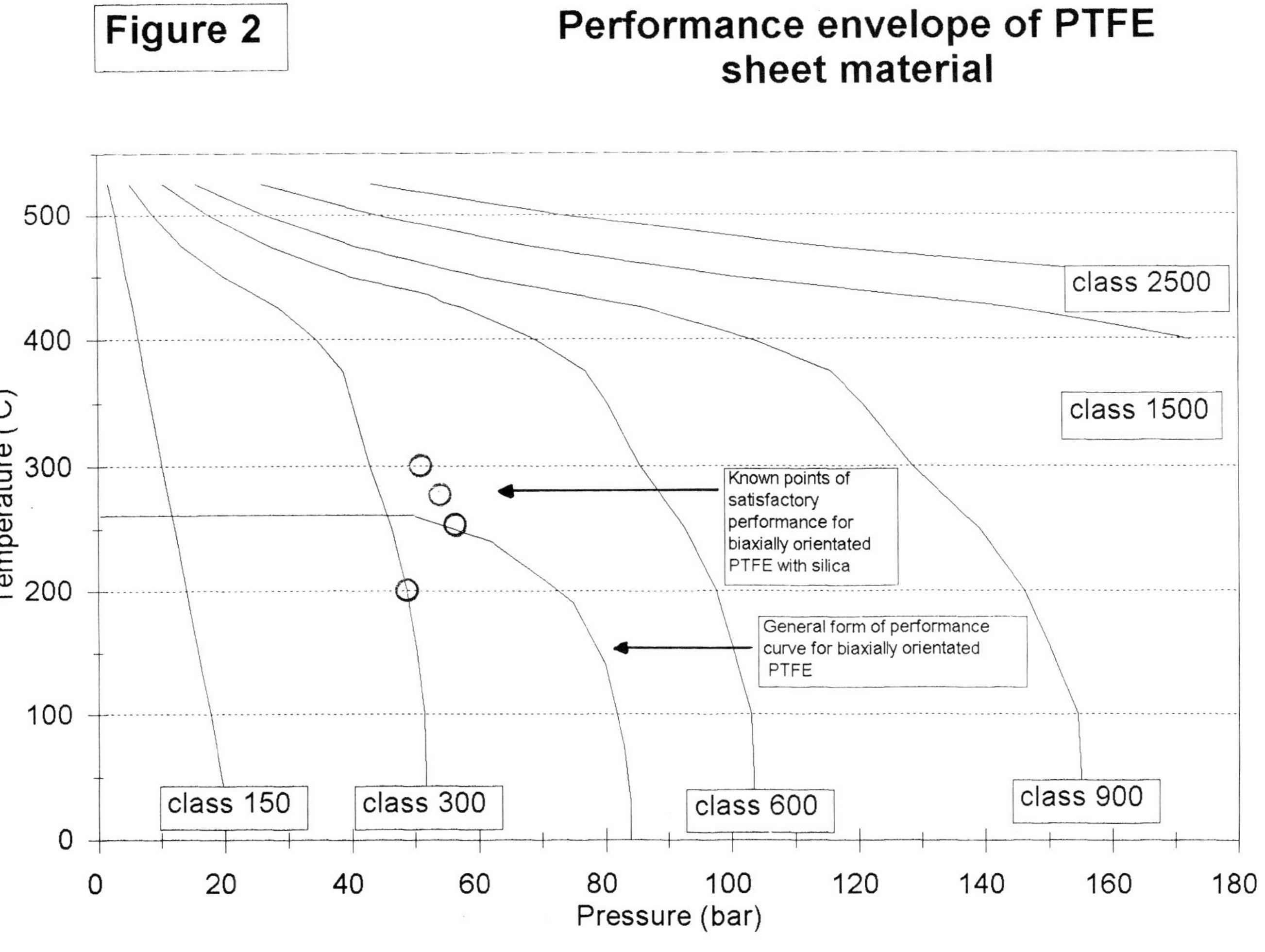

<u>Operating Envelope Increase by Boosting the Temperature Capability</u>

Two aspects limit the operating temperature of PTFE based sealing materials.

a)   Weight loss;

b)   Stress loss and creep.

On the question of weight loss with increasing temperature, it is apparent from the data presented earlier that PTFE is stable far beyond the commonly accepted operational limit of 260°C. In fact one of the major manufacturers of PTFE claims stability to 380°C, Figure 3.

Figure 3

Typical rate of weight loss of granular PTFE mouldings when heated

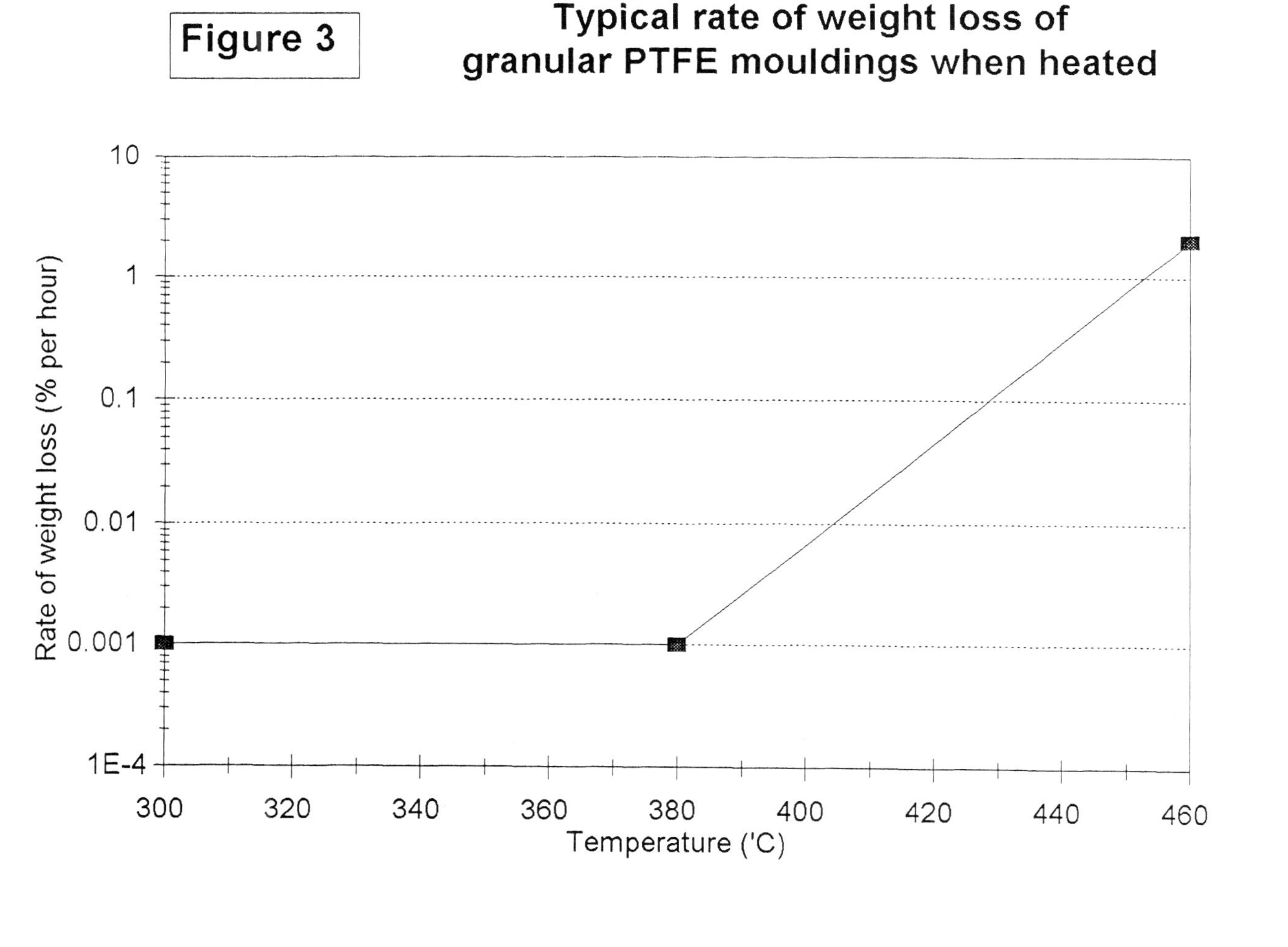

Work at TBA (Tables 2 and 3) supports the conclusion that a limit very much above 260°C could be considered to be satisfactory from a weight loss point of view.

That leaves the need to further improve the stress retention and creep resistance in order to achieve a higher functional temperature limit. Using the DIN stress retention test procedure, comparative testing of a number of relevant materials has been carried out, Table 5 and Figure 4. In parallel with this, the data shown in Figure 5 was also collected This shows how the stress on test gaskets in the DIN stress relaxation rig varied with time as the rig warmed up to 175°C at 5°C/min and then beyond that for a total period of 7 hours.

Stress retention tests according to DIN 52913 at 175°C over durations of 100 hours and 16 hours were also conducted on 2mm thick biaxially orientated PTFE with silica by MPA at Stuttgart University.

The test rigs used have been provided with transducers to allow continuous recording of the stress on the gasket.

During the course of a pair of tests of 100 hours duration the traces given in Diagram 1 were obtained. These show two things:

1.   The high level of residual stress.

2.   The fact that after the test period the residual stress appears to be very constant and not showing any progressive fall off of stress level.

The 16 hours 175°C residual stress value obtained during these tests was 25.0 MPa.

The data given in Figure 7, although different in form, supports this view. The data given is the thickness decrease as a function of temperature whilst the test specimens were held under a constant stress of 50 MPa and the temperature raised from room temperature to 320°C in 3 hours. This figure shows that for all the PTFE based materials the rate of thickness change is still very significant under this load at 320°C after 3 hours whereas for other materials, although there may have been earlier massive levels of thickness decrease, which may well have caused a failure in service, by the end of the test the rate of change of thickness is very considerably less. Once again, the excellent performance of exfoliated graphite is apparent.

Table 5

Variation of DIN Residual Stress with Temperature

Material	Residual Stress (MPa)				
	150°C	200°C	250°C	300°C	315°C
Exfoliated graphite, Ni core	46.5	42.8	45.1	47.3	45.7
Competitive non-asbestos 'it' sheet	22.4	18.9	18.3	16.9	17.1
CAF 393	7.2	2.9	4.2	3.2	4.3
CAF 399	28.8	22.3	17.6	17.1	16.9
Pure PTFE	14.9	14.2	10.6	6.2	5.0
Carbon filled PTFE	22.0	17.4	13.9	10.7	9.3
Glass filled PTFE		13.3	10.5	9.3	7.8
T1	28.6	24.9	18.2	13.1	8.8
T2	27.6	23.7	16.7	14.4	8.1
T3		23.6	20.2	16.0	10.5
T4	37.6	36.2	28.1	19.1	17.0

Key for the above:

T1	Biaxially orientated PTFE with silica
T2	Biaxially orientated PTFE with barytes
T3	High compression biaxially orientated PTFE
T4	Development material.

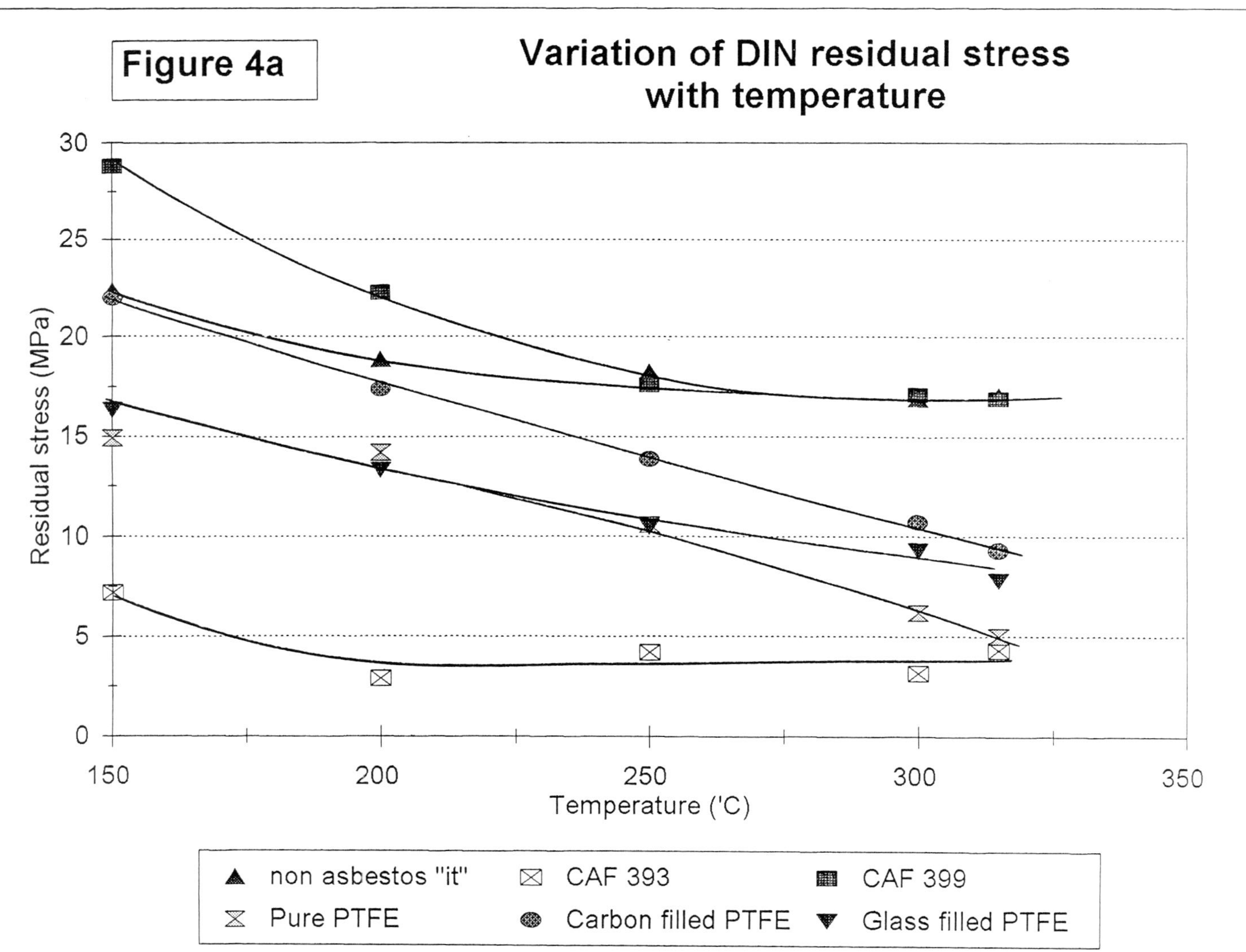

Figure 4a
Variation of DIN residual stress with temperature
Residual stress (MPa)
30
25
20
15
10
5
0
Temperature ('C)
150
200
250
300
350
non asbestos "it"
Pure PTFE
CAF 393
Carbon filled PTFE
CAF 399
Glass filled PTFE

Figure 4b

Variation of DIN residual stress with temperature

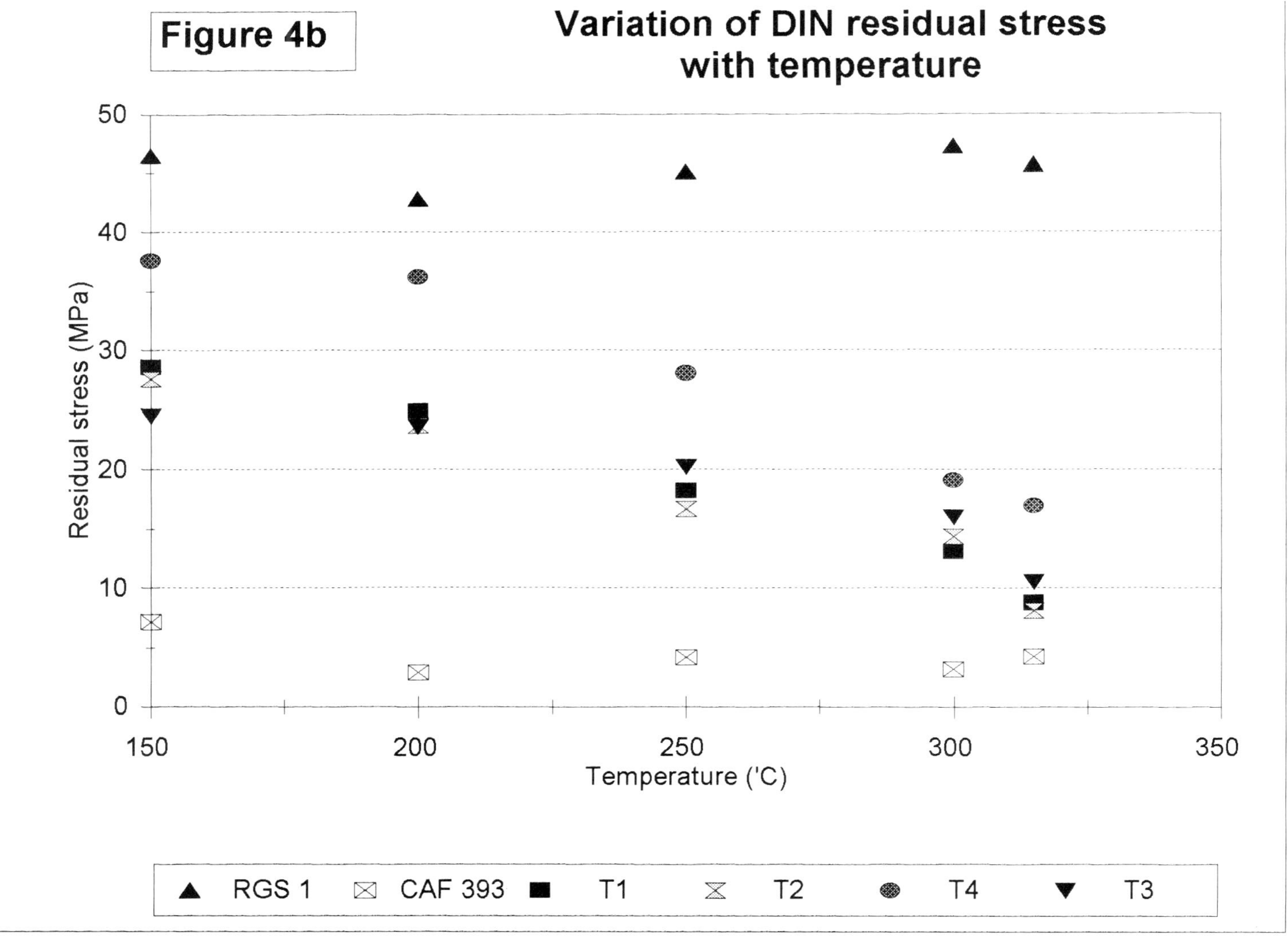

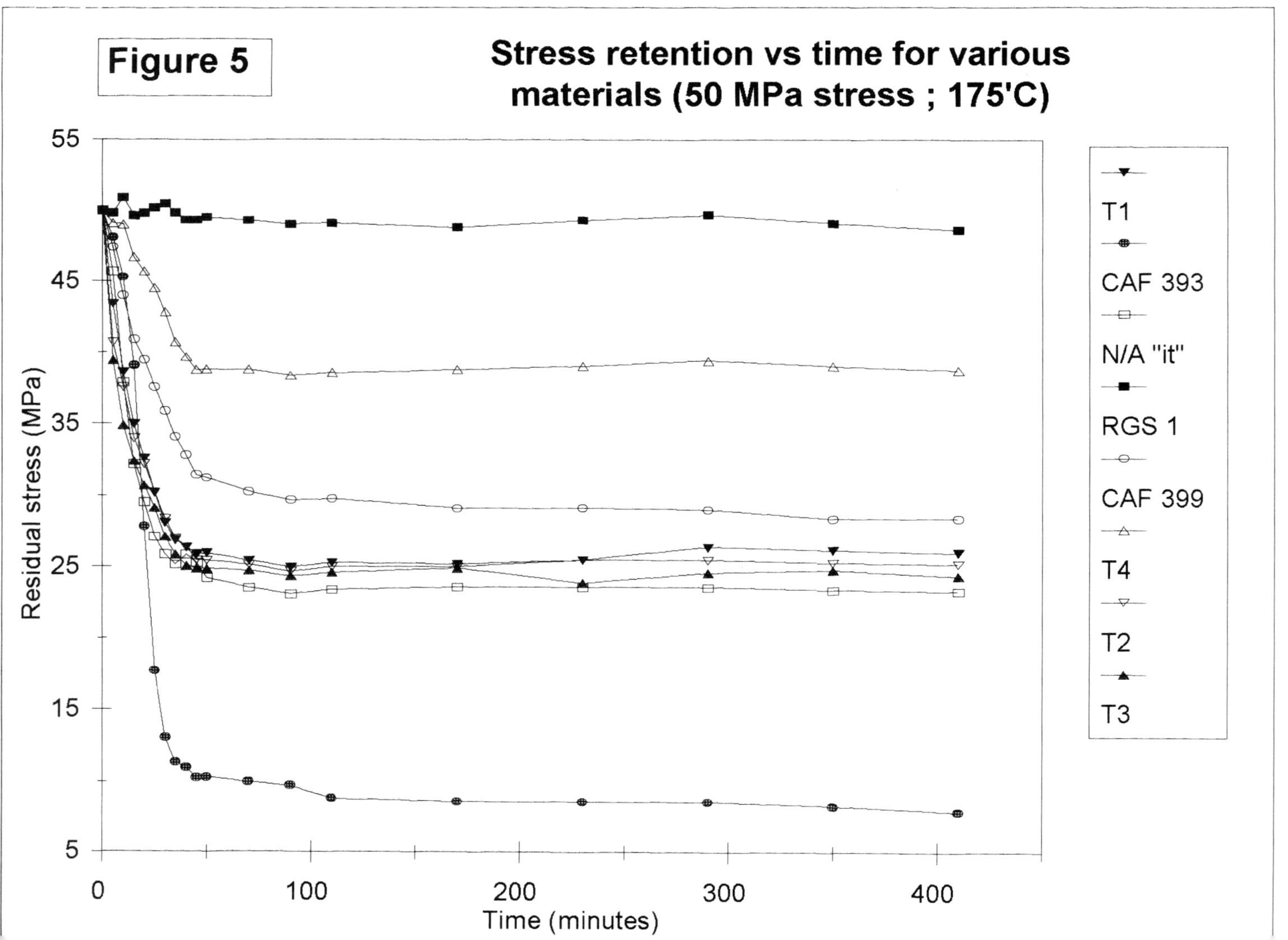

Figure 5
Stress retention vs time for various materials (50 MPa stress ; 175'C)
Residual stress (MPa)
Time (minutes)
T1
CAF 393
N/A "it"
RGS 1
CAF 399
T4
T2
T3
55
45
35
25
15
5
0
100
200
300
400

DIAGRAM 1

TRACES FROM DIN STRESS RETENTION TESTS ON SIGMA 511 (2mm,175°C) AT MPA, STUTTGART

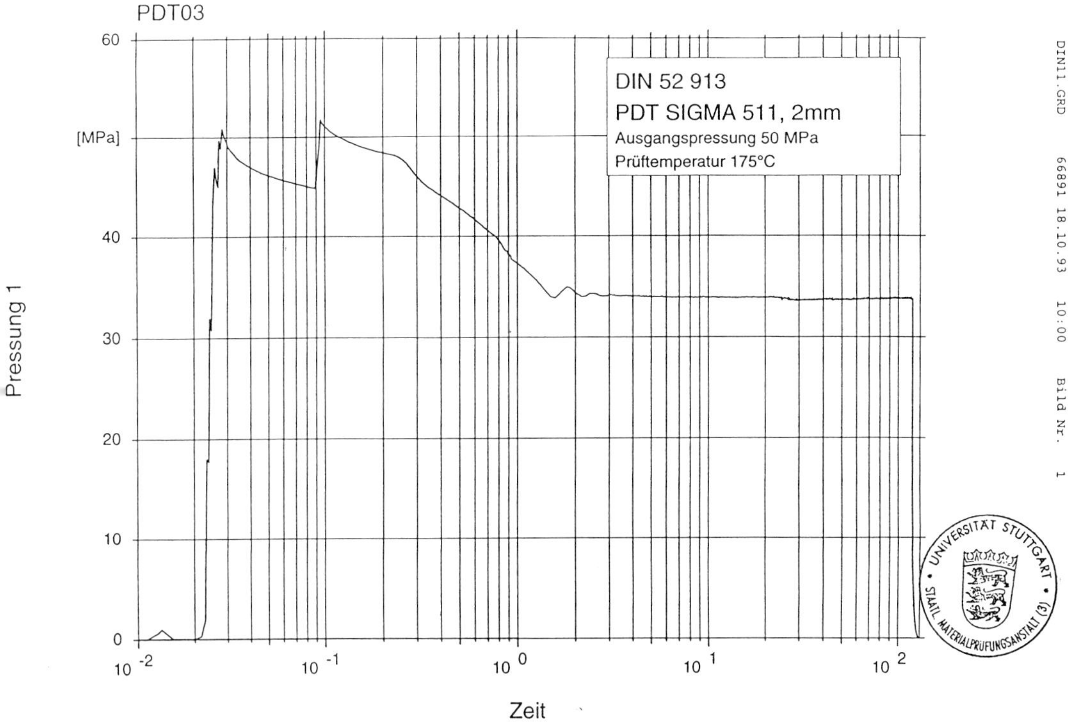

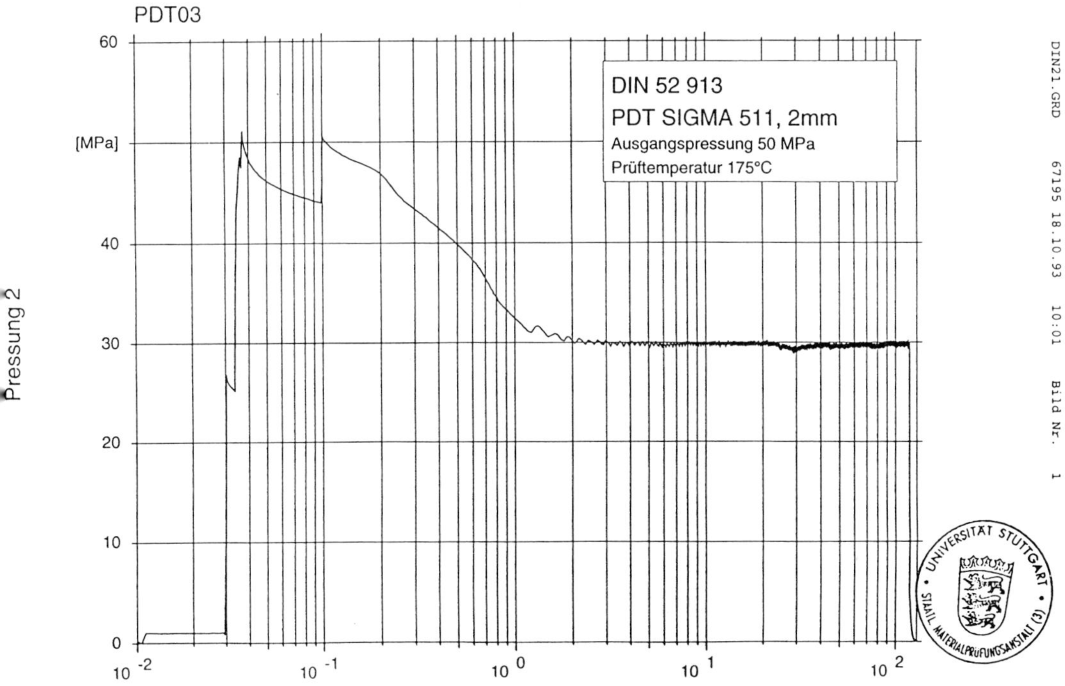

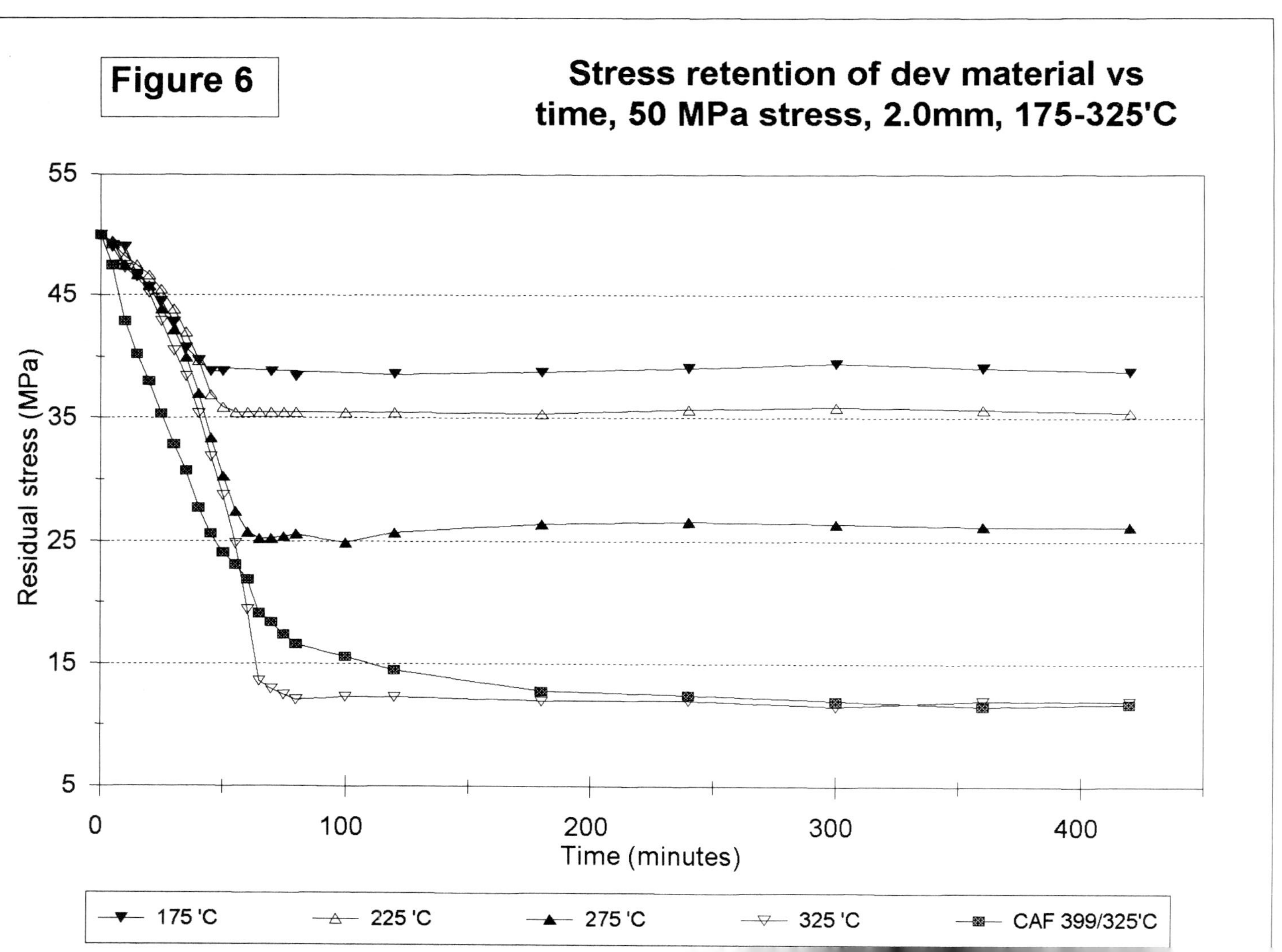

Figure 6
Stress retention of dev material vs time, 50 MPa stress, 2.0mm, 175-325'C
Residual stress (MPa)
Time (minutes)
55
45
35
25
15
5
0
100
200
300
400
175 'C
225 'C
275 'C
325 'C
CAF 399/325'C

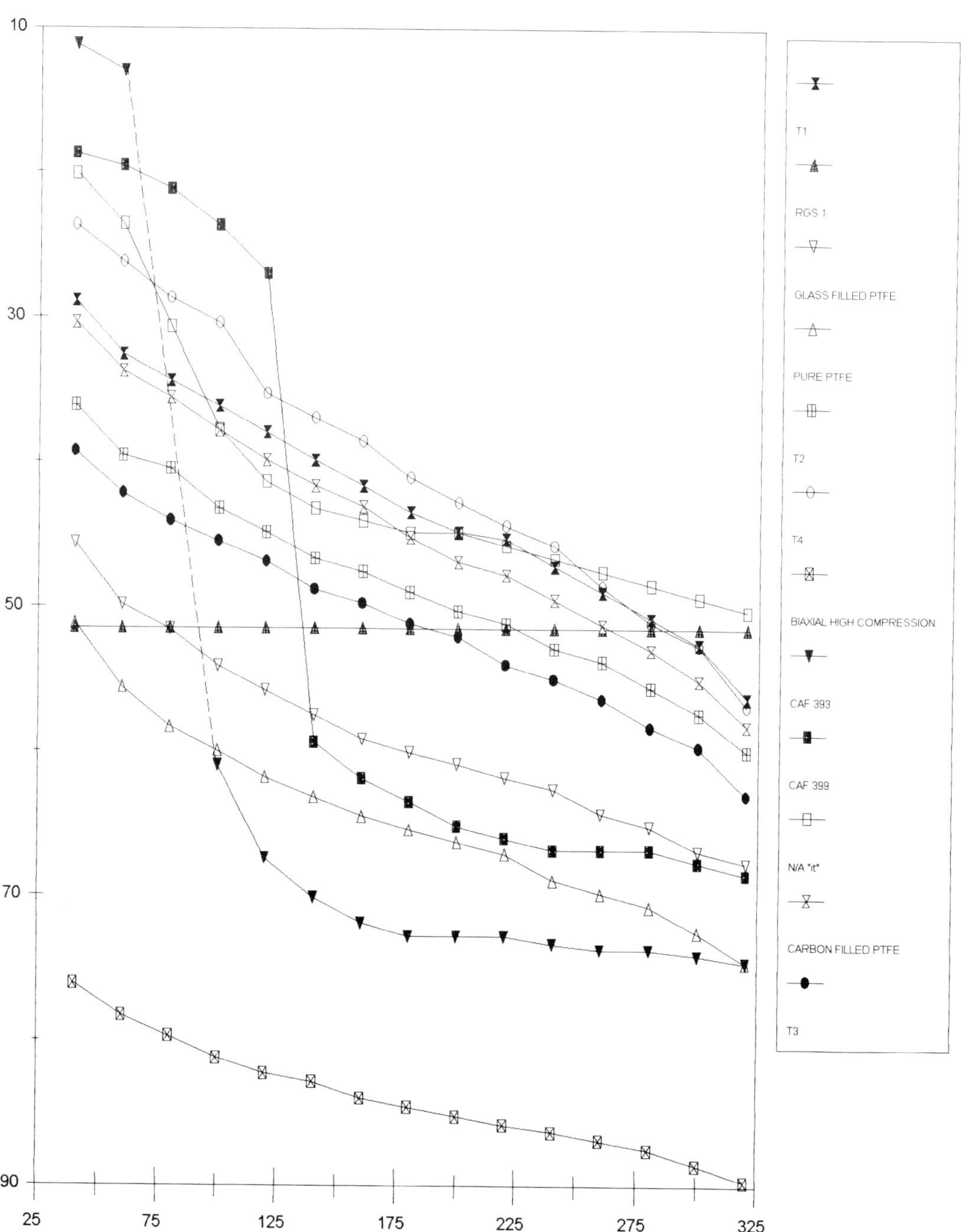
Figure 7
Thickness decrease with temperature
50 MPa stress 2.0mm initial thickness
10
30
50
70
90
25
75
125
175
225
275
325
Temperature ('C)
T1
RGS 1
GLASS FILLED PTFE
PURE PTFE
T2
T4
BIAXIAL HIGH COMPRESSION
CAF 393
CAF 399
N/A "it"
CARBON FILLED PTFE
T3

<u>Operational Envelope Increase by Enhancement of the Low Stress Sealing Capability</u>

There are many instances where good chemical resistance and sealing is needed and where either the flanges are such that only a low stress can be developed on the gasket or the flanges are so distorted, damaged or warped that areas of low gasket stress are likely to exist. In such circumstances the gasket material selected needs to be conformable enough to ensure a secure level of sealing under low stress.

Seals based upon PTFE can have a tendency to have poor surface conformability and hence require a relatively high seating stress if the flange surface finish is unsatisfactory. To attempt to overcome this situation various measures have been adopted. Those currently available are listed below.

> Compression enhancement via process improvement;
> Skived sheet incorporating microspheres;
> Biaxially orientated material including microspheres;
> Sundry forms of biaxially orientated material including
>   a filler system;
> Laminations of forms of biaxial material;
> Biaxially orientated material with no filler system.

A modification to the existing materials which was introduced in 1993, led to an increase in their compressibility as measured by the ASTM method. This had distinct advantages in terms of the seating stress required as was shown by the results of a ROTT test done on the biaxially orientated variant which includes barytes at Ecole Polytechnique, Figure 8, where the G_b value derived was 232 lbf/in^2.

FIGURE 8

GASKET CONSTANTS FOR TBA SIGMA 533 1/16"
ROTT TESTS FL15RT01, FL15RT02 & FL15RT04
NOTE: Applied inside pressure (P) for leakage measurements:
P = 800 psi points are represented by unfilled markers
P = 400 psi points are represented by filled markers
Part-A
Part-B1
Part-B2
Part-B3
a = 0.400
Gb = 232
To Gs = 2
Gasket Stress, Sg (psi)
100000
10000
1000
100
Tightness Parameter, Tp
1
10
100
1000
10000
100000

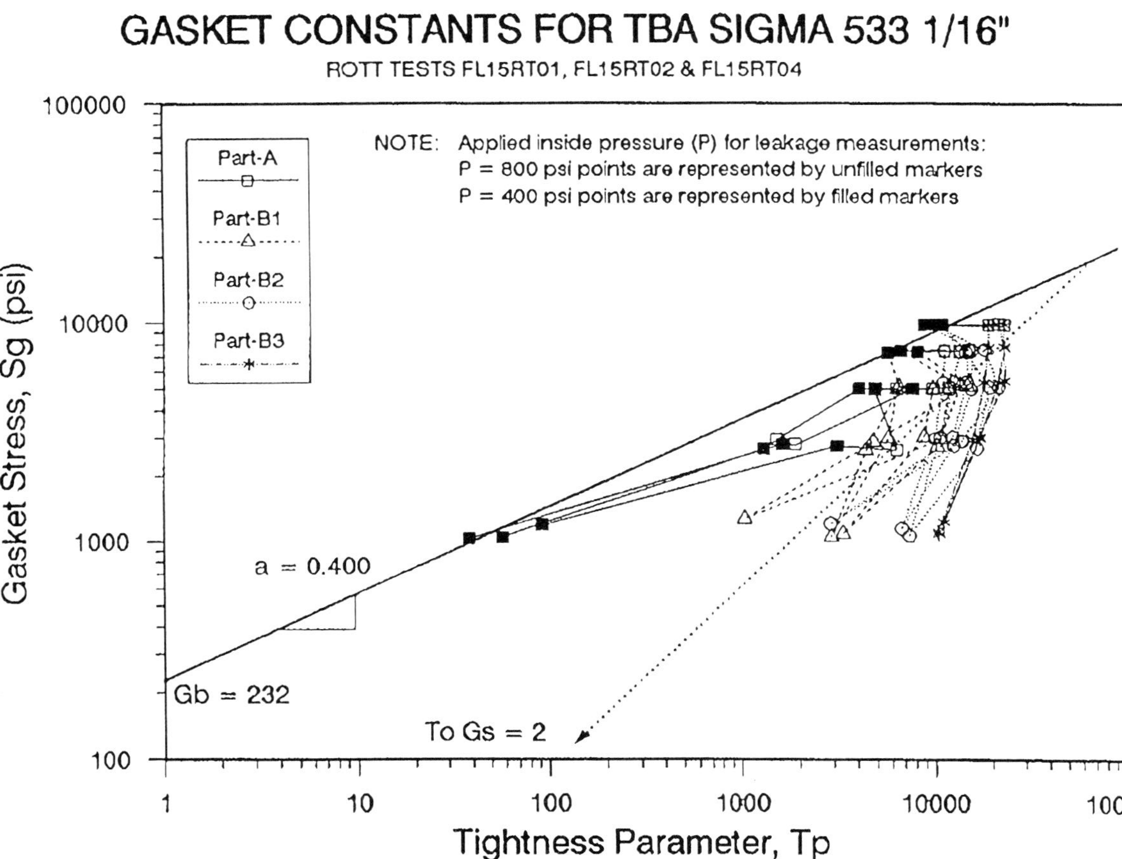

Beyond that solution, other routes have been taken. One which is quite popular is the addition of thin walled glass, or similar, microspheres which are supposed to collapse under stress thus creating a better conformability.

The concept of using collapsing microspheres in gasket materials is not new at TBA Sealing Materials. It was used and patented in the early nineteen seventies, Reference 2, in steel cored cylinder head gasket material to allow easier insertion of bore eyelet rings.

Unfortunately such a ploy has some unsatisfactory characteristics:

1.   The wall thicknesses have a distinct distribution and collapse is not entirely predictable.

2.   Those spheres which do not collapse actually tend to enhance the creep and stress loss.

3.   In PTFE matrices the level of compression achieved by their inclusion is very temperature sensitive.

Partly because of the latter point, TBA Sealing Materials, although appreciating the need for a low stress sealing material, decided that such a route was unsatisfactory. The level of the sensitivity is shown in Figure 9 together with the effect of temperature on compression of other materials.

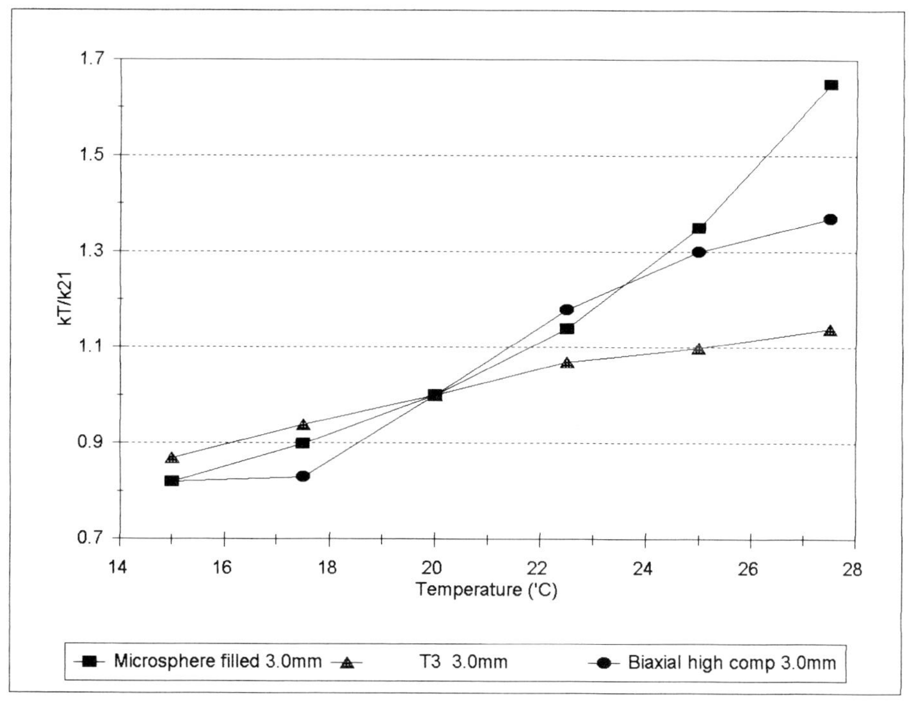

Figure 9
Normal curves of compression vs temp
kT/k21
1.4
1.2
1
0.8
0.6
14
16
18
20
22
24
26
28
Temperature ('C)
Micosphere filled 1.5mm
T3 1.5mm
Biaxial high comp 1.5mm
1.7
1.5
1.3
1.1
0.9
0.7
kT/k21
14
16
18
20
22
24
26
28
Temperature ('C)
Microsphere filled 3.0mm
T3 3.0mm
Biaxial high comp 3.0mm

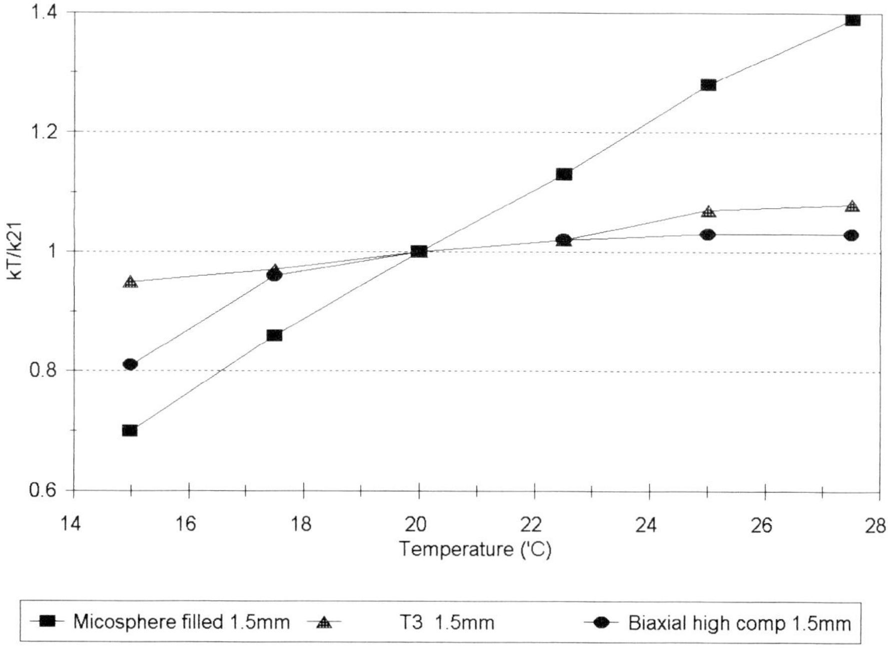

Figure 10 shows that this effect is apparent with ASTM recovery as well as compression. The consequences of season of year on the use of microsphere filled material is obvious. A fitter in winter in Alaska would get distinctly different performance from that in Italy in summer.

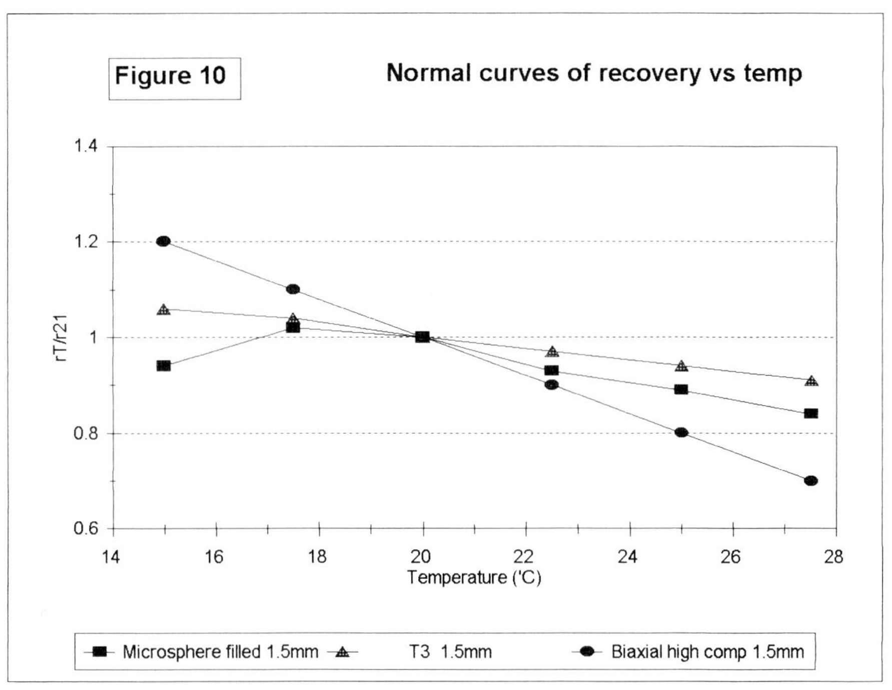

Figure 10
Normal curves of recovery vs temp
1.4
1.2
1
0.8
0.6
rT/r21
14
16
18
20
22
24
26
28
Temperature ('C)
Microsphere filled 1.5mm
T3 1.5mm
Biaxial high comp 1.5mm

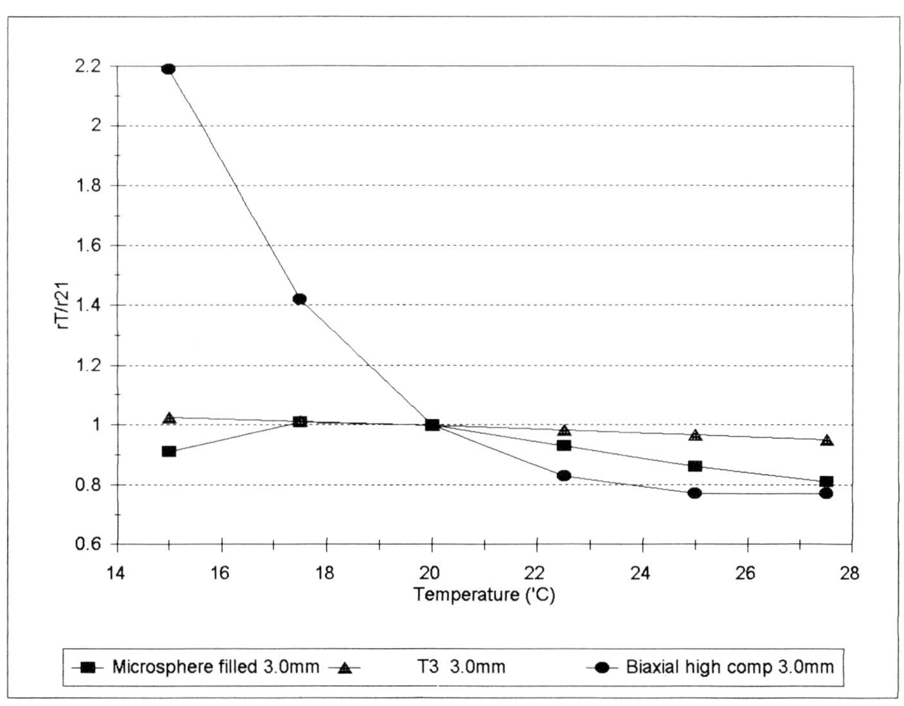

2.2
2
1.8
1.6
1.4
1.2
1
0.8
0.6
rT/r21
14
16
18
20
22
24
26
28
Temperature ('C)
Microsphere filled 3.0mm
T3 3.0mm
Biaxial high comp 3.0mm

TBA Sealing Materials decided that a different type of solution was needed. It had to have the following characteristics:

1.   Very low sealing stresses for both smooth and grooved flanges.

2.   The compression/conformability characteristics were not to be sensitive to ambient temperature.

3.   Good absolute recovery as load levels are reduced.

4.   Sufficient compression for general use.

5.   Excellent creep resistance.

The material that has been developed to these requirements has a very soft, highly conformable surface on a core of biaxially orientated PTFE with barytes. This combination of material types gives the properties required without the handling problems experienced with some low density PTFE products being sold.

Figures 9 and 10 show that the temperature sensitivity of this product is very much less than for microspshere filled PTFE or very low density PTFE with no other components.

Diagram 2 shows the compression and recovery characteristics of this product in comparison with microsphere filled biaxially orientated material and a 6mm thick envelope gasket of German origin. This work was conducted for TBA Sealing Materials by our sister company, Flexitallic.

Figure 11 shows the sealing stresses required to seal nitrogen gas at up to 600 lbf/in^2 (4.1 MPa) for smooth flanges (R_a 0.4μ inch, 16 μm) and grooved flanges (R_a 6.3μ inch, 250 μm).

SIGMA 522, MICROSPHERE FILLED & PTFE ENVELOPE GASKETS.

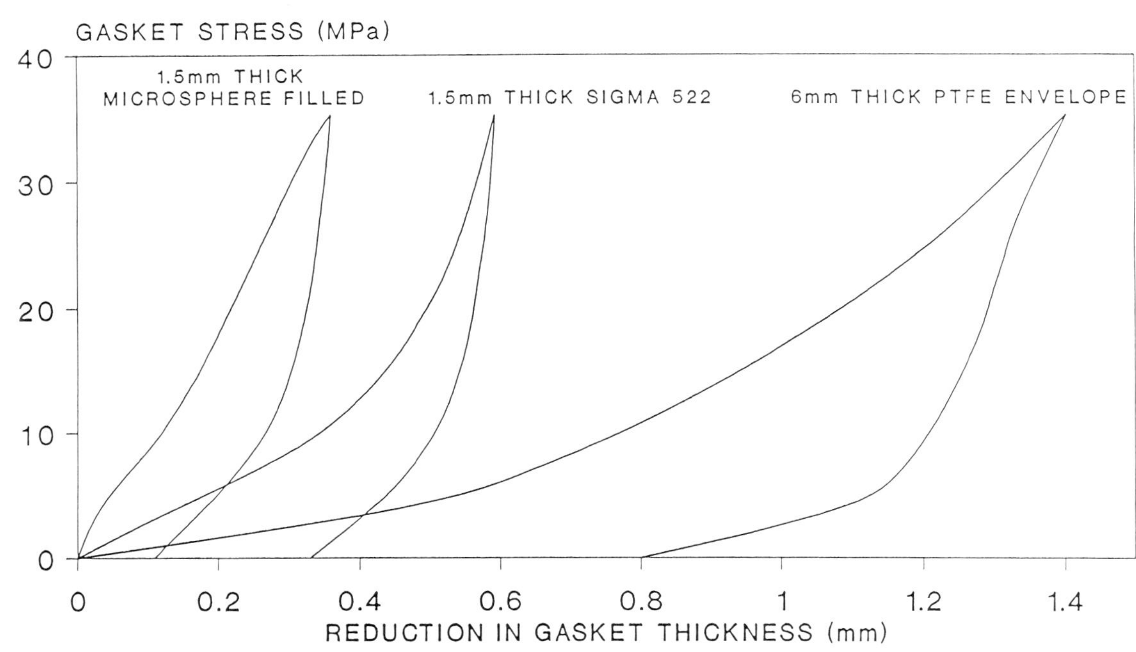

AMBIENT TEMPERATURE

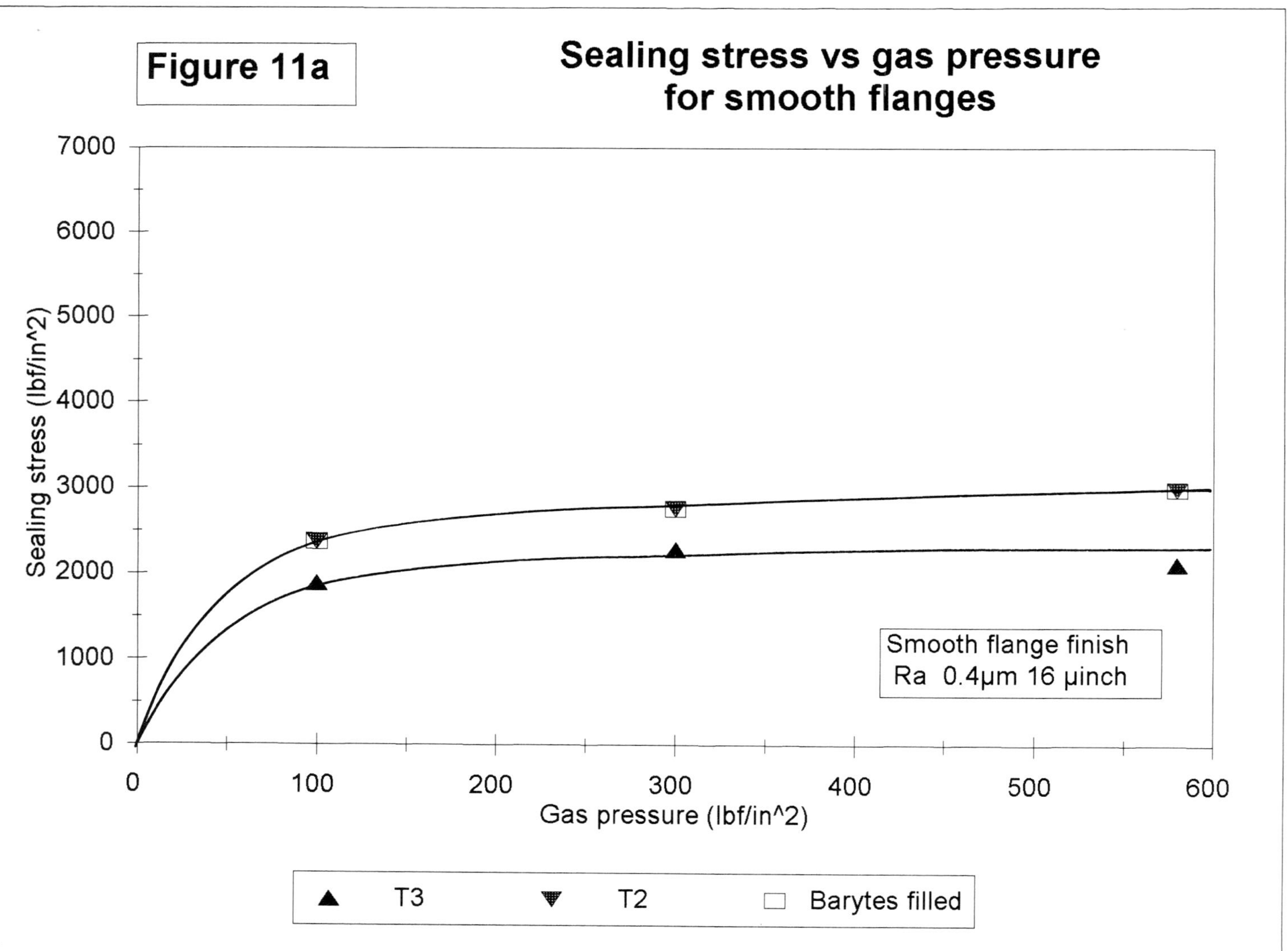

Figure 11a
Sealing stress vs gas pressure
for smooth flanges
Sealing stress (lbf/in^2)
Gas pressure (lbf/in^2)
Smooth flange finish
Ra 0.4µm 16 µinch
T3
T2
Barytes filled

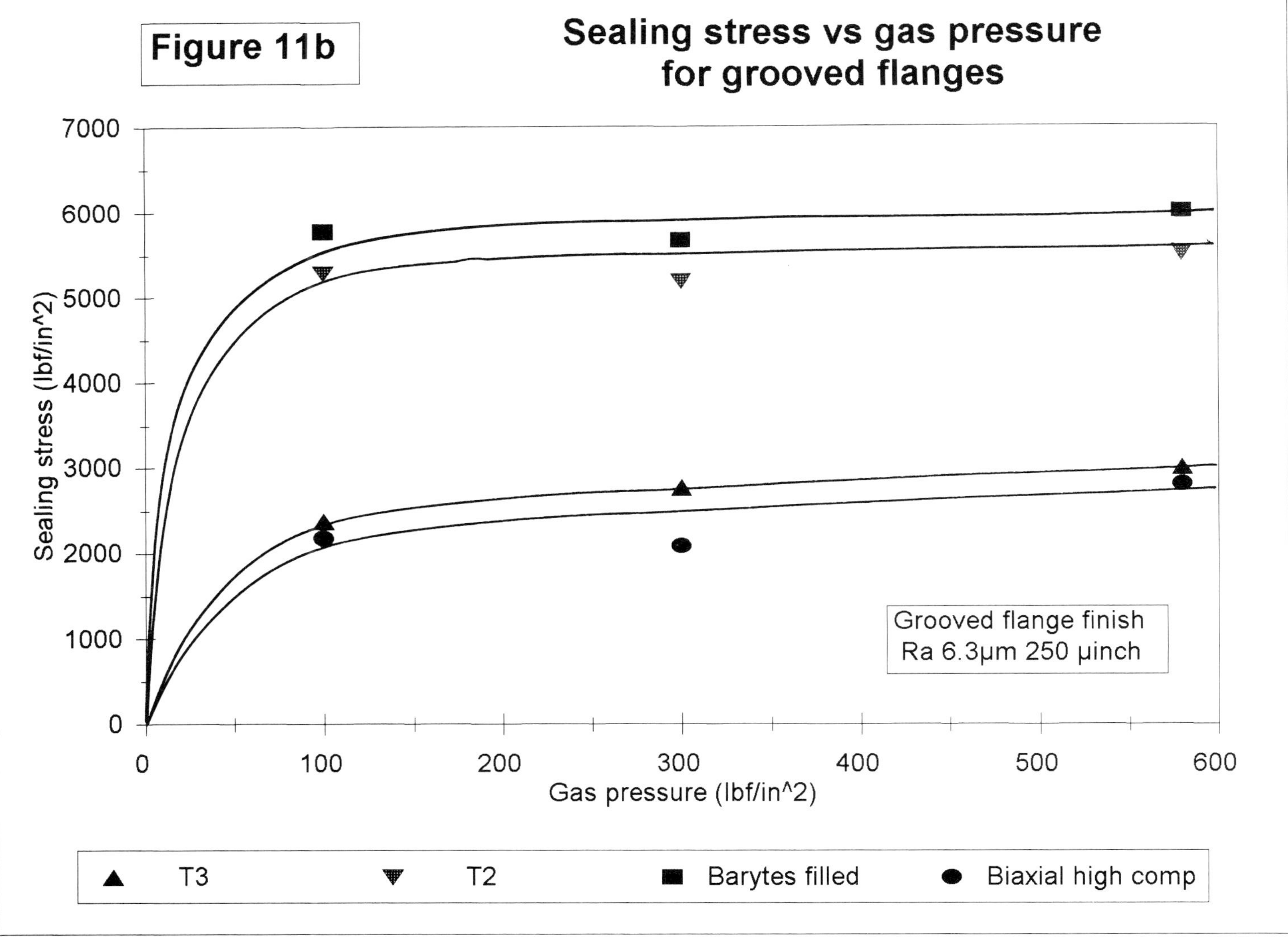

Figure 11b
Sealing stress vs gas pressure for grooved flanges
Sealing stress (lbf/in^2)
7000
6000
5000
4000
3000
2000
1000
0
Gas pressure (lbf/in^2)
0
100
200
300
400
500
600
Grooved flange finish
Ra 6.3µm 250 µinch
T3
T2
Barytes filled
Biaxial high comp

The excellent sealing characteristics of the latest addition to the product range in comparison to microsphere filled, barium sulphate filled and low density PTFE sheet are apparent. A reduction by a factor of three in the sealing stress needed over those required for microsphere and barium sulphate containing material is achieved for the same level of sealing.

In Figure 12 the absolute value of the recovery in mm is plotted as function of sheet thickness for this new product, exfoliated graphite, microsphere containing PTFE and two forms of low density PTFE sheet. The absolute level of recovery is better than for low density PTFE which is the only material to match the excellent sealing characteristics of the new product.

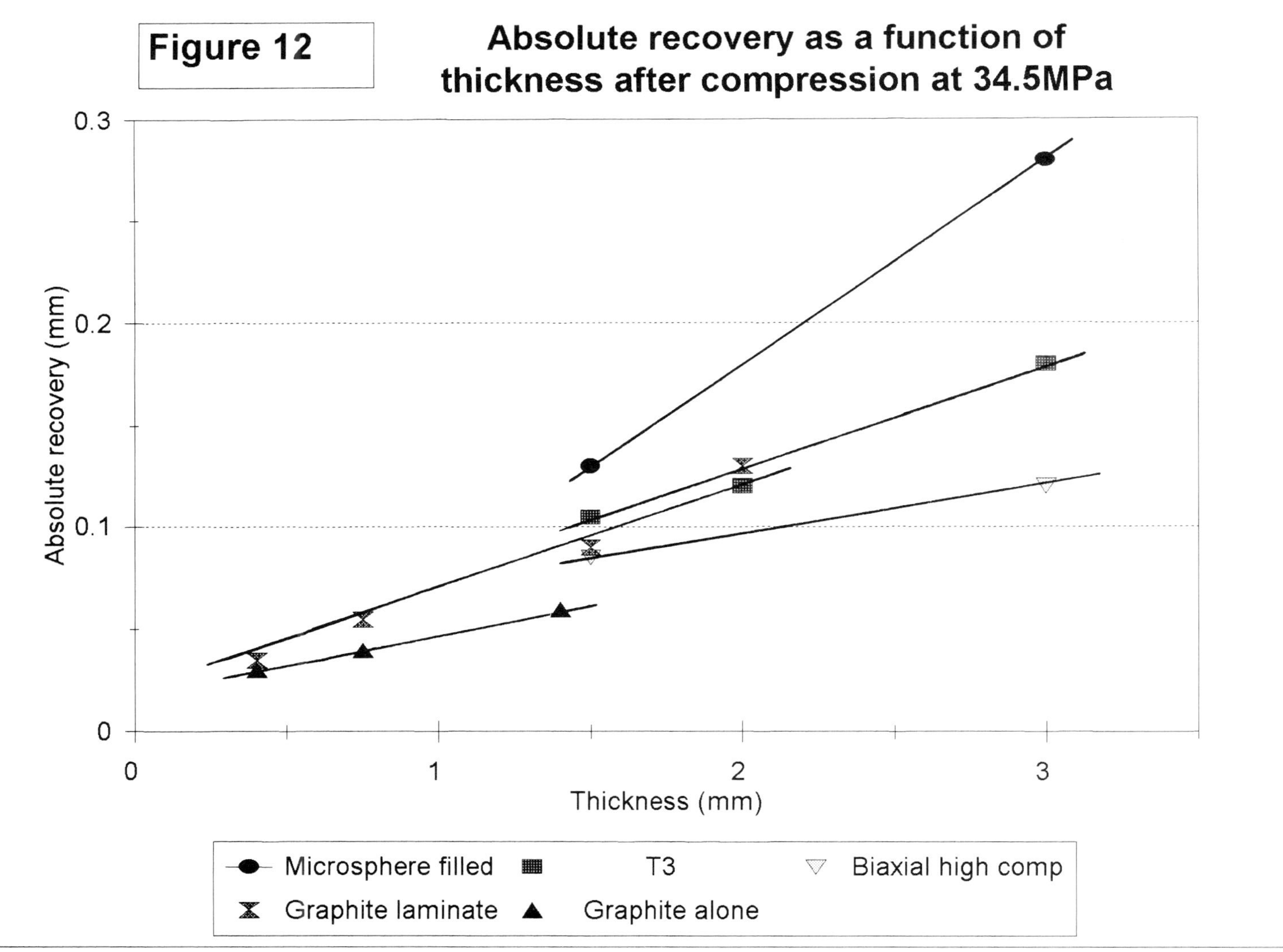

Figure 12
Absolute recovery as a function of thickness after compression at 34.5MPa
Absolute recovery (mm)
0.3
0.2
0.1
0
Thickness (mm)
0
1
2
3
Microsphere filled
T3
Biaxial high comp
Graphite laminate
Graphite alone

The general physical properties of all the material variants intended for low stress sealing applications are given in Table 6.

It is considered that the new material offers the best combination of properties:

Not sensitive to ambient temperature;

Large cut gaskets are easy to handle;

Excellent low stress sealing capability;

Better levels of absolute recovery than any other material with good low stres sealing characteristics.

Table 6

Summary of Physical Properties for Various PTFE Sealing Materials Intended to Provide Low Stress Sealing

	MANUFACTURER							
	A	B	B	B	B	C	D	T3
Thickness (mm)	1.6	1.6	1.6	3.2	1.6	1.6	2.0	1.6
Density (g/cm^2)	2.46	1.81	1.74	0.59	1.88	0.67	0.63	2.00
ASTM Compression (%)	11.4	19.3	35.9	81	39	60	63	28.5
ASTM Recovery (%)	41	48	40	6	31	8	10	25
ASTM Tensile Strength A (MPa)	15.3	19	13.8	-	-	29.7	20.1	10.2
ASTM Tensile Strength B (MPa)	11.9	17.9	24.0	-	-	17.2	26.9	10.2
ASTM Creep Relaxation (%)	67.2	55	75	44	67	51	71	52
DIN Stress Retention at 175°C (MPa)	-	22.6	20.0	-	-	37.5	28.5	28.9
DIN Gas Permeability (mL/min)	0.00	0.00	0.00	-	-	0.01	0.09	0.00
ASTM Liquid Leakage (mL/min)	-	0.5	0.5	18.7	26.9	1.8	9.9	2.3
Fillers included & comments	Glass fibre Micro-spheres	Micro-spheres	Micro-spheres Laminated	None	Not Identified	None	None	Barium Sulphate
Manufacturing	Skived	- - - - - - - - - - - Biaxially orientated - - - - - - - - - - -						

36

<u>References</u>

1.   "Biaxially Orientated Reinforced PTFE Sheet Sealing Materials",
      J.R. Hoyes, S.W. Woolfenden; 3rd International Symposium on
      Fluid Sealing, Biarritz, France; 15-17 September 1993.

2.   UK Patent 1478043 D J Adams 1973.

14th International Conference on Fluid Sealing, Firenze, Italy,
6-8 April 1994. Organised by BHR Group Limited, Cranfield,
Bedford, MK43 0AJ, UK; Tel: 0234 750422

THERMAL TRANSIENT SEALING CONCEPTS
THE "CARRIER RING" SPIRAL WOUND GASKET

Gary Briggs & Simon P Noble
FLEXITALLIC LTD.

P.O. Box 3, Dewsbury Road, Cleckheaton, West Yorkshire. BD19 5BT.
United Kingdom

INTRODUCTION

Historically, the successful selection of standard design static seals on problematic flange arrangements operating at the extremes of temperature and pressure has been limited, further heightened with the introduction of high thermal transients gradients.

Traditionally, this high specification area of static sealing has been occupied by either semi metallic or metallic gasket designs, each having their specific merits and drawbacks. Semi metalllic gaskets, such as standard spiral wound gaskets, demonstrate good sealing and general recovery characteristics, but generally have limited radial strength, offering limited recovery characteristics or possible implosion problems under critical pressure duties. Metallic gaskets, whilst exhibiting excellent strength, offer limited recovery during high thermal transient conditions.

In this paper, it is suggested that the specification for a high strength, high recovery static seal can be amalgamated into a unique gasket, the Carrier

Ring (CR) Spiral Wound Gasket offering the end user a higher level of tightness in terms of Fugitive Emissions. This concept will now be explained in detail.

BACKGROUND

In today's industry, with greater emphasis than ever before placed on fugitive emissions and joint tightness, more attention must be paid towards all variables that effect the integrity of a bolted gasket joint. Naturally, improved gasket sealing concepts will evolve to serve the needs of the user, legislative law etc.

Following discussions with the end user, it was apparent that four main features of the idealised gasket solution were considered a prerequisite to high integrity, high thermal transient sealing concepts.
* HIGH STRENGTH AND RIGIDITY - BLOWOUT RESISTANCE
* IMPROVED PERFORMANCE / TIGHTNESS
* IMPROVED THERMALCYCLIC RECOVERY CHARACTERISTICS
* IMPROVED SAFETY PROFILE
* ENDURANCE OF PERFORMANCE

The Carrier Ring was developed by Flexitallic for a particularly high pressure, thermalcyclic heat exchanger application within Western Europe, maximising the already proven recovery characteristics of the company's spiral wound gasket by means of utilising a double gasket sealing element arrangement. This "double" sealing arrangement involved the location of

arrangement. This "double" sealing arrangement involved the location of individual spiral wound sealing elements within opposing machined recesses of a rigid metallic central core. The development of this concept is seen as offering an engineered alternative to the more traditional gasket concepts, such as the standard spiral wound gasket, metal jacketed gasket and the solid metal gasket

THE DOUBLE RECOVERY CONCEPT.

As discussed, the Carrier Ring (CR) gasket consists of two fundamental elements :

1) The Metallic Body
2) The Gasket Sealing Element

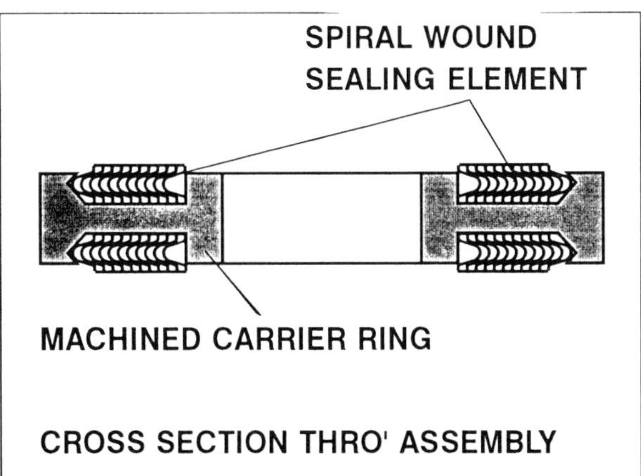

FIGURE 1

The metallic body, generally machined from either carbon steel, austenitic stainless steel or an exotic alloy (depending on service conditions), offers several benefits to the overall gasket concept besides an instrument to assist sealing element location and compression. In addition, this metallic component offers a gasket design of exceptional blow out strength and rigidity, as well as improved handleability characteristics.

The double spiral wound sealing element arrangement, totally confined within the metallic core under compression, offers superior recovery characteristics, thus ensuring the maintenance of a high integrity seal under the most exacting of thermalcyclic conditions. Each gasket sealing element is usually manufactured from precipitation hardened metallic windings (offering improved recovery characteristics to the equivalent material in the standard annealed condition) with the inclusion of soft flexible graphite filler material. By means of monitoring filler material thickness and manufacturing compression, the spiral wound gasket sealing element density (construction) is designed to match the applications available bolt load, thus ensuring that the optimum gasket compression i.e. a total metal to metal closure, is achieved under the specified bolt stress values.

DEVELOPMENT WORK

The Carrier Ring concept, first prototyped for problematic high pressure heat exchangers, has since been further developed in conjunction with the use of Finite Element Analysis(FEA) techniques, in order to analyse a whole host of problematic applications, such as high pressure Steam Heaters and high temperature Naptha Crackers. Further extensive testwork

Heaters and high temperature Naptha Crackers. Further extensive testwork has also been instigated by the Pressure Vessel Research Council who are presently developing a new approach to identifying the essential qualities of a static seal. Fundamental to this work, researchers have compiled a base of statistically reliable data points from which a new set of gaskets factors have been derived, defining the expected behaviour characteristics of a gasket design.

[Figure 2] highlights an idealised tightness curve showing the basis for the gasket constants [Gb] [a] and [Gs]. The associated numerical values associated with the gasket factors for a particular gasket style are determined by the reduction of data from [Part A] and [Part B] gasket testing. [Part A] testing provides data on leak rate vs. pressure for several levels of compressive stress, and relates to gasket seating and the "Y" factor of the ASME Boiler and Pressure Vessel Code Section VIII-1.

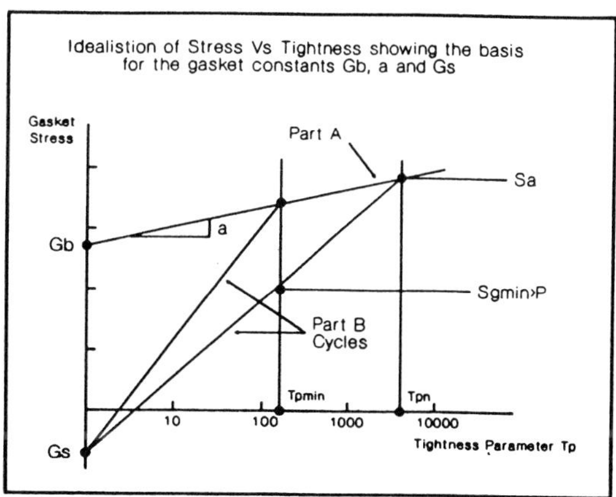

FIGURE 2

[Part B] testing provides data on the gasket operation during the unload - reload cycles at constant internal pressure and relates to the Asme Code 'm' factor. [Gb] is the stress intercept at Tp =1, and [a] is the slope associated with [Part B] tightness data. Low values of [Gb] and [a] indicate that the gasket requires low levels of gasket stress for initial seating. Low values of [Gs] indicate that the gasket requires low stress to maintain tightness during operation, an is therefore less sensitive to unloading. [Tp] is a dimensionless sealability measure used to relate the performance of gaskets with various fluids. ASTM is currently developing a standard test method for determining the gaskets constants [Gb][a][Gs] based on [Part A] and [Part B] tightness testing.

<u>Typical PVRC Published Gasket Constants</u>

Gasket Style	[Gb]	[a]	[Gs]
SWG Asbestos	3400	0.300	7
SWG Graphite	2300	0.237	13
Metal Jacketed	2900	0.230	15
CARRIER RING	1251	0.309	11

Table 1

These figures have been attained from an independent test laboratory and constitute the preliminary results of a longer term program. It is anticipated that values of [Gb] [a] and [Gs] for various types of gaskets will eventually be tabulated and listed in the revised ASME Code. Gasket manufacturers will also be expected to supply values of gasket constants for existing products as well as new gasket styles and materials.

The development of the gasket factors allows the end user to make better decisions as to the nature of the gasket chosen for a particular application. The development of the PVRC Convenient method allows for the comparison of similar gasket styles and materials.

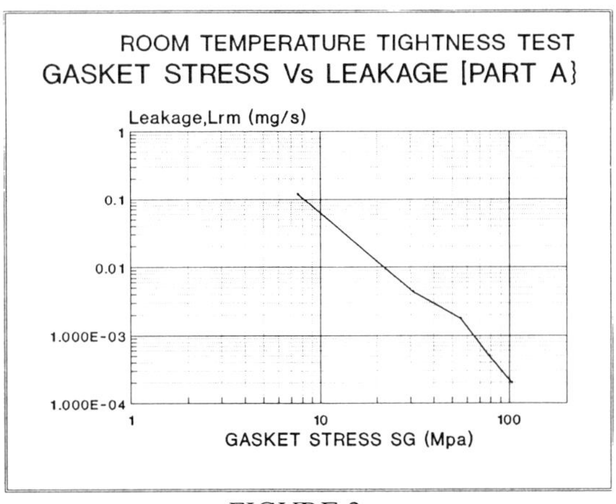

FIGURE 3

Gasket stress v Leakage.

The development of the PVRC Convenient method allows the determination of several important parameters controlling gasket performance to be identified. The methods employed in this paper was to extend the tightness class from T1 to T3 and then determine the optimum bolt stress required to meet the individual Tightness C

Class criteria. By this method, it is possible to examine the relevant tightness qualities of the chosen gasket.

Table 2 gives the example of a standard 24" Nominal Bore, Class 1500lb Carrier Ring utilising graphite filler material. This size of the gasket has been utilised to illustrate the principle that if higher levels of tightness are to be required, then progressively higher stresses will need to be applied upon the gasket face. This method of calculation offers the user an improved insight into the most appropriate gasket for a particular application, based on comparison of room temperature leakage rate performance.

PVRC CONVENIENT METHOD

GASKET TYPE	CARRIER RING GASKET				
MATERIAL	GRAPHITE FILLED				
Inside Diameter (in)	ID	4.63			
Outside Diameter (in)	OD	5.88			
Design Pressure (psi)	Pd	3750.00			
Test Pressure (psi)	Pt	5625.00			
Tp = 1 Intercept	Gb	1251.00			
Seat Loading Slope	a	0.3090			
Unload Intercept	Gs	11.0000			
Assembly Efficiency	E	0.75			
Gasket Type	(1 or 2)	1			
No of Bolts		8			
Bolt Area (Per Bolt)		0.929			
Tightness Class	T1 to T3	1.0	1.5	2	2.5	3
Ref Leak Rate (mg/sec/mm)		0.2	0.02	0.002	0.0002	0.00002
Exp Leak Rate (mg/sec)		29.9	2.99E+00	2.99E-01	2.99E-02	2.99E-03
Tightness Constant	C	0.10	0.32	1.00	3.16	10.00
Effective Sealing Width	b	0.38	0.38	0.38	0.38	0.38
Diameter of Load Reaction	G	5.01	5.01	5.01	5.01	5.01
Gasket Area (in^2)	Ag	9.18	9.18	9.18	9.18	9.18
Hydrostatic Area (in^2)	A1	19.67	19.67	19.67	19.67	19.67
Tightness Parameter	Tp min	46.61	147.40	466.13	1,474.02	4,661.25
Tightness Parameter	Tpn	69.92	221.10	699.19	2,211.03	6,991.88
Tightness Parameter Ratio	Tr	1.11	1.08	1.07	1.06	1.05
Operating Stress (psi)	Sm1	2,610	4,101	6,231	9,291	13,698
Seating Stress (psi)	Sm2	(3,909)	(2,144)	375	3,871	9,102
2P (psi)	2P	7,500	7,500	7,500	7,500	7,500
Design Factor	Mo	0.70	1.09	1.66	2.48	3.65
Design Bolt Load (lbf)	Wmo	97,723	111,411	130,954	159,036	199,471
Gasket Stress (psi)	SG	10650.00	12141.72	14271.46	17331.92	21738.58
Gasket Stress (Mpa)	SG	73	84	98	119	150
Bolt Stress (psi)	Bs	13,149	14,991	17,620	21,399	26,839
	Bs	91	103	121	148	185
Code Like Factors						
m (maintance Factor)		0.58	0.91	1.38	2.06	3.04
Ya (seating stress)		6885.46	9827.27	14025.97	20018.57	28571.50

Table 2

The resulting information was collated and plotted in graph form as a comparison of alternative gasket design that may be selected on a high pressure / temperature application. [Figure 4].

If we make a comparison between the 4 various gasket types highlighted, we can appreciate that significantly higher bolt stress values are required upon both the asbestos filled and metal jacketed gaskets than for the Carrier Ring (CR) and the graphite filled SWG to gain the same levels of tightness.

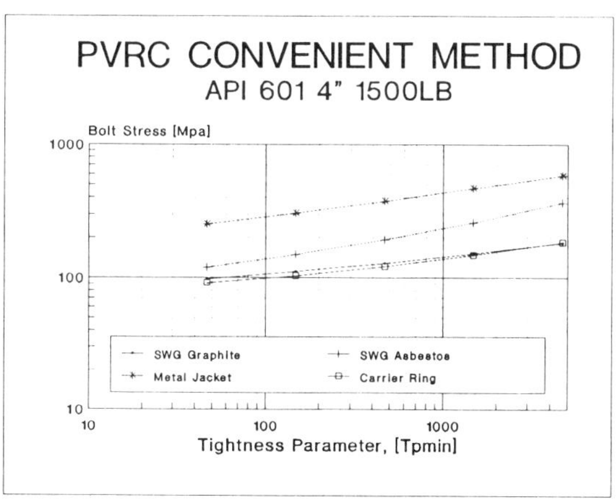

FIGURE 4

We   have now assessed the tightness performance of the Carrier Ring (CR) Spiral Wound Gasket at ambient conditions. Further to this screening process, we must now establish the gaskets ability to perform at elevated temperature conditions.

In order to evaluate performance at temperature, Finite Element Analysis techniques were introduced.

At elevated temperature or during thermal transient conditions, the behaviour of a bolted flange joint closure does not behave in a manner identical to that expected at ambient conditions. Factors such as bolt relaxation, flange distortion and differential flange expansion are significantly exaggerated, leading to possible load loss and movement within the bolted flange assembly.

In order to overcome these compounded problems, the need of a high recovery static seal is paramount, thus ensuring that the effects of excessive differential thermal expansion and potential load loss are counterbalanced by the static seals inherent mechanical features. The Carrier Ring offers these features as an integral part of the design solution.

Ensuring that the metallic body of the Carrier Ring is manufactured from similar material as that of the flanges, any expected radial expansion may be accommodated in the design, thus restricting any "slip" between sealing faces. In addition, the double spiral wound sealing elements(incorporating heat treated precipitation hardened stainless steel windings), offer recovery characteristics more than double that of a standard spiral wound gasket, approximately 0.8 to 1.0mm, ensuring the maintenance of applied surface stress upon the joint and the continuation of an high integrity seal under extreme cyclic conditions.

As part of a longer term Finite Element package investigating behaviour of bolted flange joints and gasket behaviour, a program has commissioned to evaluate the performance of a Carrier Ring gasket on a particular critical high temperature, thermal cyclic application.

CASE STUDY

A Finite Element Analysis of a flange, carrier ring and gasket assembly subjected to bolt loads, internal pressures and elevated temperatures using the Abacus finite element code was commissioned for a particular critical high temperature application.

The particular vessel in question contains gas at fluctuating temperatures up to 900 Deg C and pressures of 5 Bar. Finite Element Analysis (FEA) was performed in order to investigate the compatibility of radial expansion of the proposed Carrier Ring gasket and the mating flange assembly and the ability of the gasket sealing elements to maintain a high tightness criteria during thermal transient conditions.

The flange itself is made from a nickel based alloy, with mating Nimonic bolting. Gasket selection specified the use of an Inconel 825 metallic body with Inconel X750 HT / Graphite based sealing elements. The load deflection characteristics of the spiral wound sealing elements was represented by a hyperelastic material model giving the required stress of 68.95 Mpa (10,000 psi) when compressed to the level of the metallic body.

The assembly was represented by an axisymetric model. This modelling technique assumes that the structure and any load applied to it are constant and continuous around the circumference of the model.

RESULTS

In the analysis, the flange assembly was subjected to applied bolt loads, application of internal pressure and temperature. Figures 5 and 6 show that there is a relatively large initial clearance between the flange and gasket which is closed when bolt load is applied. The effect of the temperature is obviously evident by the extent of radial growth. This radial growth is labelled at various points around the assembly in Figure 6.

The stress nominal to the plane of the gasket for loads are shown in Figure 7. It can be seen that when the temperatures are applied the average gasket compressive stress is reduced by approximately 5%, from 68.9 MPa to 65.8MPa. This is the result of additional flange rotation induced by thermal effects.

Other effects of the increased flange rotation can be seen in the diagrams of flange and carrier ring stresses Figure 8. As the flange rotates the centre of pressure between the flange and gasket moves radially outwards. This causes a region of high contact stress to appear on the Carrier Ring at the outside diameter of the raised face on the flange

Flange Deflections

Bolt Loads, Internal Pressure and Temperature

FIGURE 5

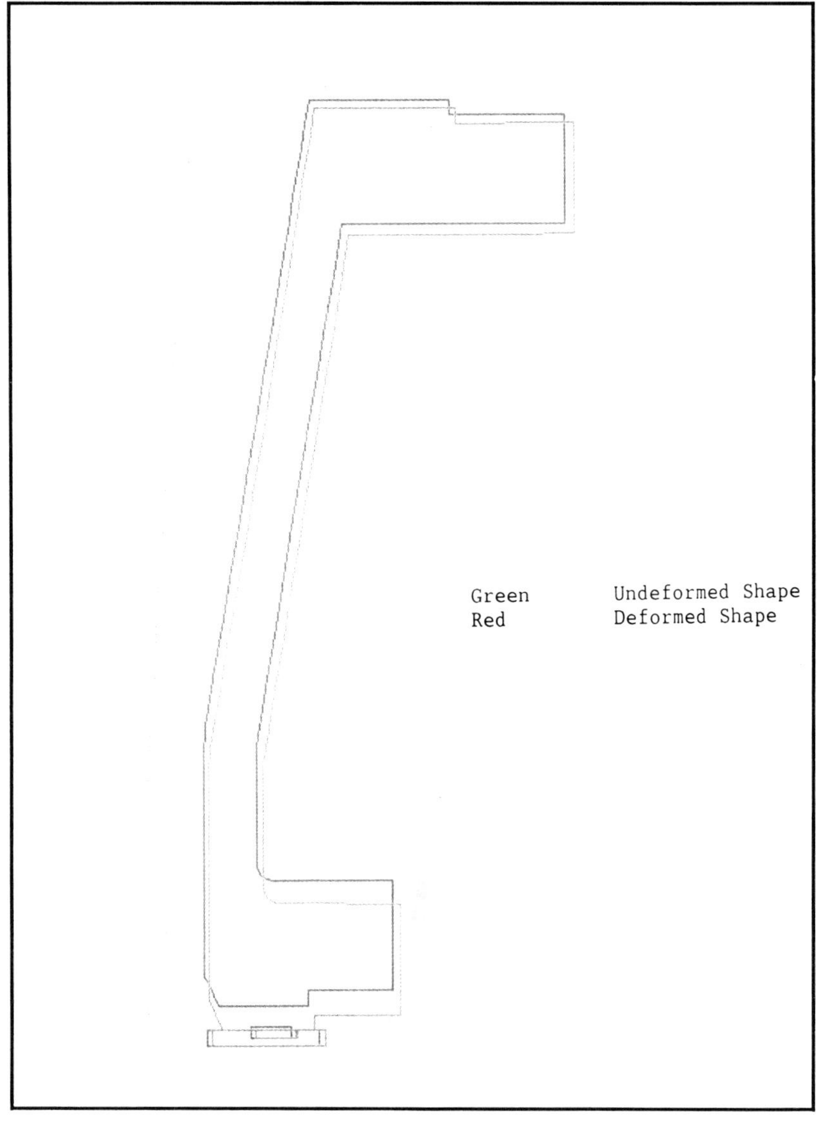

Flange Deflections

Bolt Loads, Internal Pressure and Temperature

FIGURE 6

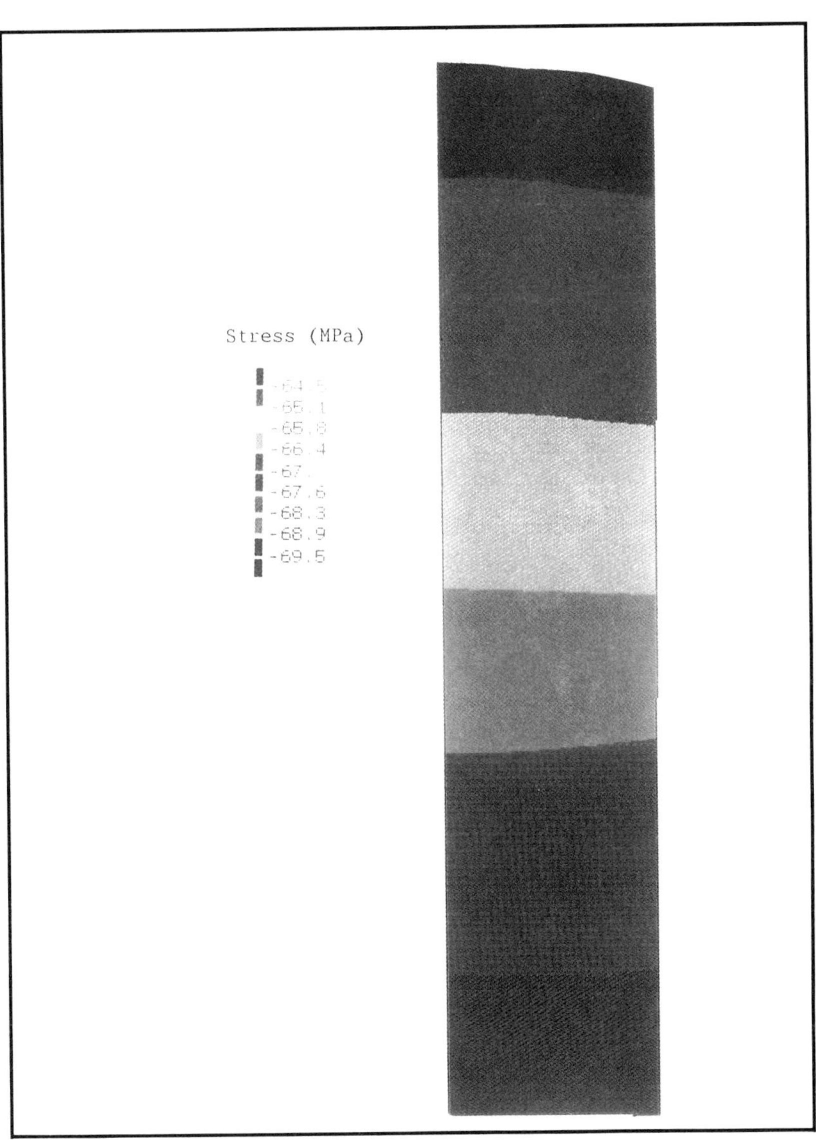

Gasket Normal Stress

Bolt Loads, Internal Pressure and Temperatures

FIGURE 7

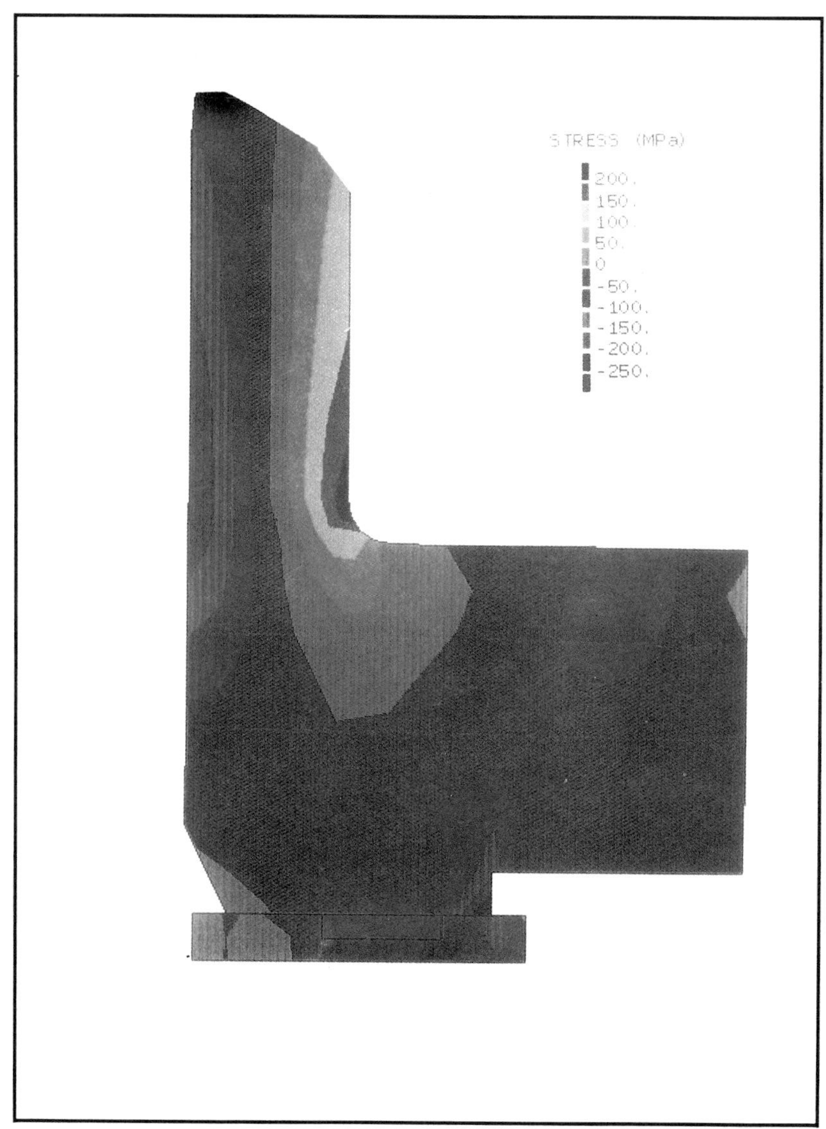

Flange, Gasket and Carrier Ring Assm - Axial Stress

Bolt Loads, Internal Pressure and Temperatures

FIGURE 8

CONCLUSIONS

From the results obtained, it can be established that the Carrier Ring Spiral Wound Gasket is suitable for the operating conditions specified. From Figure 6 it can be seen that the metallic core component of the assembly exhibits similar deflection characteristics to that of the flange, thus ensuring that radial "slip" is not allowed to impair sealing characteristics. From Figure 8 it can be established that rotation of the flange occurs 0 26.4', acting to increase contact stress,(although not of untoward concern), between the Carrier Ring and the outside diameter of the raised face thus acting to unload the gasket sealing element. However, due to the elevated recovery characteristics of the double spiral wound gasket concept, unloading is limited to an acceptable 5%,

The conclusion from the analysis, which has since been proven in field trials, is that for this particular thermal transient application, the Carrier Ring (CR) concept offered an acceptable solution, offering a high level of continued performance without inducing unnecessary stresses within the flange assembly.

DISCUSSIONS

The Carrier Ring (CR) gasket incorporates the improved features of a double high recovery spiral wound gasket sealing arrangement. A central metallic core offers a high degree of strength and rigidity minimising the risk against a blow out condition and offering excellent handleability characteristics whilst allowing a positive compression stop indicating the optimum compressed thickness has been achieved .

Long term, high thermal transient tests have indicated this concept to be capable of maintaining a high degree of tightness, even under the most enduring conditions.

As discussed earlier, this gasket concept has the ability to achieve a higher degree of static sealing performance than a standard spiral wound gasket under the same assembly conditions, offering the plant operator a higher degree of safety aimed at reducing fugitive emissions.

The Carrier Ring "CR" Spiral Wound Gasket concept may be adopted on standard pipework assemblies, heat exchanger arrangements, high pressure heaters and nuclear applications to solve persistent elevated temperature leakage problems.

REFERENCES

Note : All data presented in this paper is based on current available published information. PVRC researchers continue to refine data reduction techniques and values are therefore subject to change. The PVRC methods have not, as yet, been adopted by the ASME Code.

1) Winter J.R. " Gasket Selection - A Flow Chart Approach", 2nd International Syposium on Fluid Sealing, La Baule, France, 1990

2) Payne J.R. " Traditional Vs New Bolt Load Calculations", Panel Presentation, ASME PVP Conference, San Diego, CA, June 23-27, 1991

3) Pressure Vessel Research Commitee of the Welding Research Council, 345 E. 47th St, New York, NY 10017

4) Payne J.R., Bazergui A, Leon G.F. " Getting New Gasket Design Constants for Gasket Tightness D" Experimental Techniques, Society for Experimental Mechanics, Vol 12, No.11 1988, pg 22s-27s

5) Wright V.C " Finite Element Analysis of a Flange / Carrier Ring Assembly, T&N Technology, Cawston, Rugby, England, October 1992.

14th International Conference on Fluid Sealing, Firenze, Italy,
6-8 April 1994. Organised by BHR Group Limited, Cranfield,
Bedford, MK43 0AJ, UK; Tel: 0234 750422

Experimental Investigation Of Leakage In Static Seals Subjected To Reciprocating Vibration

Filip Rosengen
Machine and Vehicle Design
Chalmers University of Technology

Abstract

Leakage in static seals is a well known fact as well as a wide spread problem among users and manufacturers of the same. The purpose of this paper is to investigate if and under what circumstances a static seal subjected to a reciprocating vibratory motion produces leakage. Both O-rings and so called square-rings were tested with different interfacing surfaces at different operating conditions. It was found that the seal geometry as well as the ratio of stroke to seal contact length together with the surface roughness of the interface surface are important factors.

Notation

x	Coordinate in axial direction	R_a	Arithmetic mean of surface profile from mean line
q_x	Fluid flow per unit width along x-axis	R_z	ISO 10-point height
U	Relative sliding speed along x-axis	R_{max}	Maximum distance between highest peak and lowest valley within one filter cut-off length
h	Fluid film thickness		
p	Pressure distribution along seal contact	R_p	Distance between highest peak and mean line
η	Viscosity of oil		

1 Introduction

Compact seals are often used in hydraulic systems to prevent leakage. There are two main applications of a compact seal — static and dynamic. The dynamic case can be defined by the existence of relative motion between the seal and its interface.

The dynamic application is common in, e.g., hydraulic cylinders where compact seals often are used as piston and piston-rod seals.

A static seal is characterized by the absence of relative motion between seal and interface. It can be shown that such a seal is not capable of sustaining a fluid film. In other words it can not leak. According to Reynolds equation for the iso-viscous two-dimensional case the fluid flow along the x–axis, q_x, per unit width can be written as:

$$q_x = \frac{Uh}{2} - \frac{h^3}{12\eta} \frac{\partial p}{\partial x} \,, \qquad (1)$$

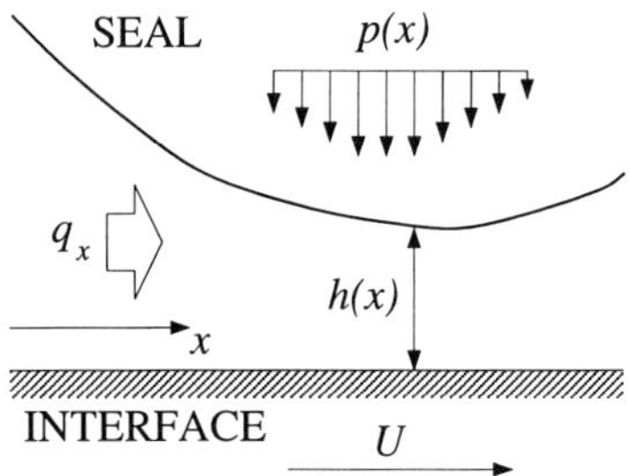

Figure 1: Fluid flow q_x through a seal contact

where U is the relative velocity between seal and interface along the x–axis, $h = h(x)$ is the fluid film thickness and $p = p(x)$ is the pressure distribution over the contact. For a static seal $U \equiv 0$ which means that q_x is a function of p and h. Most compact seals are designed so that a pressure maximum is attained when the seal is squeezed against its interface. This means that $\frac{\partial p}{\partial x}$ must change sign somewhere in the seal contact giving a fluid flow q_x, both in positive and negative direction, originating from where $\frac{\partial p}{\partial x} = 0$. Since no fluid is generated inside the contact the film thickness h will diminish until the contact is drained of fluid and the seal will not leak. This last statement is in conflict with practical experience from, e.g., the field of hydraulics where there are large problems with leakage in static seals.

In order for leakage to occur it is necessary that the pressure distribution is affected in such a way that the pressure gradient is reduced, locally or globally and/or that the static seal is disturbed by a relative motion between the sealing surfaces. A number of possible factors could cause such a situation, e.g.,

1. relative motion between seal and interface due to mechanical vibrations,
2. effects of surface roughness,
3. initial mounting squeeze of the seal,
4. choice of seal geometry,
5. choice of sealed fluid,
6. variations in the sealed pressure,
7. wear of the seal,
8. aging effects on the elastomer material in the seal,
9. particle contamination of the sealed fluid and so on.

This list could be made much longer. However the objective of this paper is to investigate the influence of the first four items, namely the effect of relative, vibratory motion between seal and interface on leakage and how it is affected by items two through five in the above list. Only axial motion, (along the x-axis in Figure 1), is considered to reduce the problem. However it is reasonable to assume that in many hydraulic applications axial vibrations are of a greater magnitude than radial ones.

2 RELATED WORK

The dynamic application of compact seals for reciprocating motion is fairly well investigated. Since Denny [1] published his pioneer work in 1950 many authors have investigated the problem. In recent years Johannesson [2] and Kassfeldt [3] have performed an extensive theoretical and experimental study of friction and leakage in hydraulic cylinder seals. Olsson [4] recently developed a theory for friction in seals acting on rough piston rod surfaces. Many others have studied various aspects of a dynamic compact seal such as the development and distribution of the fluid film and the effect of different seal geometries and/or choice of interface design on sealing capability. However, almost all investigations of dynamic compact seals for reciprocating motion are made under the assumption of constant relative sliding speed U. No published material on static seals was found by the author of this paper. There are very few examples in the literature where U is a function of time. Hirano and Kaneta however have presented one experimental and several theoretical investigations [5, 6, 7, 8, 9] where they study the effects of varying length of stroke and frequency of the reciprocating motion on friction and leakage in seals. Their investigations are concentrated on determining the stability of the periodic fluid film in the seal contact and how the stroke length and frequency of the reciprocating motion influence friction and leakage. In order to experimentally determine the stability limit of the periodic film in [9] the seal friction force was measured and a collapse of the film could be detected through an increase in friction. The investigation shows that the film collapse initially occurs at the stroke ends where the relative velocity is low. It also shows that even if the film collapses at the stroke ends it can still exist during the rest of the stroke which could make leakage possible even at relatively short strokes. However the authors of [9] do not report the existence of leakage at the stability limits of the film. Hence no conclusions can be made from [9] whether or not leakage can exist for ratios of stroke length to seal contact length near unity.

3 EXPERIMENTAL EQUIPMENT

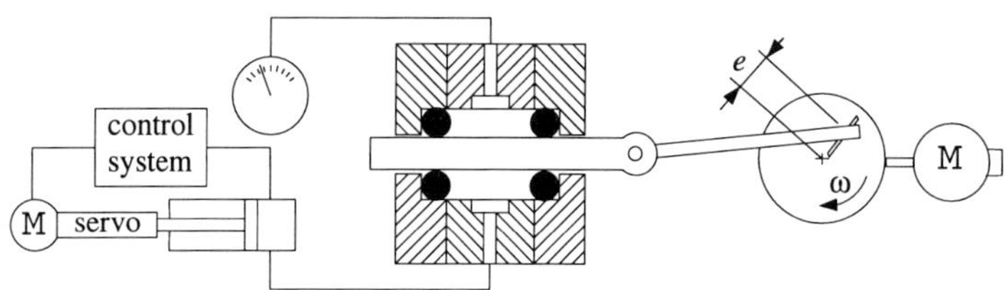

Figure 2: Schematic figure of experimental equipment

3.1 PRESSURE CHAMBER

The experimental equipment is very simple and basically consists of a cylindrical pressure chamber with two lids in which the test seals are mounted. The seals are radially squeezed against a steel shaft, Ø 20 [mm]. The play between lid and shaft is a standard H9/h8 fitting. The squeeze of the seals can be chosen by using different lids. The pressure of the oil in the chamber is set by a servo controlled hydraulic cylinder. The servo is in turn monitored by a control system keeping the pressure at a constant value with an error of less than 1%. The temperature and pressure of the oil in the chamber can be directly measured via connecting gauges.

3.2 OSCILLATOR

The shaft can be made to perform an axial vibratory motion with a stroke of zero to eight [mm] by means of a connecting rod which is eccentrically mounted to a rotating disc. The shaft is supported by a linear bearing, (visible in Photo A). The accuracy of the variable eccentricity e is 0,01 [mm]. The frequency of vibration can be varied from 1 to 8 [Hz] with an accuracy of 0,02 [Hz].

Photo A: Photo of the test rig in working condition

3.3 SEALS

Two different types of seals were tested, (see Figure 3):

- A standard nitrile, 70° shore, 19,2×3 [mm] O-ring and
- a nitrile, 70° shore, 18,7×2,5 [mm] □-ring.

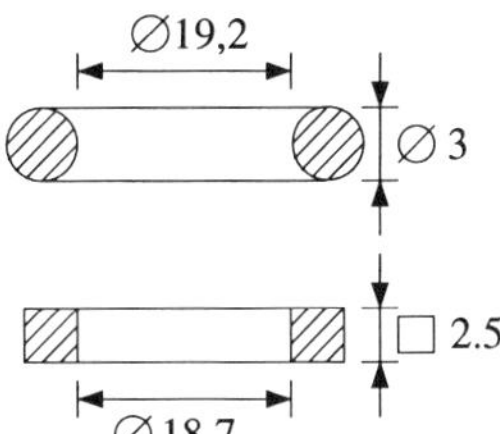

Figure 3: Different seal types used

The radial squeeze of the O-ring seal can be chosen as 10 or 20%. The radial squeeze of the □-ring is 9% according to manufacturer specifications. Both seals are mounted in an open groove, (se Figure 2).

SEAL CONTACT LENGTH

The contact length between shaft and seal was measured by using a method first presented by May [10] and will only be briefly described here. A special hollow piston, (see Figure 4) with a small hole of Ø 0.13 [mm] drilled through the envelope surface was manufactured.

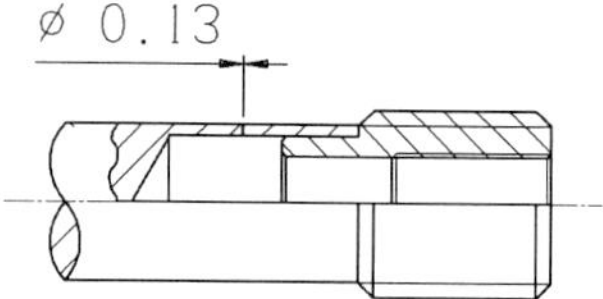

Figure 4: Schematic figure of hollow piston.

The piston has the same diameter as the shaft in Figure 2 and can be mounted in the pressure chamber. The piston can then be pressurized from the inside using a hydraulic hand pump. By using a linear actuator the piston can be axially positioned in the chamber. If the pressure inside the piston is raised above the seal contact pressure when the small hole is situated somewhere in the seal contact, oil will leak through the small hole and protrude in the direction of the local pressure gradient. If the hand pump is locked the pressure inside the piston will drop until it equals the seal contact pressure which then can be registered. It is however very time consuming to register enough points of the seal contact to get a good estimate of the whole pressure distribution since it takes considerable time for the pressure in the piston to settle. Therefore only the seal contact length was measured.

The contact length was measured for four different sealed off pressures. The results of the measurements are shown in Table II.

3.4 SHAFTS

Two shafts were manufactured with different techniques in order to investigate the effect of surface roughness. *Shaft 1* was made in a turning lathe by a manufacturer of hydraulic couplings and represents a "typical" sealing surface used for industrial purposes. *Shaft 2* was drawn, centreless-ground, chronium-plated and polished to produce a very smooth surface to be used as a reference in the experiments. A surface analysis of the shafts were made by means of a *Perthometer* profileometer. The surfaces of the shafts were only analyzed in the axial direction. The results of the analysis is shown in Figure 5 and Table I. It is noticeable that the dominating wavelength of the turned shaft, (see Figure 5C), agrees well with the pre set feed of 0,15 [mm] used in the turning lathe.

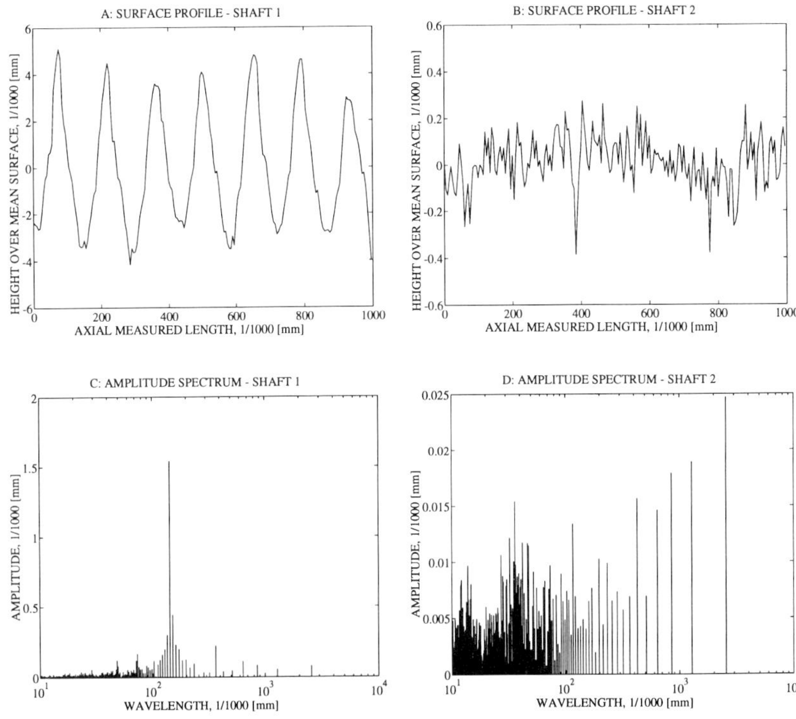

Figure 5: Measured surface profiles and amplitude spectra of the two shafts — only 1/5 of the measured lengths are shown in A and B to improve resolution.

Table I: Surface parameters of *Shaft 1* and *2* measured with a filter cut-off length of 0,8 [mm].

	Surface parameters in [μm]			
	R_a	R_z	R_{max}	R_p
Shaft 1	2,4	10,4	11,0	6,1
Shaft 2	0,10	1,19	1,81	0,57

3.5 TESTED OILS

Two different oils were tested:

Shell Tellus 68   A standard mineral based hydraulic oil with a dynamic viscosity of 0,12 [Ns/m^2] at 25° [C].

Shell Helix Ultra   A synthetic motor oil, (SAE 5W-40), with a dynamic viscosity of 0,022 [Ns/m^2] at 25° [C].

The dynamic viscosity of the oils were measured using a *Ferranti-Shirley* rotational viscometer.

4 EXPERIMENTS

4.1 CONDITIONS

The experiments were carried out at room temperature. The bulk temperature of the oil in the pressure chamber never exceeded 25° [C]. Great care was taken when mounting the seals in the pressure chamber to ensure that they did not twist or bend in their grooves. It was also made certain that the shaft and chamber were properly aligned before securing the chamber to its support. After the chamber was mounted it was pressurized and vented through a bleeder screw.

4.2 PROCEDURES

In order to determine whether a specific seal would leak under given operating conditions it was mounted in the pressure chamber and run for a fixed number of reciprocations. If leakage occurred during the specified interval it could easily be ocularly detected as a trickle of oil emerging from the clearance between shaft and lid (see Photo B). It can be argued that an ocular inspection is not sufficient to determine a very small amount of leakage since it would exist only as a very thin film on the low pressure side of the seal.

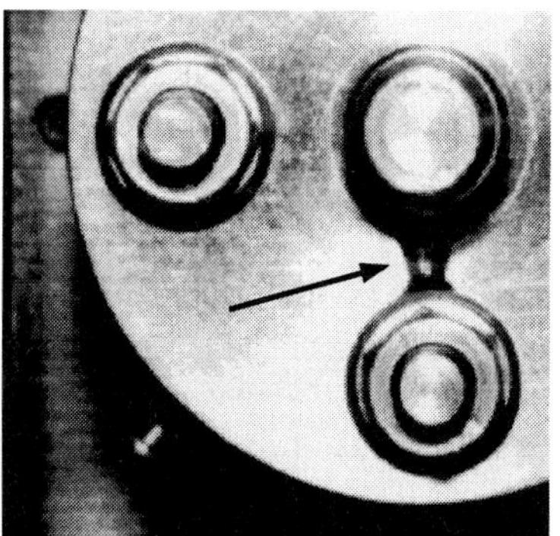

Photo B: Photo of leakage protruding between shaft and lid

Some tests with microscopic observation showed however that the ocular inspection proved sufficient. Extensive testing revealed that 1000 reciprocations were more than enough to determine if the seal would leak or not. This fact reduced the risk of long time wear which probably would interfere with the desired test situation.

MEASUREMENT OF SEAL CONTACT LENGTH

Table II: Seal contact lengths in [mm] for two O-ring cases and one □-ring case at four different sealed off pressures

Seal type	Squeeze	Sealed off pressure in [MPa]			
		6	8	10	12
O-ring	20%	2,40	2,45	2,50	2,55
O-ring	10%	1,70	1,75	1,80	1,85
□-ring	9%	2,70	2,75	2,75	2,75

The estimated error of the measurements is about ±0,05 [mm]. Three sources of error can be anticipated namely:

1. *errors due to resolution* — As the small hole approaches the seal contact edge it is only partially "covered" by the seal which makes it difficult to find the "true" edge.
2. *errors due to low pressure difference* — It is possible that the contact lengths measured at the lower pressure, 6 [MPa], suffer from errors owing to difficulties in finding the contact edge on the high pressure side of the seal. Since the pressure difference between the sealed off fluid and the contact pressure at a distance of 0,13 [mm], (Ø of the small hole), from the contact edge is relatively small oil could still emerge from the piston even if the small hole is "covered" by the seal.
3. *errors due to friction* — Friction forces will develop when the piston moves along the seal. These forces are however negligeable compared to the sealed of pressure acting on the seal.

It can be noted from Table II that the contact length of the □-ring is hardly affected by the magnitude of the sealed off pressure. This is not surprising since the □-ring conforms well with itś groove, resulting in a almost hydrostatic loading of the seal when the sealed off pressure is applied. The O-ring, on the other hand, do not conform with itś groove to the same extent, which makes the seal contact length of the O-ring more sensitive to the magnitude of the sealed off pressure.

LEAKAGE DETECTION

The experiments were carried out in the following manner for both of the two different shafts:

1. For each of the twelve cases in Table II the frequency of reciprocation was fixed at a constant value.

2. The stroke length of the reciprocating motion was then increased in steps until leakage was detected. For each step the used seals were replaced by new ones to ensure identical conditions for each test. The used seals were also checked for damage to rule out the possibility of leakage due to seal failure.

3. When leakage was detected the stroke length was fixed at the critical value and the frequency of reciprocation was varied in the same way as the stroke length to investigate its influence on leakage.

4. The stroke length was then varied by ±10% to ensure the validity of the critical length determined under item 2.

Additional testing was then performed to investigate the influence of the choice of oil.

4.3 RESULTS

SEAL GEOMETRY

The seal geometry, and hence the seal type, proved to be the most significant factor of the ones tested in this paper. There was a remarkable difference between the two seal types. The □-ring could not be made to leak even for the maximum stroke length and driving frequency of 8 [mm], (approximately three times the seal contact length), and 8 [Hz] respectively. Not even the choice of shaft had any influence in this case.

STROKE LENGTH AND SURFACE PROPERTIES

The second most important factor proved to be the stroke length in combination with the choice of shaft. Experiments with *Shaft 1* required a stroke length very close to the contact length of the seal to produce leakage. The same experiments with *Shaft 2* made it necessary to increase the stroke length to approximately 1,5 times the seal contact length in order to get leakage. The stroke in relation to the seal contact length necessary to produce leakage at the different squeezes, sealed off pressures and with the two different shafts are presented in Table III.

Table III: Ratio of stroke to seal contact lengths necessary to produce leakage using O-ring seals and *Shaft 1 / Shaft 2* respectively

	Sealed off pressure in [MPa]			
Squeeze	6	8	10	12
20%	1,03 / 1,55	1,02 / 1,52	1,05 / 1,54	1,08 / 1,53
10%	1,04 / 1,51	1,01 / 1,53	1,03 / 1,52	1,05 / 1,54

FREQUENCY OF RECIPROCATION

The frequency of reciprocation had no visible effect on the existence of leakage in the tested range.

MOUNTING SQUEEZE AND SEALED OFF PRESSURE

Neither the mounting squeeze nor the level of the sealed off pressure did directly influence the experiments. The only correlation found between these parameters and the existence of leakage was their influence on the seal contact length.

OIL

The choice of oil showed no influence on whether or not leakage could be detected for a given set of operating conditions.

SEAL DAMAGE

For some of the non-leaking O-rings a very characteristic type of damage occurred namely so called "spiral failure". This type of damage happens when a section of the O-ring is drawn into the clearance between shaft and lid due to their relative motion.

Figure 6: Sketch of an O-ring section with spiral cuts

As the shaft continues its inward motion more and more of the O-ring section will be drawn into the clearance forcing it to twist, leaving the spiral cuts shown in Figure 6. The damages only occurred when no leakage was detected, and as a rule; under conditions of a high ratio of stroke to seal contact length. Some tests showed that the damages could be avoided if the O-ring was lightly pre-lubricated with grease.

5 CONCLUSIONS AND DISCUSSION

The level of the sealed off pressure and the mounting squeeze determine the pressure distribution in the seal contact together with the material properties of the seal. Neither the squeeze nor the sealed off pressure seems to have any direct influence on the experiments. Both did however influence the seal contact length and can therefore be said to have an indirect effect on the existence of leakage.

Nor did the frequency of reciprocation seem to have any influence. This can be explained by the fact that the frequency which determines the relative sliding speed could not be set lower than 1 [Hz] in the experiments. It is very likely that the frequency would have influenced the results had this been possible since an infinitely low frequency would result in a static seal with no relative motion. Hence the pressure dependant Poiseuille term in equation (1) must be exceeded by the speed dependant Couette term on the high pressure side of the seal in order to establish a fluid flow through the contact.

This fact can also explain why there was no leakage in the □-ring case. The sharp corners of the □-ring do not provide the seal contact with a necessary inlet zone to form a fluid film and the pressure gradient on the high pressure side of the seal can not be overcome by the sliding speed. Field and Nau [11] have presented experimental results on □-rings which show that the seal edge form is of importance to conditions in the seal inlet zone. Gibson, Hooke and O'Donoghue [12] and [13] have published theoretical results that show the significance of the inlet zone and its influence together with the sliding speed on the seal pressure distribution.

Evidently the surface properties of the seal interface is of importance. The experiments show that it is necessary to increase the ratio of stroke to seal contact length by approximately 50% in order to produce leakage when using a "smooth" surface instead of a rough one. However the tests with the □-ring, which did not leak under any circumstances, implies that this phenomenon is not bound to the surface properties alone but must also be coupled with the hydrodynamic properties of the contact. If the influence of surface roughness depended only on the surface itself the □-ring would also have produced leakage. No published material on this phenomenon has been found by the author. Thus further work should be performed to investigate the influence of surface roughness and inlet zone.

ACKNOWLEDGEMENTS

This work was carried out at Machine and Vehicle Design at Chalmers University of Technology, Göteborg, Sweden. It was supervised by Dr Hans Johannesson and professor Göran Gerbert whom are thankfully acknowledged for their advice and encouragement. I would also like to express my thanks to my father as well as to my colleagues for providing valuable discussions and suggestions and to M Sc Magnus Jonasson at Manufacturing Engineering at Chalmers University of Technology for his assistance with surface measurements. The foundation for fluid system technology — IFS, (in Swedish: Intressentföreningen för Fluid Systemteknik), provided financial support.

REFERENCES

[1] Denny D. F. The friction of flexible packings. *Proc Instn Mech Engrs*, 163:98–102, 1950.

[2] Johannesson H.L. *On the optimization of hydraulic cylinder seals*. PhD thesis, University of Luleå, Sweden, S-951 87 Luleå, 1980.

[3] Kassfeldt E. *Analysis and design of hydraulic cylinder seals*. PhD thesis, Luleå University of Technology, Sweden, S-951 87 Luleå, 1987.

[4] Olsson J. Friction in seals acting on rough piston rod surfaces. lic. thesis, Machine and vehicle design, Chalmers University of Technology, S-412 96 Göteborg, Sweden, November 1990.

[5] Hirano F. Dynamic inverse problems in hydrodynamic lubrication. In *3rd international conference on fluid sealing, Cambrige, England*. The British hydromechanics research association, Cranfield, Bedford, April 1967. paper F1.

[6] Hirano F. Kaneta M. Dynamic behaviour of flexible seals for reciprocating motion. In *4th international conference on fluid sealing, Cambrige, England*. The British hydromechanics research association, Cranfield, Bedford, 1969. session 1.

[7] Hirano F. Kawahara Y. Kaneta M. A study on the characteristics of o-rings for reciprocating motion. In *26th national conference on fluid power, Chicago, Illinois*. Illinois Institute of Technology, Chicago, Illinois, October 1970.

[8] Hirano F. Kaneta M. Theoretical investigation of friction and sealing characteristics of flexible seals for reciprocating motion. In *5th international conference on fluid sealing, Cambrige, England*. The British hydromechanics research association, Cranfield, Bedford, March 1971. paper G2.

[9] Hirano F. Kaneta M. Experimental investigation of friction and sealing characteristics of flexible seals for reciprocating motion. In *5th international conference on fluid sealing, Cambrige, England*. The British hydromechanics research association, Cranfield, Bedford, March 1971. paper G3.

[10] May E.M. Pressure drop accross a packing. *Applied Hydraulics*, pages 110–114, May 1957.

[11] Field G. F. Nau B. S. An experimental study of reciprocating rubber seals. *Proc Instn Mech Engrs*, C5:29–36, 1972.

[12] Hooke C.J. O'Donoghue J.P. Elastohydrodynamic lubrication of soft highly deformed contacts. *Journal of Mechanical Engineering and Science*, 14(1):34–48, 1972.

[13] Gibson I.A. Hooke C.J. O'Donoghue J.P. The frictional behaviour of reciprocating O-ring seals under starting conditions. *Instn Mech Engrs*, C13:77–82, 1972.

14th International Conference on Fluid Sealing, Firenze, Italy,
6-8 April 1994. Organised by BHR Group Limited, Cranfield,
Bedford, MK43 0AJ, UK; Tel: 0234 750422

EMISSION CONTROL PACKING FOR HIGH TEMPERATURE/HIGH PRESSURE VALVES AND RELATED DEVICES

Takahisa Ueda, Tomikazu Shiomi, Kazuhiko Ootawa
Engineering Department, Nippon Pillar Packing Co., Ltd.
Nonakaminami 2-chome, Yodogawa-ku, Osaka, Japan

ABSTRACT

Gland packings with high reliability in emission control and maintenance-free convenience are keenly demanded in today's field of high temperature and high pressure valves. To meet this demand, various proposals have been made, including modification of valve structure such as live loading.

By braiding expanded graphite yarns compounded with reinforcing material, we have recently developed a novel packing construction. This packing features the excellent heat resistance and sealing performance attributable to the properties of the materials, and ease of handling common to all braided packings.

In this paper, the packing stress relaxation and creep properties that determine the sealing performance of high temperature and high pressure valves, especially on/off valves, are investigated, together with their behavior in high temperature conditions. In addition, the gas sealing characteristic of the packing and results of the characteristic comparison with conventional packings in actual valves are also reported.

1. INTRODUCTION

In high temperature and high pressure valves at thermal power plants, hitherto, asbestos packings, or assembled packings of asbestos and expanded graphite were used. However, the asbestos packings were noted for the major defect of large stress relaxation. Expanded graphite die molded packings are small in mechanical strength, and they are usually deformed as extrusion from the gap between the stem and the stuffing box or gland (1). The expanded graphite is also known to increase in the frictional force due to its adhering to the stem surface. Accordingly, at both ends (top ring, bottom ring) of the expanded graphite die molded packing, fiber packings other than expanded graphite, such as asbestos

72

fibers and carbon fibers are installed (2).

These conventional packings except for expanded graphite ones become hardening in the long course of use (3), or are large in the initial compression rate and fibers may fracture (4). To overcome such behaviors, the required sealing performance is maintained by retightenings of the gland studs.

As to changes during use of gland packings, stress relaxation of soft packings (5), stress relaxation due to repeated retightenings (6), and packing behavior in live loading mechanism (7) have been reported. Little has been mentioned , however, about the stress relaxation or creep behavior of packings at high temperature. As a practical evaluation, in this study, the stress relaxation and creep characteristics of packings at high temperature were investigated.

2. TESTS

2.1 Sample packings

The sample packings are expanded graphite die molded packing A, two types of expanded graphite braided packings B and C, and conventional packing D.

The structure of expanded graphite yarn is illustrated in Fig. 1. The outline of the sample packings is shown in Table 1, and the relation between tests and sample packings is shown in Table 2.

The expanded graphite yarn is a new type providing the expanded graphite with strength and flexibility by reinforcing the periphery of the narrow expanded graphite tape assembly with a small amount of metal wires.

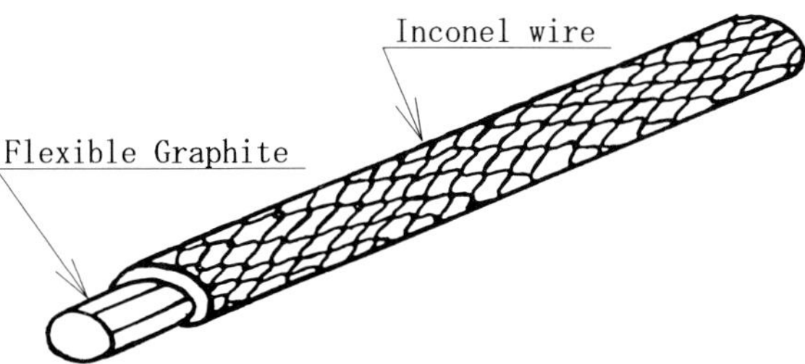

Fig. 1 Structure of expanded graphite yarn used in B and C

Table 1.　Types and structures of sample packings

Sample	Appearance	Structure
Packing A		Expanded graphite tape and heat resistant alloy mesh of high nickel content are combined, wound spirally, and pressed and formed by a die. This molded packing is easy to remove from the stuffing box.
Packing B		The packing by braiding the yarn illustrated in Fig. 1 is pressed and formed by a die. Used as top ring and bottom ring of packing A.　Not used alone because it is not treated with lubricant.
Packing C		After braiding the yarn illustrated in Fig. 1, the packing is treated with a small amount of graphite and lubricant, and pressed and formed by a die. Used alone.
Packing D (Conventional packing)		After the surface of the core made of asbestos and graphite is covered with braided asbestos fibers which are reinforced with high nickel content heat resistant wire, the packing treated with graphite and lubricating oil for impermeability is pressed and formed by a die.

Table 2.　Tests and sample packings

Test	Stress relaxation	Creep	Reciprocating	In field
Sample packing	A	A	B+A+A+A+B	B+A+A+A+A+B
	B	B	----	C+C+C+C+C+C
	C	C	----	D+A+A+A+A+D
	----	----	----	D+D+D+D+D+D

2.2 Test apparatus and test method

2.2.1 Stress relaxation test

The test apparatus is shown in Fig. 2. The apparatus is composed of the testing rig and measuring unit. The testing rig consists of a stem (32 mm in diameter), stuffing box (48 mm in inner diameter), packing retainer, tightening bolts, heater, and load cell for measuring axial tightening stress of the packing.

Test method: The one ring packing is tightened at specified tightening stress (20, 30, 40, 60 MPa) in the stuffing box. The subsequent changes in the tightening surface pressure are continuously recorded through the dynamic strain amplifier. Similarly, tightening stress changes are measured at the stuffing box temperatures of 200 and 300 ℃. The packing is tightened after temperature rise of the rig. Therefore, the effect of volume expansion of the rig at high temperature is neglected.

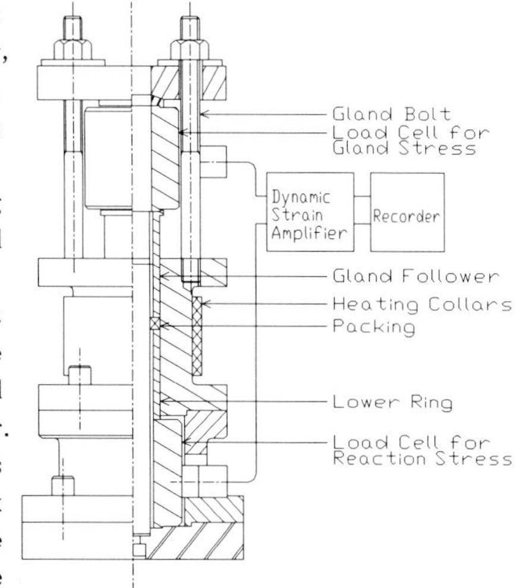

Fig. 2　Test apparatus for stress relaxation

2.2.2 Creep test

The test apparatus is shown in Fig. 3. It is composed of the testing rig, hydraulic press for always applying a constant load, and a measuring unit. The testing rig comprises, same as the testing rig for stress relaxation, a stem (32 mm in diameter) and stuffing box (48 mm in inner diameter).

Test method: Specific tightening stress (20, 30, 40, 60 MPa) are always applied to the one ring packing in the stuffing box. Subsequent changes of the compression of the packing are measured by displacement transducer, and recorded continuously by dynamic strain amplifier. Similarly, the packing compression changes at stuffing box temperature of 200 and 300 ℃ are measured.

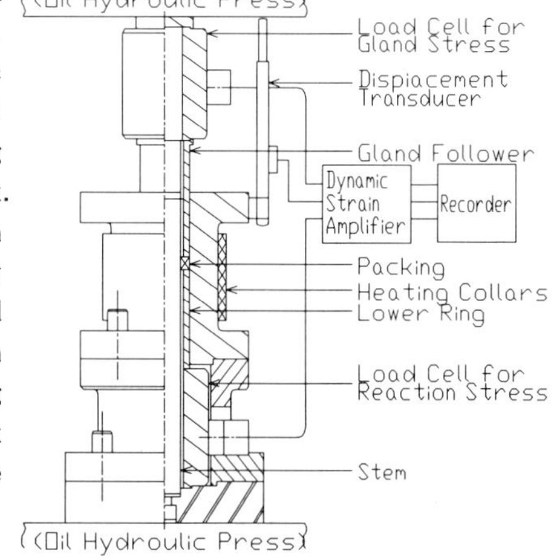

Fig. 3　Test apparatus for creep

2.2.3 Test in reciprocating

The test apparatus is shown in Figs. 4 and 5. It is composed of testing rig and measuring unit. The testing rig consists of stem(24 mm in diameter), stuffing box (37 mm in inner diameter), gland follower, gland bolt, load cell for measuring tightening stress of the packing, load cell for measuring frictional force, and hydraulic cylinder for driving stem.

Test method: At both ends of packing A of three rings, packing B of a single ring each is disposed, set in the stuffing box, and tightened up to 50 MPa at an increment of 10 MPa in tightening stress. Being loaded with He gas at 5 MPa, the stem is moved reciprocatingly at the speed 10mm/sec., and the sliding frictional force, tightening stress, and leakage are automatically measured and recorded.

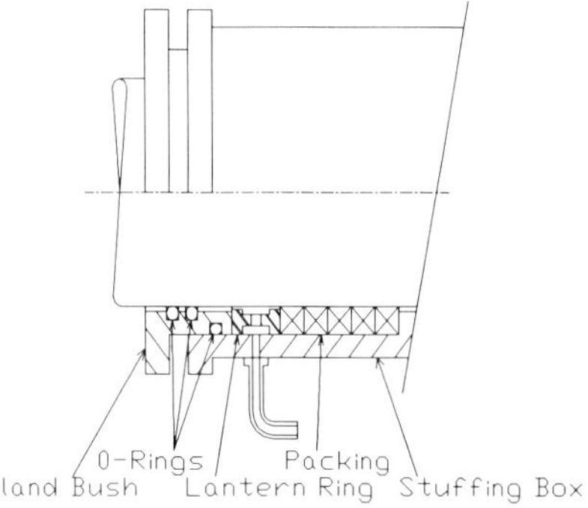

Fig. 5 Detail of stuffing box

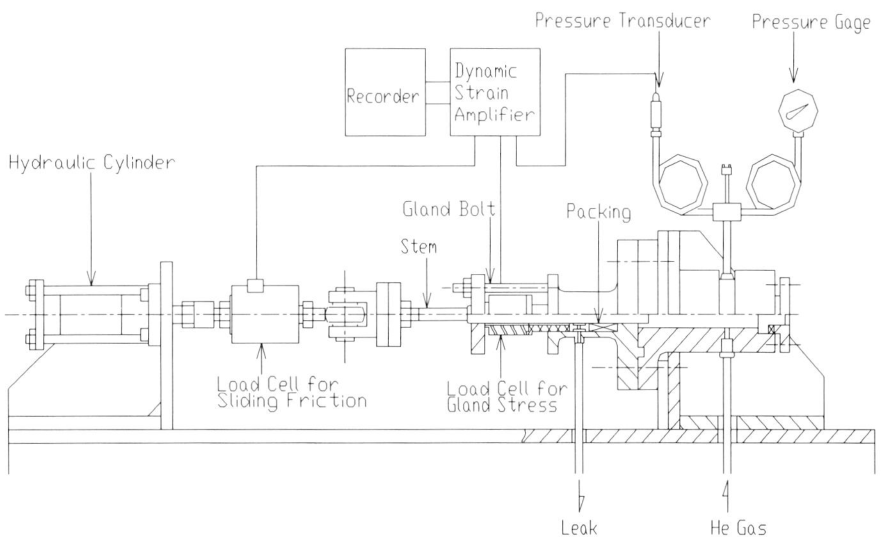

Fig. 4 Test apparatus for reciprocating

2.2.4 Test in field

Tests in field were conducted on the steam valves around the boiler in thermal power plant. The valve classes range from 1500 to 4500 Lb, and 120 valves were tested. The sample packings were packings A+B, packing C, packing D, and packings A+D. The test period was 18 months, and torque changes of gland bolts of valves in this period were measured.

3. TEST RESULTS

3.1 Stress relaxation characteristic

The stress relaxation data are shown in Figs. 6, 7 and 8. The initial tightening stress of the packing was 30 MPa, and the temperature was room temp., 200℃ and 300 ℃. Stress relaxation took place in a short time in all samples, within 10 hours after tightening in packings A and B, and within 20 hours in C. Afterwards, there was almost no stress relaxation, and it was stable.

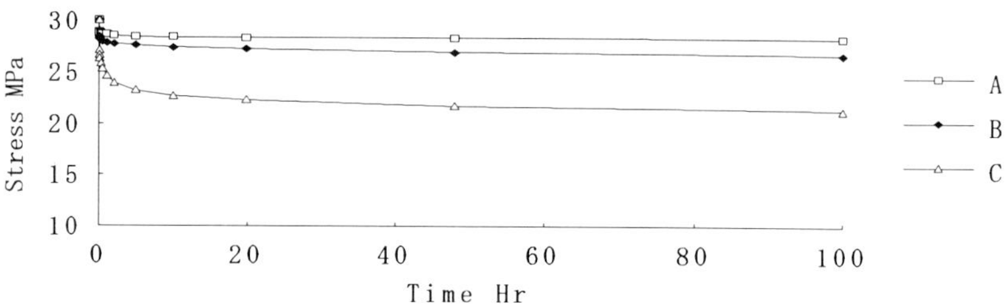

Fig. 6   Stress relaxation at RT

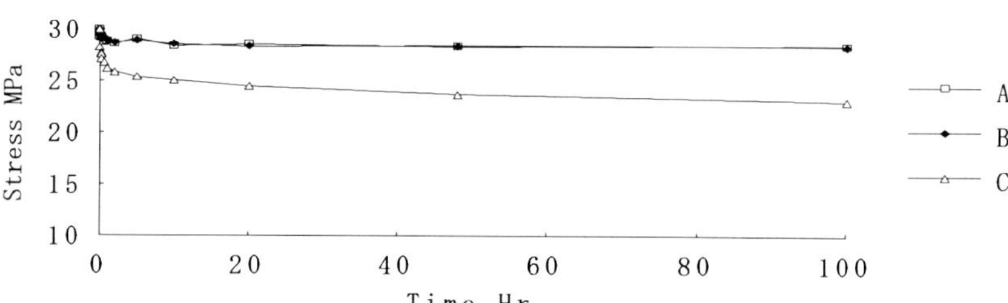

Fig. 7   Stress relaxation at 200℃

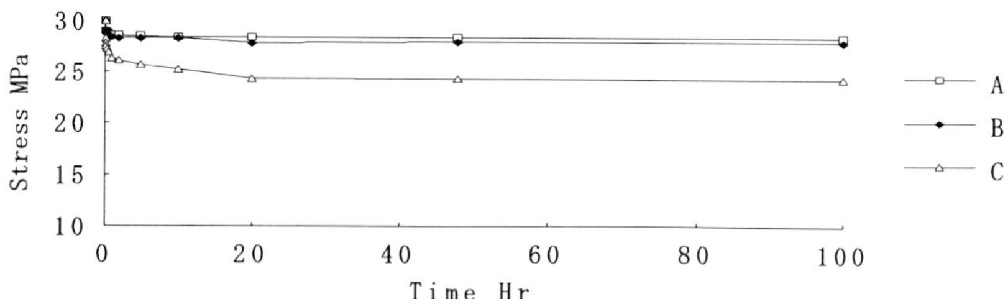

Fig. 8   Stress relaxation at 300℃

Next, another measurement was conducted after changing the test conditions of tightening stress and holding temperature. The residual rate of tightening stress after 20 hours is recorded in Figs. 9, 10, and 11. These data suggest the following.

I)   There is little effect of temperature on the stress relaxation in all of packings A, B and C.

II)   Stress relaxation characteristics of packings A and B seem to be equivalent.

III)   Packing C is greater in stress relaxation as compared with other packings.

IV)   The stress relaxation slightly increases in the condition of small tightening stress of the packing.

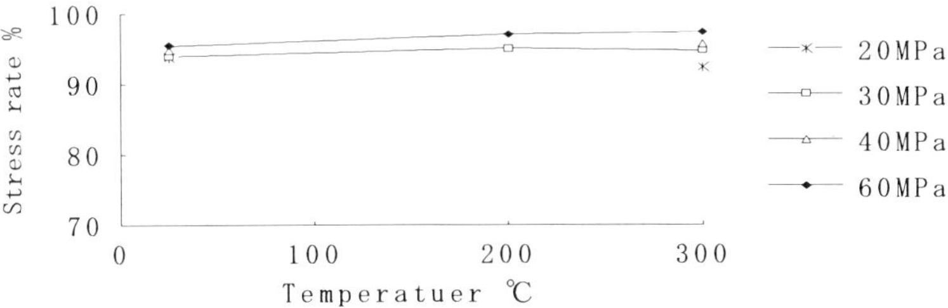

Fig. 9   Stress residual rate of packing A after 20Hr

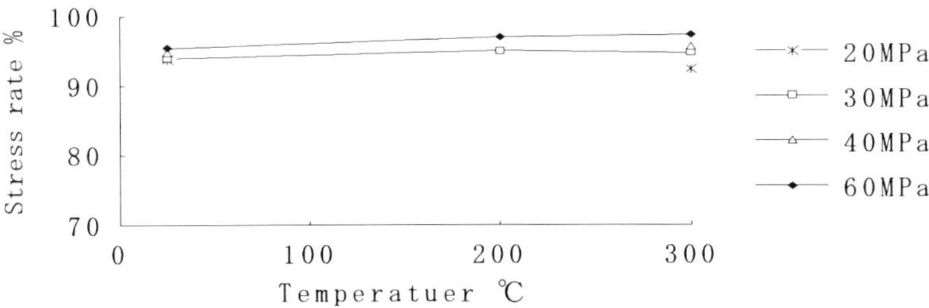

Fig. 10   Stress residual rate of packing B after 20Hr

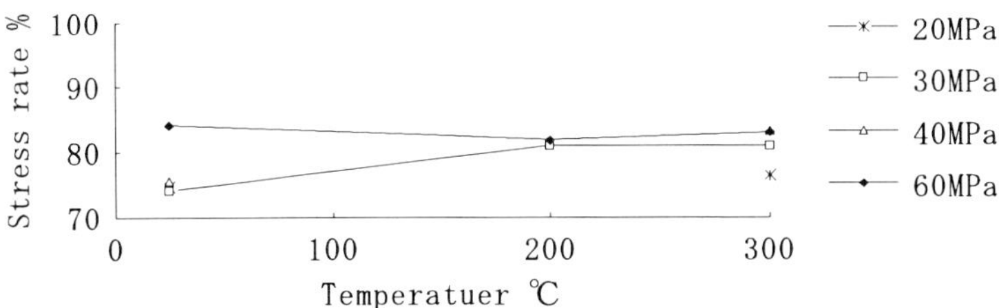

Fig. 11   Stress residual rate of packing C after 20Hr

3.2 Creep characteristic

The creep characteristic data are shown in Figs. 12, 13 and 14. The tightening stress of the packing was 30 MPa, the temperature was room temp., 200℃ and 300℃. Creep occurred in a short time, same as stress relaxation, that is, within 5 hours in packings A and B, and within 20 hours in C. Thereafter, the increase was slight and it was stable.

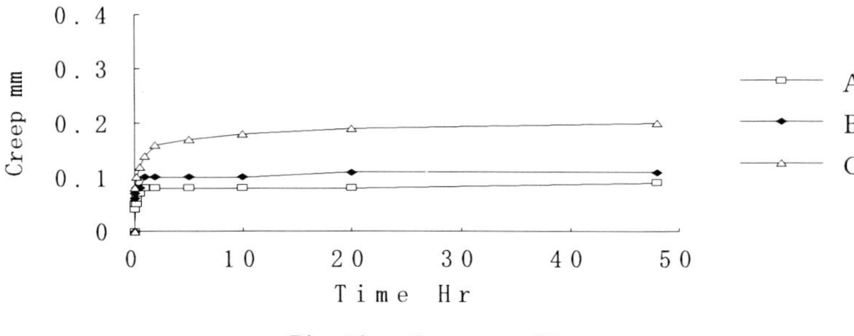

Fig. 12　Creep at RT

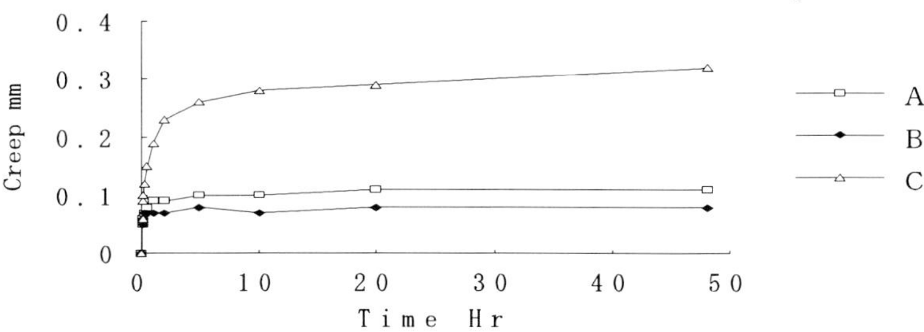

Fig. 13　Creep at 200℃

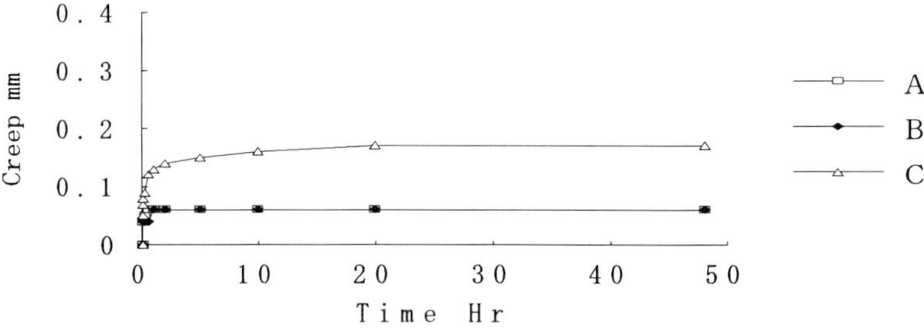

Fig. 14　Creep at 300℃

Next, the test conditions of tightening stress and holding temperature were changed, and similar measurements were taken. The creep after 20 hours is recorded in Figs. 15, 16 and 17. The data suggest the following.

I) There was little effect of temperature on the creep in all of packings A, B and C.

II) The creep characteristics of packings A and B seem to be equivalent.

III) The creep of packing C is about two times as large as that of packings A and B.

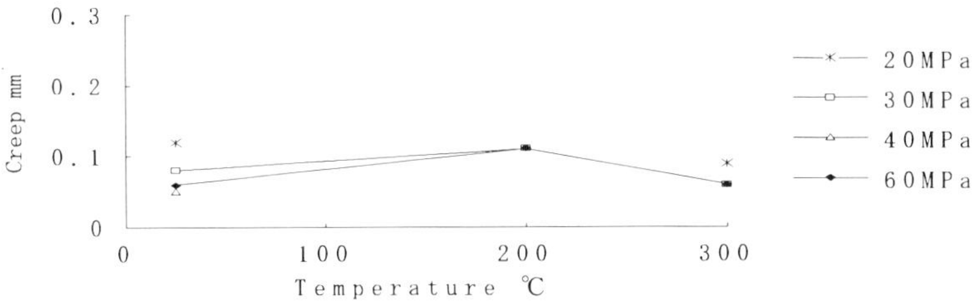

Fig. 15 Creep of packing A after 20Hr

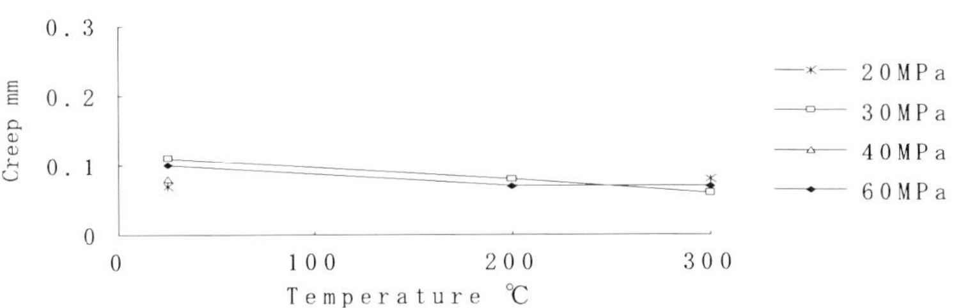

Fig. 16 Creep of packing B after 20Hr

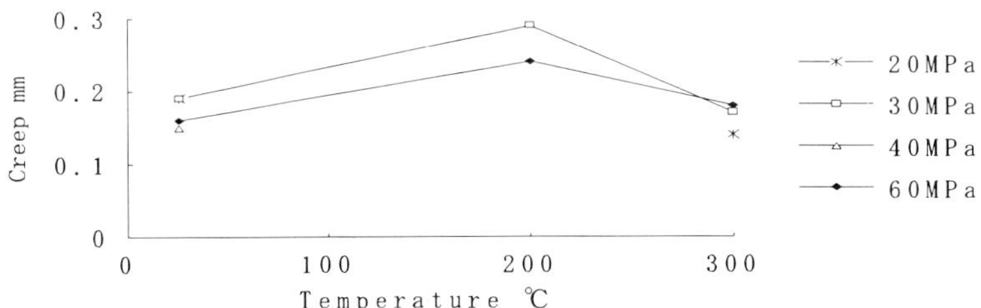

Fig. 17 Creep of packing C after 20Hr

3.3 Characteristic in reciprocating

The characteristic data in reciprocating is shown in Fig. 18. In the region of the tightening stress of 40 to 50 MPa, the leakage level was less than 0.01 mℓ/min.. This level was maintained after 300 strokes of stem.

The leakage concentration of 500 ppm in the sniffer method corresponds to a leakage of 0.3 mℓ/min. in this apparatus. In the packings after testing, extrusion of packing was not recognized.

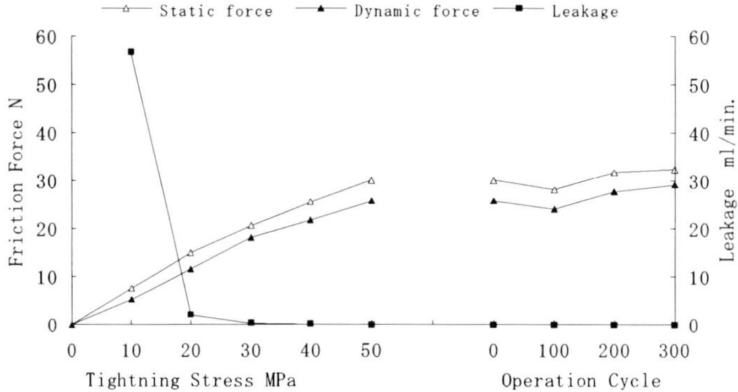

Fig. 18 Characteristic in reciprocating
Stem stroke: 90mm, Stem speed: 10mm/sec.

3.4 Changes of tightening stress in field

The time series changes of residual rate of tightening stress of the packing is shown in Fig. 19. Stress relaxation in assembled packings A+B hardly occurred, and the tightening stress was kept constant for a long time.

In other packings, meanwhile, stress relaxation occurred. The stress residual rate in packing C and the assembled packings A+D was about 40%, and that of packing D was 25%.

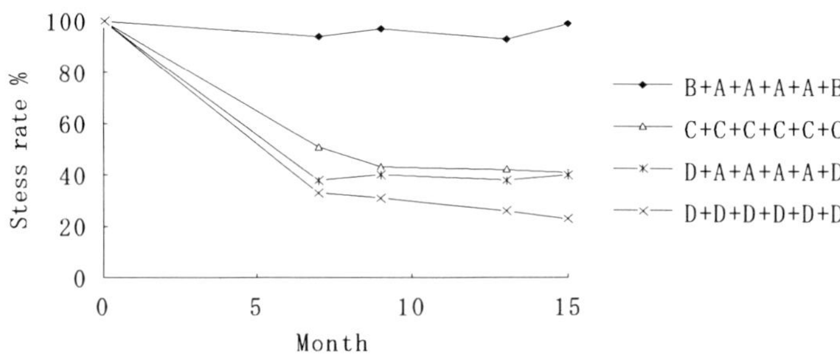

Fig. 19 Stress residual rate of packings

4. DISCUSSION

4.1 Stress relaxation characteristic

The stress relaxation rate of asbestos packing is around 40% at room temperature. In high temperature condition, this value is much larger. As a result, in the conventional packing, by retightening several times after assembling, the stress lowering was recovered to maintain the sealing performance.

On the other hand, the packings A and B are composed only of materials of high heat resistance, and the stress relaxation rate is very low, around 5 to 10%, in the temperature range from room temp. to 300°C.

The packing C contains a small amount of lubricating oil. Hence, due to the effect of this oil, the stress relaxation rate becomes about 20%. These values are 1/4 to 1/2 of the level of asbestos packings. Incidentally, even the level of packing C is similar to the level in the stress relaxation state after retightening the asbestos packing several times. This characteristic is likely to make it possible to omit the retightening conventionally practiced, and seems to be useful for alleviating maintenance.

4.2 Creep characteristic

Packings A, B and C induced creep in initial stage of retightening, and developed no creep thereafter. Therefore, all sample packings are regarded to have the heat resistance and elasticity of the expanded graphite.

Springs used in live loading valves of high temperature and high pressure are mostly belleville springs. In this type of springs, characteristically, the distortion factor is not large. When designing a live loading valve, the small creep of packing is an essential condition. The creep rate of the conventional asbestos packing is 4 to 8%.

In the present sample packings, the creep rate was about 1.3% in packings A and B, and about 2.5% in packing C. This characteristic seems to be useful for compact design of the spring structure of live loading valve.

4.3 Characteristic in reciprocating

The leakage of packings A+B is less than 0.01 mℓ/min. at the stress 40MPa. This leakage corresponds to 50 ppm of concentration in the sniffer method with CH_4 gas at 5 MPa. This value is less than 500 ppm of the CAA regulation.

After the test, there was no extrusion of packing B, and it seems to be useful as top and bottom rings.

4.4 Changes of tightening stress in field

Packing D caused an extreme stress relaxation due to deterioration of organic matter contained in the packing and hardening of asbestos.

In packings A+D, since the expanded graphite packing was combined, the stress relaxation was improved. It is this packing A+D combination that was hitherto in the mainstream of high temperature and high pressure packing, noted for high reliability.

The packing C, even when used alone, has the same stress relaxation characteristic as the packings A+D. It suggests that the conventional expanded graphite that has not been able to be used alone has a possibility of being used alone in a new form of braided packing.

The packings A+B are mainly composed of expanded graphite, and deterioration is not caused in the temperature region under $300\,°C$, which seems to be the reason of exhibiting the excellent stress retaining performance. This behavior coincides with the test data characteristic of the stress relaxation.

5.   CONCLUSIONS

The newly developed expanded graphite braided packing B and C are confirmed to possess excellent stress retaining characteristic and low creep property. Especially the assembled packings A + B is better. The new packing B and C are hence expected to be useful for emission control and maintenance-free application of on/off valves in high temperature and high pressure conditions.

6.   REFERENCES

(1)   Robert T. Wilson   How to Pack High Temperature Valves, Hydro Carbon Processing January 1978.

(2)   E. A. Bake, R. J. Gradle   Test of Asbestos-Free Stem Packings for Valves for Elevated Temperature Service, Asbestos Substitute Gasket & Packing Material Seminar August 1986.

(3)   C. W. Adey, J. J. Klein   Valve Stem Packing Improvement Study, EPRI Report August 1982.

(4)   Sentaro Yoshimoto   Present Situation and Problems of Asbestos-Free Packings and Gaskets for Valves, Valve Technical Bulletin, No. 15, January 1990.

(5)   K. Bohner, H. Blenke, R. Hinkal   Lateral Stress Ratio, Deformation, and Relaxation of Stuffing Box Soft Packing, 7th Int. Conf. on Fluid Sealing BHRA Paper No. E3 (1975).

(6)   Hisao Tashiro   Development of Optimum Maintenance and Operation Method of Gland Packings for Valves, CRIEPI Report T19, February 1991.

(7)   N. E. Pothier   Performance Testing of Prototype Live-Loaded Packed Stem Seals for Large Gate Valves in Pressurized Hot Water. 7th Int. Conf. on Fluid Sealing BHRA Paper No. G1 (1975).

14th International Conference on Fluid Sealing, Firenze, Italy,
6-8 April 1994. Organised by BHR Group Limited, Cranfield,
Bedford, MK43 0AJ, UK; Tel: 0234 750422

STRESSES AND DEFORMATION OF COMPRESSED ELASTOMERIC O-RING SEALS

by
Itzhak Green and Capel English
The George W. Woodruff School of Mechanical Engineering
Georgia Institute of Technology
Atlanta, GA 30332-0405

ABSTRACT

The sealing capability of an elastomeric O-ring seal depends upon the contact stresses that develop between the O-ring and the surfaces with which it comes into contact. It has been suggested in the literature that leakage will occur when the pressure differential across the seal just exceeds the initial (or static) peak contact stress. The stresses that develop in compressed O-rings, in common cases of restrained and unrestrained geometries (grooved and ungrooved), are investigated using the finite element method. The analysis includes material hyperelasticity and axisymmetry. Contact stress profiles, and peak contact stresses are plotted versus squeeze, up to 32 percent. The contact width, which is the length of the O-ring that touches the retaining surfaces when viewed from the cross-section, is also determined. Expressions are derived empirically to predict the peak contact stress and the contact width. These expressions are also compared to those obtained by other researchers (who assumed plain strain conditions) and conclusions to their validity are drawn.

NOMENCLATURE

b	= contact width	S	= compressive stress
d	= wire diameter	x^*	= displacement
D	= nominal (mean) diameter	x	= radial coordinate
D_{def}	= deformed mean diameter	X	= radial distance from O-ring center
E	= modulus of elasticity	y	= axial coordinate
h	= deformed O-ring thickness	Y	= axial distance from O-ring center
l	= groove width	δ	= normalized squeeze (i.e., fractional compression)
q	= chord diameter		
Q	= normalized chord diameter, q/d	δ_{ij}	= equivalent normalized squeeze

INTRODUCTION

Elastomeric O-ring seals have a broad range of service conditions that make the O-ring ideal for static and dynamic sealing functions. Its ability to seal on relatively rough surface finishes offers one of the economical solutions to sealing problems. Elastomeric O-rings are capable of undergoing large deformations under compression. Hence, grooves are often used to restrict this deformation, resulting in improved sealing capabilities and prevention of creep and extrusion. The complex geometry confederated with the deformation of restrained O-rings and nonlinear material hyperelasticity render analytical solutions infeasible. This complicated geometry and experimental inconvenience make experimental data hard to obtain. It is here where the utility of the finite element method becomes prominent. By performing a FEM analysis, comparison of the results can be made to cases where experimental data is procurable. Then, conclusions can be drawn as to the validity of FEM solutions of geometries where experimental data cannot be easily obtained.

The stiffness relationships associated with the compression of elastomeric torroidal O-ring seals have recently been studied by Green and English (1992) for the cases shown in Figure 1. That work provided empirical expressions for the prediction of compression forces and stiffnesses at squeeze levels up to 32 percent. Sealing capabilities, however, depend upon the stress related parameters at the interface. It was as early as Lindly's work (1967), who proposed that leakage onset occurs when the pressure differential across the seal, P, barely exceeds the initial (or static) peak contact stress, S_{max} (i.e., $P \geq S_{max}$). It should be noted that any increase in the contact stress, caused by the pressure loading, is ignored using this theory. Simplified expressions relating contact width to peak contact

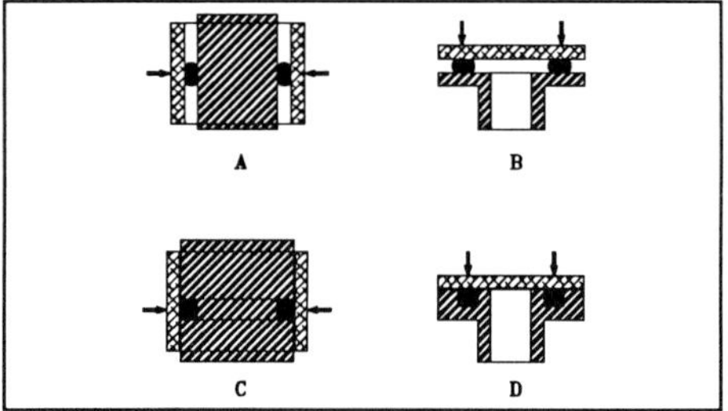

Figure 1 (A) Unrestrained Radial Loading. (B) Unrestrained Axial Loading. (C) Restrained Radial Loading. (D) Restrained Axial loading.

stress have been developed in order to predict S_{max}. Assuming unrestrained loading and plain strain Lindley (1967) obtained the contact width, b, normalized with respect to the wire diameter, d, [see Figures 2(a) and 3]

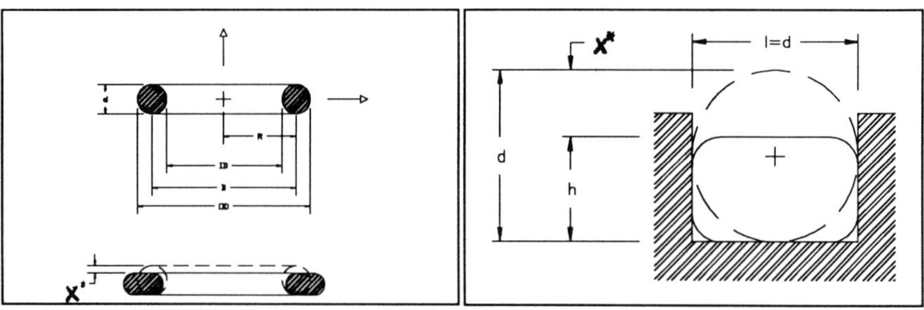

Figure 2(a) Unrestrained geometry   **Figure 2(b)** Section of a restrained O-ring

$$\frac{b}{d} = \left[\frac{6}{\pi} (1.25\delta^{\frac{3}{2}} + 50\delta^6) \right]^{\frac{1}{2}} \qquad (1)$$

and the peak contact stress, S_{max}, normalized with respect to the modulus of elasticity, E,

$$\frac{S_{max}}{E} = \left[\frac{16}{6\pi} (1.25\delta^{\frac{3}{2}} + 50\delta^6) \right]^{\frac{1}{2}} \qquad (2)$$

Here $\delta = x^*/d$, is the normalized squeeze, i.e., the compressive displacement, x^*, divided by the wire diameter, d. The first term in the equations was obtained using Hertzian theory, and the second term was added to correct for empirical data at high squeeze levels.

Wendt (1971) examined stress distributions in O-rings and X-rings, with emphasis on groove design. The most significant result of his work includes an expression for contact width of an unrestrained axially loaded O-ring. Molari (1973), who examined the stress and contact related parameters of O-rings using photoelastic techniques, lent credence to the findings of Wendt and was one of the firsts to examine the problem of restrained O-ring seals. Molari's work, however, considered one lateral wall

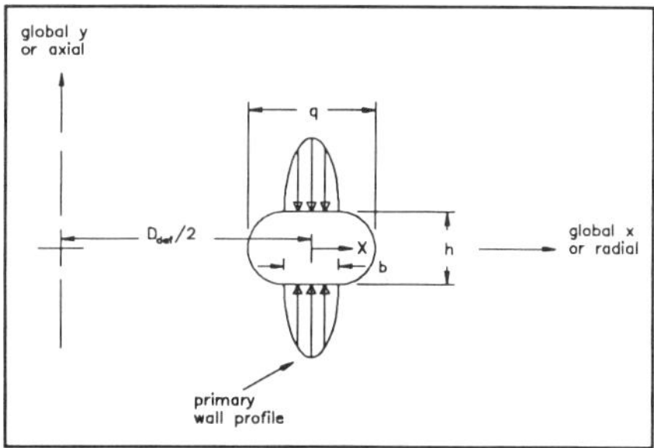

Figure 3 *Stress profile definition for unrestrained axial loading.*

only. Dragoni and Strozzi (1988) also used photoelasticity, but investigated an O-ring restrained between two lateral walls as defined in Figure 2(b). These researchers assumed that plane strain conditions were prevailing and thus did not address the condition of axisymmetric loading.

George, Strozzi, and Rich (1987) supported Lindley's results using a finite element code developed especially for the task. Experimental data taken was compared to the results obtained by numerical solution. Later Dragoni and Strozzi (1988) examined the case of laterally restrained O-ring seals in a groove using a modification of the FEM code. The results were also limited to plain strain conditions. Using Lindley's model of Hertzian contact stress, they offered an approximate analytical method for "moderately" compressed O-rings up to about 15 percent squeeze. A stress related parameter was given in terms of a normalized deformed chord diameter, $Q = q/d$ (see Figure 3). By fitting a curve to experimental results (Strozzi, 1986), they characterized Q as a function of δ

$$Q = 1 + 0.415\,\delta + 1.15\,\delta^2 \equiv f(\delta) \qquad (3)$$

where the right-hand side emphasizes the functional form of the equation, as needed for later derivations. Using only the first term of Eq. (2) the peak contact stress was given as

$$\frac{S_{max}}{E} = \sqrt{\frac{10}{3\pi}}\,\delta^{\frac{3}{4}} \qquad (4)$$

In a compromise between accuracy and simplicity they prefer Wendt's (1971) description of the contact width

86

$$\frac{b}{d} = \frac{3}{2} \, \delta^{\frac{2}{3}} \tag{5}$$

At this point Dragoni and Strozzi (1988) developed an equivalent normalized squeeze, δ_{ij}, which can be used in modeling the characteristics of restrained O-rings. The notation ij is used to denote the effects of the particular groove wall defined perpendicular to either the i or j direction. For example, for restrained radial loading the equivalent squeeze in the x direction, δ_{xy}, denotes the squeeze associated in part to the squeeze directly applied in the x direction and in part to the constraint of the walls which are perpendicular to the y direction. Alternately, for restrained axial loading, δ_{yx} is the equivalent normalized squeeze on the groove walls perpendicular to the top/bottom compressive surfaces. By definition δ_{yx} is estimated as a ratio. The numerator is the difference between two terms: (i) a virtual deformed chord diameter along the y-axis caused by the compression δ_{xy} [and is calculated by substituting δ_{xy} into Eq. (3)]; (ii) the deformed O-ring thickness, h [shown in Figure 2(b)]. The denominator is the undeformed wire diameter, d. Hence,

$$\delta_{yx} = \frac{d \cdot f(\delta_{xy}) - h}{d} = f(\delta_{xy}) - \frac{h}{d} \tag{6}$$

where f is the functional given in Eq. (3). Applying similar reasoning in the perpendicular x-direction, and using the groove width, l (as the O-ring thickness), gives the equivalent squeeze

$$\delta_{xy} = \frac{d \cdot f(\delta_{yx}) - l}{d} = f(\delta_{yx}) - \frac{l}{d} \tag{7}$$

These relationships provide estimates for any groove dimensions (allowing the possibility of a gap between the undeformed O-ring and the lateral walls, i.e., $l > d$). Next, we define the particular case (subscripted here with the letter t) where the groove lateral walls are tangent to the undeformed O-ring, i.e., $l = d$ as shown in Figure 2(b). Combination of Eqs. (6) and (7) yields

$$\delta_{tyx} = f(f(\delta_{tyx}) - 1) - \frac{h}{d} \tag{8}$$

where the functional form of Eq. (3) is used repeatedly. Eq. (8) can be solved iteratively for δ_{tyx}. Then δ_{txy} is calculated by Eq. (7), and by substitution into Eqs. (4) and (5) the normalized peak contact stress and the normalized contact width can be determined in the respective directions.

Since in all the aforementioned work plain strain conditions prevailed, it implies that no distinction exists between axial and radial loading. This was found invalid in some important loading conditions for the compression force and stiffness (Green and English, 1992). The loading cases in Figure 1 are investigated here to determine contact stresses and contact widths under axisymmetric conditions. These include a highly frictional ("unlubricated") contact where surface sliding is prevented in an unrestrained axial loading; and frictionless ("perfectly lubricated") contacts where forceless surface sliding exist in axial, radial, restrained, and unrestrained loadings. The commercial code ANSYS and the nonlinear techniques, described in Green and English (1992) and in greater detail in English (1989), are utilized. Reduced integration is exclusively applied as it was found to give most accurate results. These are best represented in normalized forms, proven indifferently to the aspect ratio, d/D. Convergence is discussed in whole in the last two references.

STRESS PARAMETERS RESULTS

Vernacular for this discussion includes "primary wall" and "lateral wall." The primary wall or walls are the surfaces which move together to force the compression of the O-ring. The lateral walls are the sides of the restraining groove. Initially all walls are tangent to the undeformed geometry of the O-ring. For the axial case the top and bottom walls are the primary walls, and for the radial case the inside and outside walls are the primary walls.

Contact stress profiles are plotted as the normalized nodal stress component, S_y/E or S_x/E, versus the normalized x or y coordinates (X/d or Y/d), respectively. For example, Figure 3 shows the X-coordinate, defined relative to the deformed nominal radius, $D_{def}/2$. X is the horizontal distance from the center of the O-ring cross-section to a node along the perimeter. The normalized stress of interest here is the y-component of the nodal stress, S_y/E. In the plots E_o is shown instead to designate neo-Hookean material representation, justified for use by Green and English (1992).

The first contact pressure profile, shown in Figure 4, was produced from a quadrilateral element mesh, using reduced integration (Green and English, 1992). Six profiles were chosen

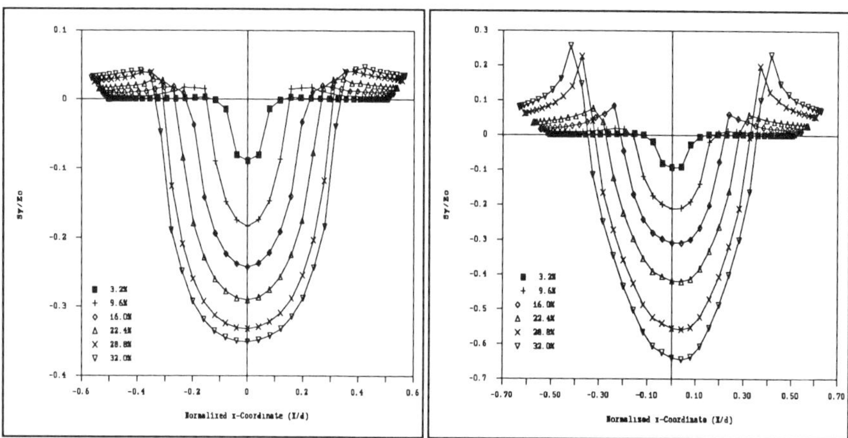

Figure 4 Primary wall contact stress profile for unrestrained - perfectly lubricated axially loaded O-ring.

Figure 5 Primary wall contact stress profile for unrestrained - unlubricated axially loaded O-ring (the only fixed case)

from ten load steps of 3.2 percent squeeze each, and are represented by the symbols given in the legend. Two other features can be extracted from the figure: (i) The normalized deformed chord diameter, Q=q/d, is the farthest distance between two opposite points on each profile, and (ii) the normalized contact width, b/d, is the distance between two points on each profile where the curve intersects the zero stress line. Notice that Q and b/d are different at each load step. Finally, there is a condition of symmetry about the x-z plane, of the global coordinate system, such that both the top and bottom profiles are identical and, therefore, only one is shown. While the profiles appear symmetric about X/d=0, close examination of the results reveals otherwise. This is due to the imposition of axisymmetric conditions upon the solution (rather than plain strain conditions).

Figure 5 shows the contact stress profile for the unlubricated case of unrestrained axial loading (where a coefficient of friction 0.9 was used). The asymmetry is much more pronounced in this case. It should also be noted that the peak value of S_y is roughly 85 percent higher than that for the lubricated case. This increase in peak contact stress can be explained by the fact that the deformed nominal diameter, D_{def}, does not expand as the load increases. Actually D_{def}

88

decreases as the loading is applied, although, only a small amount. A comparison of Figures 4 and 5, shows that friction has a dominating role in stress profile development.

Turning to radial compression, Figure 6 shows how the contact stress profiles are defined. S_x is the stress component of interest here, and Y is the vertical distance from the center of the

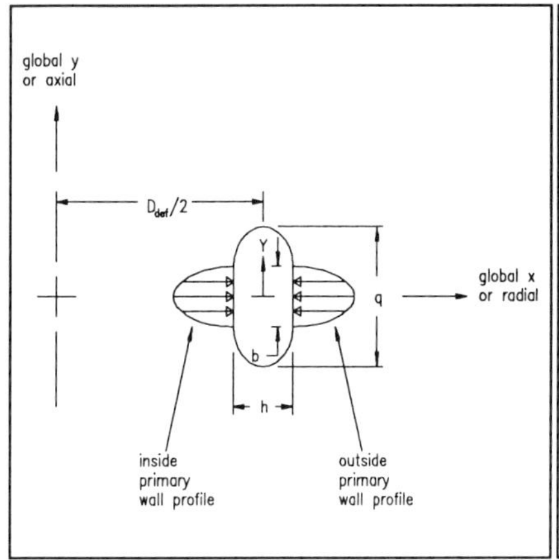

Figure 6 *Stress profile definition for unrestrained radial loading.*

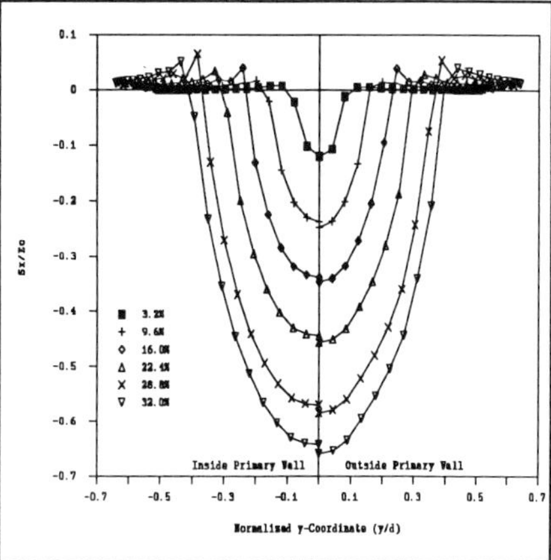

Figure 7 *Contact stress profile for unrestrained radial loading. Half profiles for the inside and outside primary walls are shown.*

O-ring cross section to a node on the perimeter. q is the deformed chord diameter parallel to the compressive surfaces. Because the model is not symmetric about the y-z plane the stresses on both walls, the inside primary wall and the outside primary wall, must be examined.

The stress profiles for unrestrained radial loading are shown in Figure 7. Here we notice that the outside primary wall profile is larger than the inside primary wall profile. This is due to nominal diameter contraction during axisymmetric loading. Figure 7 exemplifies again the necessity of using an axisymmetric model to represent the O-ring. While the difference between the inside and outside walls is minor for this particular restraining configuration it could be much more significant for a different type of loading.

Next up for attention are the restrained cases. Figure 8 gives the profile definition for restrained axial loading. Note that the profiles are symmetric about the x-z plane; however, they

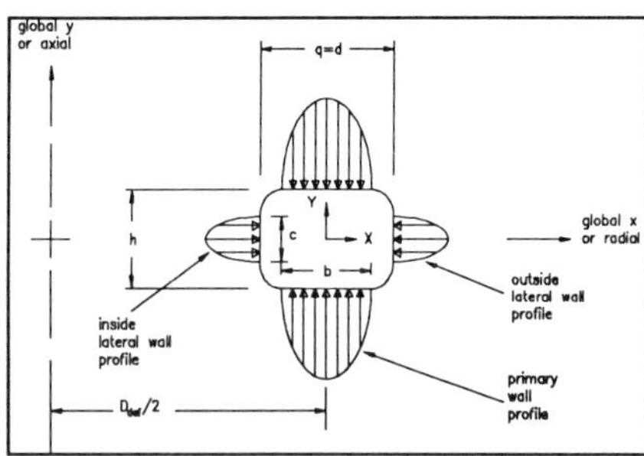

Figure 8 *Profile definition for restrained axial loading.*

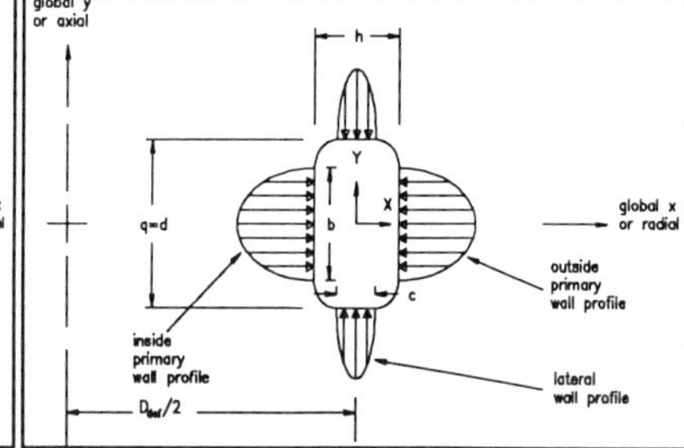

Figure 9 *Profile definition for restrained radial loading.*

are not symmetric about the y-z plane. The restrained radial profile definition can be seen in Figure 9. This profile is similar to the previous profile in that the symmetry planes are the same. However, the primary and lateral walls are different. Note the definitions for the contact widths b and c on the primary and lateral walls, respectively.

Figures 10 and 11 show the profiles for the primary and lateral walls, respectively, for the restrained axially loaded O-ring. The half profiles for the inside and outside lateral walls show

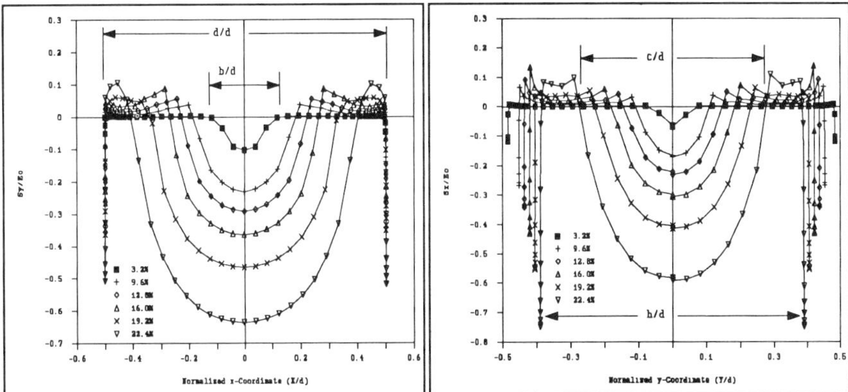

Figure 10   *Contact stress profile for primary wall of restrained axial loading.*

Figure 11   *Contact stress profile for inside and outside lateral walls for restrained axial loading.*

that there is a difference between lateral wall profiles where axisymmetric loading is concerned. Here the peak value of S/E_0 is 17 percent larger for the primary wall than that for the lateral wall. Figures 10 and 11 show an interesting formation of significant compressive normal stresses at the O-ring surface that is in contact with the respective retaining walls. Molari (1973), using bidimensional photoelastic techniques, obtained similar profiles; however, the surface stresses did not show up in his work because they were masked by the physical boundary of the test apparatus.

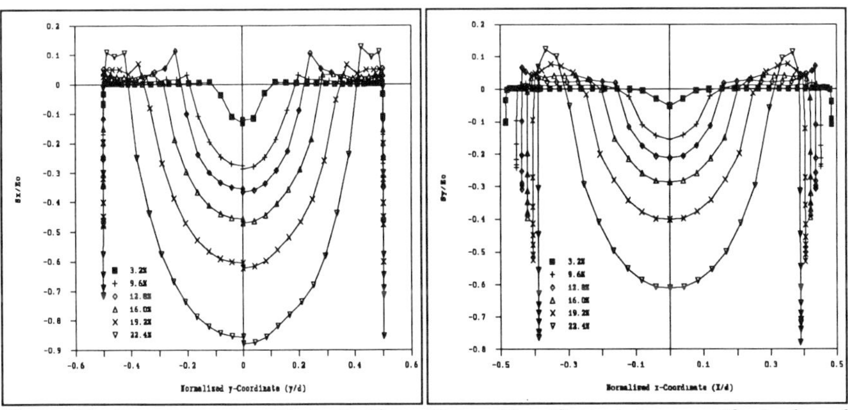

Figure 12   *Contact stress profile for inside and outside primary walls for restrained radial loading.*

Figure 13   *Contact stress profile for lateral wall of restrained radial loading.*

90

Next we consider the case of restrained radial loading. Contact stress profiles for the primary and lateral walls are given in Figures 12 and 13, respectively. In Figure 12 a difference exists in the half profiles for the inside (left) and outside (right) primary walls. Peak contact stress values for the primary walls are about 34 percent greater than those for the lateral wall. Also here there are surface stresses at the retaining walls.

The final case under investigation is that for plane strain loading. Figure 14 shows the stress profile for unrestrained plane strain loading. Plane strain loading profiles for the restrained case are given in Figures 15 and 16. In this case there is symmetry about both the x-z and y-z planes. Hence, there is no need for a half profile plot of inside or outside walls. It can be seen by comparing the stress profiles from plain strain loading to all previous cases that there is a significant difference in the profiles, especially in radial loading and peak contact stresses. It is, therefore, concluded that plain strain conditions do not commonly describe O-ring compression.

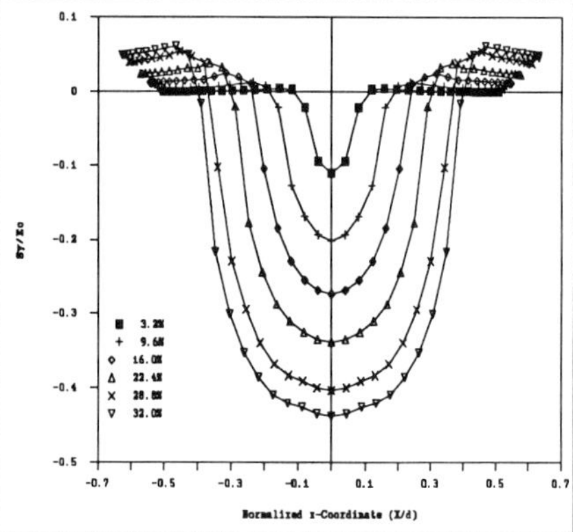

Figure 14 *Contact stress profile for unrestrained plane strain loading primary wall.*

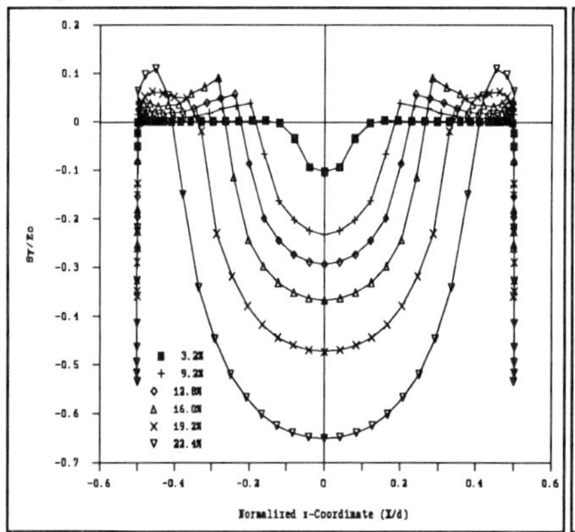

Figure 15 *Contact stress profile for restrained plane strain loading primary wall.*

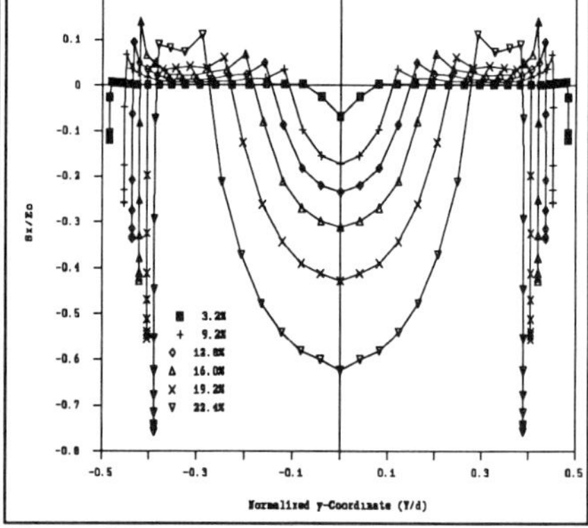

Figure 16 *Contact stress profile for plane strain loading lateral walls.*

PEAK CONTACT STRESS

Peak contact stress is of interest in order to estimate the ability of the O-ring to form a seal (Lindley, 1967). Figure 17 contains the compilation of peak contact stresses for unrestrained loadings, i.e., axial, radial, and plane strain. It can be seen that the primary wall peak contact stress response for radial loading is the greatest. This is followed by the stress response of the unlubricated axial loading, which is significantly greater than its lubricated correlative. Furthermore, we see that both Lindley's predictions underestimate the peak contact stress throughout the loading range with the exception of the plane strain case. The latter case agrees relatively well with the prediction Lindley derived from Hertzian theory (Eq. (2), but without the second correction term). Note that the empirically added correction term causes an overestimation of the peak contact stress for squeezes above 24 percent.

Lindley gives no prediction for the peak contact stress of a restrained O-ring. The comparison to Dragoni and Strozzi (1988) prediction can be seen in Figures 18 and 19. Their prediction agrees relatively well with the axial and plane strain results, but, underestimates the peak contact stress for radial loading. Where the lateral wall is concerned Strozzi's prediction underestimates the peak contact stress response for all cases. It is interesting that the lateral wall response for radial loading is less than the responses for both axial and plane strain cases. It seems

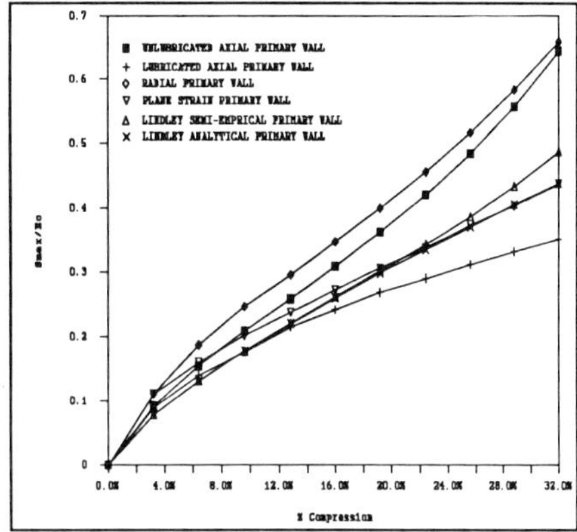

Figure 17   *Peak contact stress results for unrestrained loading.*

from this comparison, that Strozzi's model is relatively accurate for predicting the peak contact stress for the primary wall of restrained axial and restrained plane strain cases, but it breaks down in radial loading and where the lateral wall is concerned.

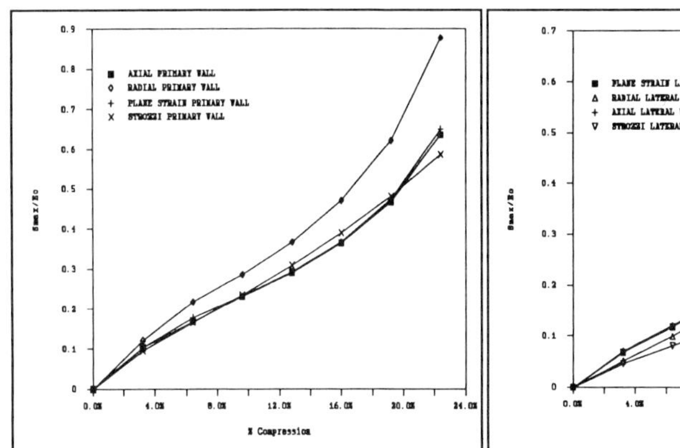

Figure 18   *Peak contact stress results for restrained loading primary wall.*

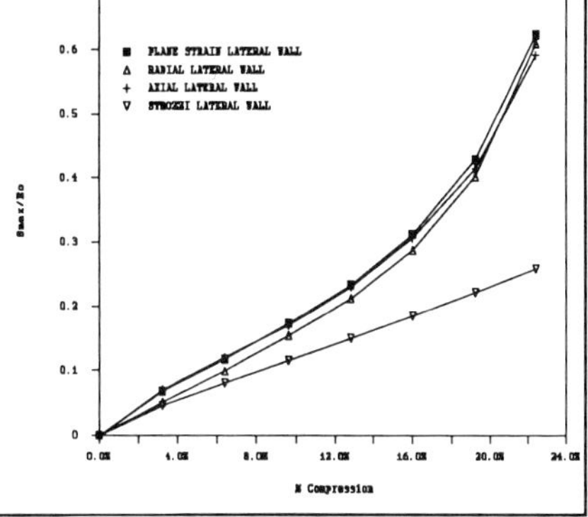

Figure 19   *Peak contact stress for restrained loading lateral wall.*

The lack of agreement to analytical work prompts the determination of the peak contact stress from the numerical data. This is accomplished by fitting a polynomial to the numerical results. Second and third order polynomials are proposed as follows:

$$\frac{S_{max}}{E} = a\delta + b\delta^2 \qquad (9)$$

$$\frac{S_{max}}{E} = c\delta + d\delta^2 + e\delta^3 \qquad (10)$$

where the coefficients are given in Table 1. While Eq. (10) produces an excellent fit, Eq. (9) may be used for simplicity with satisfactory results.

Table I *Least squares coefficients for the calculation of the peak contact stress*

Loading Case	a	b	c	d	e
Unrestrained lubricated axial primary wall	2.0572	-3.1417	2.6296	-8.8589	12.8391
Unrestrained unlubricated axial primary wall	2.0090	-0.2211	2.8383	-8.5051	18.6031
Unrestrained lubricated radial primary wall	2.4891	-1.5967	3.4591	-11.2857	21.7583
Unrestrained lubricated plane strain primary wall	2.2340	-2.8961	3.0373	-10.9192	18.0171
Restrained axial primary wall	1.9715	3.1502	3.8295	-23.0013	82.6963
Restrained axial lateral wall	1.0497	6.4631	2.6584	-16.1793	71.5999
Restrained radial primary wall	2.1587	6.7729	4.9363	-32.3232	123.630
Restrained radial lateral wall	0.5844	8.7930	2.3003	-15.3593	76.3744
Restrained plane strain primary wall	1.9933	3.2711	4.0499	-25.6765	91.5384
Restrained plane strain lateral wall	0.9400	7.5182	2.6698	-16.8290	76.9908

CONTACT WIDTH RESULTS

In Figure 3 the contact width, b, is defined as the length of the circumference of the O-ring, from a cross-sectional view, that makes contact with the compressing surface. This information is useful in calculating the total load required to compress the O-ring and in determining the contact stress profile when using Hertzian theory. Both Lindley (1967) and Wendt (1971) propose expressions which approximate the contact width as a function of compression for unrestrained loading, and Dragoni and Strozzi (1988) propose corresponding expressions for restrained loading. As illuminated in the introduction these were developed assuming plain strain conditions. This section compares results obtained in this research to those predicted by the other researchers.

The error associated with the numerical results is significant when considering the technique used to obtain the contact width. In Figure 20 there are several

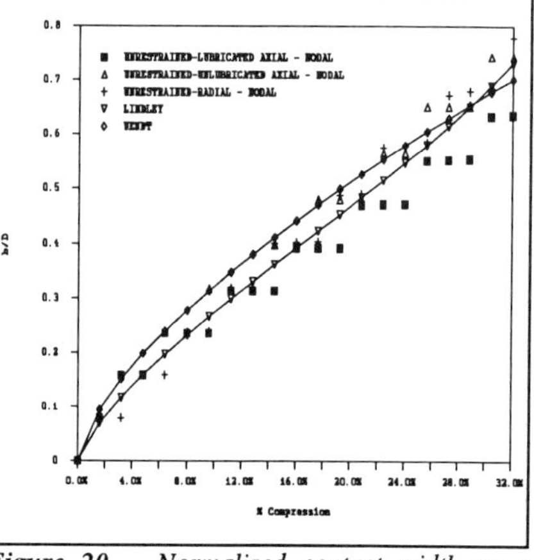

Figure 20 *Normalized contact width as a function of compression for unrestrained loading. Included are Lindley and Wendt's prediction as well as the results from nodal displacements.*

data points with approximately identical values of b/d. This comes from the discrete points used to obtain the contact width. As the loading is applied, new nodes may or may not come into contact with the compressing surfaces. Several data points with the same value of b/d imply that no new nodes have come into contact during that portion of the loading sequence. Physically the contact width response is a continuous phenomenon. By discritizing the mesh we turn this response into a discrete phenomenon. Consequently the first data point, of those which have the same magnitude of b/d, is the most accurate. Therefore, the contact width shown is an underestimation of the actual contact width.

Contact width data, for the cases of unrestrained loading can be seen in Figure 20. The numerical results from the unlubricated axial and radial cases agree well with Wendt's prediction, Eq. (5), up to roughly 24 percent compression. Lindley's prediction, Eq. (1), underestimates the numerical contact width throughout the load range. For the lubricated axial load case Wendt's expression begins to overestimate the contact width at approximately 10 percent compression while Lindley's expression underestimates, at low compression, and over estimates it at higher compressions. From this we may conclude that Wendt's prediction is good for small compressions and that Lindley's prediction should generally not be used.

Now we turn to restrained loading while probing Strozzi's approach as outlined from Eq. (3) through Eq. (8). To do so a second order polynomial [similar to Eq. (3)] must first be fitted to the data obtained from the FE analysis of the three unrestrained loading results (lubricated and unlubricated axial loading, and radial loading). By extracting nodal displacements from the output it is possible to obtain the normalized deformed chord diameter. For the unlubricated-unrestrained axial loading case the fitted polynomial is

$$Q = 1 + 0.210\delta + 0.657\delta^2 \tag{11}$$

For the lubricated-unrestrained axial loading case the polynomial obtained is

$$Q = 1 + 0.361\delta + 1.547\delta^2 \tag{12}$$

and for the unrestrained radial loading case the polynomial obtained is

$$Q = 1 + 0.355\delta + 1.626\delta^2 \tag{13}$$

The contact width results for restrained loading can be examined in Figures 21 and 22 for the primary wall and lateral walls, respectively. Looking at Figure 21 we see that the predictions

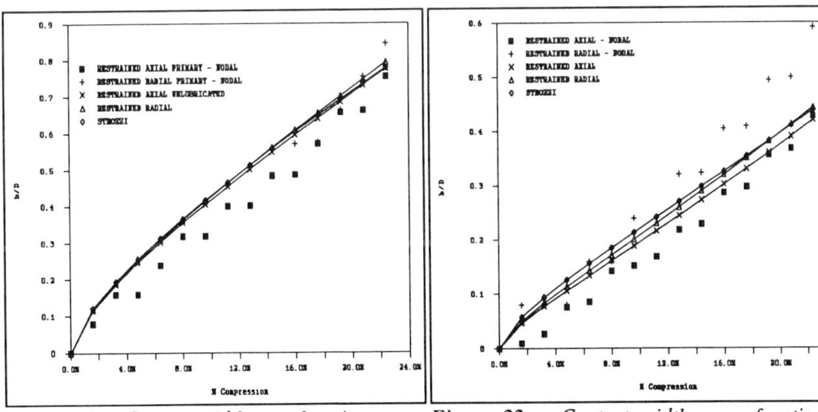

Figure 21   *Contact width as a function of compression for the primary wall of a restrained O-ring.*

Figure 22   *Contact width as a function of compression for the lateral wall of a restrained O-ring.*

in the solid lines, using Eqs. (12) and (13) as well as Strozzi's Eq. (3), produce overestimates compared to the numerical nodal contact width at the primary wall for compressions below 19 percent. At the extreme compressions the normalized chord diameter technique underestimates the contact width of the primary wall of a restrained radially loaded O-ring. However, as mentioned above, these numerical data points underestimate the actual contact width. Therefore, the prediction may be considered to give a fairly accurate prediction of contact width.

In Figure 22 we see that Strozzi's technique is a closer approximation to the lateral wall contact width of a restrained axially loaded O-ring. But, the figure also shows that the technique is a gross underestimate of the lateral wall contact width for a restrained radially loaded O-ring. Without more experimental results it is hard to make a firm statement as to the accuracy of either Strozzi's approach or the current numerical approach. It can generally be said, however, that the numerical results underestimate the actual contact width and, therefore, the method Strozzi suggests is valid, with exception to the radially loaded case at the lateral wall.

CONCLUSIONS

Acquisition of the stress parameters requires more effort in the postprocessing phase of a finite element analysis, but compared to the extensive testing equipment required for experimental stress data acquisition, this methodology is far less expensive in terms of resources and time. Another feature that FEA has to offer is the ability to examine surface stress data that is otherwise hidden by the boundary of experimental apparatus.

The most profound finding of this work is the identification of a significant difference between the peak contact stress response of plane strain models and axisymmetric models of hyperelastic O-rings (see Figs. 17 through 19, and Table 1). This is particularly true for the case of unrestrained loading . The plane strain results obtained agreed well with those predicted by Wendt. However, Wendt's prediction of peak contact stress response greatly underestimated the response generated by axisymmetric loading. Results obtained from axisymmetric modeling of the lubricated-axially loaded O-ring also indicate that the plane strain assumption is not valid for prediction of the peak contact stress response for this particular case.

Contact width examinations performed in this work yield the most inconclusive results out of all the topics investigated. This is primarily because the mesh is finite at the perimeter and only discrete information about the contact width is available. Clearly better results can be obtained with a much refined mesh at the expense of computer time. Given the results for unlubricated-unrestrained axial loading reasonable agreement exists with Wendt's prediction for compressions up to 15 percent. Lindley's prediction underestimates the contact width response for all cases except lubricated-unrestrained axial loading.

Looking at the cases of restrained loading, only Strozzi offered an analytical technique of predicting contact width and stress. To conform with that technique the numerical results were fitted to produce expressions for the normalized chord diameter in Eqs. (11) - (13). However, the use of these equations, in the procedure outlined from Eq. (3) through Eq. (8), overestimates the contact width compared to numerical data obtained from the deformed nodal coordinates. Strozzi's prediction underestimates the lateral wall stress response for the axisymmetric axial and plane strain cases, yet it gives relatively good agreement to the axisymmetric radial case for compressions below 10 percent. For compressions above 10 percent Strozzi's prediction again underestimates the peak contact stress response.

Due to the lack of consistent agreement between the results obtained here and previous analytical work an alternate empirical procedure is proposed. The peak contact stresses can be determined using Eqs. (9) or (10) for the ten loading conditions listed in Table 1. According to Lindley (1967) these equations provide estimates of the maximum pressure an O-ring can seal.

ACKNOWLEDGMENT

The authors gratefully acknowledge the support given to this work by the National Science Foundation REU Program under grant number MSM-8619190.

REFERENCES

Dragoni, E., and Strozzi, A., 1988,   "Analysis of an Unpressurized Laterally Restrained, Elastomeric O-ring," Trans. ASME, Journal of Tribology. Vol. 110, No. 2, pp. 193-199.

English, C., 1989, "Stiffness Determination of Elastomeric O-Rings Using the Finite Element Method," M.S. Thesis, Georgia Institute of Technology.

George, A.F., Strozzi, A., and Rich, J.I., 1987, "Stress Fields in Compressed Unconstrained Elastomeric O-ring seals and a Comparison with Computer Predictions with Experimental Results," Tribology International Vol. 20, pp.237-247.

Green, I., and English, C., 1992, "Analysis of Elastomeric O-ring Seals in Compression Using the Finite Element Method," STLE Trib. Trans., Vol. 35, No. 1, pp. 83-88.

Lindley, P.B., 1967, "Compression Characteristics of Laterally Unrestrained Rubber O-ring," J. IRI, Vol. 1, pp. 202-213.

Molari, P.G., 1973, "Stresses in O-ring Gaskets," 6th Int. Conf. on Fluid Sealing BHRA, pp. B2/15-31.

Strozzi, A., 1986, "Experimental Stress-Strain Field in Elastomeric O-ring seals," Experimental Stress Analysis (H. Wieringa ed.), Martinus Nijhoff Publ., pp. 613-622.

Wendt, G., 1971, "Investigation of Rubber O-ring and X-rings, 1. Stress Distributions, Service and Groove Design," BHRA Fluid Engineering, Internal Report T1115.

14th International Conference on Fluid Sealing, Firenze, Italy,
6-8 April 1994. Organised by BHR Group Limited, Cranfield,
Bedford, MK43 0AJ, UK; Tel: 0234 750422

Finite Element Analysis of Bolted Flange Connections

V. Wright   T & N Technology Ltd.
J. Hoyes   TBA Sealing Materials Ltd.
G. Briggs   Flexitallic Ltd.

Summary

This paper describes some of the work performed at T & N Technology on the development of finite element analysis applied to bolted flange connections.

T & N Technology, the research and development centre for the T & N Group, have had many years experience of applying finite element analysis to industrial applications. Recently, on behalf of Flexitallic and TBA Sealing Materials, the expertise gained has been directed at the analysis of bolted flange connections, with special emphasis on the behaviour of the gasket. A number of flange assemblies have been analysed including 24" class 150 and 4" class 300 ANSI flanges. An area of particular interest has been the consequences of thermal effects on flange behaviour. Two bolted connections operating at elevated temperature have been analysed. This paper presents an overview of the work undertaken to date and the conclusions drawn.

Introduction

Some time ago at Flexitallic and TBA Sealing Materials, it was felt that the existing methods of analysing bolted flange connections, whilst having been valuable in the past, required improvement to provide better applications engineering capability, greater information for development of gaskets and to identify improved test procedures. In addition, greater emphasis has been placed on the performance of joints from the point of view of leakage, with respect to both operating efficiency and environmental factors. It was decided to investigate the application of finite element analysis to bolted flange connections to achieve the improved analysis capability required. Finite element analysis is a well proven, general method for the analysis of structures providing all the analysis facilities required for flange connections and had been using extensively at T & N Technology on a wide variety of industrial applications, such as pistons, brake pads and automotive gaskets. A collaborative project was set up between Flexitallic, TBA Sealing Materials and T & N Technology to develop the required capability. T & N Technology is the central research and development facility for T & N plc.

The work of developing finite element analysis applied to bolted flange connections has lead to the analysis of a number of flange assemblies. These include a 24" class 150 and

4" class 300 ANSI flange. Early work compared the results from finite element analysis with published measured data for a flange on a 15 foot diameter vessel. An area of particular interest has been the consequences of thermal effects on flange behaviour. Thermal analyses have been performed on two bolted connections operating at elevated temperature, one at 900°C and the other at 180°C. Before the detailed results of these analyses are presented a brief introduction to the finite element analysis method is given.

Description of Finite Element Analysis.

Finite element analysis (FEA) is a general method for the analysis of structures and field problems, such as fluid flow and heat transfer. For structural analysis FEA takes as its inputs a description of the geometry of the structure, material behavioural data, loads and constraints, and external temperatures. It is used for determining parameters such as deflections, temperatures, strains and stresses of a structure when subjected to the applied loading, and can predict time dependent effects, for example, stress relaxation due to gasket creep.

When applied to the analysis of flange connections, the inputs to FEA are the dimensions of the flange, the pipe or vessel, the bolts and the gasket, and their material properties, from which a FE model can be generated. Bolt loads, thermal and other boundary conditions, internal pressures and external bending moments can be applied to the model. Other effects, such as thermal insulation of the flange and pipework, can be modelled. From the application of the loading and constraint conditions, the analysis provides data to calculate the following parameters;

> flange stresses and rotation,
>
> flange, bolt and gasket temperatures,
>
> variation of the bolt load due to operating loads,
>
> variation of gasket stress both across its width and between bolts, and
>
> changes due to creep of the gasket and any other components.

The change of all these parameters with time can be predicted. From this variation the service life of the joint could be determined, for example by inputting the results into leakage calculations. Transient behaviour of the flange assembly can be predicted so as to determine any detrimental effects due to start-up, shut-down or in-service variations.

Different types of FEA can be undertaken depending on the application being analysed. Figures 1 and 2 show two types of models employed. The first (fig 1.) shows an axisymmetric model. This is a relatively simple model to generate and analyse, taking a cross-section through the flange and gasket. With this type of model, the assumption is made that the structure and any loading is constant around the circumference of the flange. The loading from the bolts is spread around the flange acting at the bolt centre line. Any variation around the circumference, for example due to bolt tightening sequence or external bending moments cannot be predicted. To minimise the size of the model and hence the computing time required, only one of the pair of flanges needs be modelled. By applying a plane of symmetry through the gasket, mid-way through its thickness, the influence of the second flange can be simulated. This assumes that the second flange behaves in the same manner as the one modelled, which would be incorrect if flanges of differing materials were used, in which case both flanges would be modelled. An extension of the axisymmetric model is one that employs a similar FE model but applies any variation in loading in a harmonic manner around the circumference. A number of cross-sections are represented and the solution is interpolated between these sections. To model the complete variation of loads and structure around the circumference of the flange a three dimensional (3D) model would be employed. Figure 2 shows a 3D model of a segment of a flange assembly, although much larger models extending around half or all of the flange circumference could be generated. Conditions can be applied to the faces of 3D models to simulate the effect of the complete assembly. A 3D model is more complicated to generate and requires more computer time to perform an analysis, but provides the additional information required in some cases.

Applications of FEA to Flange Assemblies : Flange on 15 foot diameter vessel.

Before FEA can be used for the analysis of flange assemblies in a routine way the important parameters influencing the behaviour of the structure must be identified and modelling techniques developed to accurately include the effect of these parameters. Also the analysis techniques must be verified against measured data. Much testing has been performed in the past and the results that have been published were reviewed to obtain data for initial verification.

A paper was presented by N W Murray and D G Stuart [1] giving results of work performed on a flange on a 15 foot diameter pressure vessel. Tests were performed with the flange in both raised face and full face configurations. A rubber O-ring was used in each case for sealing. Longitudinal and hoop stresses on the external surface of the vessel in the region of the flange hub and adjacent vessel wall were measured during the tests. A detailed description of the flange investigated was provided in the paper enabling FEA to be applied.

An axisymmetric FE analysis of this flange was performed using the ABAQUS FE software from Hibbitt, Karlsson and Sorensen Inc. The model is shown in figure 3. For the case with the flange in the raised face configuration, the same loading conditions were applied to the model as in the tests and the predicted and measured stresses were compared. The loading from the bolts was applied as a point load at the centre line of the bolt holes. A plot of the predicted longitudinal stresses is shown in figure 4 and a comparison of the predicted and measured stresses at the surface is shown in graphical form in figure 5. It can be seen that the maximum stress occurs at the weld neck at the junction between hub and cylinder, and that there is good agreement between the results from the FEA and measurements. However, data presented in the paper indicated significant reduction in stress on the gasket due to the effect of hydrostatic end thrust. The application of the bolt loads as a point load of constant magnitude does not simulate this effect therefore an improved method was developed during subsequent analyses.

Analysis of 4" Class 300 ANSI B16.5 Flange

To establish the effect of size on the behaviour of flange assemblies a model of a 4" class 300 ANSI B16.5 flange was generated. In addition, flanges were available from which to make a test rig and testing facilities were available at Flexitallic enabling measured data to be generated for comparison with predictions.

As for the 15 foot diameter flange, an axisymmetric analysis was performed. In this case however the gasket was included in the model and an improved means of representing the bolts was used (figure 6.) The loading response of a gasket is highly non-linear and has differing loading and unloading characteristics (figure 7.) As a first approximation to this behaviour a foam material model was used within the FE analysis (figure 8), however a more complex model is required. This is discussed later in this paper. The bolts were represented by 2 lines of elements acting at the bolt centre line. In an axisymmetric FE model this equates to a cylinder, therefore the individual bolts were not modelled. However the material elasticity of the cylinder was controlled to give it the same stiffness as the bolt thereby accurately representing any off-loading of the bolts due to flange rotation. One end of the elements simulating the bolts was attached to the upper surface of the flange ring (figure 6.) with the other positioned at the plane of symmetry through the gasket. To apply the bolt load a force was applied at the lower end of the bolt elements acting downwards. Once this force had acted, inducing strain in the flange, gasket and bolting, the position of the lower end of the bolt elements was fixed, therefore any change in length would induce a change in strain and hence load.

A test rig was manufactured consisting of two flanges welded to lengths of pipes with end-caps (figure 9.) The rig was instrumented with strain gauges on the vessel and four of the 8 bolts. The gauges on the bolts were to monitor any off-loading due to flange rotation. Also special transducers were used to monitor the variation of gasket thickness during the tests. These were developed at T & N Technology initially for use in cylinder head gaskets.

The stress distribution in the flange at bolt-up and with internal pressure applied, calculated by FEA, is shown in figures 10 and 11 respectively. The same trends were seen from measurement. It can be seen that the maximum stress now occurs at the fillet radius between flange ring and hub, which is different from that seen with the 15 foot diameter flange. As the internal pressure rises the stress at the weld neck increases as flange rotation increases. From the measurements made, the mean gasket thickness at a radius of 74mm increased from 1.166mm to 1.169mm due to the application of internal pressure. The FEA calculated an increase in thickness of 0.002mm. These changes are very small and any differences between measured and calculated values can be explained by variation in the gasket material used compared to the material model employed in the FEA. The decrease of bolt load was 1.7% and 1.6% from measurement and calculation respectively.

Analysis of 24" Class 150 ANSI B16.5 Flange

T & N Technology has been sponsored by the Pressure Vessel Research Council in America to perform FE analysis of a 24" class 150 flange. This work was conducted in association with the University of North Dakota, who have a test rig featuring this type of flange. The work was part of the Flange Parameter Study of the PVRC and the aim was to investigate the application of FE to bolted flange connections in order to identify the important aspects of the analysis. Once the method is established it is intended for use in a flange re-rating program.

The test rig at the University of North Dakota had been fully instrumented with all bolts strain gauged, gauges mounted on the vessel and the latest equipment for monitoring gasket stresses. The data generated from tests has been used to verify the results from the FE analysis.

The FE modelling approach for this flange was the same as for the 4" class 300 flange. The influence of flange to gasket friction is known to affect the performance of a joint, therefore the modelling work was used to investigate the influence of this phenomenon. Flange to gasket friction was found to significantly affect the stress distribution across the gasket, and flange strains (table 1.) The effect of increasing friction is to make the gasket stiffer as the friction prevents the gasket from flowing radially outwards. In this case this results in reduced flange rotation, and hence lower flange strains, and a more even distribution of gasket stress, reducing the maximum.

Steady State Thermal Analysis

Two flange assemblies have been analysed to determine the influence of thermal effects. One case was a vessel containing fluid at 900°C at a pressure of 5 bar and the other contained steam at 180°C and 10 bar. A steady state analysis was performed on the first case to determine the effect of temperatures on gasket performance. For the second case a thermal transient analysis was performed and the results compared with measurements taken from an instrumented test rig.

Figure 12 shows a cross-sectional representation of the flange assembly operating at 900°C. The gasket proposed was a carrier ring type, supplied by Flexitallic, consisting of a solid metal carrier ring with two spiral wound gaskets acting as sealing elements (only one is seen in the model due to the plane of symmetry through the gasket.) This type of gasket gives greater recovery than a single spiral wound gasket. The temperatures used in the analysis were matched to those supplied by the customer. Important aspects of this application were how the gasket matched the flange in terms of radial growth and any potential off-loading of the gasket due to temperature effects. The radial growth of the flange and gasket are shown in figure 13. It can be seen that there was little difference for the two components. Figures 14 and 15 show the flange and gasket stresses due to bolt loads and internal pressure, and with the addition of temperature. The increase in temperature caused a threefold increase in flange rotation which resulted in a high contact stress between the outside diameter of the raised face and the carrier ring. However the stress on the sealing element was reduced by only 5%, from 10,000 psi to 9,500 psi, making an insignificant effect on sealing. The bolt loads were seen to increase from 33% of yield to 53%, which were considered acceptable. The gaskets are currently undergoing field trials.

Transient Thermal Analysis

The thermal transient analysis was performed on a 50mm DIN standard flange. The flange assembly was undergoing tests on a steam line at TBA Sealing Materials, enabling instrumented testing to be carried out.

The instrumentation used consisted of a number of thermocouples on the flange and on a bolt, (figures 16 and 17) and strain gauges on the bolts in a full-bridge configuration to eliminate the effect of bolt bending and to compensate for the effects of temperature (figures 18.) The thermocouple at the flange inner diameter was a sheaved type, fed through a hole through the flange until it was level with the surface. A thermocouple was also positioned in the centre of the pipe to measure the temperature of the steam, however this failed at the start of the test. Internal pipe pressure was also measured.

The object of the test was to obtain data over an initial start-up and through a longer steady state condition in order to supply relevant information for the development of the transient analysis and also to identify the effects of creep. Recordings of all parameters were taken every 5 seconds over the initial 20 minute start-up period, until a steady state condition had been achieved, then every 4 hours for the remaining period of the test. The test lasted 11 days.

Figure 19 shows the measured temperatures. In figure 20 the measured and calculated bolt strains through the start-up period are shown. (It is noted that bolt 4 shows a significant increase in strain at 800 seconds but returns to its value at 700 seconds after a short period.) The bolt loads, derived from the measured strains, through the entire test are given in table 2. It can be seen that the bolt stresses decrease by 10% over four hours at room temperature, and fall a further 38% over the subsequent 11 days at elevated temperature (taking room temperature values.) This was as expected. At the end of the test the flange assembly had achieved a relatively constant state with negligible changes due to creep occurring. From visual inspection the flange was completely sealed throughout the test.

The FE analysis of the thermal transient was performed in two stages; firstly to determine the flange and bolt temperatures through the test, and secondly to apply the calculated temperatures to a thermo-mechanical analysis. The analyses performed used an axisymmetric model with the bolts modelled as a line of elements as described above. In the first analysis, thermal boundary conditions were applied to the flange and gasket surfaces. Heat transfer coefficients were initially derived from literature and test experience at T & N Technology. The steam was known to be at approximately 180°C.

The heat transfer coefficients were modified, within limits, to achieve a "best fit" of temperatures to those measured. The calculated temperatures are shown in figure 21. It can be seen that they compare well with the measured values. The most significant differences are seen at the flange inner diameter within the initial 200 seconds and in the bolt temperatures. The differences at the flange i/d are due mainly to the thermal inertia of the thermocouple used causing it not to respond at the same rate as the pipe wall. The differences in bolt temperature results mainly from the method of modelling the bolt. The heat transfer mechanism between the bolt and the flange is very complex, with conduction occurring at the contact between bolt head and flange ring, and convection and radiation between the bolts and the holes in which they sit. In the FE model it was assumed that the bolts were at the same temperature as the flange at the bolt hole diameter. This is an area of modelling receiving further investigation.

From the thermo-mechanical analysis the variation in bolt strain and gasket stress was obtained. Measured and calculated bolt strains are shown in figure 20. The initial bolt loading condition was taken as that measured at the beginning of the thermal transient. It can be seen that once steady state has been achieved the calculated bolt strain is approximately 8% below the mean measured value. This is acceptable given the assumptions made in modelling the bolts and is conservative if used to calculate gasket stress. Over the first 100 seconds the bolt strains see an off-loading followed by a reloading. Subsequently they drop gradually to the steady state value. The initial drop is caused by the rapid heating of the inner wall of the flange and pipe resulting in an increase in flange rotation. As the heat penetrates into the flange this rotation is reduced, increasing the bolt strains. The gradual off-loading can be attributed to the increase in bolt temperature.

Significant variations can be seen in gasket stress through the thermal transient (figure 22.) Initial values of stress at the gasket inner and outer edges are 15 MPa and 24 MPa respectively. There is a rapid increase in stress at the gasket i/d caused by the high temperature and expansion occurring at the flange inner surface. As the heat flows into the flange this localised effect is reduced and the gasket stress reduces at the inner edge. However, flange rotation increases causing the high stress at the gasket outer diameter. The flange assembly had some amount of restraint in the direction along the pipe due to the general configuration of the test rig. Therefore, as the temperature increased, the load in this direction grew causing the overall increase in mean gasket stress.

This analysis has shown that even with a simplified modelling approach results comparable to those measured in tests can be produced. There are inevitably approximations and assumptions made when using an axisymmetric analysis. However important information can be generated to improve the understanding of flange behaviour. At the time of writing further testing and development of the analysis method are planned.

Gasket Material Modelling

As mentioned in an earlier section the behaviour of the gasket material is very complex, but it is fundamental to the analysis of flange assemblies. The loading curve exhibits highly non-linear behaviour initially but becomes more linear above given stress levels, the loading and unloading curves are different (figure 7), and from the test on the 50mm DIN flange it was seen that gasket creep is an important effect (table 2.) Research showed that existing material models available in commercial FE packages could not represent all features of this behaviour. For example, the foam model available in ABAQUS provides the initial non-linear loading response and the difference in the unloading curve. However it does not model creep effects or the linear loading response at higher stresses.

It was decided that the modelling of the gasket response was fundamental to accurately modelling and understanding the behaviour of bolted flange connections, and hence developing improved gasket materials. Therefore a research program was initiated, in association with an academic institute having expertise in the field of material modelling, to develop a comprehensive mathematical model to represent the behaviour of gasket materials. This would be programmed into the FE code being used. The material model has reached several of the early milestones in its development. Figure 23 shows the loading and unloading characteristic from the model. This compares well with the measured data, showing the initial non-linear loading region followed by the linear response, and the different unloading characteristic. The model also includes the very important effect of creep as illustrated by the horizontal portion of the graph in figure

24. The gasket model has been programmed into ABAQUS using the user definable material subroutine. When testing of this model is complete it will provide a valuable tool in the analysis of bolted flange connections.

Conclusions

It has been shown that the finite element method provides the required capabilities for the analysis of bolted flange connections. Some of the experience gained in the development of this application of FEA has been described and some important aspects of modelling have been highlighted.

A method has been developed to simulate, in a simple to generate, axisymmetric finite element model, the bolt loading and any variation due to flange rotation. The importance of flange-to-gasket friction has been illustrated. Both steady state and transient thermal analyses have been described. Comparison of measured data and calculated results has been given to verify the methods developed. The need for improved methods of modelling the gasket material behaviour has been identified and the work to develop a suitable model has been described. Future work will be to develop the existing methods to a point where they can be used on a regular basis to analyse the problems arising in the field. Modelling techniques are being developed to address the more complex situations to eliminate any shortcomings in the existing methods.

This paper has illustrated the work being performed to develop methods for the analysis of bolted flange connections to provide better understanding of their behaviour. This is leading to improved sealing products and services in the market place.

Reference

1.	Behaviour of Large Taper Hub Flanges, by N W Murray and D G Stuart. Paper 9, Pressure Vessel Research 1961, Institute of Mechanical Engineers.

Table 1: Effect of Gasket Friction on Gasket Stress and Flange Strain

FRICTION	FLANGE STRAINS AT BOLT-UP				MAXIMUM GASKET STRESS	
COEFF	RADIAL STRAIN		AXIAL STRAIN		BOLT-UP	@300PSI
	MAX	MIN	MAX	MIN		
0.3	0.045	-0.032	0.053	-0.067	70	48.9
0.6	0.039	-0.042	0.041	-0.057	57.1	38.3

Table 2: Bolt Loads Through Duration of Test

BOLT STRESSES (psi)					
	Bolt 1	Bolt 2	Bolt 3	Bolt 4	Mean
At bolt-up	25300	25600	25400	25800	25500
After 1hr @ Room Temperature	23700	24100	24100	24000	24000
After 4hrs @ Room Temperature	23000	22500	23000	22500	22750
After 11 days @ Elevated Temperature					
Hot	16300	16700	15400	14250	15650
Cold	11600	12700	14700	13200	13050

14th International Conference on Fluid Sealing, Firenze, Italy;
6-8 April 1994. Organised by BHR Group Limited, Cranfield,
Bedford, MK43 0AJ, UK; Tel: 0234 750422

UP-DATE ON HIGH TEMPERATURE LEAKAGE BEHAVIOUR OF COMPRESSED FIBRE MATERIALS AND AN ALTERNATIVE APPORACH FOR NEW JOINT DESIGN RULES

Dr. Jörg Latte + Claudio Rossi
ISTAG AG - a Klinger Group Company

Abstract

Based on leakage measurements at elevated temperature a new approach is suggested for gasketed joint design rules including the dependency from temperature, internal pressure, hydrostatic end thrust, and relaxation of the gasket. Though the work is not yet finished an outline of approach indicates an easy but elaborate access to the theory of joint design.

Introduction

On the 3rd CETIM Symposium in Biarritz Klinger-Istag presented a survey of its investigation programme on compressed fibre materials[1]. In this the characteristic differences in sealing properties between fibrous materials and flexible graphite were outlined, this leading to the conclusion that the PVRC design factors based on ROTT-Test of the Ecole Polytechnic of Montreal[2], are not applicable for compressed fibre materials - at least not at increased temperatures. The present paper now discusses this conclusion in detail and suggests a possible solution.

Basic Leakage Properties

Fig. 1 illustrates the general leakage properties of flexible graphite for a constant surface load of 20 MPa depending on the internal pressure and on temperature. As can be easily seen the leakage rate increases strongly with the increase of internal pressure at any temperature though the increase at 300 °C is only half of that at ambient temperature. But overall it stays basically in the same range.

It is reasonable, therefore, to transfer gasket design factors based on room temperature measurement to high temperature applications without making a too large error.

For compressed fibre materials this is different as illustrated by Fig. 2. At room temperature the leakage rate increases drastically as expected with each step of increase of the internal pressure. This corresponds very well with the part „A" of the ROTT test cycle. It is, therefore, evident that for ambient temperature applications the new PVRC design factors are also applicable for compressed fibre materials. But already at 100 °C the increase of leakage rates become insignificant though the internal pressure was raised in the same way as at ambient temperature. At even higher temperatures leakage rates stay basically at zero or cannot be detected with the analysing equipment used.

It is evident that for applications at increased temperatures those new PVRC gasket design factors based on room temperature measurements are misleading and could have the probable consequence of not selecting the optimal gasket material or even selecting the wrong material, this risking possibly catastrophic failure.

Fig. 3 and 4 show the corresponding graphs to Fig. 1 for surface load of 30 MPa and 20 MPa respective. They might give a better visual impression of the correlations stated above.

Design Factors for Applications at elevated Temperatures

But there might be a solution to the serious problem of this temperature induced property of all compressed fibre materials.

Taking Fig. 4 as a basis it can be seen that with a surface pressure of 20 MPa about 65 bar internal pressure can be sealed off at 100 °C (Fig. 5) without exceeding a leakage rate of 2 ml this corresponding to 0,2 mg/sec x m or PVRC tightness class 2 - 3. When a higher internal pressure has to be contained an additional surface load becomes necessary which is a function of the internal pressure and can be described by approximations as y(2) = m x p. This becomes evident when data of the sealability behaviour:

20 MPa versus 65 bar of Fig. 4 and 5 and
30 MPa versus 95 bar of Fig. 3

are transfered into a new graph (Fig. 6) now showing the necessary surface pressure σ as a function of the internal pressure Pi.

As mentioned before this is an approximation and, therefore, the straight line σ = f (Pi) was drawn through the origin of the graph.

In fact this straight line would cut the horizontal of the graph at about 5 MPa as shown in fig. 7. In this figure additional parameters were added like max. tolerable leakage rate (the sub-index stating the rate in mg/sec x m) and temperature T (the sub-index stating the temperature in °C). Further parameters are thickness and width of the gaskets - the influence of which still have to be evaluated.

Unfortunately, the test results are not yet detailed enough to state the definite functional dependency of the necessary surface load for all parameters mentioned. Also these straight lines are a simplification as they are the tangents of complex exponential functions. This means they best describe the dependency of the minimum operational surface stress from the internal pressure for higher internal pressures and surface stresses. On the other hand test results so far indicate an absolute minimum assembly stress of about 20 MPa (see later). Therefore, this simplification is justified and offers an easy approach for joint design. Additionally it gives an explanation how the «old» PVRC design factors were originally created.

Fig. 7b shows in double logarithmic scale the dependency of leakage rates from internal pressure Pi and operational temperature at constant surface stress (in this case 20 MPa). Everyone will at once recognize the analogy to ROTT test graphs displaying the dependency of leakage from internal pressure and surface stress at ambient temperature. The comparison of both graphs leads to the conclusion that for compressed fibre materials the temperature increase has the same effect on leakage behaviour as the increase of surface pressure. This is different for flexible graphite as stated above and gives compressed fibre materials a definite advantage and lead.

In analogy to ROTT test procedure it should be possible to formulate a general applicable tightness parameter Tp = f (σ,T, Pi) as a dependency from surface stress σ , temperature T and internal pressure Pi. But for this purpose some more detailed measurements have to be carried out.The outlined approach can supplement the PVRC approach for gasket design factors taking into account the different behaviour of compressed fibre materials.

The same conditions are displayed in Fig. 8 showing the necessary operational gasket load for a maximal leakage rate of λ = 0,1 mg/sec x m depending upon the internal pressure Pi and temperature T. As can be clearly seen: the surface load based on room temperature measurements would lead to too high data for the necessary stress. For instance to seal Pi = 90 bar at room temperature a surface pressure of 44 MPa would be needed while for the same sealability at an operational temperature of T = 250 °C only 8-12 MPa are necessary.

But we have still to keep in mind that surface pressures established by the outlined approach are those for operational conditions. To arrive at the necessary assembly pressure $[\sigma_{AS}]$ for the gasket the surface pressure reductions caused by the relaxation of the gasket $[\sigma_{REL}]$ and caused by the hydrostatic end thrust $[\sigma_{HET}]$ have to be taken into account. This leads to the following equation (1):

$$(1) \quad \sigma_{AS} = \sigma_{OP} + \sigma_{HET} + \sigma_{REL}$$

The relaxation is described by

$$(2) \quad \sigma_{REL} = \sigma_{AS} - \sigma_{E} \quad (\sigma_{E} = \text{residual stress})$$

and

$$(3) \quad R\,[\%] = \frac{\sigma_{AS} - \sigma_{E}}{\sigma_{AS}} \times 100\,\%$$

Replacing σ_{REL} in equation (1) we arrive eventually at:

$$(4) \quad \sigma_{AS} = (\sigma_{OP} + \sigma_{HET}) \times \frac{100\,\%}{100\,\% - R}$$

For simplification possible interrelations between operational surface stress and relaxation can to be neglected.

But nevertheless, unfortunately, these correlations are not as simple as they look as the relaxation is dependent on the assembly stress, the thickness of the gasket "d", the operational temperature and the spring constants of the bolts. Neglecting for the time being the influence of the bolts and using standard test procedures like DIN 52913, BS7531:1982 or ASTM F38B further series of measurements become necessary to establish the relaxation $R = f\,(\sigma_{AS},\, d,\, T)$ as a function of assembly stress, gasket thickness and temperature. This is just to indicate how much work has still to be done.

Leakage measurements with a surface load as low as 10 MPa (Fig. 9) indicate for practical reasons the necessity for a minimum surface load when a gasket is fitted as otherwise the compressive force would be too low to mate with flange roughness and to close the natural porosity of the sealing material. Data obtained so far from trials suggest this minimum assembly stress for this compressed glass fibre material KLINGERsil®C-4430 to be about 20 MPa which corresponds with the yield factor y of the old PVRC design rule.

Depending on the operational temperature and internal pressure, the operational surface stress σ_{op} might drop on the other hand as low as 2.5 - 5 MPa before a severe leak (blow-out) occurs. This is shown in Fig. 10a, 10b, for KLINGERsil®C4430. The first part of Fig. 10a shows at 150 °C the slight increase of gas leakage to about 2 ml/min or 0,2 mg/sec x m with the drastic increase of the internal pressure from 20 bar to 100 bar. In the second part of the graph the internal pressure is reduced to 40 bar and the surface load stepwise decreased. This second part of the graph is magnified in Fig. 10b. Surprisingly a significant leak (blow-out) occurs only at a surface load lower than 2.5 MPa.

Conclusion

As it can be seen the suggested approach for alternative design rules for compressed fibre materials have some similarity to the old ASME code design factors m and y [3]. On the contrary to these factors the new approach is based on measurements and is, therefore, more detailed and more specific for each gasket material.

Nevertheless, it is amazing to learn that the Rossheim, D.B. and Markt, A.R.C.[4] „invented" these rules for the «Boiler and Pressure Vessel Code» allegedly without performing trials which proves their deep understanding of this type of material.

References

[1]   J. Latte, C. Rossi, Publications CETIM
        3rd International Symposium on Fluid Sealing of Static Gasketed Joints,
        233 - 251

[2]   J.Payner, A. Bazergui, G. Leon
        «New Gasket factors» - A Proposal Procedure
        Proc. of the 1985 Pressure Vessel and Piping Conference - ASME - PVP
        Vol. 98.2 - pp 85 - 93

[3]   ASME Boiler and Pressure Vessel Code
        Sec VIII Div. 1

[4]   Rossheim, D.B. and Markt, A.R.C.
        «Gasket Loading Constants»
        Mech. Eng. Vol 65 (1943), 647 ff

ISTAG AG CH-5704 Egliswil	High Tempearature Gas Leakage Constant Load (Press)	Date: 26.10.93

Material	SLS 200	Sample Dimension		Apparatus	Klinger Hot Compression
Manufacturer	Klimger AG /Egliswil	D1 mm	56	Media	Nitrogen
Thickness	2.0 mm	D2 mm	92	Surface Stress MPa	20 MPa
Deliverance Date	04. Mai 93	Area under stress mm^2	5867		

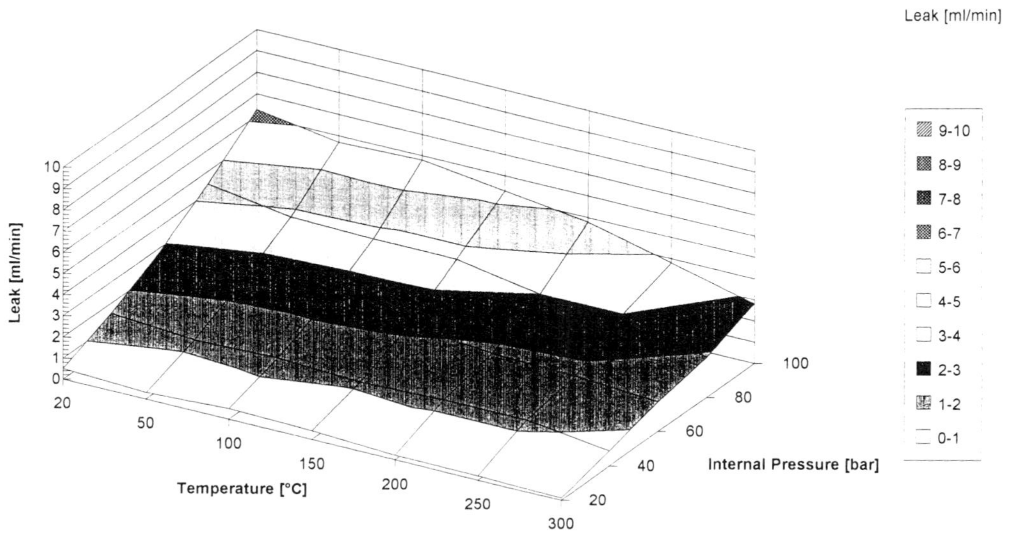

Fig.1

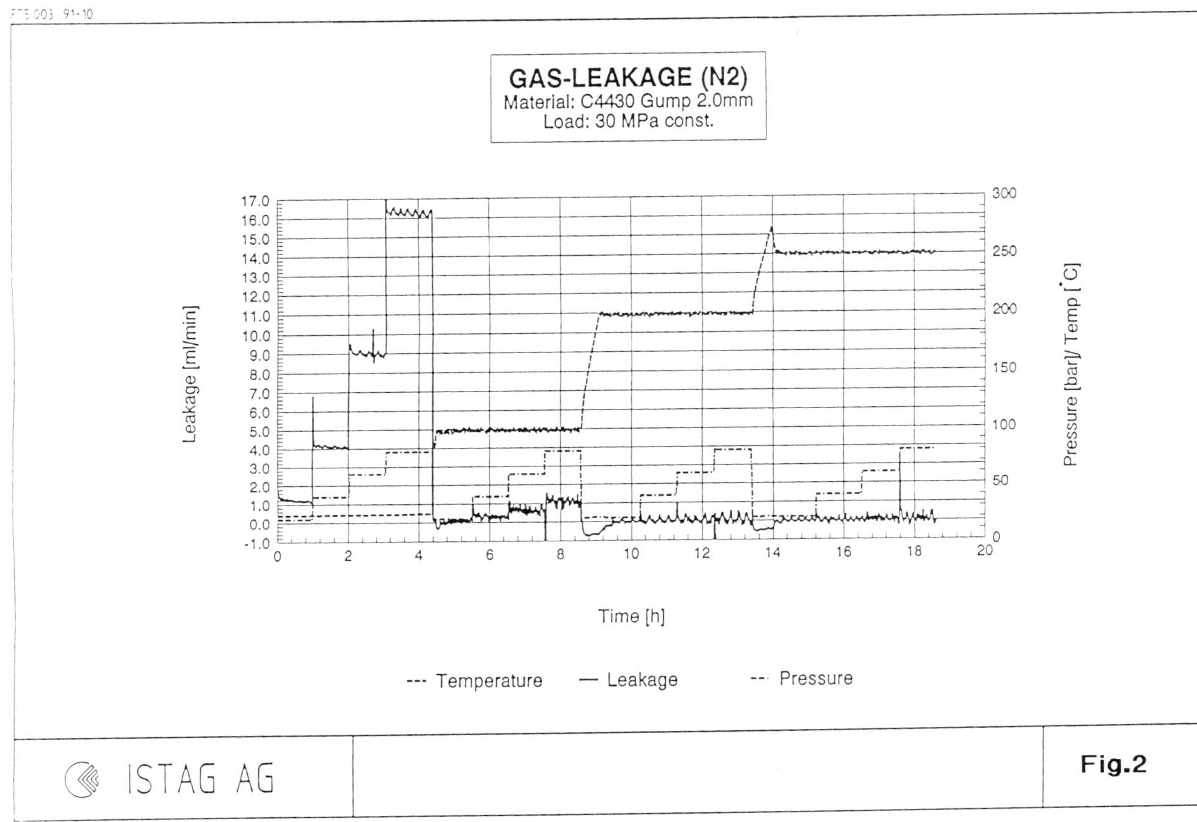

GAS-LEAKAGE (N2)
Material: C4430 Gump 2.0mm
Load: 30 MPa const.
Leakage [ml/min]
17.0
16.0
15.0
14.0
13.0
12.0
11.0
10.0
9.0
8.0
7.0
6.0
5.0
4.0
3.0
2.0
1.0
0.0
-1.0
Pressure [bar]/ Temp [°C]
300
250
200
150
100
50
0
Time [h]
0 2 4 6 8 10 12 14 16 18 20
--- Temperature
— Leakage
--- Pressure
ISTAG AG
Fig.2

ISTAG AG CH-5704 Egliswil	High Tempearature Gas Leakage Constant Load (Press)	Date: 22.09.93

Material		Sample Dimension		Apparatus	Seidner Press
Manufacturer	C 4430	d1 mm	56	Test Media	Nitrogen
Thickness	2.0 mm	d2 mm	92	Surface Stress	30 MPa
Deliverance Date	04. Mai 93	Area under stress mm2	5867		

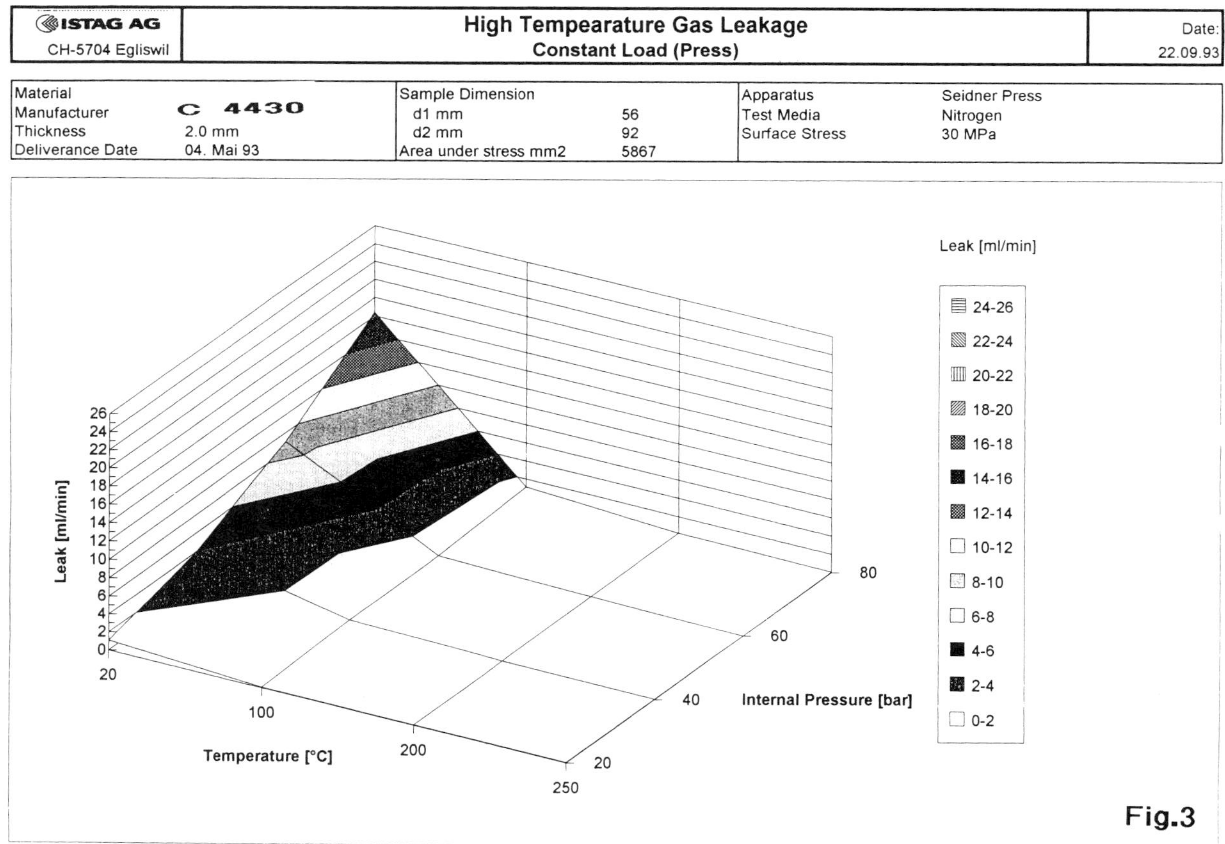

Fig.3

ISTAG AG CH-5704 Egliswil	High Tempearature Gas Leakage Constant Load (Press)	Date: 28.05.93

Material	**C4430**	Sample Dimension		Apparatus	Klinger Hot Compression II
Manufacturer	RK Gumpoldskirchen	d1 mm	56	Test Media	Nitrogen
Thickness	2.0 mm	d2 mm	92	Surface Stress	20 MPa
Deliverance Date	04. Mai 93	Area under stress mm2	5867		

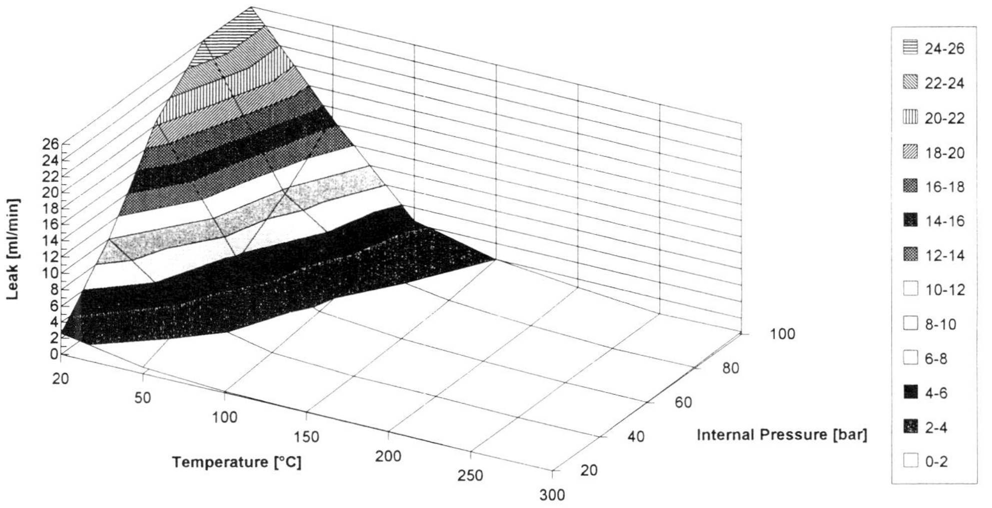

Fig.4

ISTAG AG CH-5704 Egliswil	High Tempearature Gas Leakage Constant Load (Press)	Date: 28.05.93

Material	C4430	Sample Dimension		Apparatus	Klinger Hot Compression II
Manufacturer	RK Gumpoldskirchen	d1 mm	56	Test Media	Nitrogen
Thickness	2.0 mm	d2 mm	92	Surface Stress	20 MPa
Deliverance Date	04. Mai 93	Area under stress mm2	5867		

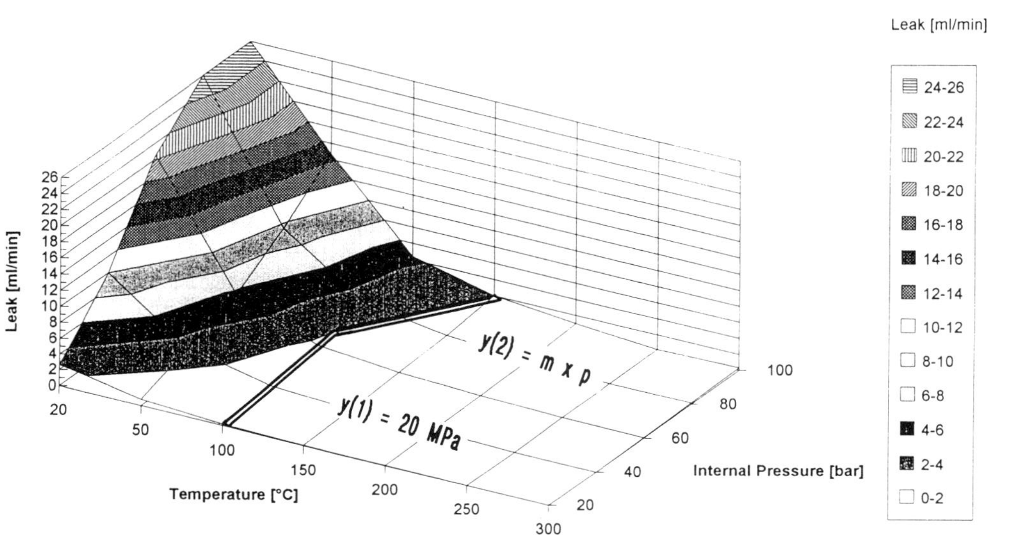

Fig.5

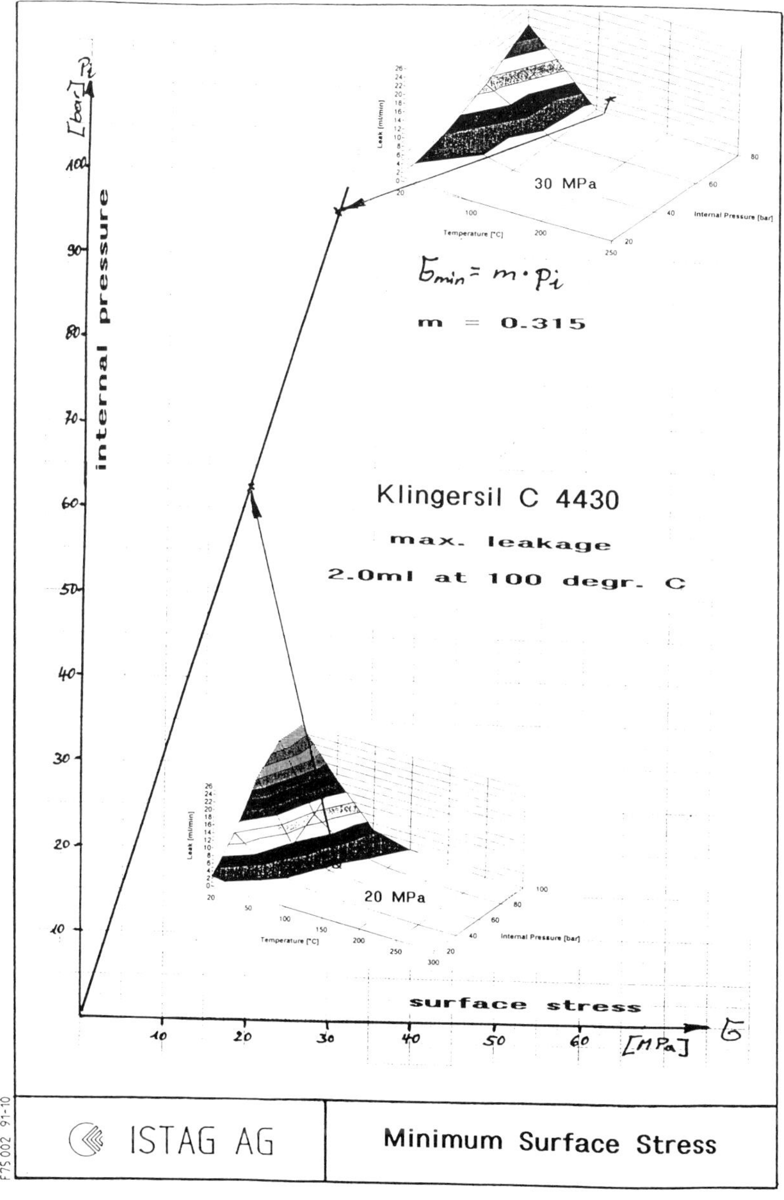

ISTAG AG

Minimum Surface Stress

F75 002 91-10

Fig.6

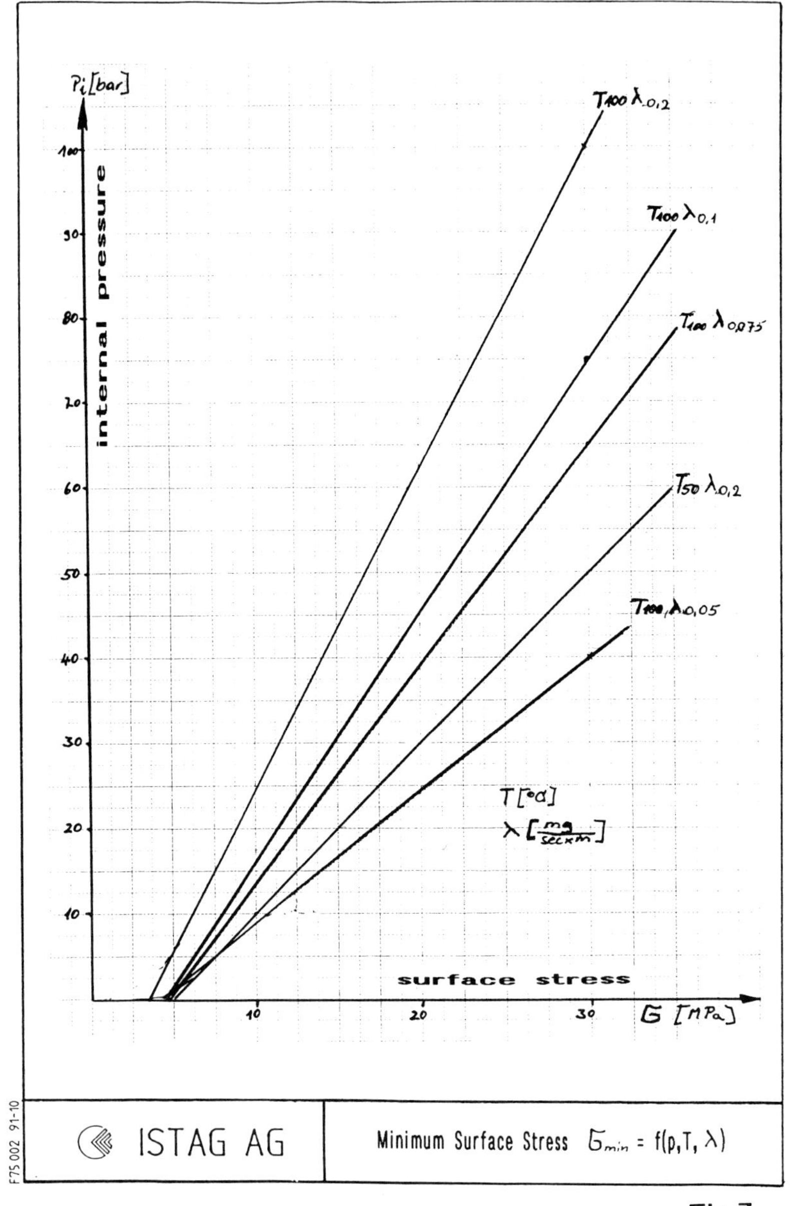

Fig.7

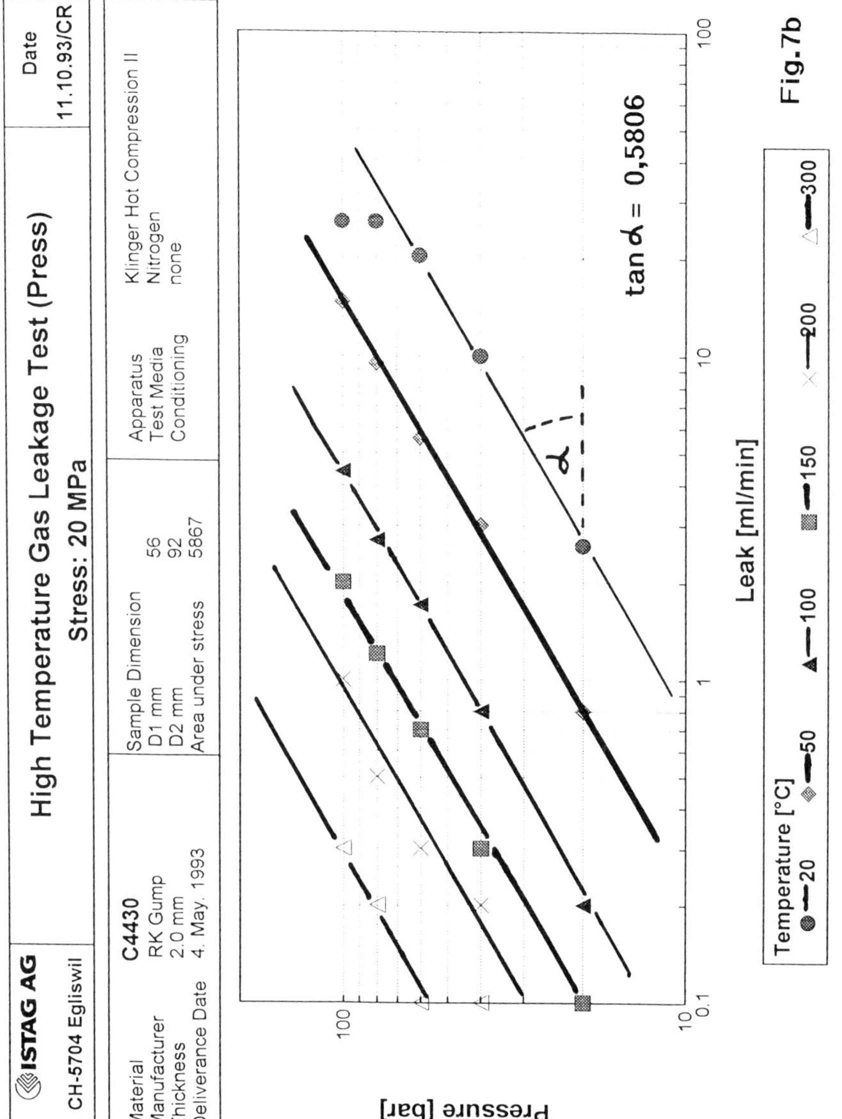

Fig.7b

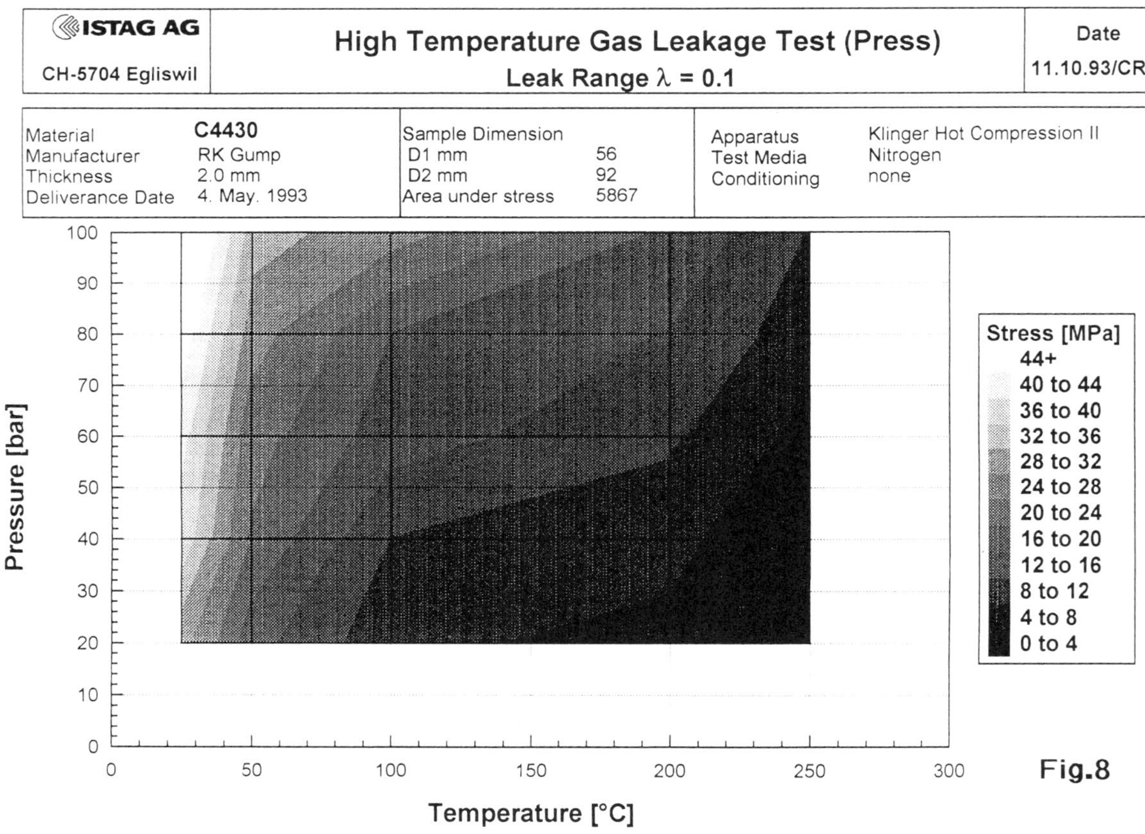 ISTAG AG CH-5704 Egliswil	High Temperature Gas Leakage Test (Press) Leak Range λ = 0.1	Date 11.10.93/CR

Material	**C4430**	Sample Dimension		Apparatus	Klinger Hot Compression II
Manufacturer	RK Gump	D1 mm	56	Test Media	Nitrogen
Thickness	2.0 mm	D2 mm	92	Conditioning	none
Deliverance Date	4. May. 1993	Area under stress	5867		

Fig.8

ISTAG AG CH-5704 Egliswil	High Tempearature Gas Leakage Constant Load (Press)	Date: 24.06.93

Material	C4430	Sample Dimension		Apparatus	Klinger Hot Compression II
Manufacturer	RK Gumpoldskirchen	d1 mm	56	Test Media	Nitrogen
Thickness	2.0 mm	d2 mm	92	Surface Stress	10 MPa
Deliverance Date	04. Mai 93	Area under stress mm2	5867	Conditioning	none

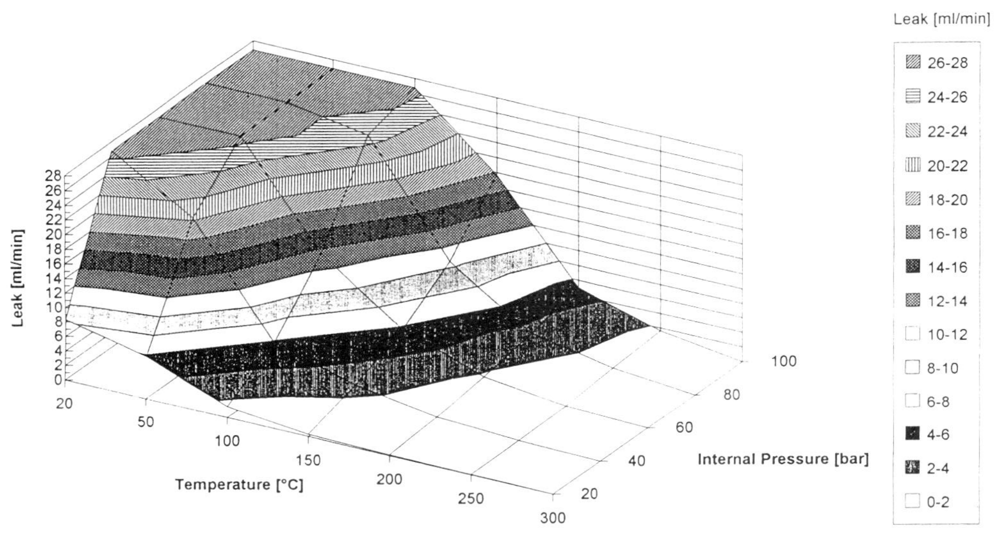

Fig.9

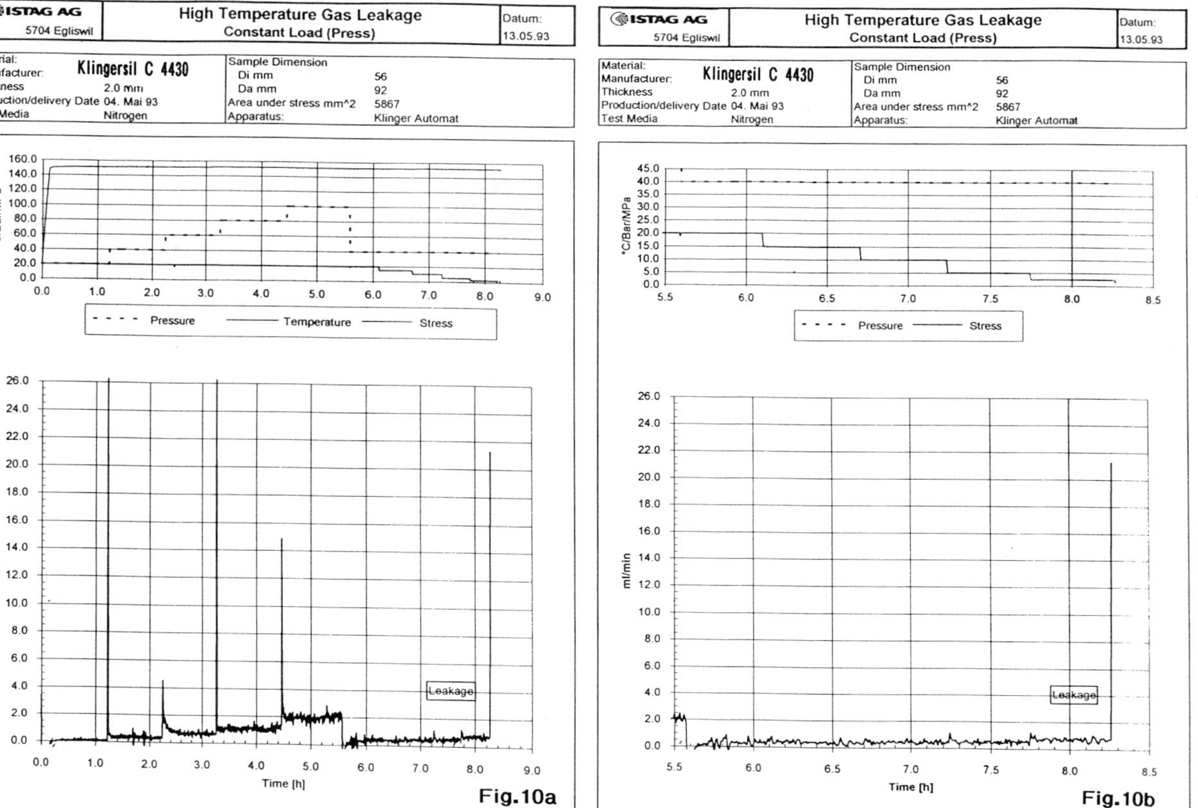

ISTAG AG
5704 Egliswil
High Temperature Gas Leakage
Constant Load (Press)
Datum: 13.05.93
Material:
Manufacturer: Klingersil C 4430
Thickness 2.0 mm
Production/delivery Date 04. Mai 93
Test Media Nitrogen
Sample Dimension
Di mm 56
Da mm 92
Area under stress mm^2 5867
Apparatus: Klinger Automat
°C/Bar/MPa
Pressure
Temperature
Stress
ml/min
Leakage
Time [h]
Fig.10a

ISTAG AG
5704 Egliswil
High Temperature Gas Leakage
Constant Load (Press)
Datum: 13.05.93
Material:
Manufacturer: Klingersil C 4430
Thickness 2.0 mm
Production/delivery Date 04. Mai 93
Test Media Nitrogen
Sample Dimension
Di mm 56
Da mm 92
Area under stress mm^2 5867
Apparatus: Klinger Automat
°C/Bar/MPa
Pressure
Stress
ml/min
Leakage
Time [h]
Fig.10b

C4430_4.XLS

14th International Conference on Fluid Sealing, Firenze, Italy,
6-8 April 1994. Organised by BHR Group Limited, Cranfield,
Bedford, MK43 0AJ, UK; Tel: 0234 750422

ELASTOMERIC-SEAL LIFE PREDICTION

E. Ho and B.S. Nau
BHR Group Ltd., Cranfield, U.K.

ABSTRACT

Elastomer seals are widely used in the petroleum industry and seal failure can be very expensive, however it is difficult to predict the life of a seal in a specific application. Computer software has been developed to address this problem, it models fluid ingress into the seal and subsequent material property degradation, due to swell and chemical reaction, these are computed using the finite element method. This permits application to a wide range of geometries. In this paper, the computer model is described and sample results are presented, using as input data material properties obtained from measurements on simple test pieces.

INTRODUCTION

Elastomer seals in oil-production equipment can be exposed to a wide variety of service fluids including water, steam, carbon dioxide, hydrogen sulphide, and hydrocarbons. Operating conditions may involve fluid pressures up to 1000 bar and temperatures from 4°C to over 200°C; during transient blow-down conditions temperature can fall to -90°C. Seal failure in such applications involves considerable costs due to inaccess- ibility and loss of production. Reliable prediction of seal life at the design stage would be a valuable aid to selecting seals and scheduling maintenance. Two failure modes are of particular concern:

     explosive decompression,
     progressive degradation.

Explosive decompression damage occurs in high-pressure gas applications when the gas pressure drops, Figure 1 illustrates typical damage. The severity

Fig 1   Example of explosive decompression damage to an O-ring

124

and extent of damage depends on the diffusion coefficient of the specific gas in the specific elastomer, the saturated concentration of the gas in the seal, the design of the seal section and housing, and the strength of the seal material. Use of the computer model in explosive decompression problems has been described in Ho and Nau (1993).

Progressive degradation is generally of a long-term nature and its extent and consequences in service can be particularly difficult to predict. Traditional approaches to this class of problem include :

(i) Reference to previous experience of 'similar' applications: but different applications differ so greatly in details of fluid composition and operating conditions that this method is often unreliable.

(ii) Functional seal tests - usually accelerated and extrapolated to the long term: to be reliable such tests are time consuming and costly.

(iii) Measurement of material property changes in test pieces, with subjective interpretation in terms of seal performance: commonly-used properties (e.g. tensile strength, hardness, swell) are difficult to translate into seal performance terms.

COMPUTER-BASED APPROACH

Within a broader project, concerned with the improvement of techniques for the prediction of seal failure due to the above mechanisms, the authors have developed a computer program which is a valuable tool for understanding and predicting the degradation of seals in service. This model is capable of analysing the following combination of factors:

seal shape and size,
time-dependent fluid diffusion into or out of the seal,
time-dependent material properties change due to elastomer / fluid interaction,
dimensional changes due to swell,
dimensional changes due to temperature change,
initial squeeze and deformed seal shape under pressure,
sealing-contact stress .

Also, account can be taken of compression set in calculating the final sealing contact stress . A colour-graphics pre-/post- processor handles input-data generation and display of results for, currently, the seal geometries in Figure 2, each of which can be analysed for user-defined values of the dimensions indicated.

Finite-element numerical analysis is used for simultaneously calculating time-dependent fluid diffusion within the seal, chemical interaction between fluid and elastomer, seal deformation under pressure, and seal contact stresses. A flow-chart of the computer program is shown in Figure 3. Six-node curve-sided axisymmetric iso-parametric elements are used, permitting accurate representation of complex seal shapes. A typical mesh in an undeformed O-ring is shown in Figure 4. The software is written in Fortran 77, it runs on Macintosh and MSDOS computers. Details of the main modules will now be described.

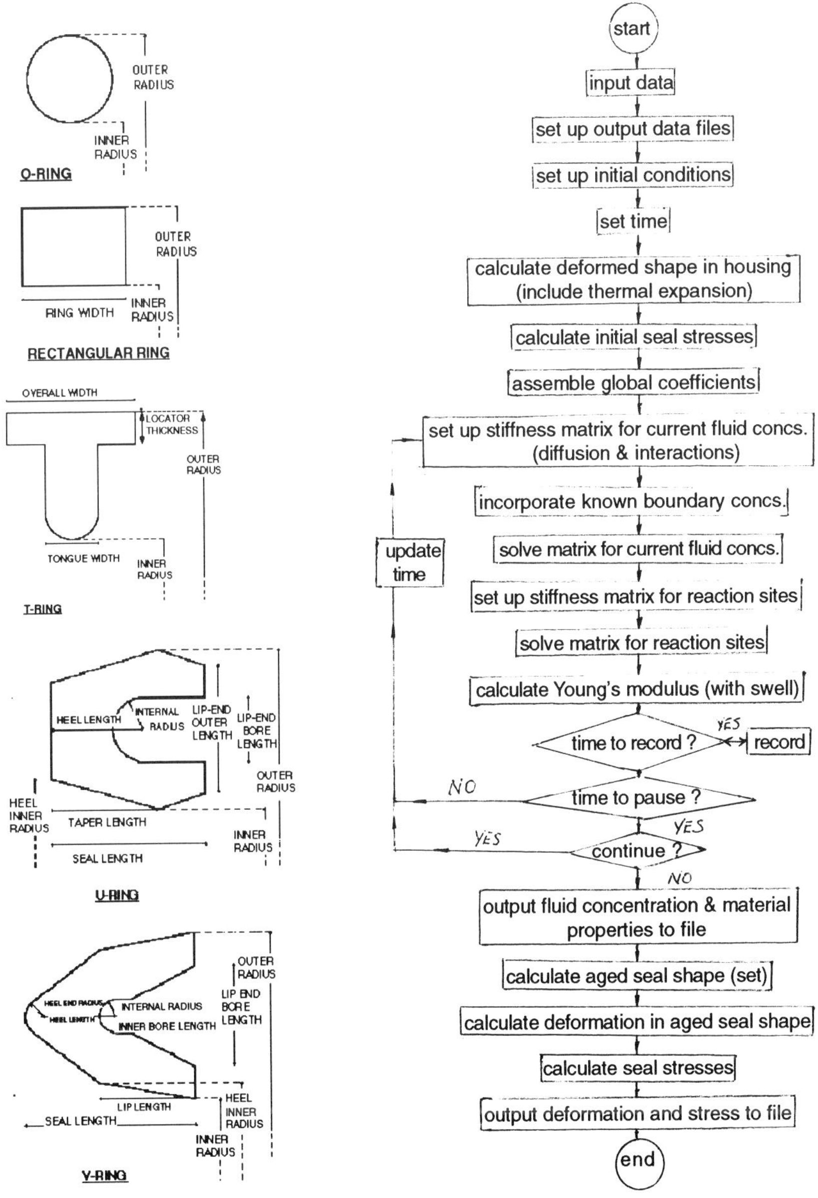

Fig 2   Preset seal geometries

Fig 3   Flow Chart for Seal Life Program

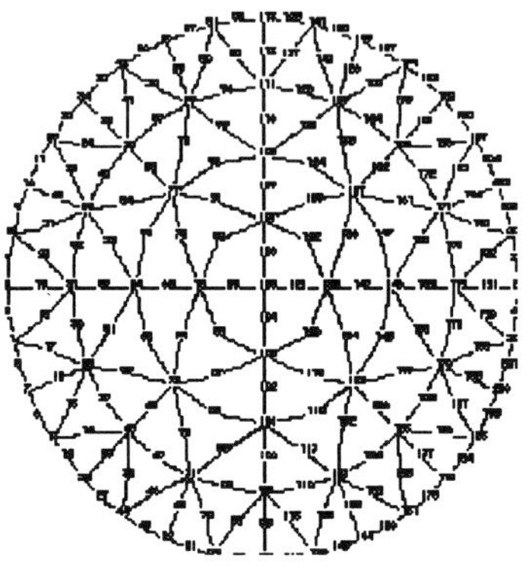

Fig 4 Finite element mesh for O-ring.

STRESS ANALYSIS

The sealing contact stress of an unaged elastomeric seal depends on the initial squeeze on the seal by its housing, the fluid pressure acting on the seal, and the material properties of the seal.   A Herrmann linear incompressible elastic approach is currently used, leading to the following variational condition:

$$\delta \sum_{n=1}^{N} \int_{B_n} \int\int \{G [\mu_1{}^2 - 2\mu_2 + 2\upsilon H \mu_1 - \upsilon(1-2\upsilon)H^2 - 6\upsilon e_T H - 2\mu_1 e_T] - F_i u_i\} d\upsilon - \int_{S_{1n}} \int P_i u_i \, dS = 0 \qquad \dots (1)$$

where G is the shear modulus (N/mm^2),

F_i is a body force,

υ is Poisson's ratio,

μ_1 and μ_2 are strain invariants,

H is a hydrostatic pressure function: $H = 3 p_m / 2G (1+\upsilon)$

σ_m is the mean pressure (N/mm^2): $\sigma_m = (p_1 + p_2 + p_3) / 3$

e_T is a thermal expansion term: $e_T = \int_{T_0}^{T} \alpha \, dT$

α is thermal expansion coefficient (K^{-1}): $\alpha = \int 1/l \, (dl / dT)$

Bn indicates volume integral and S1n indicates surface integral.

Equation (1) takes into account thermal expansion of the seal, which can significantly change squeeze and therefore sealing contact stress at high or cryogenic temperature. In matrix form, Eq.(1) becomes :

$$\begin{bmatrix} K_{11} & K_{12} \\ K_{12}{}^T & K_{22} \end{bmatrix} \begin{Bmatrix} q \\ h \end{Bmatrix} = \begin{Bmatrix} R_1 \\ R_2 \end{Bmatrix} \qquad \dots (2)$$

where K_{11}, K_{12}, K_{22} are stiffness matrices; q is the vector of nodal displacement,
h is the vector of nodal values of H, R_1 and R_2 are the load vectors obtained from external applied loads and pressures

Stresses (σ) are then calculated from :

$$\sigma = 2 G \varepsilon \qquad \ldots (3)$$

The analysis of the deformation of the seal during squeeze and pressurisation involves prediction of undetermined contact boundaries. These are found by iterative application of boundary constraints on the meshes (Salita 1987). Initially, all seal nodes crossing the wall of the housing are constrained to the wall. A fluid-pressure deformation calculation is then performed. If further nodes cross the walls, they too are constrained to the wall and the calculation repeated. This is straightforward to implement but demanding of computer time, therefore deformation and stress calculations are performed for initial squeeze and final aged condition only.

FLUID DIFFUSION AND INTERACTIONS WITH ELASTOMER

Sealing-contact stresses of an elastomer seal in service are affected by fluid absorbed by the elastomer. In the short term, fluid absorption can cause physical swell, which can increase the sealing stress due to dimension changes and decrease it due to the change of Young's modulus. In the longer term, the absorbed fluid can react with the elastomer, changing various material properties. Fluid uptake and subsequent interaction with the elastomer can be considered in the following stages.

In the first stage, adsorption onto the seal surface, liquids and gases behave differently . Liquid adsorption is dominated by the balance of chemical potentials of the bulk liquid and liquid dissolved in the elastomer. For gases, on the other hand, the adsorbed surface concentration is dominated by the gas pressure, unless gas density approaches the liquid range. According to Henry's law, the gas concentration (c) on the polymer surface is directly proportional to pressure (p);

$$c = sp \qquad \ldots (4)$$

where constant s is the solubility coefficient (bar^{-1}). Surface adsorption can be regarded as instantaneous for practical purposes, the surface concentration of adsorbed fluid (liquid or gas) then determines the mathematical boundary conditions for the computation of diffusion in the interior of the seal elastomer.

The next stage, the diffusion of fluid into the interior of the seal is driven by concentration gradient, for both liquids and gases. Eventually an equilibrium concentration distribution is reached in the elastomer, this may require a considerable time. Diffusion is governed by Fick's first and second laws of diffusion, Crank (1975) gives the time-dependent axisymmetric form of the differential equation governing Fickian diffusion:

$$\partial c/\partial t = \partial/\partial z \, [D \, (\partial c/\partial z) \,] + 1/r \, \{\partial/\partial r \, [\, D \, r \, (\partial c/\partial r) \,] \, \} + S_i \qquad \ldots (5)$$

where D is the diffusion coefficient (mm^{2}s^{-1})

r is the radial coordinate (mm),

z is the axial coordinate (mm),

S_i is a sink or source term (s^{-1}).

Values of the diffusion rate for gases in elastomers can be derived from permeation measurements on thin diaphragms (Campion and Morgan 1992), using Fick's 1st law of diffusion and the relationship between D, s, and permeation coefficient P :

$$P=Ds \qquad \ldots (6)$$

For liquids, values of the diffusion coefficient can be obtained from mass uptake measurements during immersion tests. Values used in this paper were obtained using these methods, by MERL our partners in the broader project.

128

The third stage to consider is the interaction between diffusing fluid and elastomer, in the interior of the seal. The sink (or source) term in Equation (5) represents this interaction. The nature of such interaction varies depending, *inter alia* , on the nature of the fluid and the elastomer. Examples of the interactions under consideration are cross-link formation and polymer chain-scission. For the present, it is assumed that there is a single dominant reaction and that this can be represented as follows. The fluid concentration at time t, at any point, is denoted by $c(t)$; the corresponding concentration of potential reaction sites is denoted by $Y(t)$; these two unknowns are related by the following simultaneous equations :

$$\frac{d\,c(t)}{dt} = \frac{d}{dz}\,D\frac{d\,c(t)}{dz} + \frac{1}{r}\,\frac{d}{dr}\,D\,r\frac{d\,c(t)}{dr} - \sum k_c\,Y(t-\delta t)\,c(t) \qquad \ldots\,(\,7\,)$$

$$\frac{d\,Y(t)}{dt} = -\,(n_1/n_2)\,k_c\,Y(t)\,c(t) \qquad \ldots\,(\,8\,)$$

where: k_c is the reaction rate,

n_1/n_2 is the ratio of fluid consumption to reaction site consumption.

As the reaction proceeds, Young's modulus of the seal material changes and the relationship with reaction site concentration is assumed to be linear:

$$E\,(t) = a\,[\,Y(t) - Y(0)\,] + E(0) \qquad \ldots\,(\,9\,)$$

Swell caused by fluid uptake will also change Young's modulus, the relationship assumed is given by Treloar (1975) :

$$G' = v_2^{\,1/3}\,G \qquad \ldots\,(\,10\,)$$

where v_2 is the volume fraction of polymer in the swollen elastomer,

G' is the shear modulus of the elastomer in the swollen state,

G is the shear modulus of the elastomer in the unswollen state.

SOLUTION OF THE SYSTEM OF EQUATIONS

Equations (7) and (8) can be written in matrix form :

$$[K]\,\{c\} = [B]\,\{\partial c/\partial t\} + \{F\} \qquad \ldots\,(\,11\,)$$

Non-dimensionalising time and then discretising the time variation, at time-step 'n' write equation (11) in the form:

$$[K]\,\{c\}_\xi = [B]\,\{\partial c/\partial t\}_\xi + \{F\}_\xi \qquad \ldots\,(\,12\,)$$

where

$$\xi = (t - t_{n-1})\,/\,\Delta t_n \qquad \ldots\,(\,13\,)$$

$$\Delta t_n = t_n - t_{n-1} \qquad \ldots\,(\,14\,)$$

Next introduce the time-discretisation of $\{c\}_\xi$, $\{\partial c/\partial t\}_\xi$ and $\{F\}_\xi$ using:

$$\{c\}_\xi = (1-\xi)\,\{c\}_{n-1} + \xi\,\{c\}_n \qquad \ldots\,(\,15\,)$$

$$\{F\}_\xi = (1-\xi)\,\{F\}_{n-1} + \xi\,\{F\}_n \qquad \ldots\,(\,16\,)$$

$$\{\partial c/\partial t\}_\xi = (\{c\}_n - \{c\}_{n-1})\,/\,\Delta t_n \qquad \ldots\,(\,17\,)$$

Equation (12) becomes :

$$([K] \xi - [B] / \Delta t_n) \{c\}_n = - ([B] / \Delta t_n + [K] (1- \xi)) \{c\}_{n-1} + (1- \xi) \{F\}_{n-1} + \{F\}_n$$

$$\dots (18)$$

Setting $\xi = 0$ gives an explicit forward difference method which is computationally fast but can be unstable. Setting $\xi = 0.5$ gives a stable time-centred method which can suffer from oscillation. Setting $\xi = 1.0$, gives a backward difference method which is "well behaved" and reliable, but slow. The best compromise is a value of ξ between 0.5 and 1.

COMPRESSION SET

An elastomeric seal maintained in a strained condition does not necessarily return to its original shape, such loss of memory is compression set and can adversely affect the sealing contact stress. The computer software allows the user to specify the final compression set in an analysis, and takes this into account when calculating the sealing stress. Figure 5 shows the computed shape of an O-ring in the installed condition, in a recess between flanges, and free with two levels of compression set.

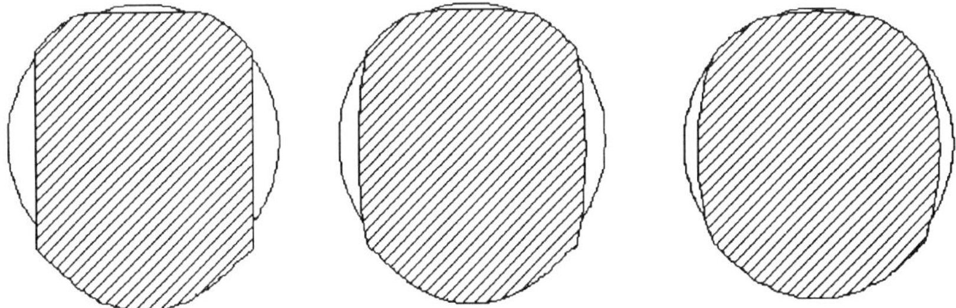

**Fig 5 Computed shape of O-ring with compression set:
a) installed condition; b) free ring, 75% set; c) free ring, 50% set.**

VALIDATION

The software validation has necessarily been mainly on a module by module basis. The time-dependent diffusion module has been validated (Ho and Nau 1993) against exact solutions for thermal diffusion in simple geometries (Carslaw and Jaeger 1959), gas permeation in a thin sheet, and liquid uptake in rectangular slabs. The chemical reaction module was validated for a simple chemical analogue (Fagg 1977), Figure 6, and for ageing data on nitrile rubber made available by the Ministry of Defence, e.g.Figure 7.

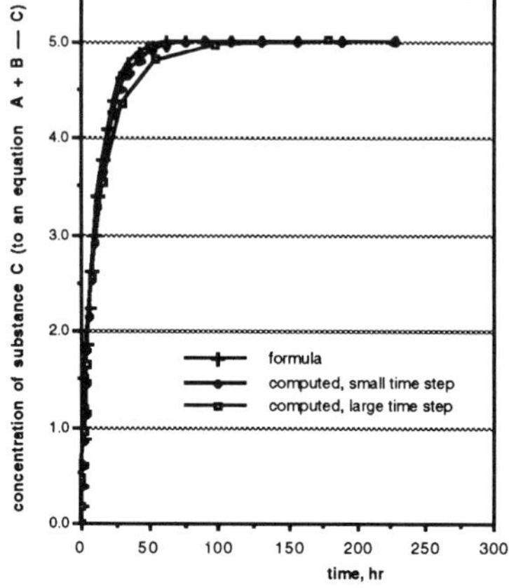

Fig 6 Comparison of computed and anaytical (Fagg1977) reaction product predictions.

Fig 7 Comparison of computed and measured Young's modulus of air-aged NBR (MoD data).

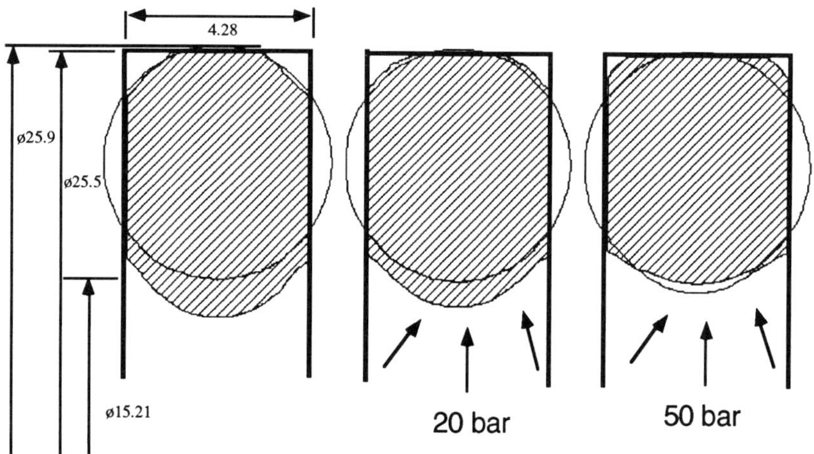

Fig 8 Computed effect of fluid pressure on O-ring shape:
a) installed condition; b) 20 bar; c) 50 bar.

EXAMPLES OF COMPUTATIONS FOR SEALS

To illustrate the application of the software to seals, examples will now be given of the capabilities of some major modules.

Fluid-pressure deformation

Figure 8 illustrates the computed effect of fluid pressure on the shape of an O-ring in a flange recess. The left-hand diagram shows the unpressurised O-ring with radial interference on the curved upper wall and lateral compression between the two flange faces. The effects of two levels of fluid pressure are shown in the remaining diagrams.

Sealing stress of an aged O-ring

Within the broader project (Stevenson 1992), MERL performed various laboratory tests on sample pieces of elastomer. Tensile tests after ageing in a synthetic oil at 100°C gave the rate of change of Young's modulus: 6.4×10^{-7} MPa /s. Liquid uptake tests gave a diffusion coefficient: 9.32×10^{-8} cm^2/s, and equilibrium mass uptake: 6.4%. Also, from the preceding data, it can be calculated that the ratio n_1/n_2 (in eq.8) is 0.0024. Using these data values as input, the computer program calculates change of Young's modulus and hence change of sealing-contact stress with time. Computed results for a size 312 NBR O–ring are presented in Figure 9. This shows the initial contact stress, the value after applying pressurised fluid, and the value after additionally heating to 100°C; after two months at these conditions set reduces the contact stress at the elevated temperature as shown, then when fluid pressure is removed there is a very large drop in contact stress. Finally the diagram shows the contact stress when cold pressurised fluid is reapplied.

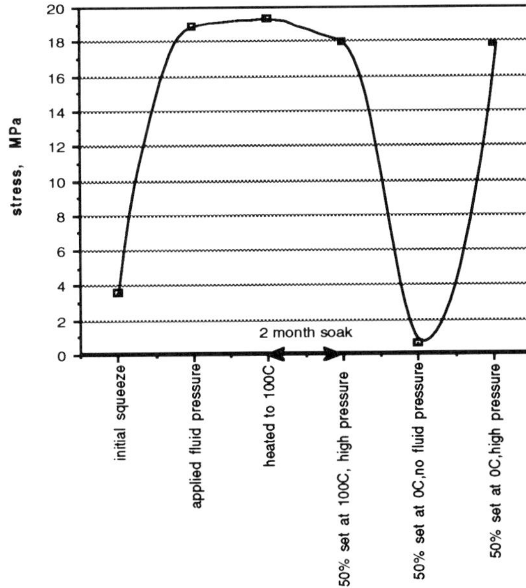

Fig 9   Computed sealing-contact stress (maximum) for various conditions.

132

Explosive decompression

An important factor affecting the severity of explosive decompression damage is the magnitude of the *difference* in concentration of dissolved gas, between the interior and surface of a seal. If the depressurization rate of the sealed fluid is low enough, most of the dissolved gas in the interior of an elastomeric seal can diffuse out and there is less potential for creating large pressure gradients, and consequent structural damage, due to gas coming out of solution from the elastomer in the interior of the seal. To study the effect of depressurization rates, the diffusion of dissolved carbon dioxide *out* of a fully saturated 312 O-ring at 100°C has been computed using the present software. Examples of computed results are shown in Figure 10. With instantaneous decompression of the bulk gas, the concentration differences rises to a high peak instantaneously. With decompression spread over a 2-hour period, the concentration difference rises to a, delayed, lower peak. This is only 2/3rds of that due to instantaneous decompression and is further reduced by a 4 hour decompression, significantly reducing the risk of explosive decompression damage. This has, of course, been generally understood, however only by the use of the present approach has it been possible to quantify such effects. Interestingly, the computed results show that the effect of a 15 minutes decompression, for conditions modelled, has little benefit over an instantaneous decompression. Figure 11 illustrates the typical gas concentration distribution in an O-ring several hours after half of the seal surface contacting gas.

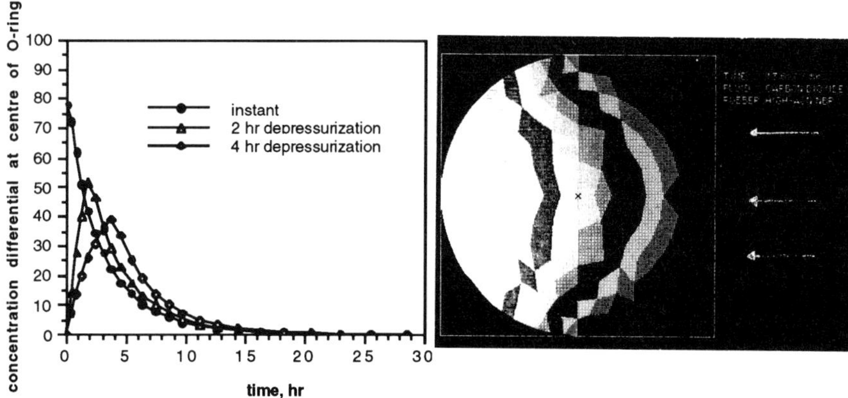

Fig 10   Effect of decompression time on concentration differential in an O-ring.

Fig 11   Gas concentration distribution in an O-ring

Permeation emission

In most elastomeric seal installations there is, on the low pressure side of the seal, a narrow clearance between the sealed parts, i.e.the extrusion gap. Gas which has diffused through the body of the seal evaporates here from the exposed surface of the elastomer and escapes to the surroundings and constitutes leakage.This form of gaseous emission can be quantified using the present software. Steady-state emission rates for carbon dioxide and methane at 100°C, 2500 psi through a square-section elastomeric seal (Figure 12a) with different extrusion gaps have been computed and are presented in Figure 12b. For both gases, leakage increases rapidly with extrusion gap for small gaps, below 5% of the seal section. Notice that the carbon dioxide emission rates are over 300% those of methane, this is mainly due to the high solubility of carbon dioxide in the elastomer.

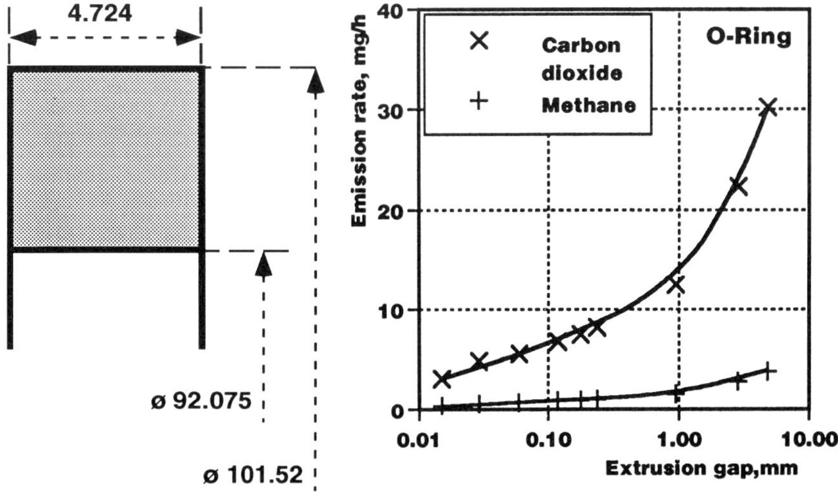

**Figure 12   Emission due to permeation through a square section seal.
a) dimensions; b) emission rates.**

CONCLUSIONS

It is now possible to combine computer-modelling techniques with data measurements on simple test pieces, to predict various aspects of elastomeric seal behaviour. This has application to prediction of ageing processes, change of sealing-contact stress with time, swell, permeation emissions, and factors affecting explosive decompression damage.

ACKNOWLEDGEMENTS

The support of the BHRG/MERL International Seal Life Prediction Project sponsors : Amoco Corp., Baker Oil Tools, British Gas plc, BP International Ltd, Conoco (UK) Ltd, DuPont (UK) Ltd, Exxon Production Research Co., HSE, Norsk Hydro a.s., Saga Petroleum a.s., Shell International Petroleum, Statoil, and James Walker & Co. Ltd, and their permission to publish this paper, is gratefully acknowledged. The authors also thank the Ministry of Defence for supplying valuable ageing data, and MERL for measurements of permeation and ageing.

REFERENCES

Ho, E. and **Nau, B.S.**, 1993, "Prediction of fluid behaviour in elastomeric seals", *Proc. ASME Arctic and Offshore Engineering Conf.*, vol.IIIA, 407-412 .

Ho, E., Flitney, R.K. and **Nau, B.S.**,1991, "Elastomeric Seal Life - Prediction", *Proc. Seminar on Elastomers in Oil Recovery*, RAPRA, Shrewsbury.

Herrmann, L.R., 1965, "Elasticity Equations for incompressible and nearly incompressible materials by a variational theorem", *AIAA Journal*, Vol.3, no.10, pp1896-1900.

Salita, M.,1987, "A simple finite-element model of O-ring Deformation and activation during squeeze and pressurization", *AIAA/SAE/ASME 23rd Joint Propulsion Conference*, AIAA-87-1739,San Diego, California, June 1987.

Crank, J.,1975, Mathematics of Diffusion, 2nd edition, book publ. by Clarendon Press, Oxford.

Campion, R.P. and **Morgan, G.J.**, 1992, "High pressure permeation and diffusion of gases in polymers of different structures", *Journal for Plastics Rubber & Composites : Processing & Applications*, 17, No.1, 51-58.

Treloar, L.R.G., 1975, *"The Physics of Rubber Elasticity"*, 3rd Edition, Clarendon Press, Oxford.

Carslaw, H.S. and **Jaeger, J.C.**, 1959, *"Conduction of Heat in Solids"*, 2nd edition, book publ. by Oxford University Press, Oxford, pp101-102

Fagg, S. V.,1977, *"Differential Equations"*, book publ. by Hodder and Stoughton, pp27-29.

Stevenson, A., 1992, "An Overview of the Requirements for Seal Life Prediction", *Proc. Corrosion 92*, Paper 65, NACE, Houston.

14th International Conference on Fluid Sealing, Firenze, Italy,
6-8 April 1994. Organised by BHR Group Limited, Cranfield,
Bedford, MK43 0AJ, UK; Tel: 0234 750422

Fibre Combinations for use in Asbestos-Free 'it' Calendered Sealing
Materials

By: D.A. Thomas
     J.R. Hoyes

Of: TBA Sealing Materials Ltd
    PO Box 21
    Rochdale
    OL12 7EU
    England

Introduction

When asbetos-free 'it' calendered sheet sealing materials were being
developed in the late 1970's, the reinforcement used was either
aramid or glass fibres. Since then other fibres have been
incorporated into such products amongst which cellulose, carbon and
mineral wool are the most common. Most of this latter group of
fibres are not used as the sole fibrous reinforcement but are present
in blends of two or more fibre types.

In this paper the advantages of using fibre blends are discussed as
are the properties and performance of gasket sheets containing such
blends.

Types of Fibre Combinations

a)   Aramid/Mineral Wool Blends

      Most gasket sheets containing a single reinforcing fibre are
      based on aramid.

      The main disadvantage of aramid based gasketing, as compared to
      CAF*, is its high cost. Aramid fibres are typically 25-30
      times the cost of good quality asbestos fibres and, in order to
      keep sheet costs as low as possible, manufacturers have had to
      learn how to make the most effective use of low fibre contents.

      A second disadvantage of these fibres is that they embrittle at
      elevated temperatures (e.g. 200-250°C). A consequence of this
      embrittlement is that the resilience or recovery of the sheet
      is reduced thus reducing the ability of the gasket to follow
      any movements in the flange assembly such as may occur as a
      result of thermal cycling or vibration.

* Trade mark of TBA Sealing Materials

These dual disadvantages have been mitigated by the inclusion of mineral wool as a secondary reinforcement. This enables the aramid content to be reduced (thus lowering the cost of the product) and at the same time introduces a higher degree of resilience to the matrix. The term "mineral wool" is intended here to cover a wide range of fibrous types such as are manufactured from molten rock, usually by blowing air through the melt, or by introducing the melt into a current of air which causes fine fibrils of wool to be spun off. Such processes result in the production of fibres having a wide range of lengths and diameters - some of which may be of respirable size. Also formed in this process are fibres which have a "head and tail" appearance where the "head" is a lump of non-fibrous rock, often referred to as "shot" and the "tail" is a low diameter fibre. In subsequent processing the head and tail separate due to mechanical action leaving pieces of shot mixed in with the fibres. This shot is not reinforcing and usually has to be removed (at least in part) by an expensive separation or washing process. The degree of shot removal has a direct effect on the cost of such mineral wool fibres but those currently used in the gasket industry cost 10 to 20% of the cost of aramid.

With combinations of aramid and mineral wool fibre it is possible to meet the requirements of BS 7531 Grade Y at an acceptable cost. This specification grade covers asbestos-free material intended to be used to seal water, steam and gases at temperatures up to 200°C and for oils, solvents and inert liquids at temperatures up to 350°C.

Typical properties of such a material, TBA's SFS 2800, are given in Table 1.

TABLE 1

SFS 2800 Aramid/Mineral Wool Reinforced Gasket Sheet v. BS 7531 Grade Y

		Result	Requirement
Actual Thickness	mm	1.5	1.5 ± 0.15
Residual Stress	MPa	24	22 Min
Thickness Increase in Oil	%	2	10 Max
Thickness Increase in Water	%	2	15 Max
Compressibility	%	8	6 Min
Gas Permeability	mL/min	0.2	1 Max

SFS 2800 has been tested for its ability to maintain a seal with nitrogen gas at elevated temperature (200°C) using the Shell test rig. This rig consists of a pair of 4" Class 300 lb raised face flanges between which the gasket (1.5mm thick) is clamped. Full details are given in Reference 1. The requirement for gaskets for 200°C service is that the initial nitrogen pressure of 33 bar shall not be reduced by more than 1 bar after 3 thermal cycles from room temeprature to 200°C. SFS 2800 easily met this requirement with a pressure drop of only 0.3 to 0.4 bar. As a consequence of this test performance, this material has been approved by Shell for service up to 200°C.

Whilst nitrogen gas has only a very small molecular volume and may thus be considered as difficult to seal, a more practical sealing problem at elevated temperature occurs with steam. The steam sealing performance of SFS 2800 was assessed on the BHRG steam joint test facility using 4" NB test flanges of Class 300 lb rating. Two different points on the 300 lb flange service envelope were evaluated (see Fig. 1).

These were:

a)   45 bar/256°C
b)   30 bar/400°C

Note that in both cases the temperature is in exces of that given in BS 7531 Grade Y for steam sealing (200°C maximum).

These tests were carried out on 1.5 and 2.0mm thick material to reflect the different usages of gasket thickness in UK/USA and Continental Europe. The full test details are given in Reference 2. Each test was of seven days duration and included three thermal cycles from room temperature to test temperature.

During the tests, any steam leaking from the joints was measured by a sensitive pressure transducer measuring the height of condensed leakage in a collector tube: this method gives a resolution of 10 mg/hr which, for the gasket dimensions used, equates to the PVRC "Tight" reference leakage.

In all four tets carried out, no measurable leakage occurred, i.e. performance of these gaskets was better than the PVRC "Tight" classification. During the tests the bolt stress was also monitored and it was found that for the 45 bar/256°C tests, 86% of bolt stress was retained while at 30 bar/400°C the retention was 83%. These results indicate only a small change in gasket thickness due to compaction - a highly desirable feature.

138

b)   <u>Aramid/Glass Blends</u>

Glass fibres which have a high softening point (e.g. greater
than 600°C) are ideal candidates for use in sheet sealing
materials. In this paper glass fibre is restricted to E-glass
manufactured by the bushing route in which molten glass is
drawn through tiny, accurately drilled holes in a heated block
of platinum so that fibrils of a precise, controlled diameter
are obtained. The glass fibrils are then passed through a bath
of dressing prior to drying and chopping. By carefully
monitoring the rate of drawing and of the diameter of the
holes, the diameter of the glass fibrils can be controlled so
that none have a diameter which may be respirable (i.e. none
below 3 micron). Fibre diameters of 6 or 10 micron are most
beneficial for use in gasket products: this contrasts with the
coarser diameters usually used for most glass reinforced
plastics (15-25 micron).

The degree of reinforcement achieved is a function of the
specific surface area of the fibre and the fibre content.
Specific surface area is defined as the total surface area of a
given weight (usually 1 gramme) of the fibre. For fibres
having a high aspect ratio (fibre length ÷ fibre diameter),
surface area is proportional to the product of fibre diameter
and fibre length. Thus, for a given fibre length, the degree
of reinforcement achieved with 6 micron diameter fibres will be
double that of 12 micron diameter fibres of the same fibre
type. In other words, to achieve the same reinforcement as a
given weight of 6 micron diameter glass fibres, twice the
weight of 12 micron diameter glass fibre of the same length
would be required.

Because glass fibres have surfaces which are smooth in both a
physical and chemical sense they are coated with surface
dressings to promote compatibility with the matrix in which
they are being incorporated. Dressings for glass fibres
intended for gasket service often include coupling agents which
enable chemical cross links to be produced between the glass
fibre and the rubber matrix in which they are dispersed.

Glass fibres are non-fibrillating, i.e. they will not sub-
divide longitudinally into finer, possibly respirable fibrils
when subjected to mechanical action - they can only truncate to
give shorter fibrils or, eventually, glass dust. This
potential shortening of fibres means that special precautions
must be taken in mixing and sheet manufacturing processes to
ensure that fibre length, and hence matrix reinforcement, is
maintained.

When glass fibres such as discussed above are used as the sole
reinforcement for a gasket sheet produced by the 'it'
calendering process it is observed that the fibres are
preferentially aligned in the calendering direction to a much
greater degree than when asbestos fibres are used. For
example, the ratio of tensile strength in the calendering
direction to that in a direction at right angles to it may be
as high as 4.5 to 5 for glass fibres compared with the more
familiar 2 to 2.5 with asbestos.

When a blend of aramid and glass fibres is used to reinforce a
gasket sheet, the fibrillating aramid fibres entangle with the
glass fibres and trap more of them in the cross-grain direction
thus giving a tensile ratio of around 3, i.e. much closer to
that of asbestos than when glass fibres alone are used.

With these blends asbestos-free material has been produced
which comes close to the performance of a top range CAF. This
material, SFS 3300, readily complies with the highest, X-grade,
classification of BS 7531 and typical properties are given in
Table 2

TABLE 2

SFS 3300 Aramid/Glasss Reinforced Gasket v. BS 7531 Grade X

		Result	Requirement
Actual Thickness	mm	1.5	1.5 ± 0.15
Residual Stress	MPa	29	25 Min
Thickness Increase in Oil	%	3	10 Max
Thickness Increase in Water	%	2	15 Max
Compressibility	%	12	6 Min
Gas Permeability	mL/min	0.1	1 Max

Gaskets of SFS 3300 have, like the aramid/mineral wool product
described earlier, been subjected to the Shell Thermal Cycling
Test (Reference 1) but this time at the more stringent test
temperature of 350°C. Once again excellent high temperature
sealability was demonstrated and Shell have approved this sheet
for use in services up to 350°C.

When subjected to testing on the steam joint test facility at
BHRG excellent performance resulted on Class 300 and 600 lb
flanges even at 2mm thick. In these tests the level of sealing
achieved was better than the PVRC "Tight" level (Reference 3).

c)   <u>Aramid/Carbon Blends</u>

Sheet sealing materials based on blends of aramid and carbon
fibres have been introduced into the market over the last few
years and have been found to give performances similar to that
of aramid/glass blends with exceptionally good service in
strong alkaline conditions. The carbon fibres used are
typically 10-13 micron in diameter and can have carbon contents
which vary from 65 to 95+%. These fibres are, like glass,
smooth and straight and do not fibrillate when subjected to
mechanical action.

The effect of alkali on aramid/carbon reinforced gasket sheet
is shown in Table 3.

Table 3

The Effect of Alkali on Aramid/Carbon Reinforced Gasket Sheet

	Competitive Material A			Competitive Material B			TBA Sealing Materials Development Material			SFS 3300		
Days in 20% NaOH	0	2	7	0	2	7	0	2	7	0	2	7
Tensile Strength with grain (MPa)	26.1	11.4	10.5	32.0	14.9	12.0	51.7	25.7	21.2	56.9	14.3	13.5
Loss (%)	-	56	60	-	54	63	-	50	59	-	75	76
Tensile Strength cross grain (MPa)	8.0	3.4	3.1	12.3	6.0	4.7	21.0	8.9	7.9	20.5	4.7	4.7
Loss (%)	-	58	61	-	52	61	-	58	62	-	77	77

It is worth noting that from the viewpoint of the user of these sheets, alkaline resistance is most readily assessed by comparing the percentage reduction in tensile strength after free immersion in caustic soda solution rather than by functionally based tests. From Table 3 it can be seen that if the magnitude of the residual strength was used as the comparison then aramid/glass blends would have equivalent performance to some of the aramid/carbon blends currently in use.

The development material referred to in Table 3, XJ 569, has been evaluated worldwide in industries where strong alkaline conditions are frequently encountered such as pulp and paper mills and in aluminium extraction. This product also has good sealing properties in steam and in general chemicals and complies with the requirements of BS 7531 Grade X. Typical physical properties are shown in Table 4 against the BS 7531 Grade X requirements.

TABLE 4

<u>XJ 569 Aramid Carbon Reinforced Gasket Sheet v. BS 7531 Grade X</u>

		Result	Requirement
Actual Thickness	mm	1.5	1.5 ± 0.15
Residual Stress	MPa	26	25 Min
Thickness Increase in Oil	%	3	10 Max
Thickness Increase in Water	%	2	15 Max
Compressibility	%	7.5	6 Min
Gas Permeability	mL/min	0.2	1 Max

The steam sealing performance of XJ 569 was assessed on the BHRG steam joint test facility using 4" NB test flanges of Class 600 lb rating. Two different points on the 600 lb flange service envelope were evaluated (see Figure 2):

a)	77 bar/290°C
b)	40 bar/450°C.

Note that in both cases the temperature is in excess of that given in BS 7531 Grade X for steam sealing (250°C maximum).

These tests were carried out on 1.5mm thick material and were
of 7 days duration, including 3 thermal cycles from room
temperature to test temperature. Full test details are given
in Reference 2.

As for the tests described earlier with SFS 2800, no measurable
leakage occurred during either test, i.e. the performance of
these gaskets was better than the PVRC "Tight" classification.

Retained bolt stress was also monitored in these tests and
excellent figures of 95% retained stress were recorded, i.e.
even better than previously mentioned.

d)   <u>The Ideal Fibre for use in Asbestos-Free 'it' Calendered
      Sealing Materials</u>

The fibre blends discussed so far have given materials with
acceptable performance properties. However, for enhanced
performance we need a new web-forming fibre with a higher
temperature capability. Such a fibre will almost certainly be
respirable.

We are looking to those companies skilled in appropriate areas
to develop such a fibre and will be pleased to take part in
joint evaluation of any likely contenders.

For sheet sealing materials containing no respirable fibre,
glass fibre can be used. Whilst blends of glass fibre with
aramid have already been discussed in b) above, the use of
glass fibres alone will now be considered.

e)   <u>Glass Fibre</u>

Sheet sealing materials reinforced with glass fibre have been
available for more than 10 years following extensive
development at TBA Sealing Materials covered by worldwide
patents (Reference 5). Initially, 10 micron fibres were used
but it was soon recognised that, for enhanced reinforcement, a
finer diameter was required and 6 micron chopped glass strand
was introduced. Gasket sheets reinforced with 6 micron glass
fibres can give excellent service performance, particularly in
steam, but only if used correctly.
Whilst the properties of sheets reinforced with aramid/glass
blends can approach those of conventional CAF, sheets
reinforced with glass fibres alone give a different property
profile. These differences are shown in Table 5.

TABLE 5

Comparative Physical Properties of AFS 2100 and SFS 3300 Sheets

		AFS 2100 Glass Reinforced	SFS 3300 Aramid/Glass Reinforced
Actual Thickness	mm	1.5	1.5
Density	g cm^{-3}	1.6	1.5
ASTM Compressibility	%	18	12
ASTM Recovery	%	70	63
ASTM Tenisle Strength			
- with grain	MPa	50	60
- cross grain	MPa	12	20
BS Residual Stress	MPa	20	29
Thickness Increase ASTM Oil 3	%	3	3
Thickness Increase ASTM Fuel B	%	6	4
DIN Gas Permeability	mL/min	0.01	0.1

The main functional properties are compressibility, recovery, sealability and stress retention: tensile strength is not considered a prime functional property although it can affect the cuttability and handleability of the sheet and may have an influence on blow-out resistance.

The compressibility of the glass based product shown in Table 5 is 50% higher than that of the aramid/glass product. Thus for equivalent actual compression a thinner gauge can be used (e.g. 1.0mm instead of 1.5mm).

The higher recovery of glass based products mean that they are better able to follow any flange movement such as may take place due to thermal cycling or vibration - this is clearly advantageous - moreover, the glass fibres are less prone to embrittlement at temperatures of 200-250°C than aramid fibres and thus the enhancement in recovery is maintained even at elevated temepratures.

The excellent nitrogen gas sealability of AFS 2100 gaskets assessed under DIN 3754 test conditions gives leakage levels amongst the lowest ever recorded for any sheet material.

The main apparent disadvantage of glass based products is in the area of stress retention where, at a nominal thickness of 1.5mm, the value obtained is significantly lower than that routinely measured for aramid/glass blends (or for CAF).

However, by taking advantage of the higher compressibility of glass based products and using these materials at a reduced thickness of 1mm stress retention values equivalent to those of a 1.5mm CAF can be obtained (see Figure 3).

Evidence of the steam sealing ability of such materials was obtained during testing on the steam joint test facility at BHRG. This work was carried out as part of the work of a consortium of gasket manufacturers and end users and is fully reported in Reference 3.

The salient facts emerging from these tests are:

1.   Glass fibre based gaskets behaved satisfactorily within the pressure/temperature envelopes of Class 150 and Class 300 lb flanges. This applied at both 0.75 and 1.5mm thick.

2.   All 0.75mm glass based gaskets operated satisfactorily in Class 600 lb flanges within the pressure/temperature envelope from 70 bar/300°C to 30 bar/485°C, including operation in saturated steam. There was no measurable leakage in any of these tests.

3.   Because the medium term performance at 0.75mm was so good an endurance test was carried out. This encompassed 35 pressure/temperature cycles over a total period of 5,657 hours (nearly 8 months). The results showed that long time sealing performance was satisfactory - there was no measurable leakage.

AFS 2100 and AF 2200, glass based gaskets,were also included in a test programme carried out by the US Chlorine Institute (Reference 4). In this work 13 different asbestos free sheet sealing materials were evaluated for service in gaseous and liquid chlorine in 2" Class 150 lb flanges: the gasket thickness was 1.5mm.

Evaluations were carried out at four different chemical plants covering a total of 5 different temperature/pressure combinations as shown in Table 6.

146

<u>TABLE 6</u>

<u>Temperature/Pressure Combinations for US Chlorine Institute Tests</u>

	Temperature		Pressure	
	°F	°C	PSIG	Bar
1	-20	-29	0	0
2	160	71	110	7.5
3	100	38	85	6
4	223	106	110	7.5
5	122	50	170	12

There were two series of tests, one of 6 months duration, the other of 12 months. At the end of these tests a panel of users reviewed the Evaluation Sheets completed by the four test centres and also examined the used gaskets. For each temperature/pressure combination the panel gave the material a mark based on the following scale as a comparison with the performance of CAF:

A	-	Improved Performance
B	-	Comparative Performance
C	-	Inferior but Acceptable Performance
D	-	Inferior Performance
E	-	Unacceptable Performance

The overall performance (i.e. a combination of the marks awarded at each temperature/pressure combination) was reported for the 6 months and 12 months tests.

After the 6 months tests 4 of the 13 materials gave ratings of B, including both TBA Sealing Materials' products. After the 12 months tests only one material was deemed to have given performance comparable to that of CAF - this one, which gave a B ranking at all 4 temperature/pressure combinations, was AF 2200.

<u>Conclusions</u>

Sheet sealing materials reinforced with blends of fibres can give properties and performance which approach those of CAF. Some fibres in these blends may be respirable.

For yet higher temperature applications a new web-forming, thermally resistant fibre is required. Such a fibre will probably be respirable.

When a calendered product, free of respirable fibre is required, the TBA Sealing Materials AF series is recommended, but, for optimum performance a thinner gauge of material should be used.

References

1.   "Requirements for Asbestos Substitutes for Jointing, Packing
      and Sealing Materials"; May 1992; M. Dol, C. Robbe & L.Voogd;
      Shell Internationale Petroleum Maatschappij, The Hague, The
      Netherlands.

2.   "Report on Steam Leak Tests conducted on XJ 569 and SFS 2800
      Gasket Materials"; October 1993; M.D. Reddy; BHR Group. BHR
      Group Project No. 71611.

3.   "Gasket Performance and Stress Requirements in Standard
      Joints"; February 1993; B.S.Nau and M.D. Reddy; BHR Group.
      BHR Project No. 5433.

4.   "Gaskets for Chlorine Service"; October 1992; Chlorine
      Institute Pamphlet 95.
      The Chlorine Institute, Inc., 2001 L Street N.W.
      Washington D.C. 20036, USA.

5.   UK Patent 2094821.

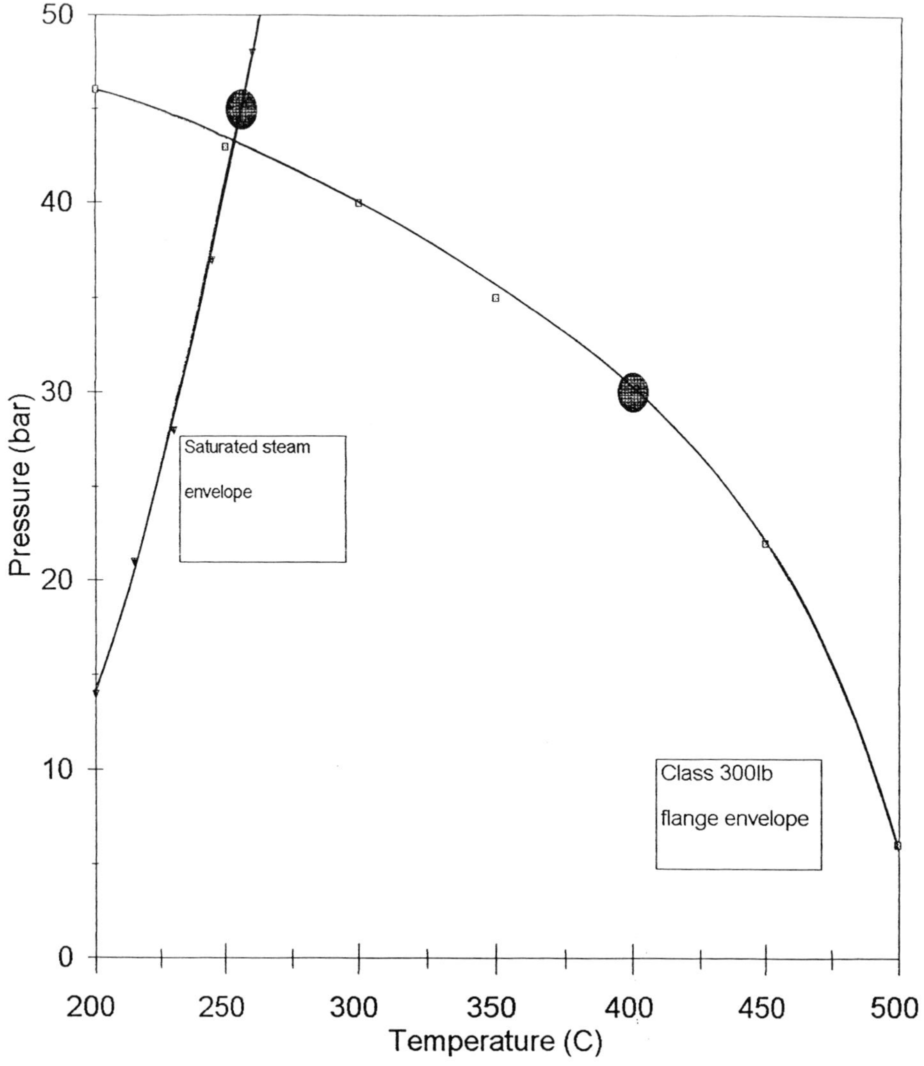

Figure 1

Test conditions for SFS 2800 1.5 and 2mm thick on BHRG steam test
facility using 300 lb flanges.

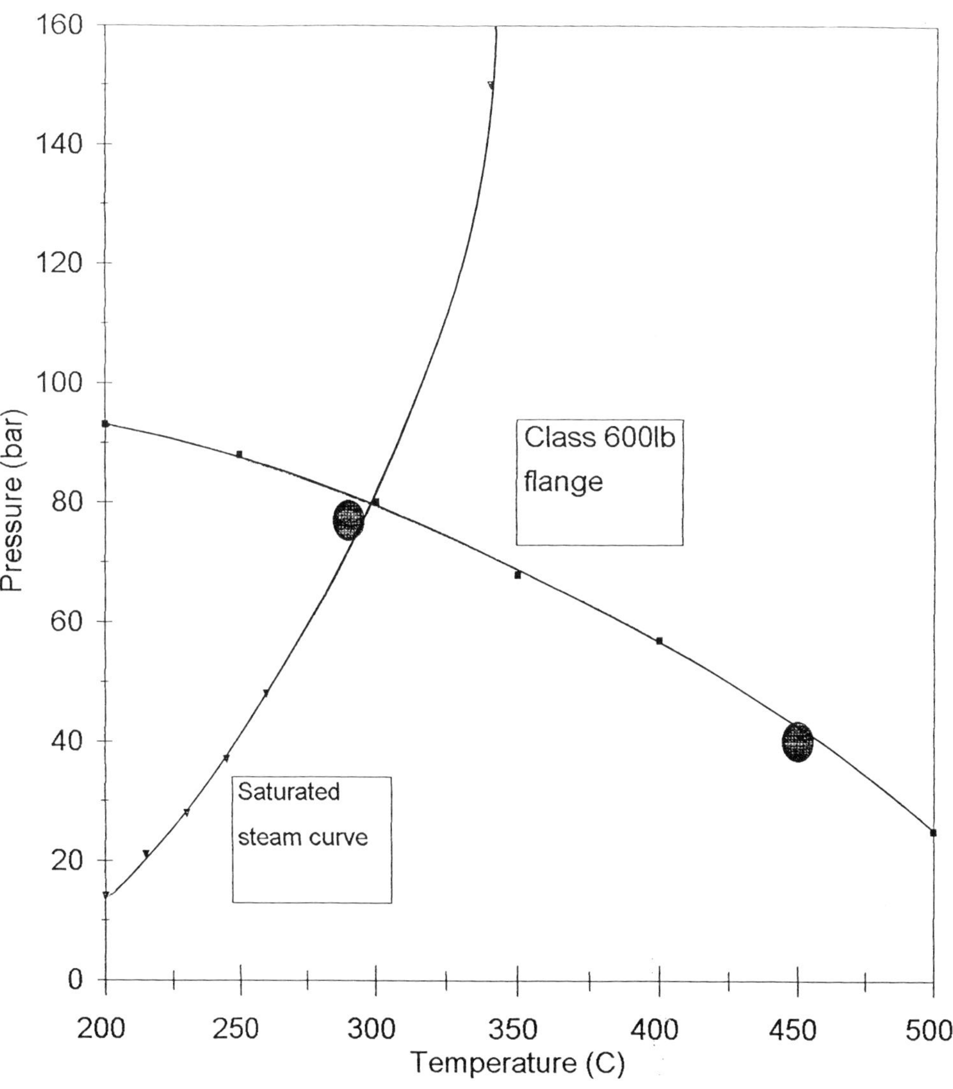

Figure 2

Test conditions for XJ 569 1.5mm thick on BHRG steam test facility
using 600 lb flanges.

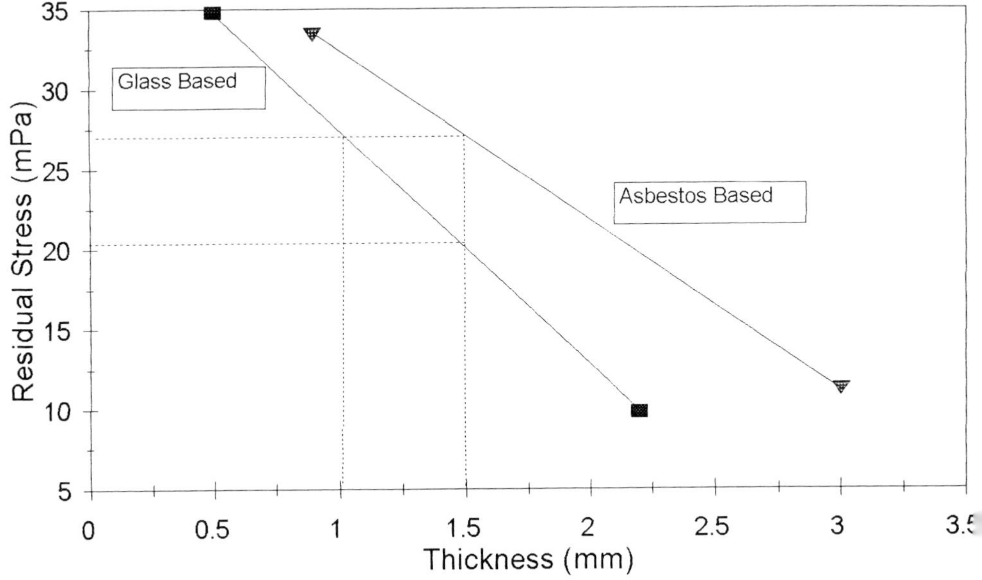

Fig 3 Stress retention v Thickness
Glass and Asbestos based materials

BS Residual Stress of a glass based gasket at 1.5mm thick is typically 21 MPa: at 1.0mm thick a value of 27 MPa is obtained – the same as for CAF at 1.5mm thick.

The higher compressibility of the glass based product allows this thickness reduction to be made.

14th International Conference on Fluid Sealing, Firenze, Italy,
6-8 April 1994. Organised by BHR Group Limited, Cranfield,
Bedford, MK43 0AJ, UK; Tel: 0234 750422

A Review of Corrosive Processes in Sealed Joints

Alfred Hirschvogel
SIGRI Great Lakes Carbon GmbH, Germany

ABSTRACT

Corrosive processes at statically and dynamically sealed joints in pipelines, pumps, valves and vessels cause avoidable costs and reduce operational safety and service life of plants.

Using typical examples, the author describes the major causes of corrosion and corrosion mechanisms. The manifestation of corrosion have been divided into three main types: crevice corrosion, contact corrosion and electrochemical corrosion.

By selecting appropriate sealing materials and sealing designs corrosion at sealed joints can be considerably reduced.

When mounting the sealing material care should be taken to ensure sufficient and uniform contact pressure between metallic sealing surface and sealing material as well as the use of sealing auxiliaries, eg. adhesives, "release pastes" or lubricants, without corrosive secondary reactions under industrial conditions.

Some corrosion problems could be avoided by using high purity flexible graphite.

INTRODUCTION

Sealed joints in pipelines, vessels, pumps, fittings and at the junctions between individual components are points where corrosion usually occurs to a greater extent because of the combined effect of various factors.

The following describes manifestations of corrosion at metallic sealing faces (flanges, spindles and housings of fittings), which may occur by their interaction with sealing materials or sealed joints. As the effects described are usually very complex chemical or electrochemical processes, the author tries to make the subject easy to understand by describing the major processes in simplified terms.

Although different cases of corrosion often present a similar picture at first glance, a distinction should be made between different corrosive mechanisms with a variety of causes.

The most important types of corrosion occuring in conjuction with sealing materials are:

- °   **Contact Corrosion (galvanic corrosion)**
- °   **Crevice Corrosion**
- °   **Electrochemical Corrosion**

As in works practice these types of corrosion rarely occur in an isolated form, it is often difficult to find the exact cause of the damage.

CONTACT CORROSION

This type of corrosion occurs in particular with sealing materials containing substances that induce a chemical reaction when in contact with a metallic sealing face. This often occurs to a greater extent in connection with crevice corrosion.

The damage most frequently observed (at the spindles of fittings and at flanges) with asbestos-containing seals such as CAF gaskets and asbestos stuffing box packings was pitting of Cr/Ni steel grades as caused by the high chloride content of these sealing materials (typically between 200 and 1000 ppm).

By switching to low-chloride sealing materials on the basis of PTFE or high purity flexible graphite, instances of corrosion caused by chlorides can be substantially reduced (depending on the degree of purity, the chloride content of flexible Graphite is typically between 5 and 50 ppm).

Corrosion may also be induced by specific coatings or impregnations of the sealing materials, which result in the release of corrosive decomposition products at elevated temperatures. In the case of organic compounds containing chlorine, this is usually HCl. With PTFE, highly corrosive compounds of fluorine are released. As the decomposition process already starts at temperatures around 300 °C, PTFE-containing sealing materials should not be used at a continuous service temperature above 250 °C, despite the PTFE's short-term resistance of approx. 400 °C.

The same applies to lubricating or "release" pastes, which are frequently emploed as mounting aids or sealing auxiliaries in flat gaskets or as lubricants in stuffing box packings. If such auxiliaries cannot be dispensed with, special attention should be paid to their composition and possible decomposition products.

CREVICE CORROSION

This is probably the most frequent type of corrosion found in conjunction with sealing materials in electrolytic media. It is mainly caused by physical factors, which are attributed, among other things, to poor seal design or errors in selecting and mounting the sealing materials. Severe effects of crevice corrosion may even be observed with high-purity, electrochemically inactive as well as PTFE if these are not properly mounted.

Crevice Corrosion

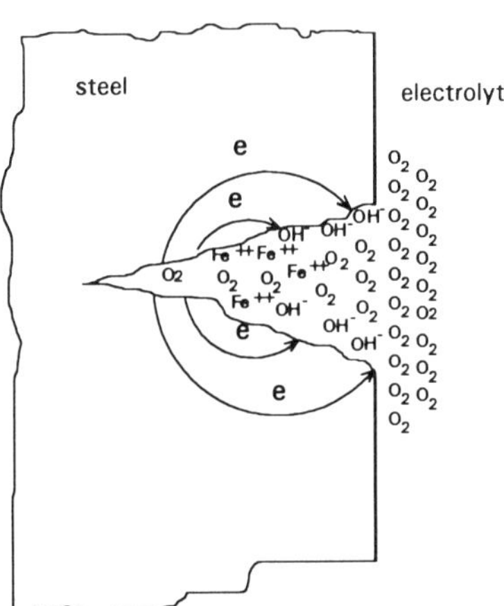

In such cases, the corrosive mechanism is due to the concentration differences (eg of oxygen) often found in the narrow crevices of seals, and the resulting potential differences. The effects of crevice and contact corrosion mostly occur concurrently. Owing to the potential differences, electrochemical corrosion currents flow and metal atoms dissolve by redox processes to form ions. The concentration of corrosive products then produces a self-intensifying effect so that eventually a considerable amount of metal may be dissolved.

Such effects are particularly marked in stuffing box packings, where a more or less pronounced sealing crevice always occurs as a result of the design. The service pressure drops off over the entire length of the packing, in which process the concentration of operating medium in the packing itself and the sealing faces gradually declines from the service to the atmospheric pressure side. As a stuffing box packing nearly always allows a certain degree of leakage, a build-up of substances dissolved in the operating medium also occurs, particularly in the region of the outermost packing ring. By mounting the packing rings properly, these effects can at least be alleviated. In the case of flexible graphite packings (frequently referred to as high-purity graphite packing), for instance, it is recommended as a general rule that the packing rings has to be precompressed one by one in the housing using a two-part tube.

This ensures that each packing ring is uniformly and radially prestressed prior to applying the stuffing box gland pressure. Thus, the bottom ring contributes substantially to the sealing effect and the risk of a wide sealing gap is reduced.

Stuffing box sealing system

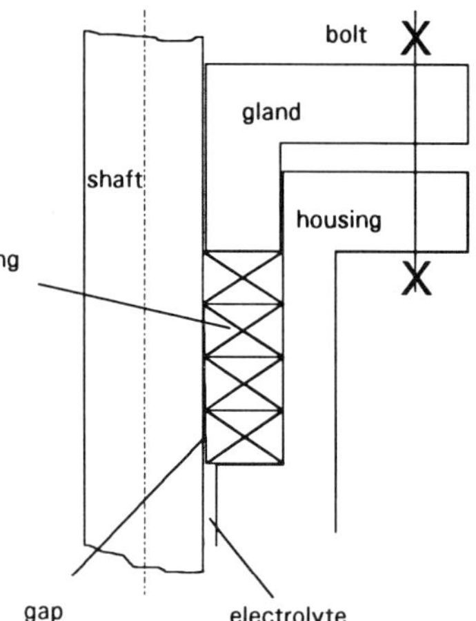

Flange sealing system

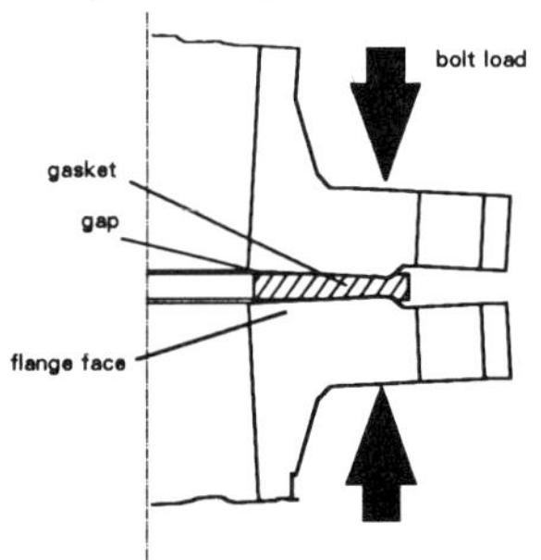

In static gaskets joints, crevice corrosion processes are likely to occur if the sealing faces are not properly adjusted to the flange sealing faces. In works practice, instances of severe corrosion to the flange sealing faces have often been observed and erroneously attributed to an insufficient resistance to the operating medium. If the flange faces are slanted, i.e. by flange rotation, or the flanges are distorted, sections of the relatively hard CAF seals and CAF substitute seals in particular may not be properly pressed against the flange faces (especially along the inner perimeter) so that a seal crevice can form.

These effects become more critical with reducing gasket thickness i.e. due to intent improvement of creep relaxation. To avoid such problems, it is recommended to use gaskets with enough compressibility and adaptability to create enough gasket stress at all areas of the sealing surfaces.

The high chloride content especially with CAF is another factor of increased corrosion. According to the favourable experience gained over many years with relatively flexible and adaptable materials such as flexible graphite based seals, this type of corrosion can be largely avoided.

ELECTROCHEMICAL CORROSION

(eg.

This type of corrosion is characterized by electrochemical processes occurring exclusively in the presence of an ion-conducting phase. It is not necessarily induced by direct electrolytic removal of metal, but may also be caused by a reaction with an intermediate electrolytic product (eg. atomic hydrogen).

A characteristic feature of electrochemical corrosion processes is their dependence on the electrode potential or a current flowing through the interface of the material and medium (as defined in DIN 50900, section 2).

By the formation of galvanic elements, i.e. by the specific arrangement of base elements and nobler elements, especially metals, and in the presence of an electrolyte (or an aqueous solution made conductive by dissolved ions), the base element is dissolved in a redox process (electrons emitted by the base element and taken up by the noble element). As is well known, such processes are utilized in electric batteries and accumulators.

Element	Normal potential $E^\circ H$ [Volt]
Mg/Mg^{2+}	-2.40
Al/Al^{3+}	-1.70
Mn/Mn^{2+}	-1.70
Cr/Cr^{3+}	-1.00
Zn/Zn^{2+}	-0.76
Fe/Fe^{2+}	-0.44
Co/Co^{2+}	-0.29
Ni/Ni^{2+}	-0.22
Sn/Sn^{2+}	-0.14
Pb/Pb^{2+}	-0.12
$H_2/2H^+$	0.0
Cu/Cu^{2+}	$+0.35$
O_2/OH^+	$+0.40$
C/C_2^+	$+0.72$ *
Ag/Ag^+	$+0.80$
Hg/Hg^{2+}	$+0.86$
Cl_2/Cl^-	$+1.36$
Au/Au^{3+}	$+1.50$
Pt/Pt^{2+}	$+1.60$

* Values calculated for:
$$C + 1/2\ O_2 \rightarrow CO + 2e$$

In sealing joints, galvanic elements are often produced by the combined action of different materials, metals or alloys. This may be conditioned on the one hand by different flange materials (eg. connection of pipelines to the equipment) or different materials used within the component (materials of the spindle and housing of fittings), and on the other hand also by the sealing material itself. This latter may be a metallic seal made from a material unsuitable for the application in point or a metal reinforcement in the flexible material. Finally the flexible material itself may become electrochemically active. **In any case an electrolyte must be present to ensure the transport processes and the relevant electrochemical reactions.** In pure vapour systems, however, the risk of this type of corrosion is usually very small because of the lack of electrolytically active ions.

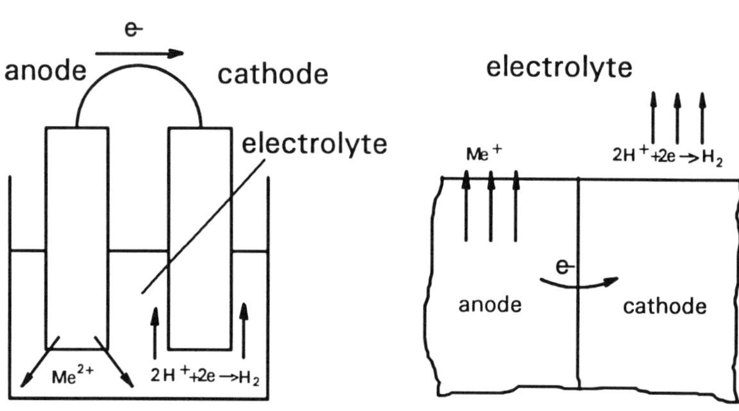

Since graphite is classified as a relatively noble element in the electrochemical series, but is unable to produce ions or take up electrons, redox processes are not directly possible with carbon, but take place only in the presence of other reducible elements such as oxygen.

The material combination of a Cr/Ni steel spindle with a grey cast iron housing is widely used in the manufacture of fittings, but different materials for both components are used as well. If materials with different electrochemical potentials are combined, the electrochemical potentials are superimposed, depending on the material combination. It is therefore often impossible to determine the exact cause of any specific damage. More than 20 years' experience in the field of graphite stuffing box packings has shown, however, **that instances of corrosion sharply decline after replacing asbestos-containing packings by high-purity graphite material** (see also the section headed "Contact corrosion").

If graphite packings are employed, corrosion may occur at the valve spindles in certain isolated cases, as is, incidentally, also the case with PTFE packings. Extensive investigations have revealed that the causes of corrosion are extremely varied (crevice corrosion, build-up of contaminants and corrosion products from the pipework, **low-purity** graphite, structural disorder in poor-quality steel grades, chloride-containing media) and cannot always be exactly determined because of the complex processes within the packing crevice. However, if the corrosion process is already under way, for whatever reasons, the graphite is thought to have a certain supportive effect.

Graphite packings made from molybdate-inhibited graphite foil should be used in such applications. More than 10 years favourable experience with this "Anti Corrosion" grade, especially in atomic power plants, exists.

One should be careful when using inhibitors based on the principal of sacrifical anode like metallic zinc. Although this inhibitor works well in some cases, especially at elevated temperatures it shows a contary effect and causes stronger corrosion. Furthermore, due to generation of zink oxide, some other negative effects for the function of gland packings have to taken into acount.

Although graphite gaskets have been employed for many years for the flange joints in nuclear power stations and industrial chemistry, no major corrosion problems have so far been reported.

Owing to the geometrical surface design (small electrochemically active cathode face of the seal (inner circumference x seal thickness) compared with the total inner surface of the pipework with an anodic action) and the associated very low density of the corrosion current, **corrosive processes** induced by seals in **carbon steel** flanges and piping are **negligible** and usually less than other forms of erosion that occur during plant operation.

On flanges manufactured in **austenitic steel**, electrochemical corrosion is particularly likely in the presence of electrolytic media. To date, only a few cases of pitting have become known and these occurred in very strong chlorine-containing electrolytes such as monochloroacetic acid. It is advisable to look out for possible pitting, especially in the case of chloride-containing aqueous media. In some cases it may be necessary to bear in mind that certain intervals occur between inspections. Corrosion specialists try to avoid damage by using resistant materials for the design.

For cases where serious problems occur or are likely to do so, a PTFE covering can be recommended for the graphite gasket (**special grade without adhesive and stainless steel reinforcement and a PTFE envelope**).

The majority of stainless steel flanges in the chemical industry, however, are not affected as most of the media are organic.

ROTARY SEALS

14th International Conference on Fluid Sealing, Firenze, Italy,
6-8 April 1994. Organised by BHR Group Limited, Cranfield,
Bedford, MK43 0AJ, UK; Tel: 0234 750422

Impact Tests of Rubber Materials for Seals

FUJIO HIRANO

Visiting Professor, Oita University

Professor Emeritus, Kyushu University

4-10-12 Takamiya, Minami-Ku, Fukuoka-Shi, 815 Japan

HIROOMI MIYAGAWA

Professor, Department of Engineering , Oita University

KEIJI IMADO

Assistant, Department of Engineering, Oita University

700 Dannoharu, Oita-Shi, 870-11 Japan

and

YOSHIO KAWAHARA

NOK Corporation, 4-3-1 Tsujido-Shinmachi, Fujisawa-Shi, 251 Japan

ABSTRACT

Using an impact testing apparatus originally designed by the authors, which enables accurate measurements of repeating collisions and rebound of a ball fixed to the end of a hammering arm. Viscoelastic properties of rubber materials for flexible seals were investigated. Four kinds of representative materials different in chemical composition were experimented: acrylonitrile-butadiene rubber (NBR), silicone rubber (VMQ), acrylic rubber (ACM), and fluoro rubber (FKM).

Displacement of the ball while indenting or rebounding from surfaces of rubber sheets was recorded with sufficient accuracy, from which the instantaneous ball velocity can be derived by means of data-processing. In order to facilitate analysis of viscoelastic behaviour of the rubber materials so that to clarify its difference distinctly, the velocity-displacement diagrams were plotted automatically. From these results qualitative information of viscoelastic properties, e.g. dynamic elastic moduli, damping factors, and effects of swelling can be immediately and clearly obtained.
For quantitative treatments of the velocity-displacement curves, a differential equation was

established based upon the two-element Voigt model in Hertzian contact and numerically solved, The equation consists of inertia of the ball, gravitational, non-linear spring and viscous damping terms.   This theoretical treatment was proved to be sufficient for the estimation of a spring and dashpot terms of the rubber materials by comparing the numerical solution with the experimental results.   The damping factors were estimated in order of magnitude: FKM > ACM > VMQ > NBR.

Furthermore, the change of the viscoelastic properties of the rubber materials was easily observed and the glass transition nearly at -25°C was found by cooling with dry ice. Thus, much information is available through the observation of repetition of bouncing on a rubber material in contrast to a single collision as in the Shore hardness testing.

INTRODUCTION

In the authors' previous investigation [1] behaviour of traction oils at high contact pressures as the squeezing film effect or their solidification was observed by means of the impact testing apparatus which was originally designed and developed at Oita University together with its measuring and data-processing system: impulse and proximity sensors, an electric circuit for detecting contact voltage and a group of amplifiers.   Here, synchronized records of the motion of the ball, intensity of the impacts, and contact voltage enabled observation of the phenomena within extremely short duration of each contact, e.g. 0.3 ms with resolving power of a few microseconds as well as in sequence of contacts and rebounds for long periods.

Thus observed coefficients of restitution serve more accurate measures of hardness in comparison with the Shore hardness which utilizes only first collision and rebound.   The sequential behaviour of repeated collisions and rebounds bring about further information as to changes of surface layers of materials or oil film and surface coating film.
Afterwards, by improving this apparatus the scope of its applicability was widely and successfully extended to estimation of load carrying capacity of oils and greases, hardness of ceramics, reliability of bolted joints etc.

The present investigation was carried out for further extension of its applicability to elastomer, in particular, to rubber materials for flexible seals. So far a number of standards were prescribed for estimating elastic and viscoelastic properties of rubber, e. g. by ISO. In this connection, the impact resilience test is used to estimate internal friction of rubber. However, its physical meaning and technological applicability are not always fully clarified. Concerning them, the majority of investigators dealt with rolling friction and wear of rubber [2,3,4]. For example, Engel summarized investigations of impact phenomena [5].

Moreover, since this test utilizes intentionally only the first collision and rebound measurement, thus information obtained is of limited nature. In contrast to this, more information is available through observation of repeated impact and rebound behaviour. The elastic and viscoelastic material constants of rubber can be estimated by means of data processing, as will be shown in the following report.

NOTATION

g	: acceleration of gravity		, ms^{-2}
h	: displacement of ball, + indentation, − rebound, m		
h_∞	: depth of static indentation $(mg/k)^{2/3}$		, m
H	: initial height of ball bottom		, m
k	: non−linear spring constant		, $Nm^{-3/2}$
m	: mass of hammer		, Ns^2m^{-1}
t	: time		, s
v	: velocity of ball		, ms^{-1}
V	: initial collision velocity of ball		, ms^{-1}
X	: dimensionless velocity v/V		
Y	: dimensionless displacement ξh		
Y_∞	: dimensionless depth of static indentation ξh_∞		
λ	: non−linear damping constant		, $Nsm^{-3/2}$
Λ	: dimensionless damping constant $\lambda/(mV\xi^{3/2})$		
τ	: dimensionless time ξVt		
ξ	: parameter group $(4k/5mV^2)^{2/5}$		, m^{-1}

EXPERIMENTAL METHODS

Test Apparatus

Figure 1 (a) shows the diagrammatic view of the test rig used in the authors' previous investigations [1]. Here, a hammering arm of mild steel is supported by a deep-grooved ball bearing as a hinge. The contact element applying the impact is a bearing steel ball 13.49 mm in diameter mounted near the arm end. The equivalent weight of the hammering head is 7.15 N.

A proximity sensor is attached on the base plate to detect the accurate position of the arm , i.e. the ball motion by means of a magnetic field change induced by a high-frequency oscillator. Its output voltage was calibrated for estimation of the spacing between the ball and the test piece on the base plate. For this purpose the spacing was measured with the aid of a precision digital length gauge, by which the dropping height of the ball was also adjusted prior to each experiment. Satisfactory linearity was proved between the measured voltage and the position of the ball.

Besides these sensors the electric circuit shown in Fig.1 (b) was connected between the arm-ball system and the insulated test piece. By applying a voltage difference 30 mV under the open condition of the circuit, the moment and duration of collision, i.e., of direct contact between them were estimated. Since rubber shows inherently an insulating property, in case of necessity the test pieces sputtered with gold films 60 nm in thickness were specially prepared. Electrically conductive rubber was also available for this purpose and actually compared with them, considering some difference in viscoelastic properties.

Test Pieces

Table 1 shows main properties of experimented rubber materials used generally for flexible seals: **A**, acrylonitrile-butadiene rubber NBR, **S**, silicone rubber VMQ, **T**, acrylic rubber ACM, and **F**, fluoro carbon rubber FKM. The test pieces were cut off in a form of narrow pieces 10 mm in width from vulcanized rubber sheets 2 mm in thickness. Each test piece was fastened on a base plate with a cover steel plate and set screws at both ends.

Test condition and procedure

The dropping height of the ball was H=1 mm and the corresponding impact velocity was 0.14 m/s. Ball motion itself triggered records of ball motion and contact voltage.

From the records of the ball motion, i.e. indenting into the test piece or rebounding from it, the instantaneous velocity was obtained through numerical differentiation. In order to facilitate an intuitive grasp of the viscoelastic behaviour of the rubber materials, the velocity–displacement diagram was depicted automatically, as will be demonstrated later.

The experiments were performed at room temperature 20 to 23°C. In particular, low temperature properties of ACM rubber were observed by cooling it with the ball at the same time with circulating alcohol cooled with dry ice.

EXPERIMENTAL RESULTS

Variation of displacement

First, the sheets of the rubber materials **A**, **S** and **F** coated with the sputtered films of Au, 60 nm in thickness, were tested. Figure 2 shows variations of the displacement of the ball dropped from a height of H=1 mm. After repetition of collision and rebound the variation is damped and the displacement reaches stationary. Figure 3 is the corresponding changes of the contact voltage, where the upper level, 100% separation, stands for the ball is off the rubber surface, and lower levels means direct contact between the test pieces and the ball respectively. By superposing these records on their records of ball motion the moments of collisions and separation can be estimated. The behaviour of the rubber material **T** is shown in Fig.4 together with the results of the low–temperature experiments.

Velocity–displacement diagram

From these time–displacement relations the damping characteristics of the rubber materials are qualitatively compared. For qualitative treatment, however, representation as the velocity–displacement diagram, i.e. the phase plane is considered to be more suited. An example of the rubber material **A** is shown in Fig.5 for the explanation of dynamical behaviour of the rubber material and the ball.

Here, the velocity of the ball is estimated by differentiating the time–displacement relation. Plotting h, displacement of the ball in the direction of indentation, against v, velocity of the ball, the velocity–displacement diagram is obtained characterized with a spiral converging finally to the stationary point $v = 0$, $h = h_\infty$, as illustrated in Fig.5.

The initial point V shows the point of first collision: $v = V$, $h = 0$, where V denotes the first collision velocity $V = (2gH)^{1/2}$ and $h = -H$ is the initial dropping height of the ball. From the moment of collision the ball is decelerated from acceleration of gravity due to resistance of rubber, By way of point A, the maximum velocity point, it decreases with an increment in depth of indentation. The maximum depth is at the point B : $v = 0$. Then the displacement h turns to decreasing, which means recovering from indentation, the direction of the velocity is reversed. After reaching the maximum reversed velocity at the point C, the contact force becomes zero at the point D. Here, the ball jumped off the rubber surface. The ball rebounds freely with the initial velocity at the point D. After reached the maximum rebound height at the point D' it begins to fall and the second collision occurs at the point V_1.

The progress of the second collision $V_1 \rightarrow A_1 \rightarrow B_1 \rightarrow C_1$ is similar to the first one. In most case the separation hardly occurs in the second collision, as shown in this example. Then the minimum indentation takes place at the point D_1. The third collision progresses as $D_1 \rightarrow A_2 \rightarrow B_2 \rightarrow C_2 \rightarrow D_2$. The behaviour of silicone rubber **S** is little different from **A**, as shown in Fig.6. The behaviour of fluoro carbon rubber **F** shows no rebound even in the first collision owing to its high hysteresis loss, as shown in Fig.7. Figure 8 shows the effect of adhesion character. Figure 9 shows the results of swelling test. Clear difference was brought about through swelling. Furthermore, property at low temperature are also tested. Samples are shown in Fig.10 and Fig.11. Around glass transition point – 25°C, a number of rebounds were observed.

DISCUSSIONS

Fundamental equation

The experimental results proved to be sufficient to obtain qualitative information as to the dynamic and viscoelastic behaviour of the representative rubber materials, including effects of low temperatures, swelling, surface adhesion, etc. For the purpose of the quantitative analysis, however, a differential equation describing the behaviour should be established, considering inertia, gravity, restoring and viscous damping terms. According to theory of Hertzian contact, the normal displacement h of a spherical solid of radius R on a half plane and the radius of a contact area are written as follows[5]:

$$h = \frac{3P}{4aE^*} \quad and \quad a = \left(\frac{3PR}{2E^*}\right)^{\frac{1}{3}} \qquad (1)$$

where P and E^* denote the normal load and the effective modulus of elasticity defined by

$$\frac{2}{E^*} = \frac{1-v_1^2}{E_1} + \frac{1-v_2^2}{E_2}$$

Subscripts 1 and 2 express the rubber and the spherical body, respectively. In the present case the modulus E_1 of rubber is much lower than E_2 of the metallic ball and its Poisson's ration v_1 is nearly 0.5. Hence,

$$E^* = \frac{2E_1}{1-v_1^2} = \frac{8}{3}E$$

Denoting E for E_1 hereafter. The relation between the normal load P' and the displacement of the ball, i.e., its depth of indentation becomes as follows:

$$P' = \frac{16E}{9} R^{\frac{1}{2}} h^{\frac{3}{2}} = kh^{\frac{3}{2}} \qquad (2)$$

168

Here, the spring constant k is introduced in the non-linear spring characteristics (2) for convenience as:

$$k = \frac{16E}{9} R^{\frac{1}{2}} \qquad (3)$$

Next, to express the viscoelastic resistance P'', i.e., viscous damping of rubber, the two-element Voigt model is assumed for simplification:

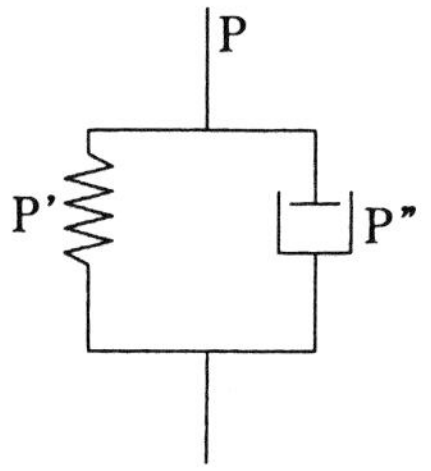

$$P'' = \lambda \frac{dh^{\frac{3}{2}}}{dt} \qquad (4)$$

$$P = P' + P'' \qquad (5)$$

Hence the equation of the ball motion is established as:

$$m \frac{d^2h}{dt^2} + kh^{\frac{3}{2}} + \lambda \frac{d}{dt} h^{\frac{3}{2}} = mg \qquad (6)$$

This differential equation is written in the dimensionless form.

$$\frac{d^2Y}{d\tau^2} + \frac{5}{4}(Y^{\frac{3}{2}} - Y_\infty^{\frac{3}{2}}) + \frac{3}{2}\Lambda Y^{\frac{1}{2}}\frac{dY}{d\tau} = 0 \qquad (7)$$

where

$$\left.\begin{aligned}
\tau &= \xi V t \\
Y &= \xi h \\
\xi &= \left(\frac{4k}{5mV^2}\right)^{\frac{2}{5}} \\
Y_\infty &= \xi h_\infty \\
h_\infty &= \left(\frac{mg}{k}\right)^{\frac{2}{3}} \\
X &= \frac{dY}{d\tau} = \frac{v}{V} \ , \quad v = \frac{dh}{dt}
\end{aligned}\right\} \qquad (8)$$

where V and h_∞ denote the velocity of initial collision of the ball and the stationary depth of indentation, respectively. The parameter ξ with the dimension of reciprocal length was generally used in the most simples theory of collision. Duration of contact under $Y_\infty = 0$, $\Lambda = 0$ is calculated from following equation [6].

$$t^* = \frac{2.943}{V\xi} \qquad (9)$$

Comparison with the observed result

In order to compare with the experimental results, which are expressed in the form of the velocity–displacement diagram with the aid of data–processing, the equation (6) and (7) are transformed into the following relations, respectively:

$$v\frac{dv}{dh}+\frac{k}{m}(h^{\frac{3}{2}}-h_{\infty}^{\frac{3}{2}})+\frac{3\lambda}{2m}vh^{\frac{1}{2}} = 0 \qquad (10)$$

$$X\frac{dX}{dY}+\frac{5}{4}(Y^{\frac{3}{2}}-Y_{\infty}^{\frac{3}{2}})+\frac{3}{2}\Lambda XY^{\frac{1}{2}} = 0 \qquad (11)$$

The equation (11) was numerically solved, taking account of the condition explained in Fig.5. An example of the solution is shown in Fig.12. The correspondence with the experimental results is facilitated by comparing the relation $r_X=X_2/X_1$, and $r_Y=Y_2/Y_1$, as demonstrated in Fig.12. Figure 13 shows the relation between r_X and r_Y. This chart was obtained by solving (11) under certain Y_{∞} and Λ then evaluate r_X and r_Y for each solution. Experimental data were also plotted on this chart. Y_{∞} and Λ. Figure 14 also shows the chart for low temperature test. This diagram serves the estimation of the actual values of Y_{∞} and Λ. These estimated values are listed in Table 2.

Estimation of Viscoelastic properties

Using the values of Y_{∞} and Λ listed in Table 2, actual viscoelastic properties of the rubber materials are estimated with the aid of relations derived from (10) and (11):

$$k=mg\left(\frac{4g}{5V^2}\right)^{\frac{3}{2}}Y_{\infty}^{-\frac{15}{4}} \qquad (12)$$

$$\frac{\lambda}{k}=\frac{V}{g}\Lambda Y_{\infty}^{\frac{3}{2}} \qquad (13)$$

These are also shown in Table 2. In the case of linear viscoelasticity, the equation of motion is as follows.

$$\frac{d^2h}{dt^2}+\frac{\lambda}{m}\frac{dh}{dt}+\frac{k}{m}h=f\sin\omega t \qquad (14)$$

The ratio of the viscosity term $\lambda\,\omega$ to the elasticity term k in the equation of motion is defined as a loss tangent.

$$\tan\delta = \frac{\lambda\omega}{k} \qquad (15)$$

In the present non-linear case, it is difficult to define the loss tangent strictly. For simplicity (15) is also employed here for estimation of loss tangent. As a rough measure of circular frequency ω, it is estimated by using the relation (9) in an ideal case $Y_\infty = \Lambda = 0$,

$$\omega = \frac{\pi}{t^*} = \left(\frac{\pi}{2.943}\right)\xi V \qquad (16)$$

Introducing this into (15), and considering $\xi = (4g/(5V^2))Y_\infty^{-3/2}$, the loss tangent becomes as:

$$\tan\delta = \frac{\lambda\omega}{k} = 0.854\Lambda \qquad (17)$$

In fact, however, k and λ are considered to be markedly influenced by frequency, as generally shown by master curves of rubber elasticity. Moreover, in the case of impact, behaviour is transient phenomenon. The duration of contact t^* varies with the parameter Y_∞ and Λ. Therefore, the estimation by using (17) is considered to be of tentative nature. During repeated collisions as shown in Fig.5, the estimated values r_Y decease considerably compared with those of r_X as shown in table 3. On the other hand, in the linear viscoelastic results in the simple relation $r_X = r_Y$ as shown in Fig.15. Observed r_Y approaches to this relation with progress of collisions as shown in Table 3. These facts are attributed to increasing conformity being formed in preceding collisions. The first collision is subjected to the non-linear relation between displacement and contact pressure characterized by Hertzian theory. Increased conformity due to the elastic after effect of the indentation produces nearly parallel surfaces without changing the contact area and results in linear relation between them. The fact as in the table 2, that r_Y is generally much higher than r_X is considered to be evidence of the Hertzian theory.

Next, the coefficient of restitution is difficult to estimate strictly due to difficulty in finding a point of falling on or jumping off without using a sputtered rubber sheet. Practically it is approximately represented by the ratio r_X, the ratio of maximum upward velocity to the maximum downward velocity. Hence, the coefficient of restitution is conveniently estimated by r_X. Figure 16 shows the relation between the coefficient of restitution r_X and the loss tangent.

As far as the effect of surface friction is concerned, the effect of an oil between the ball and rubber surface is investigated. The results are as follows:

Experimental condition	Y_∞	Λ
Dry condition	0.275	0.34
With PWF60 ($\nu = 9.4 \text{mm}^2/\text{s}$ at $40\,^\circ\text{C}$)	0.269	0.30
With BS ($\nu = 405 \text{mm}^2/\text{s}$ at $40\,^\circ\text{C}$)	0.255	0.353

Here, it is suggested that by lubricating with low viscosity oil easily spreading by impact its damping coefficient decreases. Hence, the effect of surface friction should not be neglected. These estimated values in Table 2 show some discrepancy compared with those calculated based upon their master curves applying WLF formula [8]. At the present stage, more accurate observation should be carried out to remove the cause of discrepancy experimentally as well as theoretically. The cause is considered to be due to incomplete treatment such as:

(a) Assumption of the two–element of the Voigt model
(b) Disregarding of the normal component of the friction or the effect of microscopic slip between the ball and the rubber surface
(c) Neglecting of the elastic after effect or its accurate observation
(d) Master curves corresponding to the contact condition should be established
   These will be treated in detail in the future, to utilize fully effective information obtained by this simple method.

CONCLUSION

Using a simple impact apparatus wide information of basic viscoelastic properties of rubber materials practically used as a flexible seal was obtained.

(1) By expressing experimental results in the form of the velocity–displacement diagram, the different viscoelastic behaviour of rubber materials could be easily analyzed and clarified not only qualitatively but also quantitatively.

(2) In order to derive practical properties, an equation of motion consisting non–linear reaction and damping terms was established. Experimental results were compared with the solution of the equation.

(3) Dynamical and viscoelastic properties of the actual rubber materials obtained are sufficiently to explain their different behaviour.

(4) In low temperature experiment, the change of the viscoelastic behaviour was easily observed and succeeded in finding the glass transition temperature.

(5) The estimated damping factors were in order FKM > ACM $\gtrless$ VMQ > NBR.

ACKNOWLEDGEMENTS

The authors wish to acknowledge Atuyoshi Miura, technical staff of our laboratory and Sigeaki Ishii for their support of this research.

REFERENCES

1. Imado, K., Miyagawa, H., Miura, A., Ueyama, N. and Hirano, F., Behaviour of Traction Oils under Impact Loads, STLE 48th Annual Meeting Calgary Technical Preprints No.93–AM–3C–3

2. Bowden, F.P. and Tabor,D., The Friction and Lubrication of Solids, Part I, Clarendon Press, Cambridge(1980).

3. D.F.Moore, The Friction & Lubrication of Elastomers, Pergamon Press, Oxford, 1972

4. G.M.Bartener & Yu.F.Zuyev, Strength and Failure of Viscoelastic Materials, Pergamon, 1968

5. Peter.A.Engel, Impact Wear of Materials, Elsevier, 1976

6. S.P.Timoshenko and J.N.Goodier, Theory of Elasticity, McGrow–Hill Kogakusha, Ltd, 421, 1970

7. Standard Practice for Rubber and Rubber Latices, ASTM Designation D 1418–85, 1985

8. Ferry, J.D., Viscoelastic Properties of Polymers, 3rd Edition, Wiley, New York, 1980

TABLE 1. Property of rubber materials

Abbreviation	A	S	T	T'	F
Designation[*]	**NBR** Acrylic nitrile butadiene rubber	**VMQ** Silicon rubber	**ACM** Acrylic rubber	**ACM** Acrylic rubber	**FKM** Fluoro rubber
Hardness Hw (IRHD)[**]	75	75	73	72	70
Young's modulus MPa	4. 5	4. 1	6. 6	5. 0	4. 6
Tensile strength MPa	15. 8	5. 0	7. 3	7. 9	9. 3
Elongation %	550	180	140	220	360

[*] ASTM Designation D 1418–85:

Standard Practice for Rubber and Rubber Latices [6]

[**] International Rubber Hardness Degree

TABLE 2. Viscoelastic properties of rubber materials

	r_x	r_y	Y_∞	Λ	$k\,(\times 10^7)^*$	$\tan\delta$
A	0.598	1.362	0.275	0.340	0.72	0.29
A ZnO	0.590	1.606	0.238	0.355	1.25	0.30
S	0.439	1.155	0.197	0.593	2.53	0.51
F	0.344	0.504	0.210	0.810	1.99	0.69
T (25°C)	0.386	0.748	0.212	0.705	1.92	0.60
T ZnO	0.410	0.810	0.222	0.645	1.62	0.55
T'	0.284	0.223	0.365	0.995	0.25	0.85
T' ZnO	0.287	0.313	0.209	0.980	2.03	0.84
T(−4°C)	0.221	0.660	0.102	1.403	30.0	1.20
T(−15°C)	0.372	1.789	0.119	0.78	16.8	0.67
T(−25°C)	0.571	2.961	0.149	0.395	7.20	0.34

* unit $N\,m^{-3/2}$

TABLE 3. Change of r_x and r_y in repeated collisions

	r_x	r_y	r_{x1}	r_{y1}	r_{x2}	r_{y2}	r_{x3}	r_{y3}
A	0.598	1.362	0.631	0.774	0.621	0.524		
A ZnO	0.590	1.606	0.623	0.947	0.613	0.766	0.724	0.785
S	0.439	1.155	0.613	0.764	0.678	0.588		
F	0.344	0.504						
T(23℃)	0.386	0.750	0.440	0.490				
T ZnO	0.410	0.810	0.470	0.530				
T'	0.284	0.223						
T' ZnO	0.287	0.313						
T(−4℃)	0.221	0.660						
T(−15℃)	0.372	1.789						
T(−25℃)	0.571	2.961	0.665	1.418	0.704	0.960		

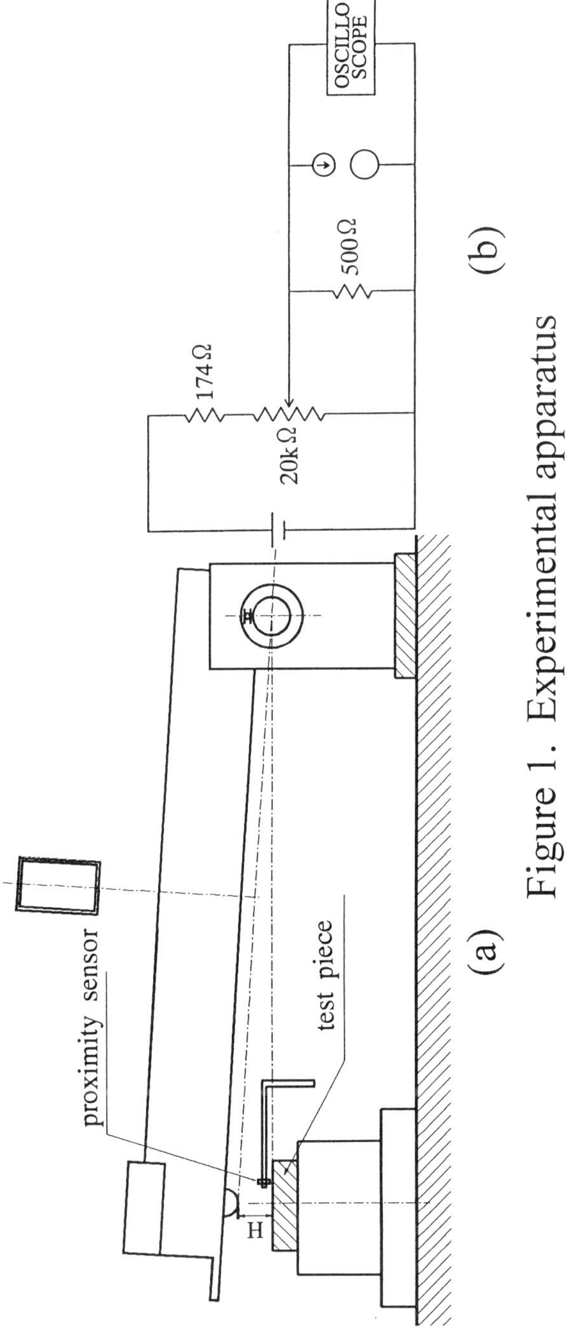

Figure 1. Experimental apparatus

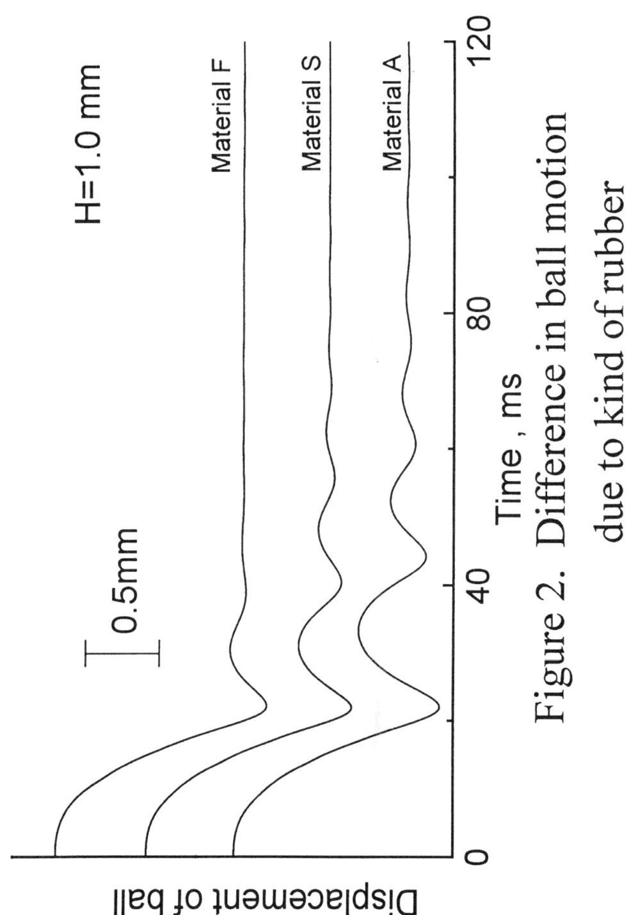

Figure 2. Difference in ball motion
due to kind of rubber

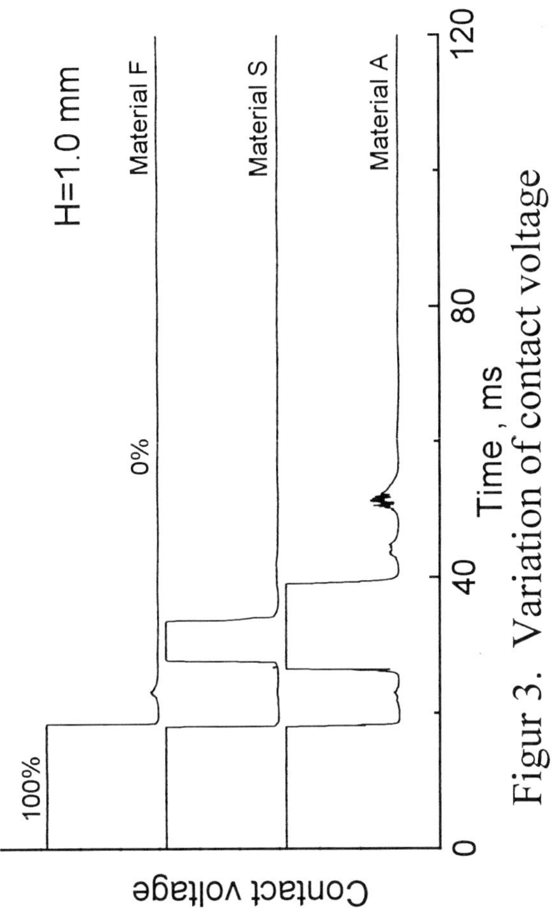

Figur 3. Variation of contact voltage

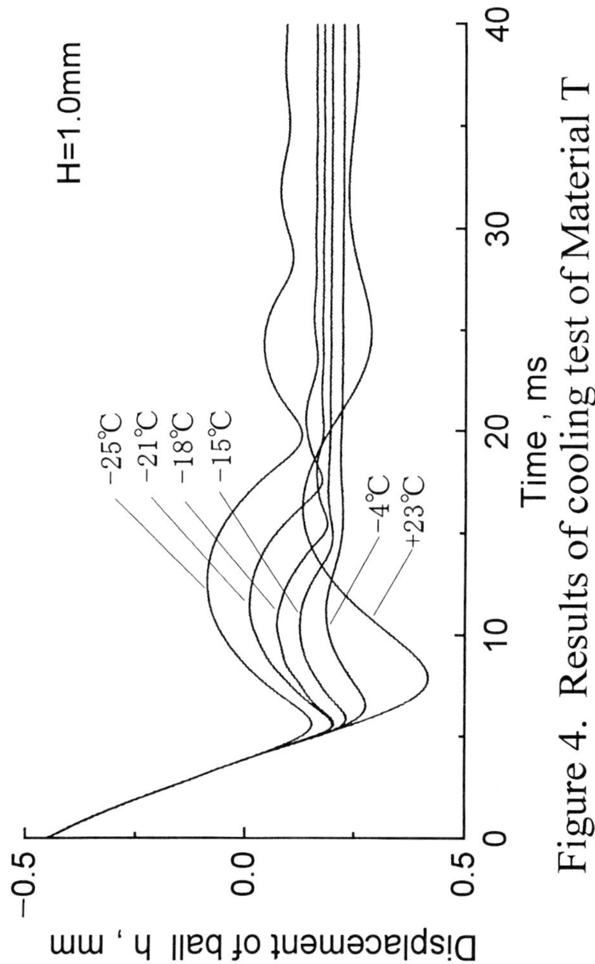

Figure 4. Results of cooling test of Material T

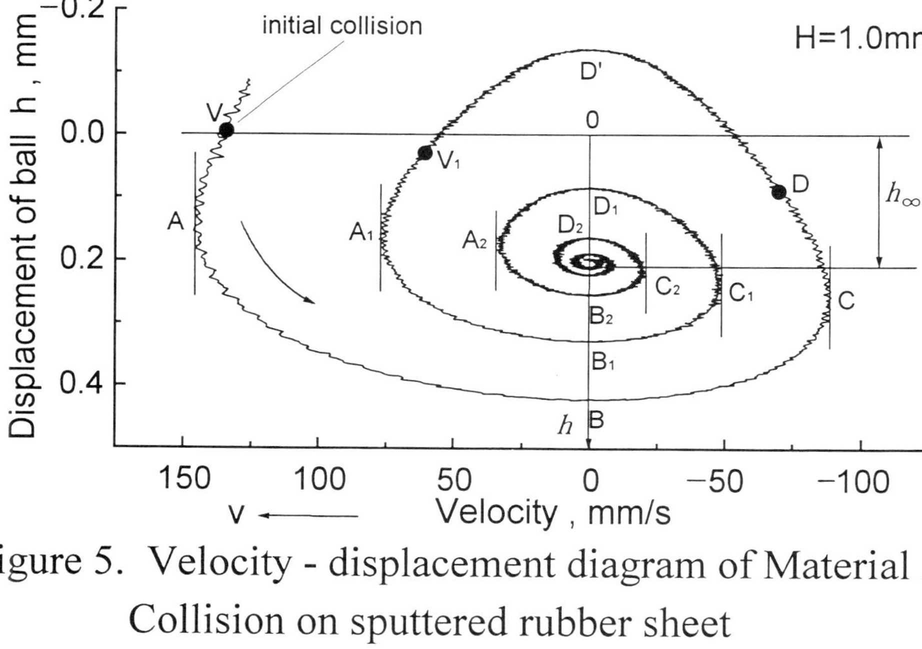

Figure 5. Velocity - displacement diagram of Material A
Collision on sputtered rubber sheet

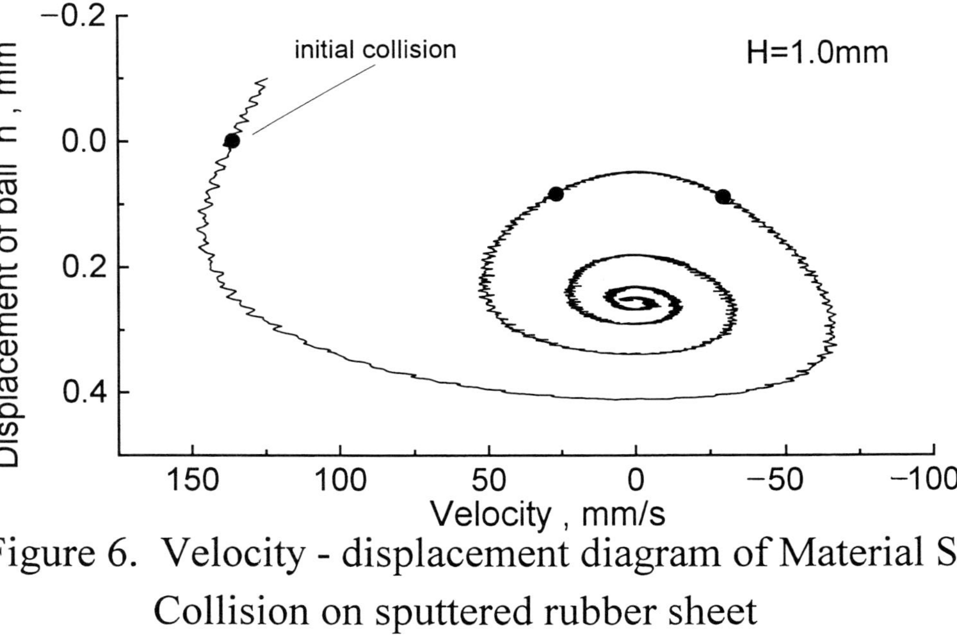

Figure 6. Velocity - displacement diagram of Material S
Collision on sputtered rubber sheet

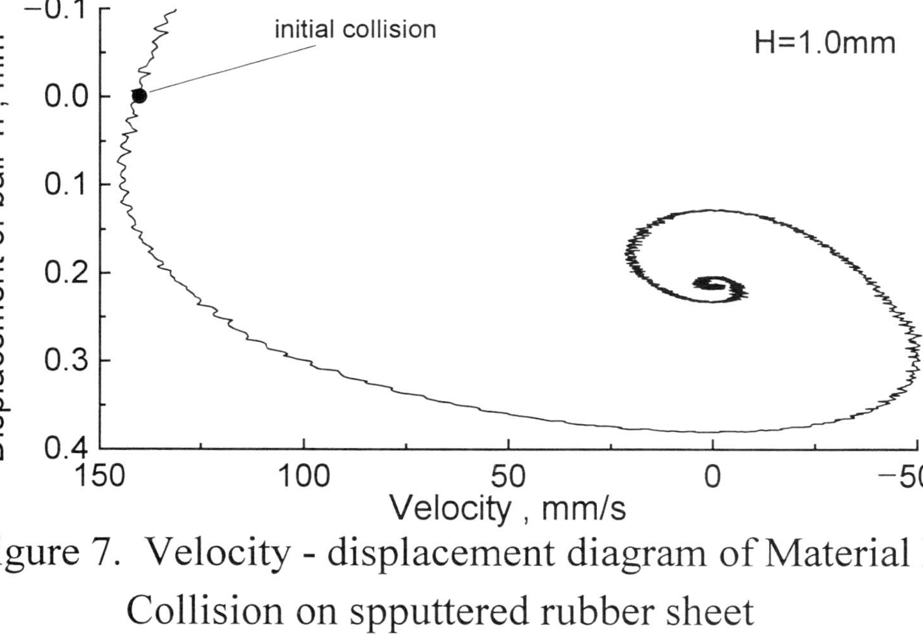

Figure 7. Velocity - displacement diagram of Material F
Collision on spputtered rubber sheet

183

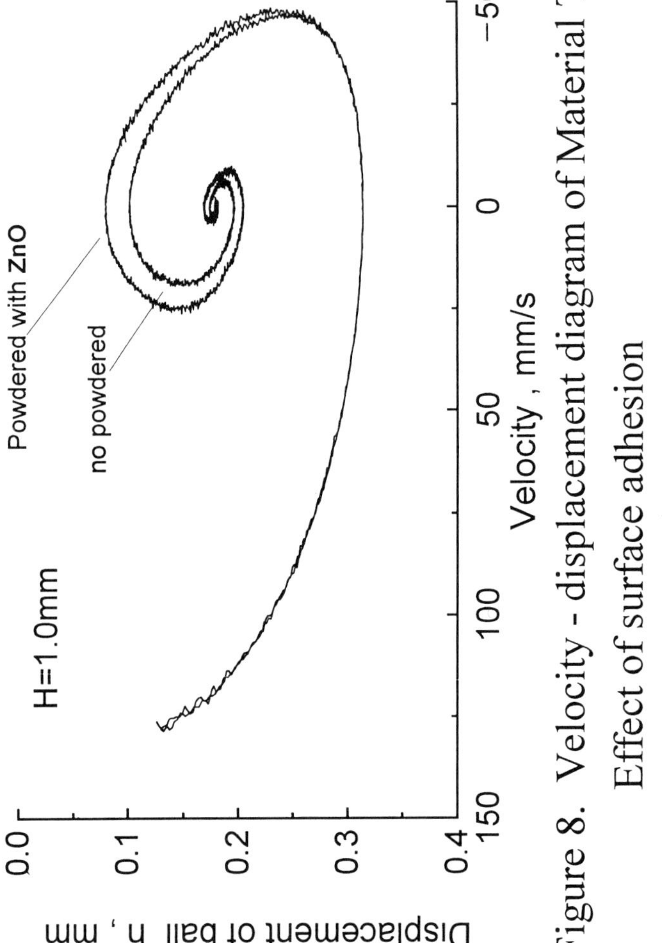

Figure 8. Velocity - displacement diagram of Material T
Effect of surface adhesion

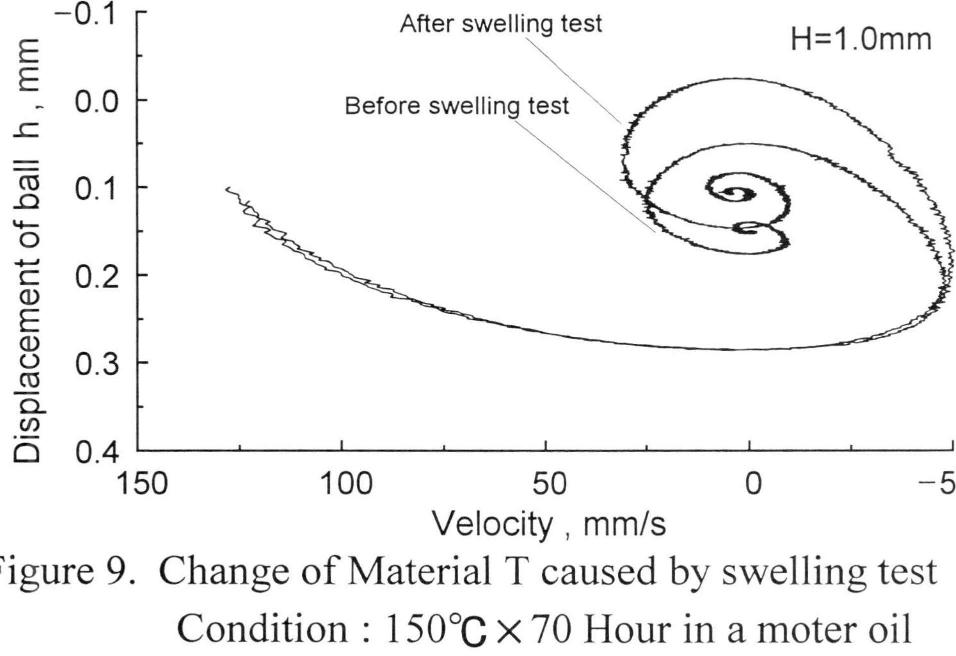

Figure 9. Change of Material T caused by swelling test
Condition : 150°C × 70 Hour in a moter oil

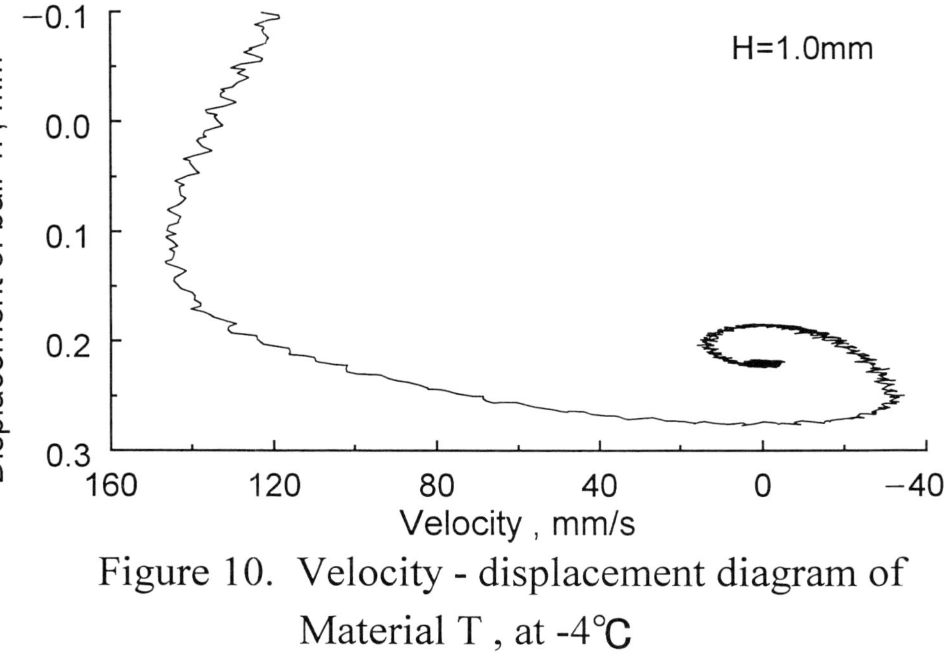

Figure 10. Velocity - displacement diagram of
Material T , at -4°C

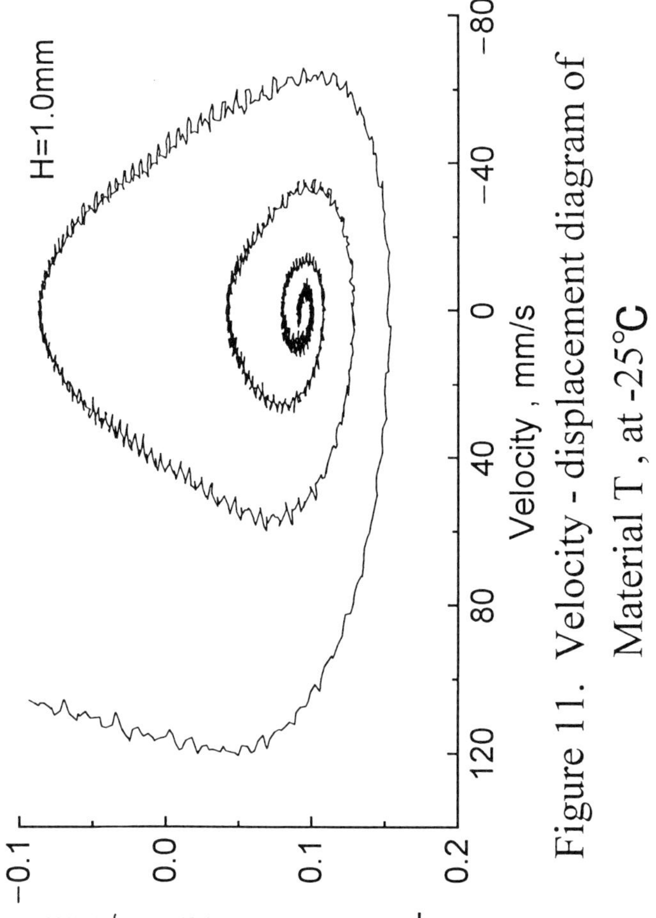

Figure 11. Velocity - displacement diagram of
Material T , at -25°C

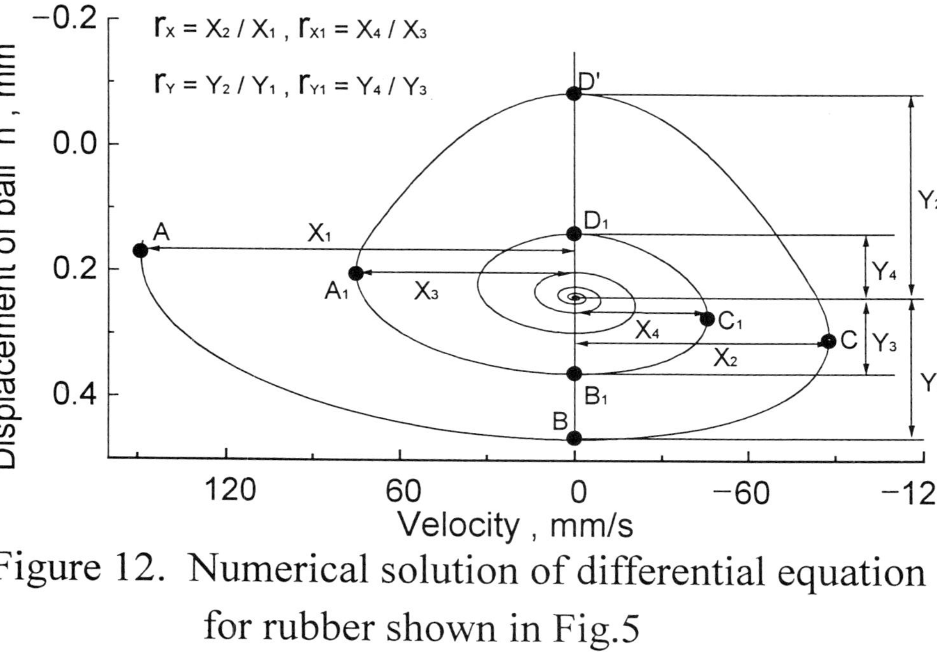

Figure 12. Numerical solution of differential equation
for rubber shown in Fig.5

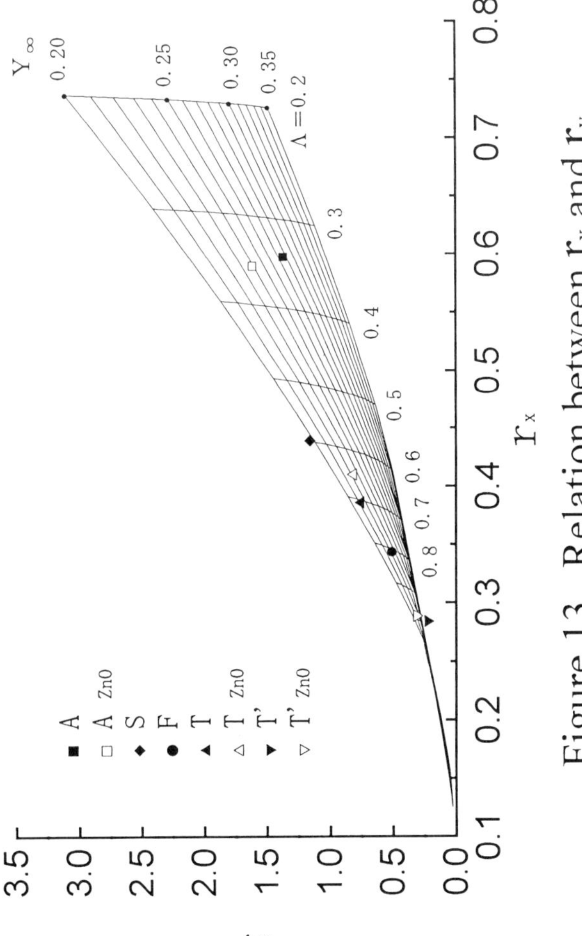

Figure 13. Relation between r_x and r_y

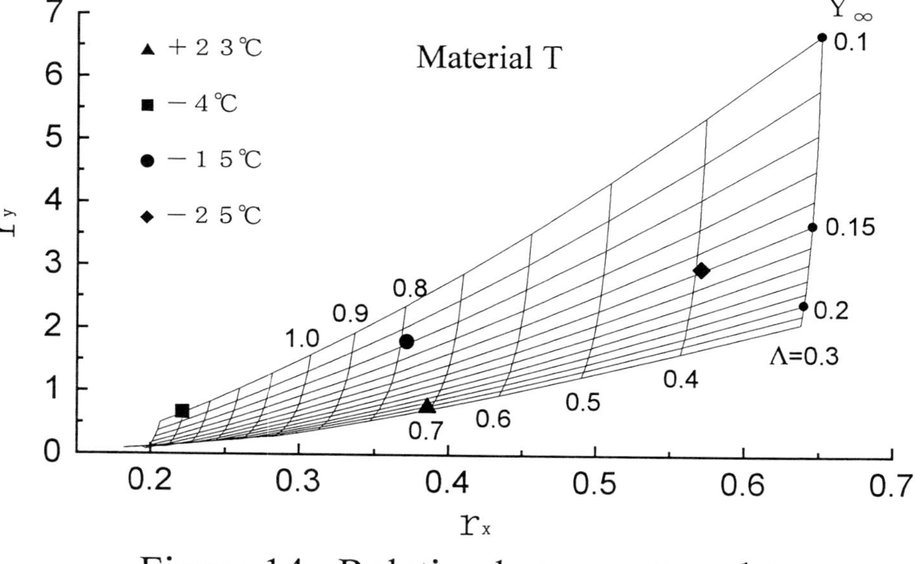

Figure 14. Relation between r_x and r_y

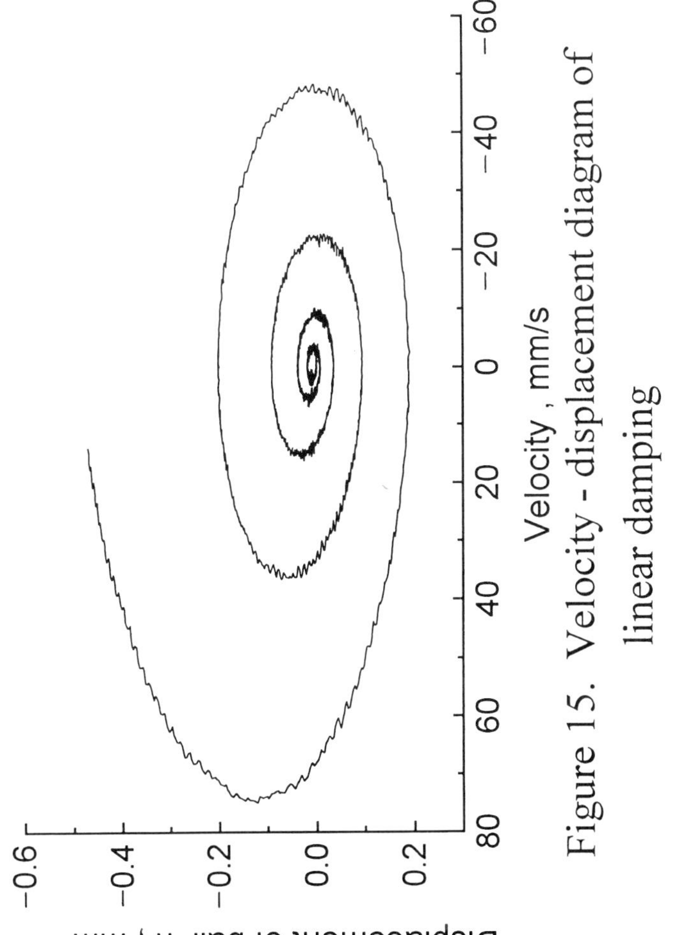

Figure 15. Velocity - displacement diagram of linear damping

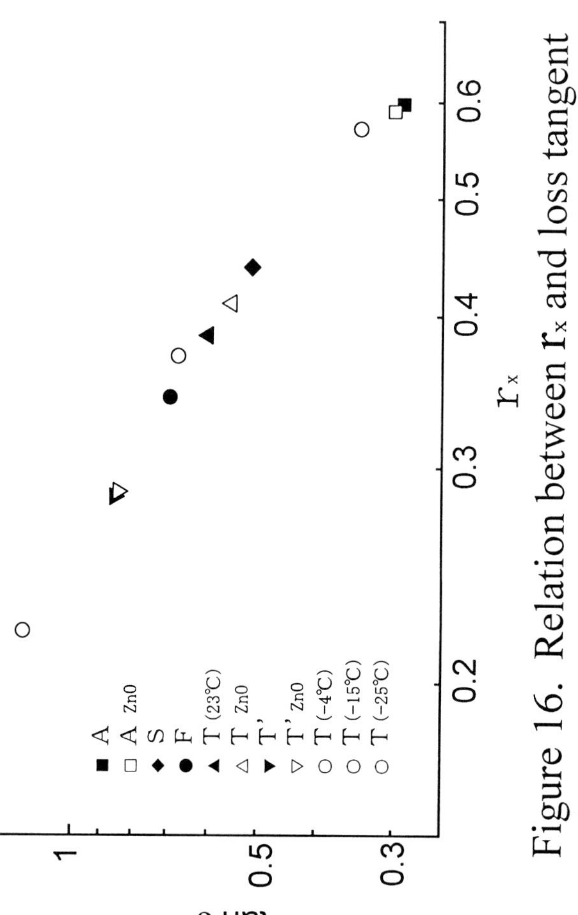

Figure 16. Relation between r_x and loss tangent

14th International Conference on Fluid Sealing, Firenze, Italy,
6-8 April 1994. Organised by BHR Group Limited, Cranfield,
Bedford, MK43 0AJ, UK; Tel: 0234 750422

Sealing liquid with air

Norbert Stanger, Werner Haas, Heinz K. Müller

Institut für Maschinenelemente

Universität Stuttgart

Air-barrier seals are compact non contacting seals offering the possibility to seal fast running machine tool spindles which might permanently change their position in space. Hitherto the efficiency of such seals depends on the scattered know-how of the designers. In the paper the basic functions of air-barrier seals are described. In an air-barrier seal air is injected into narrow clearances and creates a barrier pressure p_a. Cooling liquid which at high velocity splashes onto the entrance of the outer clearance builds up pressure p_{fl} at this point. The air-barrier prevents liquid to leak into the bearing area as long as p_a is higher than p_{fl}. For various configurations of the seal the barrier pressure p_a is calculated and compared to experimental results. The influence of eccentrical or inclined position of the spindle is demonstrated. In view of the economy of sealing the conditions of minimum air consumption are discussed. Finally, based on experience and systematic testing, the paper presents design directives for air-barrier systems designated to seal a splashing liquid.

Introduction

The shaft seals of high speed machine tool spindles are expected to run at low friction and without wear. Under all circumstances the bearings must be protected against the cooling lubricants which are splashed onto the working area of the machine tool. Liquid-collecting labyrinth seals /1,2/ have been applied successfully to solve this problem on condition that enough space is available to install a number of comparatively large internal chambers and channels to collect and drain the liquid which found its way into the sealing system. Because draining depends on gravity the functioning of liquid-collecting labyrinths is questionable if the spindle axis changes its position in space. Therefore, if variably positioned spindles have to be sealed both compact and contactless air-barrier seals may be a suitable answer.

An air-barrier is created by injecting pressurised air in a sealing system with narrow clearances.

194

Basic performance of an air-barrier

In an air-barrier seal air is injected into a buffer area between narrow clearances. The flow branches off, on the one side towards the bearing and on the other towards the atmospheric side where cooling liquid splashes on the seal entrance.

Due to the flow resistance of the clearances the injected air in the buffer area builds up the internal barrier pressure p_a. The cooling liquid which approaches the entrance at high velocity builds up the stagnation pressure p_{fl} at the end of the outer clearance. The

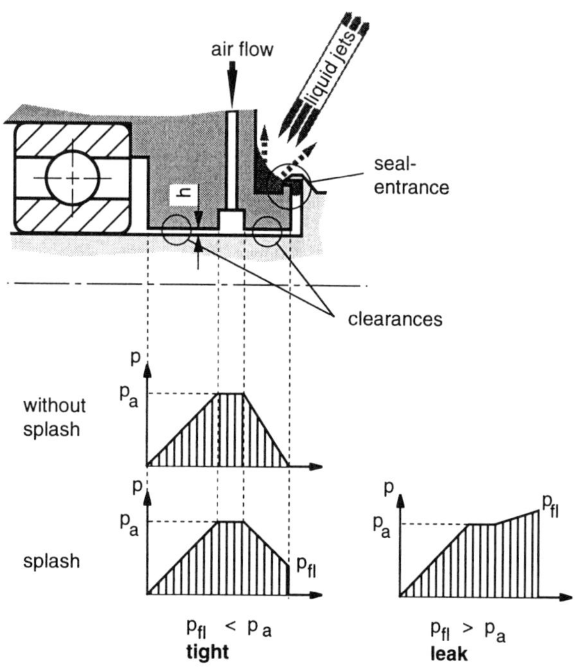

Figure 1: *Schematic view and pressure distribution of an air-barrier seal*

pressures p_a and p_{fl} are defined as pressure differentials against atmospheric pressure. Fig. 1 shows a schematic view and pressure distribution of an air-barrier seal. The air-barrier prevents liquid from leaking into the bearing area as long as the air pressure p_a is higher than the pressure p_{fl} of the liquid. In an experiment the air-barrier pressure was lowered to allow liquid to pass through the barrier area and to penetrate the bearing area. Fig. 2 for example shows the amount of leakage as a function of the air flow rate at variable height of the clearance. At a specific clearance the seal to be tight requires a certain value of air flow. The air consumption of a leakage-preventing air-barrier is linearly proportional to the length but proportional to the cube of the height of the clearance. Therefore, an effective and economical air-barrier seal is achieved by providing a small clearance height. Allowing larger clearances at the expense of a high air consumption is not only a question of energy costs but additionally is a risk for the life of the bearings because a high air flow in the bearing area accelerates drying of the lubricating grease. Therefore, it is recommended to make the clearances smaller than 0.07 mm.

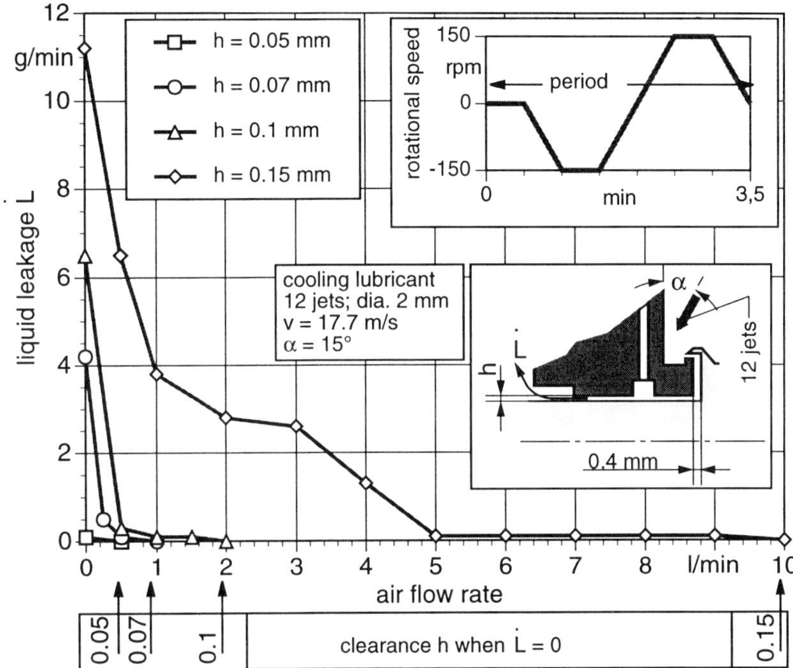

Figure 2: Leakage of an air-barrier seal dependent on air consumption and the height of the clearance

Another measure to reduce the required air-pressure and hence the air consumption is to reduce the velocity at which the cooling lubricant can approach the entrance of the seal. For two different designs of the entrance area Fig. 3 shows the air pressure required to prevent liquid to get over the barrier. The parameter in Fig. 3 is the velocity of the liquid which, in the experiments, was squirted from 12 jets towards the shaft seal entrance. Clearly superior is the type 2 entrance design, which is characterised by a rotating splashboard with an axial "nose" to shield the entry section of the clearance. The nose forms a short axial gap open towards a stationary circular rim which collects and deflects the cooling liquid squirting onto the radial wall of the spindle housing. Due to this precautions the stagnation pressure p_{fl} at the entrance practically remains very small, in the range of some millibars. In contrast, if high speed cooling jets hit an unshielded entrance, as is the case with the type 1 seal, the stagnation pressure p_{fl} is high and hence a high air pressure is necessary to prevent leakage. Therefore, in order to increase the sealing reliability and to save energy at the same time a properly shielded entrance of an air-barrier seal is highly recommended.

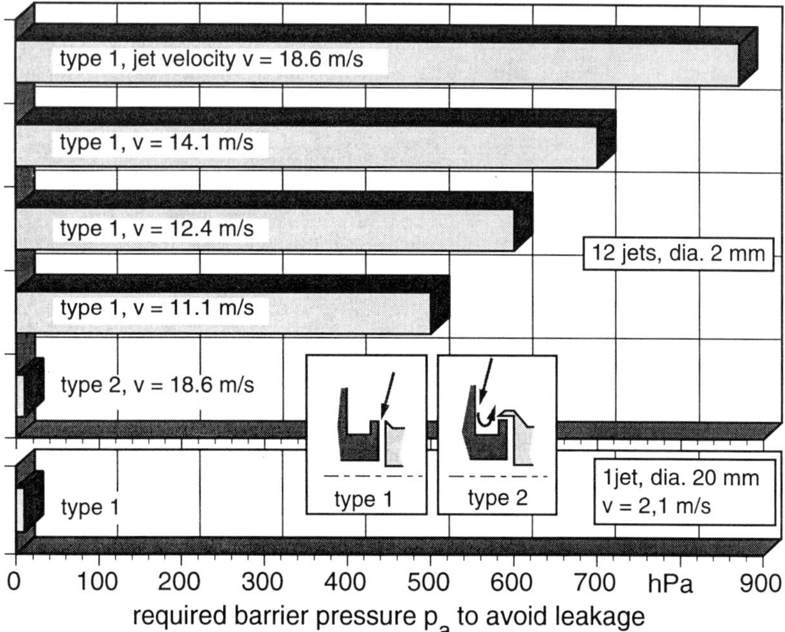

Figure 3: Required pressure p_a to avoid leakage as a function of the seal-entrance geometry and velocity of the approaching liquid

Considerations on the design of air-barrier seals

In air-barrier seals the direction of the clearances can be radial and/or axial. Figure 4 shows a number of basic designs with the entrance properly shielded. In Figs. 4a and 4b the barrier chamber is a rectangular groove followed by axial or radial clearances and Fig. 4c shows a seal with an axial clearance towards the bearing and a radial at the outside. Figs. 4d to 4f represent air-barriers with variable clearances formed between relatively floating or resilient components. Because they allow to adjust very small clearances the latter systems are most attractive. Fig. 4f shows a seal having a spring-loaded splashboard. The clearance is self-adjusting according to the air flow rate, the spring load and the dimensions of the clearance. Such design may be suitable for radially deflected spindles. Fig. 5 shows the measured height h of the radial clearance as a function of the air flow for various values of spring force F. Reliable operation requires a clearance of at least 20 μm. To stabilize the clearance, the sealing system works like an air bearing. Inclining of the splashboard, which might result in contact and wear, is avoided by injecting the air through a number of small throttling bores or through porous sintered material. The stagnation pressure p_{fl} of the liquid at the entrance has to be kept low. Otherwise the clearance may be blown open by the liquid itself.

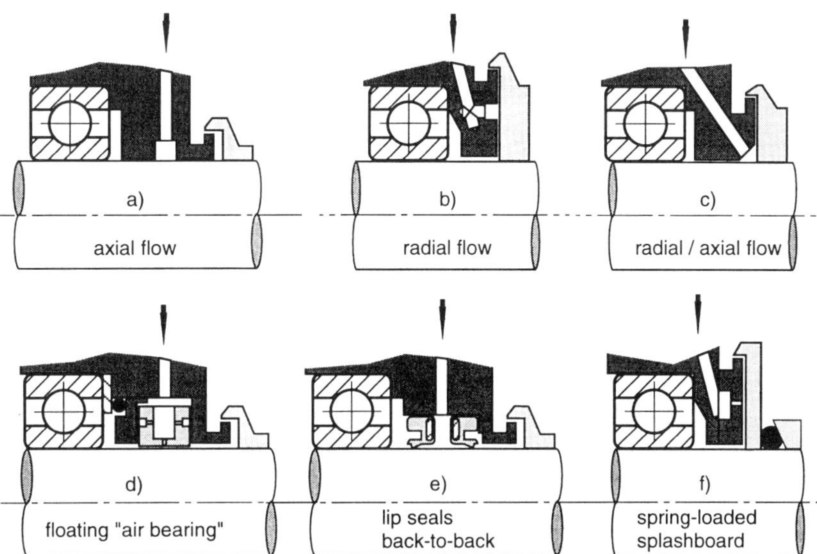

Figure 4: Basic design concepts of air-barrier seals with shielded entrance

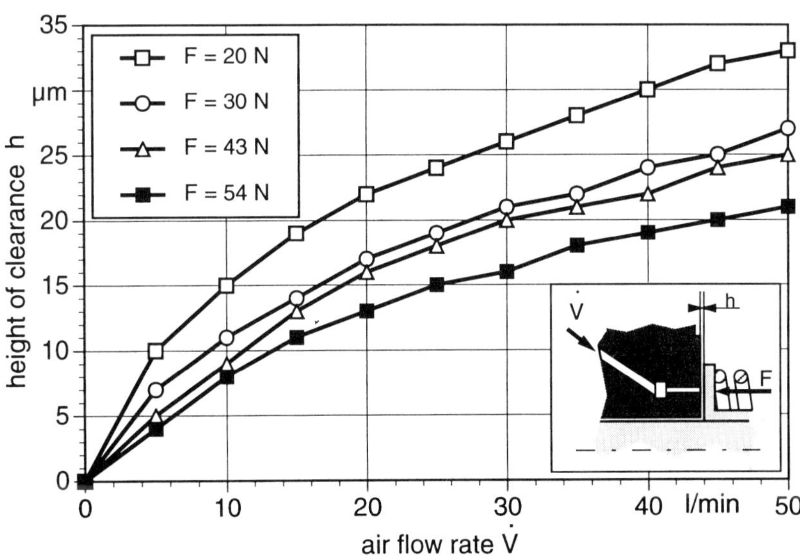

Figure 5: Height of the clearance dependent on the air flow rate

198

Fig. 6 shows the results of experiments with two different design concepts. Leakage occurred when the entrance of the clearance was in line with the front wall and, therefore, the splashing liquid directly penetrated the clearance. As already shown in Fig. 3, under equal conditions also the seal with the self-adjusting radial gap did not show any leakage when it was equipped with a circular rim and shielded by the nose of a rotating splashboard.

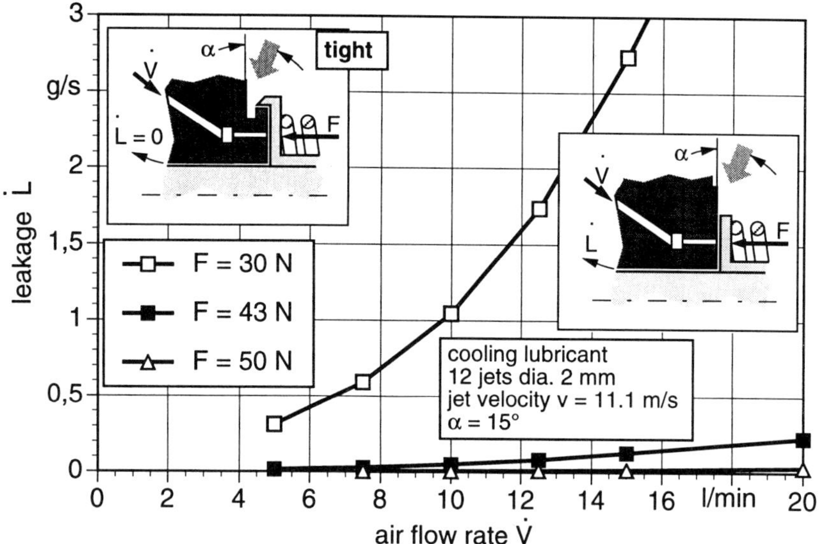

Figure 6:   Leackage of an air-barrier seal dependent on the air flow rate

Fig. 7 in more detail shows the type of air-barrier of Fig. 4d. Sealing is provided by a self adjusting "air-bearing" ring. One of the side walls is axially movable and spring-loaded. Air is injected into the clearances between the ring, side walls and shaft through a number of small bores. The pressure in the clearance rises where the height of the gap decreases. Therefore, the resulting forces F_{ax1} and F_{ax2} stabilize the seal ring. The lowest pressure maximum in the axial clearance is important for successful sealing of the liquid. Experiments on a 35 mm dia. shaft showed that this kind of sealing system combined with a splashboard and a circular rim is reliably non-contacting and liquid-tight at air flow rates of approximately 30 l/min (jet velocity v ≈ 18 m/s).

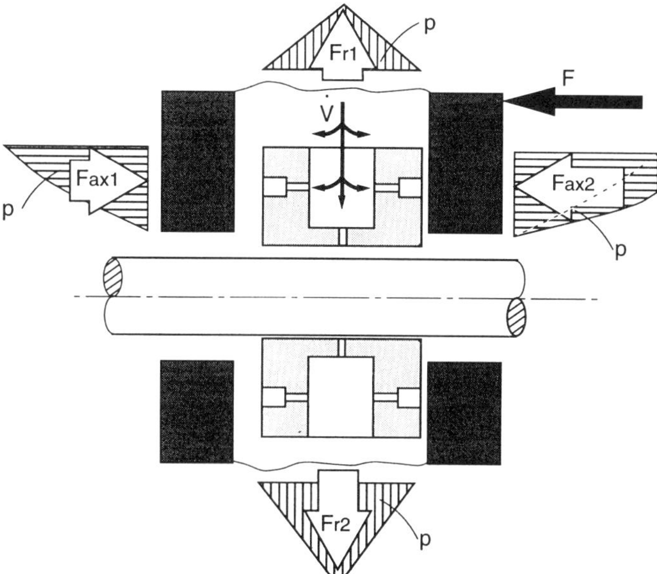

Figure 7: Air-barrier seal with self-adjusting "air-bearing"-ring

Most of the air-barrier seals used in practice are based on small clearances between non-contacting rigid components where the air pressure decreases almost linearly in the direction of flow. Another concept of air-barrier, shown in Fig. 4e, is to inject the air between two back-to-back mounted lip seals. Here, the air pressure has to be as high as to relief the radial force of the lip-seals until the lip lifts-off and the air pressure decreases suddenly to ambient in the short clearance under the lip. Due to the very small clearances the air consumption of this type of air-barrier is similar to that of the other types discussed. A disadvantage of this type of air-barrier, however, is that the lip lifts off at several spots only and the sealing edge of the rubber lip seal remains in contact with the shaft at other spots. The experiments showed that this results in lip abrasion and time dependent increase of air consumption. Fig. 8 shows the measured values of friction torque and pressure p_a versus the air flow of a lip type air-barrier seal.

Generally, a successful operation of air-barrier seals requires the air flow to be maintained at all time otherwise liquid droplets can be sucked in during the cooling period of the spindle.

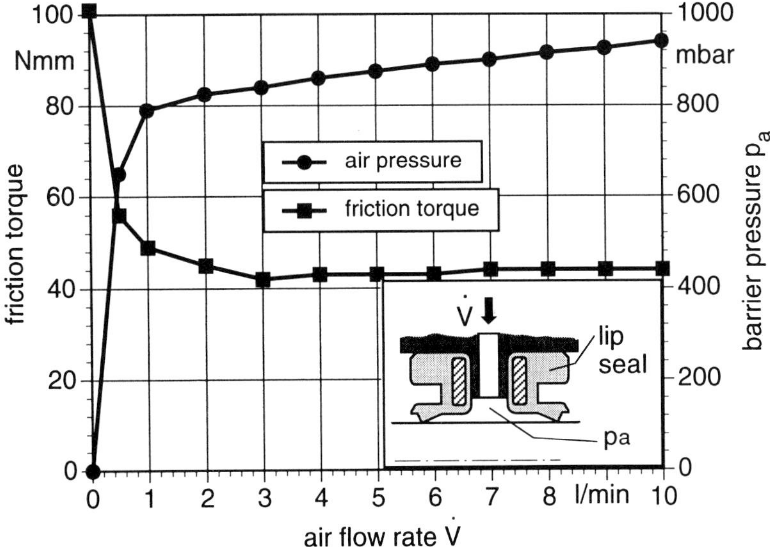

Figure 8: Air-barrier established by back-to-back lip seals: Friction torque and barrier pressure as function of the air flow rate $\dot{V}$

Calculation of the air flow rate

The air flow to establish the value of air pressure p_a required for leak-proof sealing is calculated by applying the formulas for laminar flow in an annular axial clearance.

Applying the symbols given in Fig. 9 the flow rate is given by

$$\dot{V}_0 = \frac{\pi \cdot D \cdot \Delta p \cdot h^3}{12 \cdot \eta \cdot l}$$

For compressible laminar flow the flow rate is given by

$$\dot{V}_L = \dot{V}_0 \cdot \frac{1+\beta}{2\beta}$$

For parallel eccentric position of the spindle the flow rate is given by

$$\dot{V}_{Le} = \dot{V}_L \cdot (1 + 1{,}5 \cdot \varepsilon^2)$$

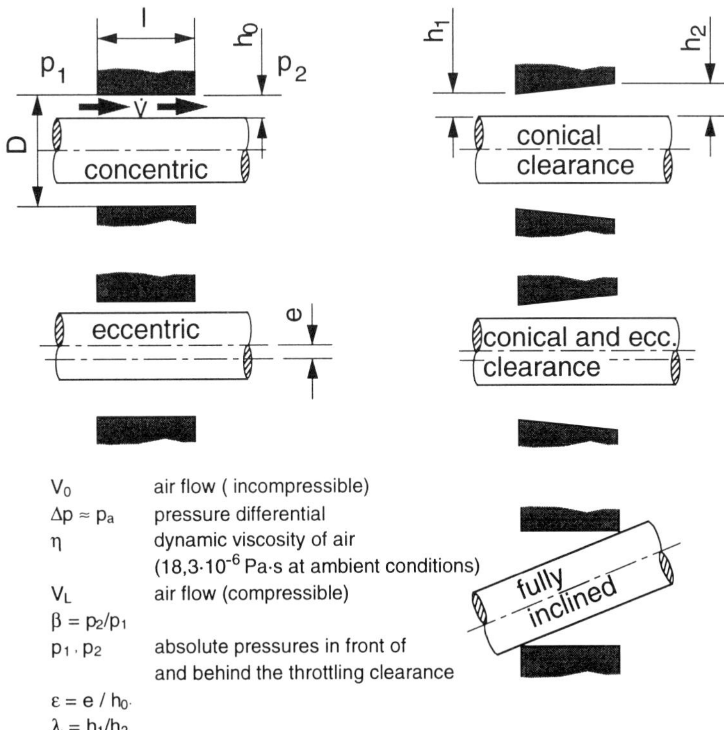

$\dot{V}_0$ air flow (incompressible)

$\Delta p \approx p_a$ pressure differential

η dynamic viscosity of air
(18,3·10⁻⁶ Pa·s at ambient conditions)

$\dot{V}_L$ air flow (compressible)

$\beta = p_2/p_1$

$p_1 \cdot p_2$ absolute pressures in front of
and behind the throttling clearance

$\varepsilon = e / h_0$

$\lambda = h_1/h_2$

Figure 9:   General shaft positions and clearance geometry

For a conical spindle or a conical hole in the housing the flow rate is given by

$$\dot{V}_{Lk} = \dot{V}_L \cdot \frac{2 \cdot \lambda^2}{\lambda + 1}$$

For a parallel eccentrical spindle position and a conical clearance ($\varepsilon < 0.5$, $\lambda < 2$) the flow rate approximately is given by

$$\dot{V}_{Lke} = \dot{V}_L \cdot \left(\frac{2 \cdot \lambda^2}{\lambda + 1} + 0.75 \cdot (1 + \lambda) \cdot \varepsilon^2 \right)$$

Most astonishing, no formula was found in the literature to calculate the flow rate when the shaft is inclined in the bore. Only Trutnovsky /3/ described some experimental results obtained with inclined shafts. In practice inclining very frequently is present and a normal kind of deflection. For example, in machine tools lateral deflection caused by cutting forces will create inclining of the spindle relative to a bore fixed to the structure of the machine tool.

In our experiments by increasing the inclining of the spindle the air flow decreased appreciably. Along a fully inclined shaft (diameter of shaft 58 mm, diameter of bore 58.16 mm, length of bore 9 mm) - oppositely contacting at both ends of the bore - the air flow rate was approximately 40 per cent of the flow rate with a concentric annular gap. Practically the same reduction was found at other shaft/diameter ratios and different height of the clearance. Fig. 10 compares the measured air flow rate through an annular clearance when the shaft is in concentric and in inclined position respectively. Also shown is the calculated flow rate for the concentric position. The upper right hand diagram shows the air velocity measured around the circumference at the exit of the clearance and the diagram below shows the pressure distribution measured along particular axial surface lines of the bore.

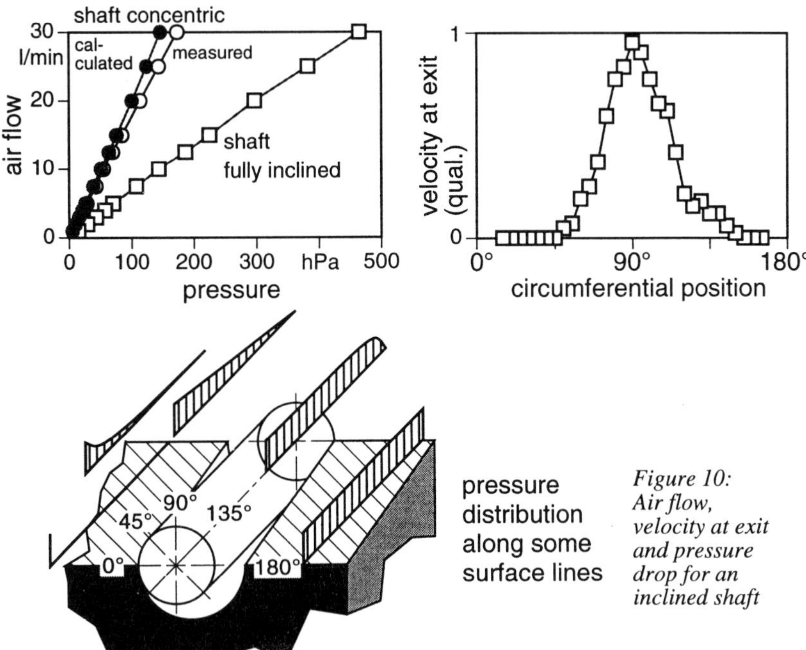

Figure 10: Air flow, velocity at exit and pressure drop for an inclined shaft

The air flow through short radial clearances is calculated approximately by the formulas as given above by replacing the diameter D by the medium diameter of the radial clearance. The air flow calculated in this way is somewhat higher than the correct value and the error depends on the radius ratio r_a/r_i, but, even at $r_a/r_i = 2$ the error is approximately 4 per cent.

Conclusions

The sealing of machine tool spindles which are heavily splashed by cooling lubricants requires a multistaged sealing system. If limited space necessitates a compact design a favourable solution is the combination of an outboard shielded clearance and an inboard air-barrier. Based on experience and systematic testing the paper presents some figures of the required air pressure and air flow rates and also some recommendations on the proper design of such sealing systems. With the advantage of higher reliability of opertation properly shielded air-barrier seals in many applications can replace contacting seals or conventional labyrinth seals.

Acknowledgements

The work presented was supported by VDW (Verband Deutscher Werkzeugmaschinen-fabriken e.V.) and AIF (Arbeitsgemeinschaft industrieller Forschungsvereinigungen).

References

/1/ Fritz E., Haas, W., Müller, H. K. Liquid-collecting labyrinth seals for machine tool spindles 13 th International Conference on Fluid Sealing, organised by BHRG, Brügge, 1992

/2/ Fritz, E., Haas, W., Müller, H. K. Konstruktions-, Richtlinien- und Lösungskatalog "Berührungsfreie Spindelabdichtungen im Werkzeugmaschinenbau". Institutsbericht Nr. 39, Institut für Maschinenelemente, Universität Stuttgart, 1992

/3/ Trutnowsky, K., Berührungsfreie Dichtungen, Düsseldorf 1973 (VDI-Verlag) pp. 43/44

/4/ Müller, H.K., Abdichtung bewegter Maschinenteile, Medienverlag Ursula Müller, Waiblingen, 1990

14th International Conference on Fluid Sealing, Firenze, Italy,
6-8 April 1994. Organised by BHR Group Limited, Cranfield,
Bedford, MK43 0AJ, UK; Tel: 0234 750422

Finite-Element-Analysis of a PTFE-Shaft-Seal

Guido Wüstenhagen, Heinz K. Müller, Klaus-Dieter Meck

Institut für Maschinenelemente, Universität Stuttgart, Germany

Sometimes the designer is doubtful whether the fascinating coloured pictures presented by a Finite-Element-analysis do reliably agree with reality. In particular this is becoming a problem when constructional elements are concerned which, for example, consist of PTFE-compounds the suppliers of which are unable to procure the relevant material properties. More often than not there is a lack of information on the stress-strain behaviour within the intended operating temperature range. In this case the user of a FE-program by himself and sometimes by costly experiments has to establish the necessary material properties. In this process it proved most effective on the one hand to experiment with a simply shaped specimen under load conditions similar to those of the original component and, on the other hand, to apply the material properties measured by setting up a FE-model of the specimen and performing an equivalent calculation. If there is satisfactory agreement of the results a reliable calculation of more complex constructional components is justified as well. In the paper the authors show how this procedure was applied to calculate the radial sealing force of a complex PTFE-seal ring under realistic mechanical and thermal conditions.

Introduction

A Finite-Element-analysis provides an efficient tool to the designer of PTFE-Shaft-seals. Advanced Finite-Element-programs with their captivating user interfaces allow an immediate access to this tool enabling the designer to get short-term results, for example with regard to the deformation or the reaction forces of resilient seals. Generally the user immediately can recognize a false outcome if it is a result of any coarse input errors, for example, wrong geometric values or material properties. Much more difficult, however, is to detect errors which are due to unsufficient experience of appropriate FE-modelling. To simulate different kinds of loading and working conditions the FE-programs offer various calculating options [1]. For instance, various options are available for the elasto-plastic behaviour of materials or different kinds of friction models. Furthermore, by applying user subroutines the user can include calculation options of his own. In view of the applicability to the actual problem the user has to make a competent choice out of the available options and has to decide how far the model can be simplified not to produce misleading conclusions.

To solve the particular problem of calculating a complex PTFE-seal the material properties of which were unknown it proved useful to start with an FE-analysis of a simply shaped specimen and, parallel, to measure the stress-strain relationship of an equally shaped specimen hardware in order to get the relevant material properties under load and temperature conditions similar to those of the original seal. It is important in the measuring process that the characteristic mechanical load conditions present in the complex seal are being put into effect and measured separately on the specimen level. Thereby too, it is possible to study the influence of essential parameter variations and, by comparing the results of FE-analysis and experiment to draw useful conclusions about the adequate modelling of the material. Using the FE-program MARC-MENTAT in this way, in consideration of actual mechanical and thermal load conditions the authors have calculated a complex *Balanced Shaft Seal* (*Entlasteter Wellendichtring* "EWDR") [2].

Radial force of the Balanced Shaft Seal

Figure 1 shows a cross section of the *Balanced Shaft Seal* made of carbon-filled PTFE compound which radially contacts the shaft surface with a sealing edge [3]. Because of the interference of the circular edge and the shaft initially a sealing radial force is created [4]. Due to the particular design of this seal the radial force is but weakly increased when the fluid pressure rises. Due to the high thermal expansion coefficient of PTFE the seal expands appreciably when the temperature increases during operation. Once thereby the radial interference gets lost the seal will leak excessively.

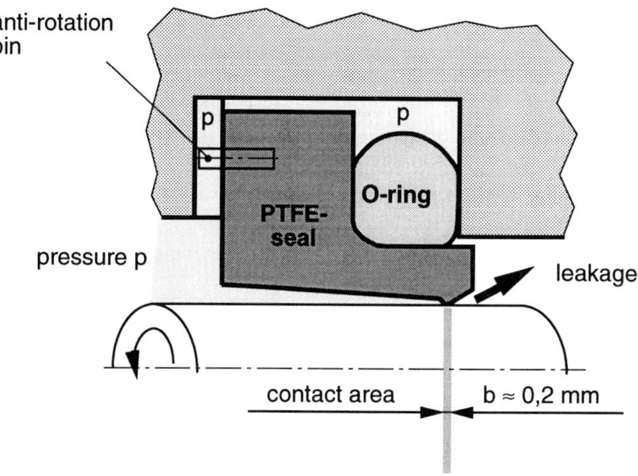

Figure 1:   Balanced Shaft Seal

Figure 2 shows the most important parameters which influence the radial force. The initial contact pressure resulting from interference and supported by the fluid pressure is relieved by creeping and thermal expansion of the PTFE material. In a practical seal a temperature rise of only 80K may result in a total loss of radial force and excessive leakage. Cooling down the seal will restore its performance but in practice an intermediate thermal leakage is not acceptable. In the course of the development of the *Balanced Shaft Seal* this unwelcome effect was eliminated by the implementing into the seal the Plastic-memory-effect [5].

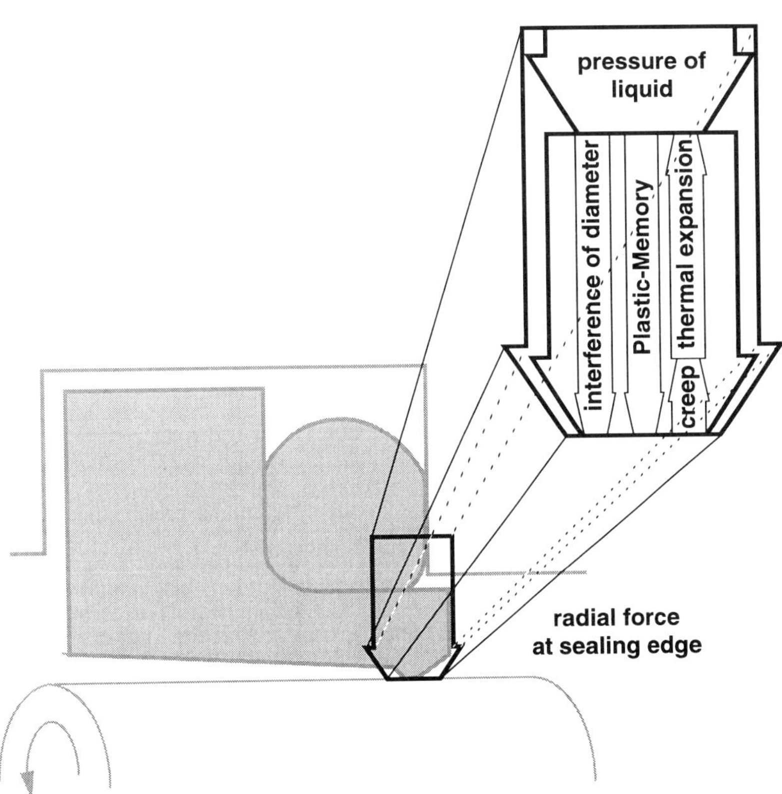

Figure 2:   Parameters influencing the radial force of a Balanced Shaft Seal

The Plastic-memory-effect is characterised by the habit of certain materials after an initial plastic deformation to "remember" their initial shape when they are heated again. This effect is applied in a PTFE-seal in the manufacturing process by radially stretching the sealing edge beyond the yielding point. A temperature increase of the seal in operation now activates the Plastic-memory-effect which results in a higher radial force which in turn counteracts the thermal expansion. It is within the scope of the optimization of the *Balanced Shaft Seal* to obtain a maximum compensation of the thermal expansion. Therefore the radial force is taken as the characteristic parameter for the measuring and calculating procedures. To measure, in the relevant temperature range, the radial load of an original *Balanced Shaft Seal* a measuring device has been developed. In addition an FE-analysis of the sealing system is performed simultaneously simulating the influences on the radial force, except the Plastic-memory-effect.

To achieve realistic results of the FE-analysis the following parameters are required with reference to the variation of operating temperature:

      Young´s-modulus,

      Thermal expansion coefficient,

      Yielding point and

      Creeping-behaviour.

The suppliers of the PTFE-compounds used were not able to procure this data. As an example, in the following the work of the authors to determine the relevant values of Young´s-modulus is described.

Developing the FE-model

For a given interference Young´s-modulus in the first place determines the magnitude of the radial force and hence the initial sealing pressure. In the tubular membrane as well as the sealing edge circumferential elongation creates tangential stress while the radial force results in radial compressive stress. Basically, the state of stresses is biaxial and the problem is similar to that of a thin-walled internally pressurized tube. Therfore, it appeared reasonable to measure Young´s-modulus by a specimen and reference-model in the form of a thin-walled PTFE-tube. Figure 3 in principle shows the corresponding experimental arrangement. With this arrangement the outer diameter of the tube was measured as a function of internal pressure and system temperature, the variables which represent the load conditions of the seal ring in practical operation. According to the readings Young´s modulus E was calculated from:

$$E = \frac{p\, d_a^3}{\Delta r \,\, (d_a^2 - d_i^2)}$$

where p is the fluid pressure, d_a and d_i the external and internal diameters of the tube and Δr the increase of the tube radius as a result of pressure p. The formula is valid if the ends of the tube are free of clamping and other forces. This is adequately satisfied by sufficient length of the tube as well as by the free floating ends and the kind of sealing. Figure 4 for a PTFE-Compound MT12/006 [6] shows Young´s modulus within the temperature range of 20° to 120°C. With the same experimental arrangement the coefficient of thermal expansion and the yielding point were measured within the same temperature range. The creep is measured separately using conventional tensile test procedures.

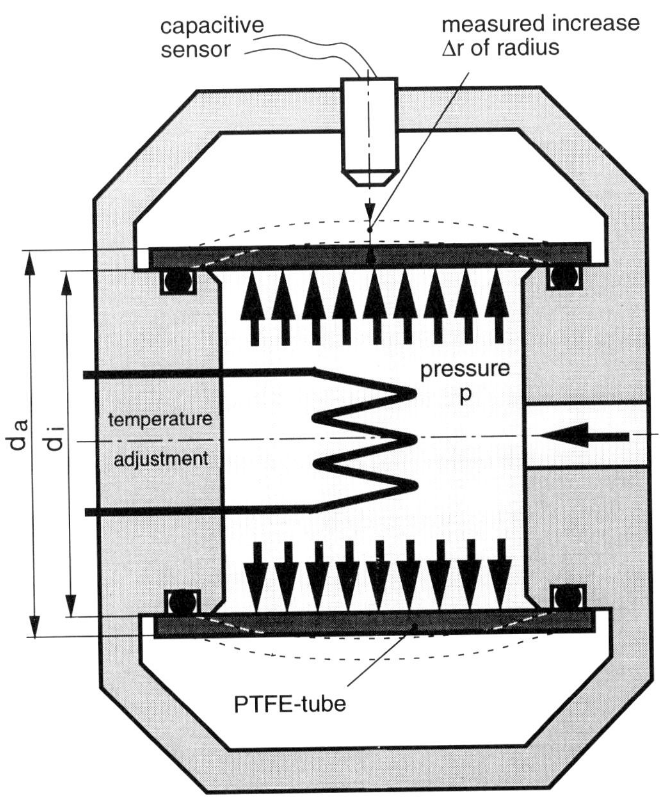

Figure 3:   Experimental arrangement to establish Young´s-modulus

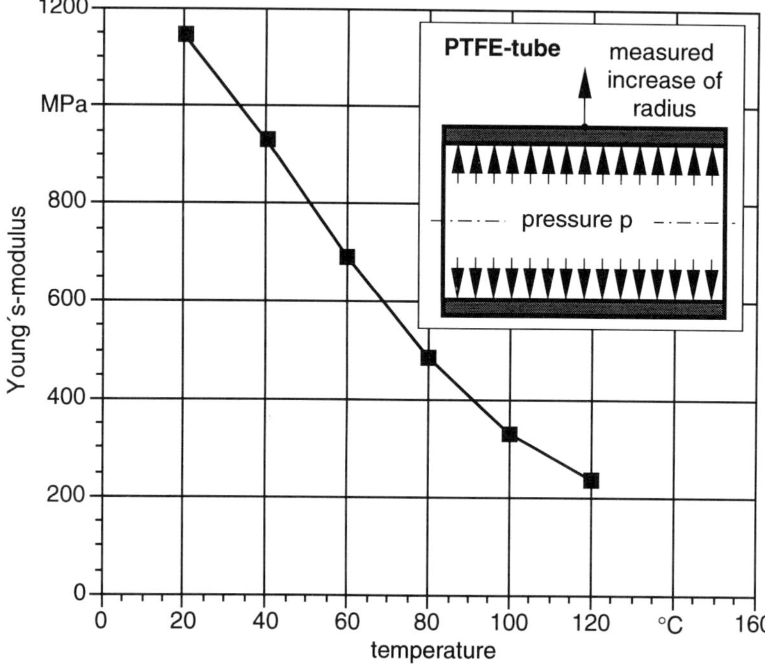

Figure 4:   Young´s-modulus of a PTFE-compound versus temperature

The material properties derived from the experimental readings were implemented in a FE-calculation modelled for the experimental PTFE-tube with the corresponding load parameters. The results found justify reliable application of the established material properties in a FE-analysis of the real *Balanced Shaft Seal*.

Calculation and measuring of the radial force

The mesh of a "21/2dimension"-FE-calculation is shown in Figure 5. Besides of the forces, stress and deformation in the viewing plane the selected finite ring-elements also consider peripheral stress. Except for the creeping values which have not yet been investigated satisfactorily, Figure 5 shows qualitatively the influence of the material properties.

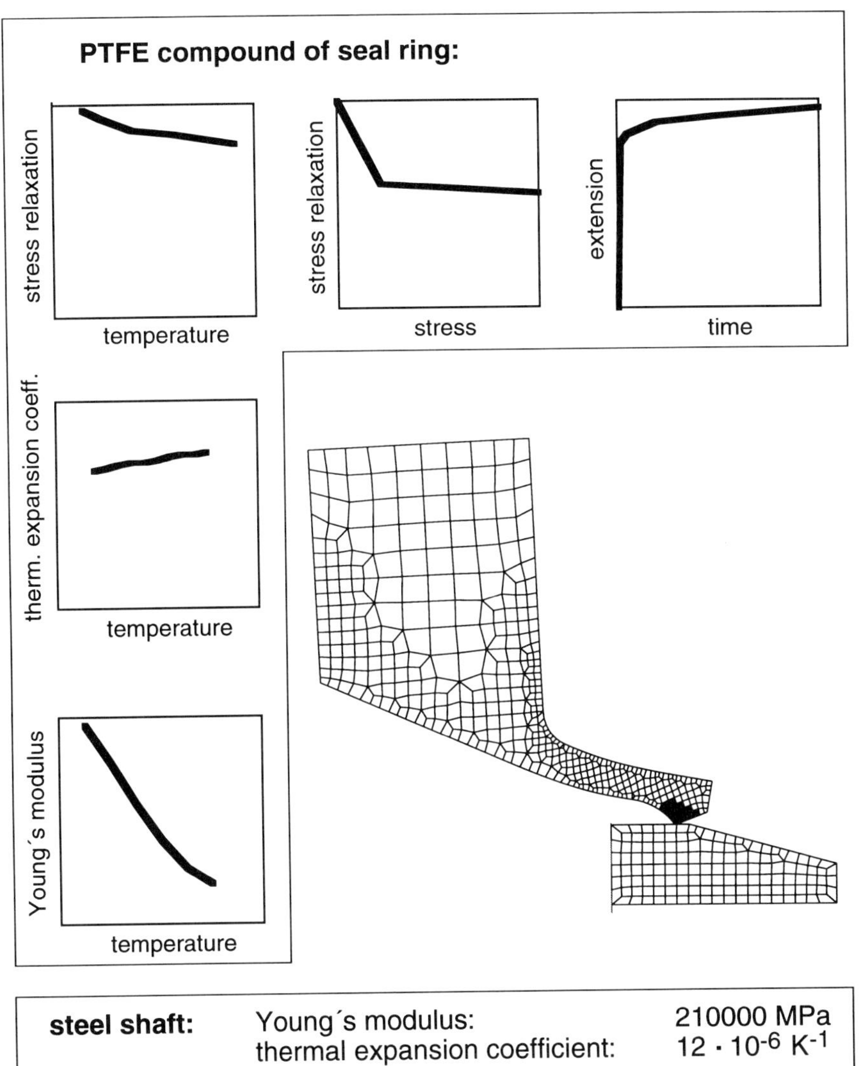

Figure 5:   FE-mesh and material properties of a Balanced Shaft Seal

Figure 6 in principle shows a heated "radiameter" developed for measuring the radial force of shaft seals in the temperature range up to 200°C. It consists of two half-shafts (1), (2), bridged by a pre-loaded strain gauge force transducer (3). When the seal is mounted on the shaft the signal of the transducer (3) changes as a result of the radial force. A second force transducer (4) and the adjusting screw (5) allow to readjust the initial distance of the both half-shafts to compensate the initial change caused by the radial force. After readjustment of the distance the transducer (3) signals the real value of radial force.

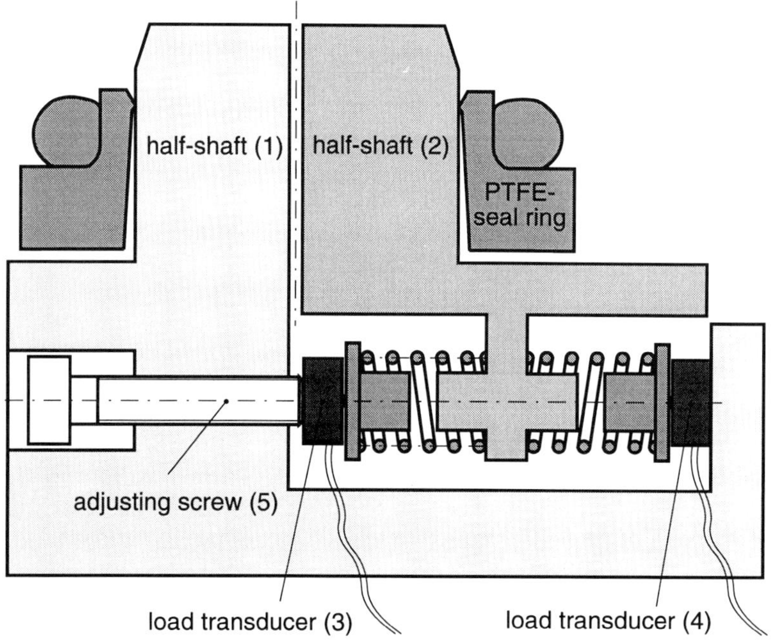

Figure 6:   Heated radial force measuring device

Figure 7 shows in normalized form the measured radial force of a *Balanced Shaft Seal*, type EWDR9, and also the calculated FE-results, the plastic-memory-effect being unconsidered. The values shown refer to ambient temperature. The slope of calculated and measured values is similar. The difference of calculated and measured values is predominantly a result of the creep-behaviour missing in the FE-analysis. The measurement makes distinctly visible the complete loss of radial force at 100°C.

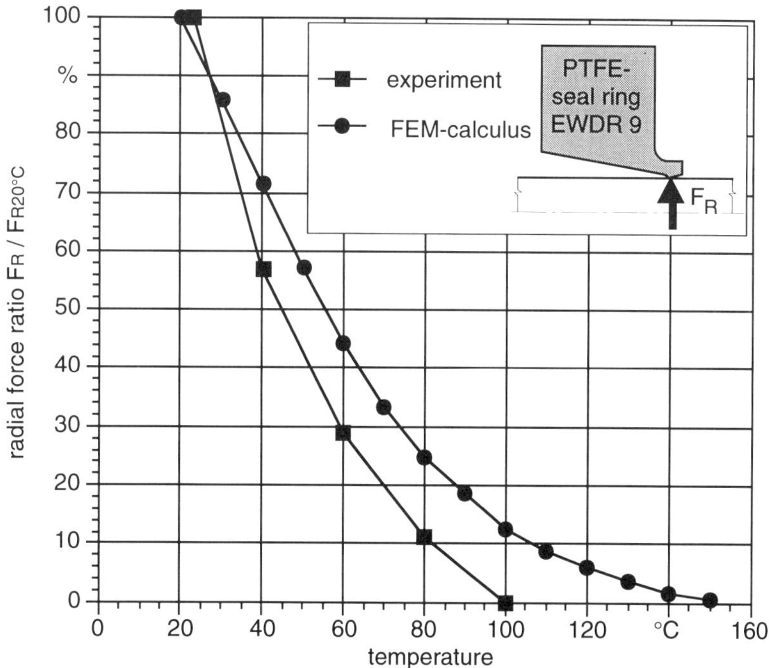

Figure 7:   Radial force versus temperature of a Balanced Shaft Seal

Plastic-Memory-Effect in Balanced Shaft Seal

The idea of implementing the Plastic-memory-effect to counteract the loss of radial force created a new geometry of the *Balanced Shaft Seal* the sealing lip of which allows more plastic deformation. Figure 8 shows the new seal "EWDR-PM2" together with the results of the calculated and measured radial force. Except the geometry all parameters in the calculation are the same as with the type "EWDR9", shown in Figure 7. According to the

FE-analysis which did not account for the Plastic-memory-effect the radial force should completely vanish at 160°C. This result shows that the increased interference of the type "EWDR-PM2" seal is not sufficient to avoid thermal leakage. The measured values, however, show definitely a higher radial force than it has been calculated by neglecting the Plastic-memory-effect. The conclusion therefore, is that the difference of the measured and calculated radial force - as shown by a separate curve in Figure 7 - predominantly represents the positive influence of the Plastic-memory-effect. Once more reliable quantitative values of the creep-behaviour will be available the Plastic-memory-effect will appear to be even stronger than it seems from the difference-curve of Figure 7 The experiments have verified that the implementation of the Plastic-memory-effect is an efficient method to avoid any thermal leakage of a *Balanced Shaft Seal*. It is within the scope of further investigations to establish isolated values of the intensity of the plastic-memory-effect. The goal is to finally establish a user-subroutine for the FE-program accounting for the Plastic-memory-effect.

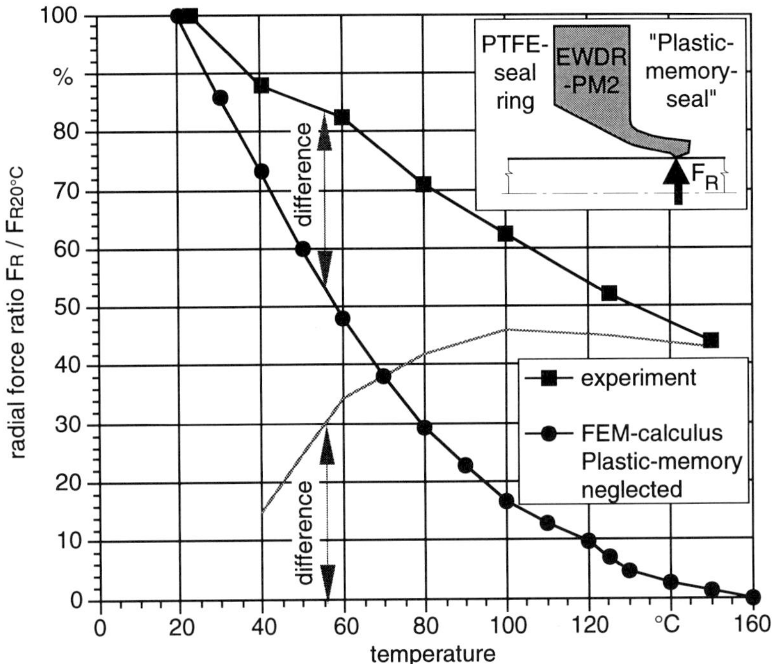

Figure 8:   Radial force versus temperature of a Plastic-memory Balanced Shaft Seal

Conclusions

Finite-Element-Analysis is a powerful tool for the optimization of PTFE-seals. The key for a successful application of the tool, however, is a detailed knowledge of the relevant material properties. To establish these properties for a particular PTFE-compound the authors carried out experiments with a specimen in form of a thin-walled PTFE-tube and thereby measured the relevant properties within the pressure and temperature range of real seal operation. The material properties obtained were then applied in a FE-calculation of the specimen. Owing to the accordance of the FE-calculated and measured values the calculation of more complicated seals became possible.

Earlier geometries of the *Balanced Shaft Seal* exhibited thermal leakage which is now eliminated by implementation of the Plastic-memory-effect. The magnitude of the Plastic-memory-effect was shown by comparison of the results of FE-calulation and experiment.

References

[1]   User information MARC Vol. A-C, Marc Analysis Research Corporation, Revision K.5, 1992

[2]   Müller, H. K.: Dichtung. Patent appl. DE 3616689 C1

[3]   Wäschle, P.; Müller, H. K.: PTFE-Shaft Seal for higher pressures. 13th International Conference on Fluid Sealing, (BHRG), Brugge, Belgium, 1992

[4]   Müller, H. K.: Abdichtung bewegter Maschinenteile. Medienverlag Müller, 1990

[5]   Wüstenhagen, G.; Müller, H. K.: Plastic-Memory-Effekt contra Wärmedehnung bei Dichtungen. Antriebstechnik 32 (1993), Nr. 6, S. 61-63

[6]   Werkstoff Polytetrafluoräthylen (PTFE). Merkel GmbH, Hamburg, 1989

MECHANICAL SEALS I - practical

14th International Conference on Fluid Sealing, Firenze, Italy,
6-8 April 1994. Organised by BHR Group Limited, Cranfield,
Bedford, MK43 0AJ, UK; Tel: 0234 750422

TRENDS IN MECHANICAL SEAL PERFORMANCE AT THREE PROCESS PLANTS IN THE OIL INDUSTRY

by

P.A.Conner, Conoco

M.T.Thew, Mech Eng Dept, University of Southampton, UK

ABSTRACT

The paper discusses the monitoring of mechanical seal performance
on approximately 4200 centrifugal pumps, in three process plants
within the oil industry. There is information on approximately
6200 seal lives. Preliminary information was discussed in a paper
at the 13th ICFS, which also surveyed previous relevant
literature.

A Weibull analysis covering 7 types of liquid with a total seal
population of 930, suggests that a generalised failure
distribution is followed. This, if a valid concept, allows a
statistical predictive estimate for such fluids.

Quantitative links are made between various causes for failure and
seal life, and also between causes of failure and seal face
material combinations.

The restricted range of liquids for which the populations are
statistically adequate illustrate how the Weibull index changes
with failure mode.

The data sets range provides the period between 1967 and 1993, and
provide broad evidence that a significant minority of seals are
still failing with expensively short lives. However the latest
data indicates that efforts to increase seal life show a clear
improvement for installations that used to produce a reasonable
life and now exhibit lives of several years.

NOTATION LIST

To	Origin of a failure mode on a Weibull distribution.
h	Characteristic life for a Weibull distribution (the life when 63% of the population of a single failure mode has failed).
t	Life at which failure occurs.
B	Weibull index.
MTBF	Mean time between failures.
N	Population size.
Ho	Characteristic life of a Weibull distribution (the life when 63% of the total population has failed - including all failure modes).

INTRODUCTION

Plant failure data (1) suggests that perhaps 60-65% of all maintenance costs on centrifugal pumps are attributable to mechanical seal failure. Increasing pressures to reduce maintenance operating costs, especially in the oil industry where traditionally the cost of maintenance has been less significant, have resulted in a great deal of interest in improving mechanical seal performance. Due to the poor reliability of mechanical seals (and interrupted service life of most centrifugal pumps), most pumping duties have a spare pump to avoid a loss of throughput in the event of the primary pump failing.

In terms of sealing efficiency and versatility of application, mechanical seals represent a significant improvement over any of the other shaft sealing techniques. A glance through a seal manufacturer's catalogue would give the impression that most mechanical seals should perform without problems, assuming the correct selection. Unfortunately the reality is very different. Mechanical seals remain a major reliability problem. Admittedly the mean time between failures has been significantly increased through improved seal face materials and improved seal designs. However mechanical seals remain a major problem at most (if not all) process plants. As far as plant operators are concerned, mechanical seals will remain a reliability problem as long as the majority of seals cannot survive, consistently, the periods between major plant shutdowns (often 3-6 years).

There is very little published data on mechanical seal performance in the field. The BHRA carried out a collaborative research project in the 1980's (2), but little has been published since then. This paper presents details and analysis of seal life from 3000 centrifugal pumps, between 1967 and 1993, from three different refineries/ petrochemical plants. The aim of this paper is to highlight and publish real data on seal performance, and provide a yardstick by which improvements in seal performance can be traced. All seal lives relate to installed time, not running time.

Description of the Data Sources

Plant A

Plant A is a large petrochemical plant and oil refinery. Some of the plant is over 30 years old, but considerable investment during the 1980's has resulted in a large amount of new equipment. The site operates 2600 centrifugal pumps fitted with mechanical seals. Mechanical seal failures have been recorded on the plant's computerised maintenance record system since 01/01/88. The database contains details of the date and cause of each seal failure. The quality of the seal data is high because failed seals are analysed and recorded by Engineers, who are part of a team dedicated to rotating equipment reliability.

This data set contains the failure of 2457 mechanical seals, and
records the trouble-free operation of a further 1300 mechanical
seals. The data covers the period 01/01/88 to 01/04/91.

<u>Plant B</u>

Plant B is an oil refinery about one third the crude throughput of
Plant A. Most of the refinery is at least 30 years old. This is
broadly reflected in the mechanical seal designs and seal face
materials. Plant A and Plant B are operated by the same Company.
Plant B operates about 700 centrifugal pumps fitted with
mechanical seals. Mechanical seal failures were recorded manually
from 1967 to 1983. A computerised maintenance record system (same
as Plant A) was introduced in 1984.

This data set contains 1364 mechanical seal failures, and covers
the period from 1967 to 01/01/88.

<u>Plant C</u>

Plant C is a modern oil refinery, built in the late 1960's, and
has undergone continuous expansion ever since. Crude throughput at
Plant C is about half that of Plant A. The majority of mechanical
seals on the site now use faces of carbon against tungsten or
silicon carbide. Two hard faces are used where conditions require
their use. The refinery operates 900 centrifugal pumps fitted with
mechanical seals. Mechanical seal failures have been recorded
manually since 01/01/91 as part of a rotating equipment
performance improvement programme. The majority of process pump
mechanical seals (> 95%) are supplied by a single manufacturer.
Plant C has a close working relationship with this seal
manufacturer, to investigate and solve mechanical seal problems.

The data set records the failure of 700 mechanical seals, and
trouble-free operation of seals in a further 400 pumps. The data
covers the period 01/01/91 to 31/7/93.

Weibull Analysis

Weibull analysis utilises a flexible empirical function to
graphically plot trends enabling rapid interpretation of data. The
basis and derivation of the Weibull distribution function is well
covered in the literature (3). Discontinuities in a Weibull
distribution indicate a change in failure mode. This is perhaps
the most important feature of a Weibull analysis. The Weibull
index (B) and distribution origin (To), provide a good indication
of the failure mechanism (Table 1).

$$R(t) = \exp \left[-((t - T_o)/h)^B \right]$$

Weibull analysis has been frequently used to analyse mechanical
seal data and is probably the most popular statistical analysis

tool used upto now. Gu and Wang (4) have made the most extensive use of Weibull analysis in relation to mechanical seals. Lines shown by Gu and Wang are short compared to the data presented in this paper.

In almost all of the literature on mechanical seals, a best-fit line is used to describe the Weibull distributions. This practice may have been used because of insufficient data points. However a real failure distribution will usually contain several failure mechanisms which dominate the failure of components at different lengths of life. When real data is displayed as a Weibull distribution the data will not plot as a straight line. Discontinuities in the gradient of the Weibull distribution mark the change between two failure mechanisms. As a rough guide a minimum of 50 data points is necessary to produce a Weibull distribution which can indicate changes in failure mechanism.

SEAL LIFE v SEALED FLUID

Weibull analysis has been used to compare failure distributions of mechanical seals operating on a wide range of fluids at Plant B. Typical fluid properties are shown in Table 2. The aim was to see whether there are any fundamental differences in the failure distributions of mechanical seals operating on hydrocarbon, water, and chemical duties. Significant differences would indicate that the factors affecting seal life and reliability are closely associated with the properties of the sealed fluid. The barrier fluid is the fluid at the seal faces in double seals, and the outer seal of a liquid tandem arrangement. Barrier fluids are chosen to be non-hazardous, compatible with the sealed fluid, and provide the best possible conditions at the seal faces. LPG, HF acid, and high temperature hydrocarbon pumps are generally fitted with (liquid barrier fluid) tandem or double seals.

Sealed Fluid Properties

Hydrocarbons

Liquefied petroleum gas (LPG) is made up of propane and butane. LPG is a gas at atmospheric pressure, but is usually pumped under pressure in a liquid form. Double or tandem seals are normally used due to the hazardous nature of LPG (highly inflammable). Plant B uses double and tandem arrangements with a barrier fluid. At Plant C most seals on light hydrocarbon duties use tandem arrangements with a dry running back-up seal. These have proved extremely reliable.

Gasoline contains a blend of hydrocarbons. Although gasoline has a broad boiling range, it is generally regarded as a non-arduous sealing duty. Although gasoline is highly inflammable, it cannot form an explosive gas cloud like an LPG leak, so simple single seal arrangements are usually sufficient.

The loss of vapour from seals is increasingly being regarded as environmentally detrimental, even if there is no immediate safety hazard. Dry running back-up seals are being used in increasing numbers to reduce fugitive emissions from seals, especially on light hydrocarbon duties.

Naphthas are the major constituent of gasoline, and generally need processing (Reforming process) to make suitable quality gasoline. The performance of seals on naphtha and gasoline duties would be expected to be very similar.

The lighter hydrocarbons (eg LPG, gasoline, and naphtha) have fairly poor lubricating properties. Operating experience indicates that the most common failure mechanisms are chipping, vaporisation, and excessive wear. These mechanisms are all associated with poor fluid lubrication, and insufficient cooling.

The term "heavy hydrocarbons" has been used to describe all the vacuum distillation products. These include gas oil and residue. Vacuum distillation takes place at high temperature (390-50 deg.C) and low pressure (0.03 - 0.05 bar abs.). These conditions create a very arduous environment for mechanical seals to operate in. Operating experience at all three plants suggests that the most common failure mechanisms are hang-up and coking. Double and tandem seals with a relatively cool barrier fluid supply are often used to reduce the seal face operating temperature.

<u>Water</u>

Water is a poor lubricant, but also has a very narrow boiling range. Water sealing duties are not generally considered arduous. Simple single seals are normally specified for this type of duty. Water is used in oil refineries both as a process fluid, and to generate steam in boilers. Sour water is the name given to water containing dissolved hydrogen sulphide and ammonia. In a refinery, the pH of the sour water system can vary from 9.5 to slightly acidic. Sweet water is used as boiler feed water and cooling water; it does not contain dissolved hydrogen sulphide or ammonia.

<u>Hydrofluoric (HF) Acid</u>

HF acid is an extremely hazardous fluid. The slightest contact will cause extremely severe burns. Full protective suits, complete with hoods and gauntlets are required by personnel working on these units. HF is used as a catalyst in the alkylation process. Double or tandem seals with a liquid barrier fluid are normally used. Plant C has shown excellent reliability using double seals with a clean barrier fluid circuit. The process normally takes place at ambient temperature and moderate pressures.

Linking Weibull Index to Failure Modes (Table 3)

<u>Short Life Failure Modes</u>

The first mode on all the duties has a weibull index greater than
1.0, which indicates an increasing failure rate. This contrasts
with the conventional idea of a high initial failure rate, which
decreases with time as infant mortality failures are overcome (ie
the "bathtub" curve). The first failure mode accounts for seal
lives upto about 50 days, for between 4% and 20% of seal failures
on the different duties.

Data from Plant A indicates that poor shaft-seal alignment,
failure of auxiliary seal systems, and hang-up are the most
significant failure mechanisms. The Weibull index indicates a
wear-out characteristic. This is consistent with the failure
mechanisms described. The severity of misalignment will affect the
time to cause fatigue failure in metal bellows, excessive wear of
elastomeric secondary seals, and fatigue of springs. Failure of
auxiliary systems occurs most frequently on start-up, as a result
of a blockage (debris left in the pipework, or hardened product)
or incorrect pump operating procedures. Hang-up occurs as a result
of crystallisation or coking deposits, which build-up and prevent
free movement of the floating seal face. There is a very distinct
change of slope on all the weibull distributions after this first
failure mode.

<u>Mid-life Failure Modes</u>

The weibull distributions indicate a clear change in failure
characteristic, starting at seal lives of 20 to 50 days. This
signifies the end of the first failure mode, and onset of the
mid-life failure modes. The mid-life failure modes have a Weibull
index less than 1.0 (indicating a decreasing failure rate with
time). The majority of seals fail into the mid-life range, which
extends to seal lives of 1600 days or more. The Weibull
distribution characteristics support the idea that mid-life
failures are related to face break-up due to thermal/mechanical
fatigue (Table 1).

The majority of seals (80-96%) on the hydrocarbon duties fall into
a single mid-life failure mode, with a Weibull index of between
0.72 and 0.84. Most seals (90%) on the sweet water duty fall into
two distinct failure modes in the mid-life range, with a Weibull
index of 0.88 and 0.62. There is very little difference between
the failure distribution characteristics of sweet water and the
hydrocarbons. The distribution for the sour water duty is slightly
more complex, with three mid-life failure modes with Weibull index
ranging from 0.63 to 1.11, but is basically quite similar to the
hydrocarbon distribution. The HF acid distribution has two
mid-life failure modes, with a Weibull index of 1.17 and 0.89.
These indicate an almost constant failure rate after initial
start-up. The general characteristics of the distribution for
sealing HF acid is very similar to the distributions for sealing

water and hydrocarbons.

<u>Long Life Failure Modes</u>

If mechanical seals conform to the classic "bathtub" failure
distribution then we would expect to find a long-life failure mode
corresponding to wear-out. It is well known that only a very small
percentage of seals actually wear-out in service.

A long life wear-out failure mode is evident only for the LPG
distribution (Fig.1). This mode has a weibull index of 1.29,
starts at around 1600 days, and accounts for about 10% of seals on
LPG duties. The other distributions show vague signs of long-life
failure modes corresponding to wear-out, starting at between 5000
and 8000 days, but accounting for less than 2% of seals in each
distribution.

A Generalised Failure Distribution

The Weibull analyses for data at Plant B indicate that mechanical
seals on a wide range of fluids have very similar failure
distribution characteristics (Table 3). There are two or more
distinct failure modes for each distribution. The Weibull index of
each of these failure modes is similar for each of the fluids.
Discontinuities between the failure modes occur at a similar
percentage of the population survival. Experience has shown that
certain fluids suffer greater mechanical seal problems (5). This
is reflected by considerable variations in the characteristic life
of the distributions.

All the seal lives were plotted onto a single Weibull distribution
using a dimensionless age parameter, to determine if a common
distribution curve could be obtained.

$$\text{Dimensionless Age} = t \ / \ h$$

A Weibull plot is produced (Fig.2) with a narrow scatter of
points. This distribution provides a simple method of predicting
seal life against probability of seal survival, for seals
operating over a wide range of fluids, eg water and hydrocarbons
(light and heavy). It is then possible to use this single graph as
a design tool for plant risk assessments where mechanical seals
are used. No such tool exists at present. The authors have not yet
been able to confirm whether this distribution is specific to
Plant A, or can be applied more generally.

The distribution confirms that mechanical seal life is dependent
on the fluid being sealed, but also indicates that failure
mechanisms are common to mechanical seals operating on a wide
range of fluids.

SEAL LIFE v CAUSE OF FAILURE

Plant A provided record sheets for 98 mechanical seal failures, with sufficient information to establish the probable cause of each failure. Twelve basic causes of failure were established (Table 4). Despite the limited amount of data, trends do emerge to link cause of failure with seal life. The information is displayed graphically in Fig.3.

Misalignment and embedded metal particles are the most significant causes of short seal life (less than 50 days). The link with high carbon wear is clearly shown.

Hang-up due to coking and crystallisation are a very significant cause of failure in the mid-life range (50 to 300 days). Dry running was by far the greatest single cause of seal failure at Plant A (20% of all the failures). The majority (70%) of these failures occurred in the mid-life range. The data suggests that dry running failures on start-up are not very common, but usually the result of prolonged exposure to dry running conditions rather than isolated incidents (eg plant upset).

Long life seals are most likely to fail due to an external (ie non-seal) component/system failure, or secondary seal failure (eg embrittlement of elastomers through extended exposure to high temperatures).

SEAL LIFE v SEAL FACE MATERIALS

The 98 record sheets from Plant A provided details of the seal face materials, so it has been possible to look at a link with seal life. Weibull distributions have been plotted for six different seal face material combinations. Due to the limited number of data points only the general slope has been calculated.

The stellite and Niresist v carbon indicate a Weibull index of 0.87 and 0.92; agreeing closely with the parameters established for Plant B, where the majority of seals use these 'older' material combinations.

The Weibull distributions for the other material combinations are significantly different (Table.5), suggesting that there is clear evidence in the field that seal face materials can fundamentally change the distribution of seal failures.

The combination of SiC/SiC appears much less sensitive to failure mechanisms on start-up, and exhibits an almost random failure distribution. The Al/C distribution confirms operating experience of high sensitivity to start-up failure mechanisms (especially dry-running, and thermal shock).

The relationships between cause of failure and seal face materials are represented in Fig.4.

AN UPDATE ON MECHANICAL SEAL LIFE IN PROCESS PLANTS

Our previous paper (5) presented a discussion on the seal life
distributions at Plant A and Plant B. Data from another similar
process plant, Plant C, adds to the total published data on actual
seal performance in Industry. The data from Plant C is especially
important in that it establishes current mechanical seal
performance in what is considered to be a modern, and highly
successful Oil Refinery.

Mechanical seal performance at the Refinery was the subject of a
study in December 1991, carried out by a consultant working for
the main seal vendor. The study was based upon less than 12 months
of actual plant data, and concluded that the overall MTBF was 3.47
years, placing it in the top half of refinery performance in the
U.K. Leaders in the field claim a MTBF in excess of 6 years. It
would be interesting to see evidence of this published.

During August 1993 the database at Plant C has been analysed, to
provide a better measure of seal performance at Plant C, compare
with the 'competition', identify where resources need to be spent
most urgently, and assess whether the performance improvement
programme has had any impact on seal life.

Overall Seal Life Distribution at Plant C (Fig.5)

Seal failures on or shortly after start-up (less than 1 month)
account for 7% of the total number of failures. This is a
significantly smaller percentage than for Plant A or Plant B
(Fig.7). Experience indicates that most of these failures are
due to incorrect assembly, installation, or pump start-up
procedure.

Hang up due to coking is perceived as the most common single cause
of seal failure. About 20% of the seals failed within 12 months.
This is dramatically better than Plant A and Plant B where 50% had
failed after only 7 months (500 hours).

Seals failing after 12 months would appear to be due to hang-up,
secondary seal degradation, or an external random cause (eg
bearing failure, plant upset). Almost 64% of the seals had not
failed in the period between January 1991 and August 1993
(32 months=23,360 hours).

It is evident that mechanical seals at Plant C perform
considerably better than at Plant A, Plant B, and the plants
studied by the BHRA(2). However the shape of the distribution
curves are common to all, and appear remarkably similar.

The overall MTBF (Fig.6) has improved significantly over the
period 1991-1993, which suggests that the plant improvement
programme has had some impact. The refinery's current approach is
to analyse each seal failure and engineer out recurrent problems

228

in conjunction with the seal vendor. The MTBF of just under 4
years indicates a major improvement in mechanical seal performance
in recent years, and vindication of the refinery's approach to
improving seal reliability.

Seal Life v Seal Type

The majority (80%) of seals at Plant C are single seals, in line
with most other process plants. The other 20% are split equally
between double and tandem seal arrangements. Tandem seals are used
only on light hydrocarbon duties, and many of these are using a
dry-running tandem arrangement. Double seals are used on all of
the critical pumps on heavy hydrocarbon duties (high temperature),
and all of the hazardous (containing HF acid) pumping duties on
the alkylation plant. Plant C has installed ring mains for
supplying barrier fluid to the double seals.

The MTBF by seal type (Fig.8, Table 6) is very interesting. The
tandem seals are over twice as reliable as single seals, and
almost 3 times as reliable as the double seal arrangements. This
indicates clearly that initial seal cost cannot be used as an
indicator of potential seal reliability. The performance of tandem
seals is excellent, and provides a good indication that dry
running arrangements do exhibit good reliability on light
hydrocarbon duties. The performance of single seals is approaching
a level where it is not the most significant reliability problem.

At Plant C duties using double seals represent the area of most
concern. In fact no seals had failed on the alkylation plant in
the period January 1991 to August 1993. The 'problem seals' are
located on two heavy hydrocarbon processing units within the Black
Oils Division. Over the past two years considerable effort has
been put into engineering new seals, refining existing designs,
and training technicians in fitting and installing these seals; in
close collaboration with the main seal manufacturer. Specific
efforts have been made to simplify seal installation and design
out hang-up due to coking. The effect of this is clearly
demonstrated by the near 30% improvement in seal MTBF on the Black
Oils Division (Fig.9) since 1991.

CONCLUDING COMMENTS

(1) Weibull analysis has been used to identify different seal failure
     mode characteristics, in a wide range of fluids. The analysis
     indicates that there are distinct transition characteristics
     between infant mortality (typically life less than 50 days),
     premature failure, and wear-out failure.

(2) A Weibull plot of 930 mechanical seals on a wide range of fluids,
     using a dimensionless age parameter (ratio of actual life to
     characteristic life), produces a distribution with very little
     scatter. This suggests that predominant failure mechanisms are

common to a wide range of sealed fluids.

(3)   The dimensionless Weibull plot could form the basis of a
        statistical predictor of mechanical seal performance, as it is
        applicable to seals over a wide range of fluids. However a note of
        caution; the majority of seals at Plant B used Stellite/C or
        Niresist/C face materials. The weibull plots at Plant A indicate
        significant differences in the weibull index for other material
        combinations.

        It would be valuable for other authors with access to large
        databases to publish their data for the common good.

(4)   New data presented from Plant C indicates that seal performance
        has improved very dramatically since the BHRA survey in the early
        1980's (1). Indeed, mechanical seals are no longer the greatest
        source of poor pump reliability on significant areas of the plant.

(5)   Plant C provides a good indication of current seal performance, in
        a modern, well maintained Oil Refinery.

REFERENCES

1.   NAU, B.S., FLITNEY, R.K.
       Reliability of mechanical seals in centrifugal process pumps.
       Proc. 11th ICFS, A2, P17-45, 1987.

2.   NAU, B.S., FLITNEY, R.K.
       Seal survey: Part 1 - rotart mechanical face seals.
       BHRA, Confidential Report CR1386, Dec 1976.

3.   CARTER, A.D.S.
       Mechanical Reliability.
       Pub. Macmillan, 2nd Edition, ISBN 0-333-40586-2, 1986.

4.   GU, Y.Q., WANG, R.M.
       Trouble-shooting of process pump mechanical face seals with
       Weibull distribution statistical analysis.
       Proc 12th ICFS, G2, P329-341, May 1989.

5.   CONNER, P.A., THEW, M.T.
       A statistical review of mechanical seal failures in process plant
       pumps.
       Proc. 13th ICFS, P347-368, April 1992.

6.   CONNER, P.A.
       Mechanical seals in process plant: a critical review of seal life
       and seal failure.
       Thesis (M.Phil), Dept. of Mechanical Engineering,
       University of Southampton, April 1992.

ACKNOWLEDGEMENTS

The authors acknowledge the help and cooperation of personnel at
the three plants, and their permission to use data has been
invaluable in the compilation of this paper.

The Normal Weibull Distribution

		Failure Mode	Causes of Failure
To = 0		Failure mechanisms begin as soon as the component enters service	
	B < 1	Decreasing failure rate. A premature failure mode, typical of "infant mortality".	Low safety margin (eg stress rupture), overloaded seal faces.
	B = 1	Constant failure rate. Random failures. There is no dominant failure mechanism.	External component failures, and random operational mishaps.
	B > 1	Increasing failure rate. Premature wear-out.	Excessive wear rate
To > 0		Components are intrinsically reliable until time To	
	B < 1	Decreasing failure rate after a sudden onset of failures at time To. An acceptable wear-out failure mode if To is beyond the expected design life. There is a sudden sharp drop in reliability at To.	Fatigue (eg metal bellows), thermal distress due to excessive thermal cycling.
	B > 1	Increasing failure rate starting at time To. This is the ideal wear-out failure mode providing To is sufficiently close to the design life. There is a gradual loss of reliability after To.	Erosion, corrosion, build-up of deposits which lead to hang-up.
To < 0		Failures start to occur before the component enters service	
	B < 1	Decreasing failure rate. Premature failure mode.	Handling damage during transit or installation, manufacture error
	B > 1	Increasing failure rate, with some failures before the component enters service.	Limited shelf life, poor storage.

The Modified Weibull Distribution

To = Tc			
	B < 1	Increasing failure rate, until Tc, after which no failures occur. A premature failure mode.	Excessive wear
	B > 1	Decreasing failure rate, reaching zero at Tc. A premature failure mode.	Installation and assembly errors.

<u>Table 1: Interpretation Of Weibull Distribution Parameters (6)</u>

Sealed Fluid	Temperature (Deg.F)	Vap. Press. (psia)	Boiling Temp. @14.7 psia (Deg.F)	Spec. Gravity (60/60 Deg.F)	pH
LPG	100	800 to 5000	-40 to 32	0.300 - 0.356	
Gasoline	100	5 to 190	-40 to 155	0.507 - 0.664	
Naphtha	100	0.06 to 15	86 to 400	0.631 - 0.734	
Heavy Hydrocarbon	100	< 0.06	> 450	> 0.734	
Sweet Water	100	0.95	212	1.00	7
Sour Water	100	0.95	212	1.00	6 to 9.5
HF Acid	100	27	67	0.88	< 1

Table 2: Typical Properties of the Sealed Fluids

Sealed Fluid	Failure Modes																OVERALL	
	First Mode				Second Mode				Third Mode				Fourth Mode				Ho (days)	MTBF (days)
	N	To	B	h	N	To	B	h	N	To	B	h	N	To	B	h		
LPG	11(6)	0	2.61	18	160(84)	20	0.84	327	20(10)	1640	1.29	2300					530	545
Gasoline	18(20)	0	1.28	37	72(80)	50	0.72	453									457	550
Naphtha	7(4)	0	1.28	12	172(96)	15	0.77	506									420	535
Heavy Hydrocarbons	39(18)	0	2	21	177(82)	30	0.75	298									413	500
Sweet Water	7(11)	0	1.06	50	20(32)	80	0.88	145	36(57)	245	0.61	910					325	420
Sour Water	6(12)	0	3.45	24	6(12)	25	0.63	45	16(32)	65	1.11	122	22(44)	160	0.81	864	270	340
HF Acid	12(8)	0	3.27	17	97(69)	25	1.17	176	32(23)	400	0.89	975					248	305

Table 3: Weibull Distribution Parameters for Mechanical Seals at Plant B

A	Hang-up (coking)
B	Hang-up (crystallisation)
C	Shaft and seal face plane misalligned
D	Auxiliary seal system failure (eg quench, cooling, recirc., flush)
E	Incorrect seal spring compression
F	External system or component failure (eg bearings)
G	Dry-running
H	Secondary seal failure
I	Seal component failure (other than the faces or sec. seals)
J	Metal particles embedded in the carbon
K	Highly worn carbon (more than 4mm of wear)
L	Chemical attack

Table 4: Codes For Causes Of Mechanical Seal Failure At Plant A

Seal Face Materials			MTBF (Days)	Weibull Distribution			
				N	To	B	h
Silicon Carbide	v	Silicon Carbide	290	14	0	1.07	350
Silicon Carbide	v	Carbon	91	6	0	0.86	122
Tungsten Carbide	v	Carbon	290	13	0	1.69	75
Stellite	v	Carbon	250	16	0	0.87	180
Alumina	v	Carbon	233	11	0	0.49	170
NiResist	v	Carbon	190	39	0	0.92	120

Table 5: Failure Statistics for Seal Face Material Combinations At Plant A (6)

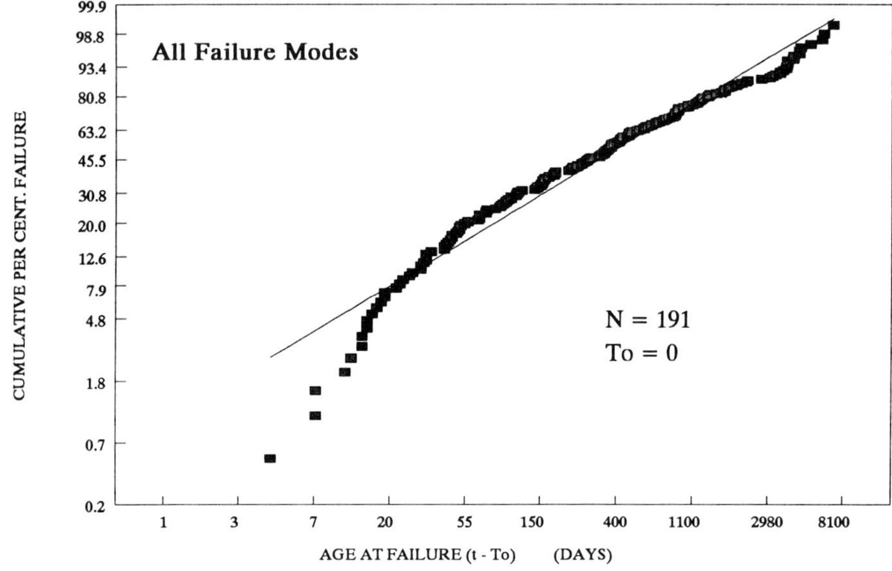

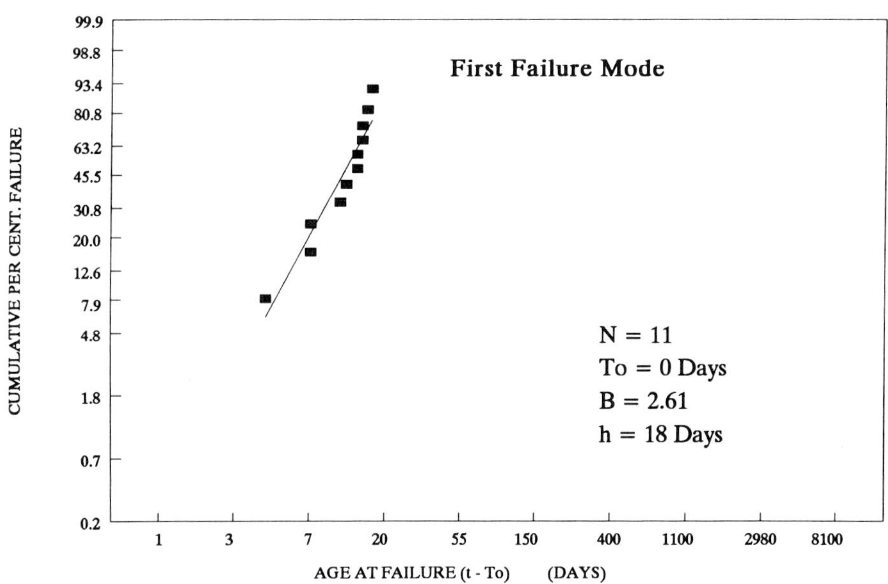

Figure 1: Weibull Plot Of Mechanical Seals On LPG Duties At Plant B (6)

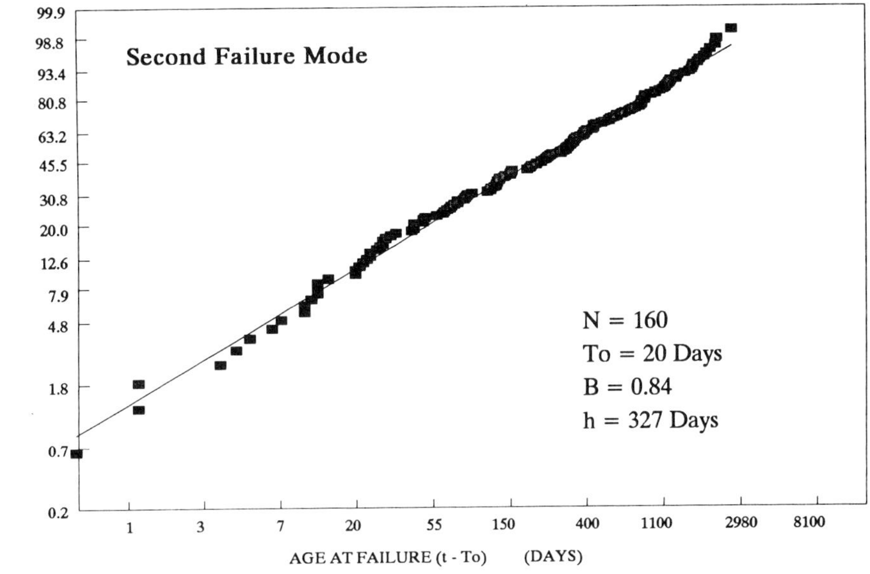

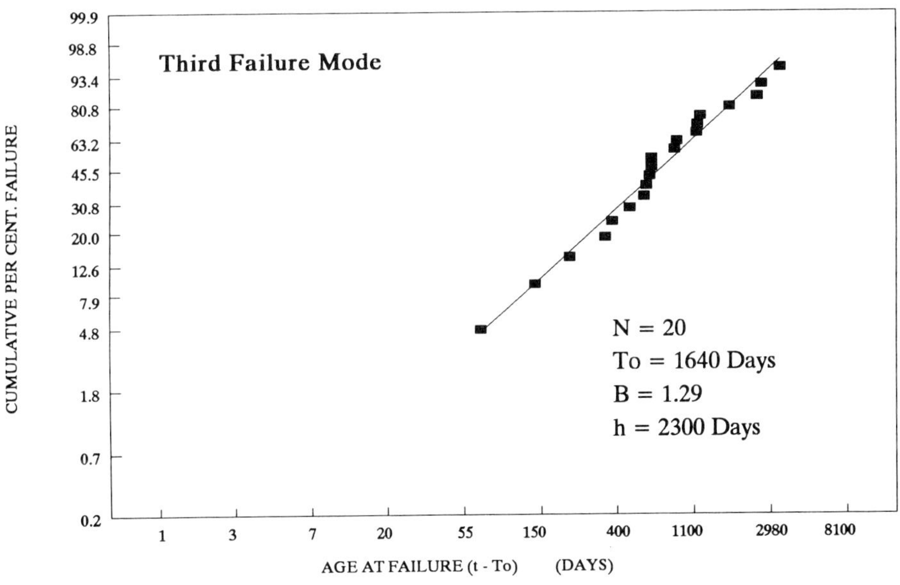

<u>Figure 1 (Cont/): Weibull Plot Of Mechanical Seals On LPG Duties
At Plant B (6)</u>

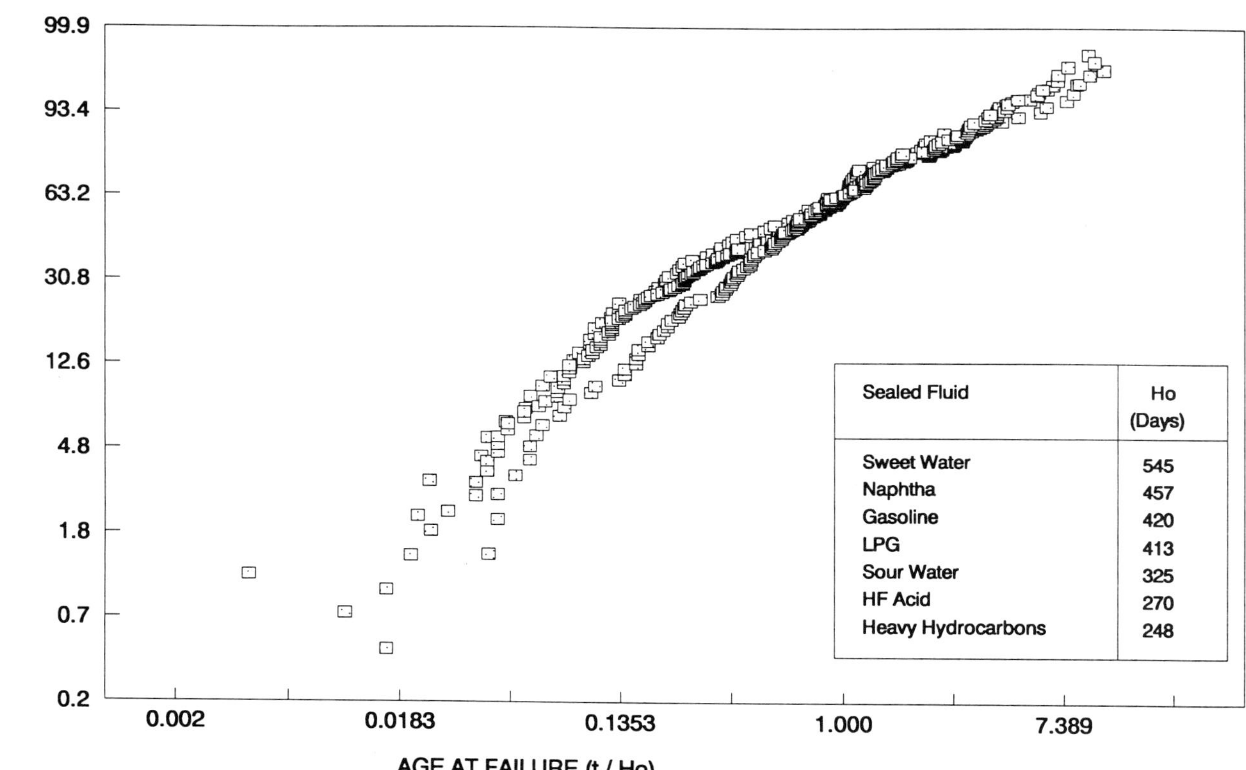

Figure 2: Dimensionless Weibull Plot for Mechanical Seals at Plant A (6)

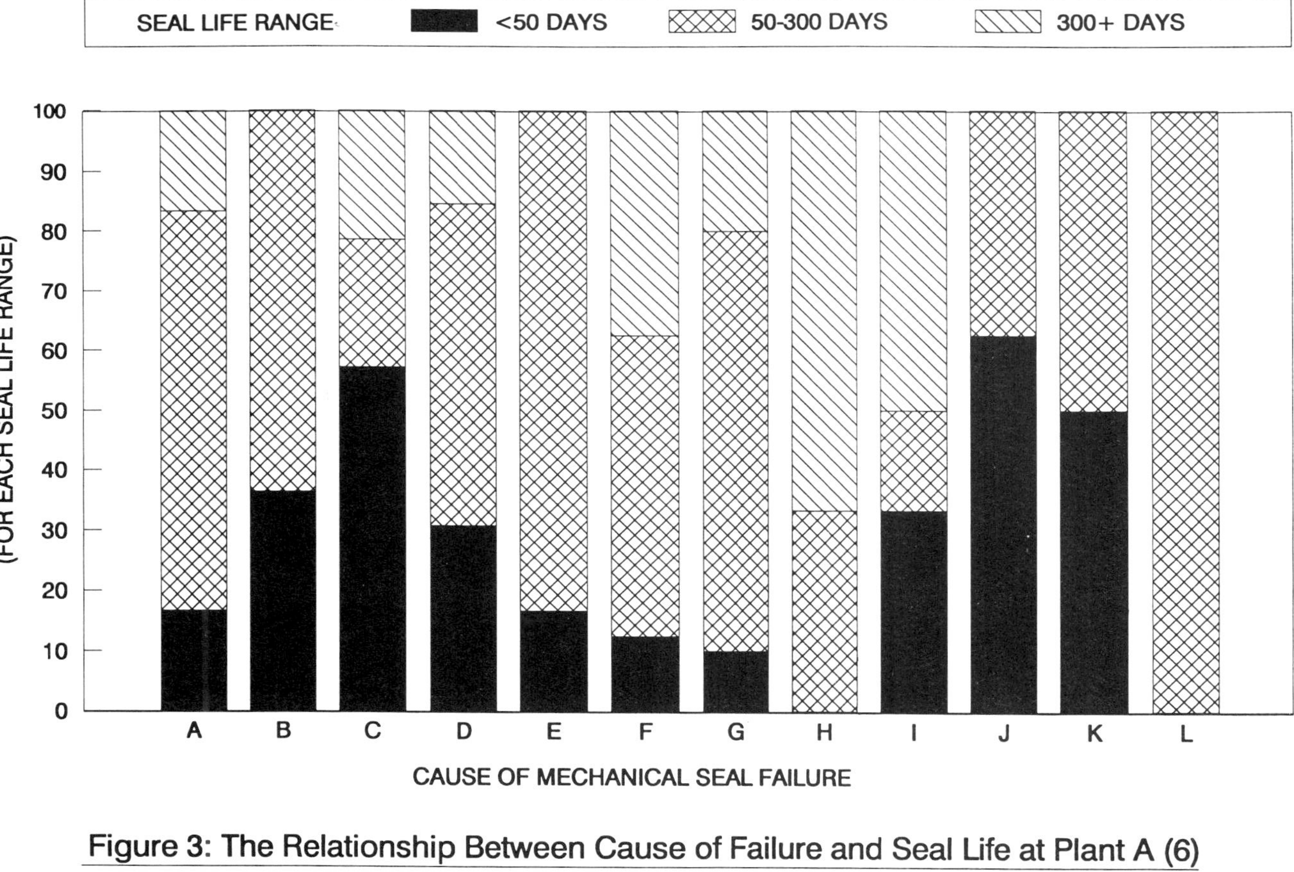

Figure 3: The Relationship Between Cause of Failure and Seal Life at Plant A (6)

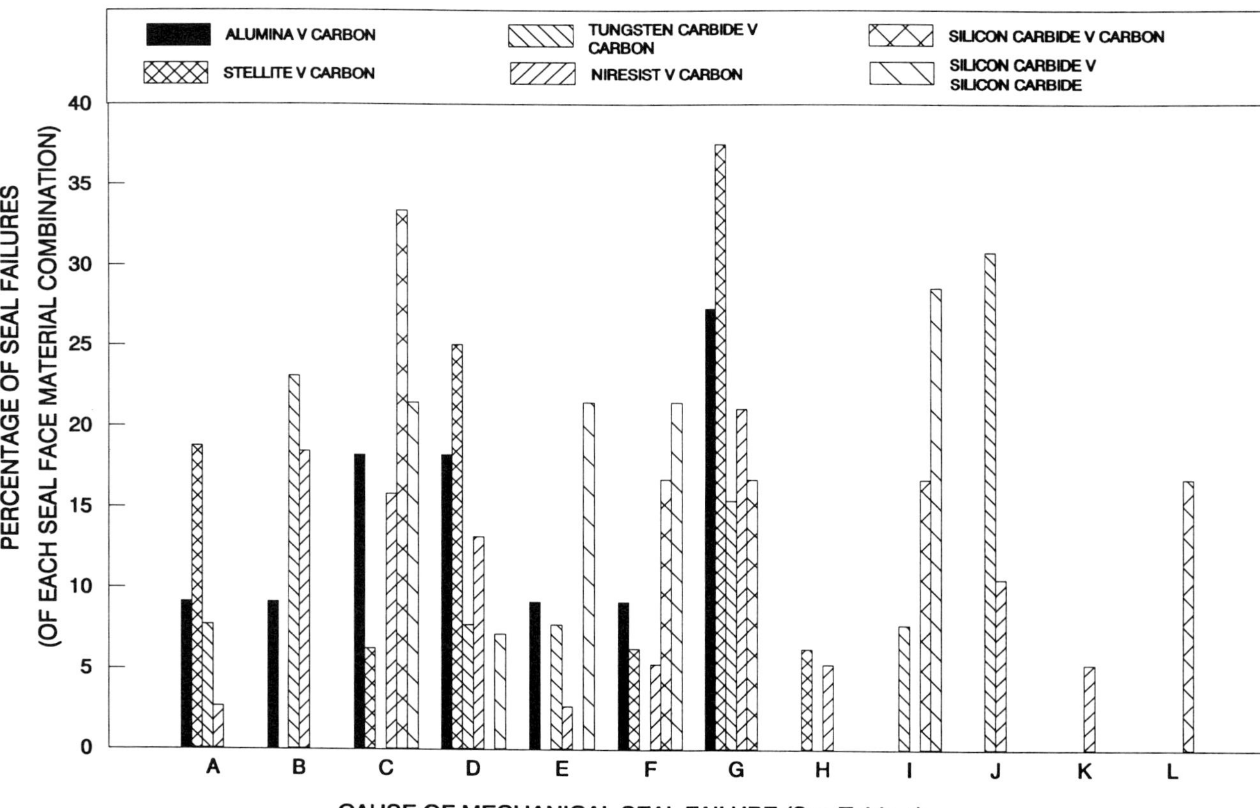

Figure 4: The Relationship Between Cause of Failure and Seal Face Materials at Plant A (6)

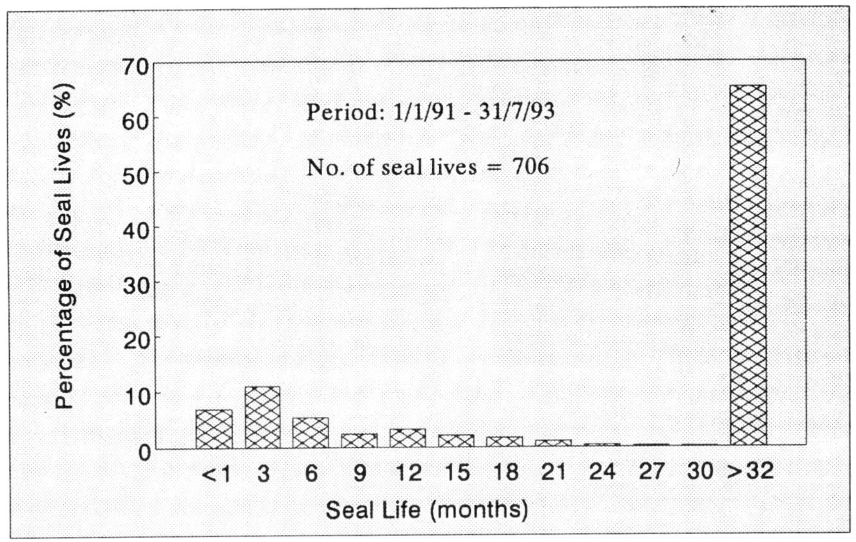

Fig.5: Overall Mechanical Seal Life Distribution at Plant C

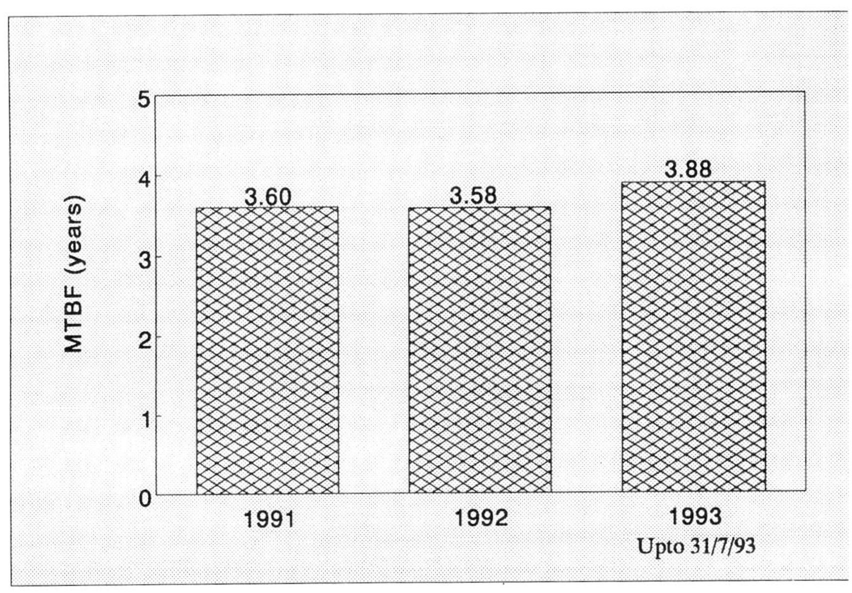

Fig.6: Overall Mechanical Seal MTBF at Plant C

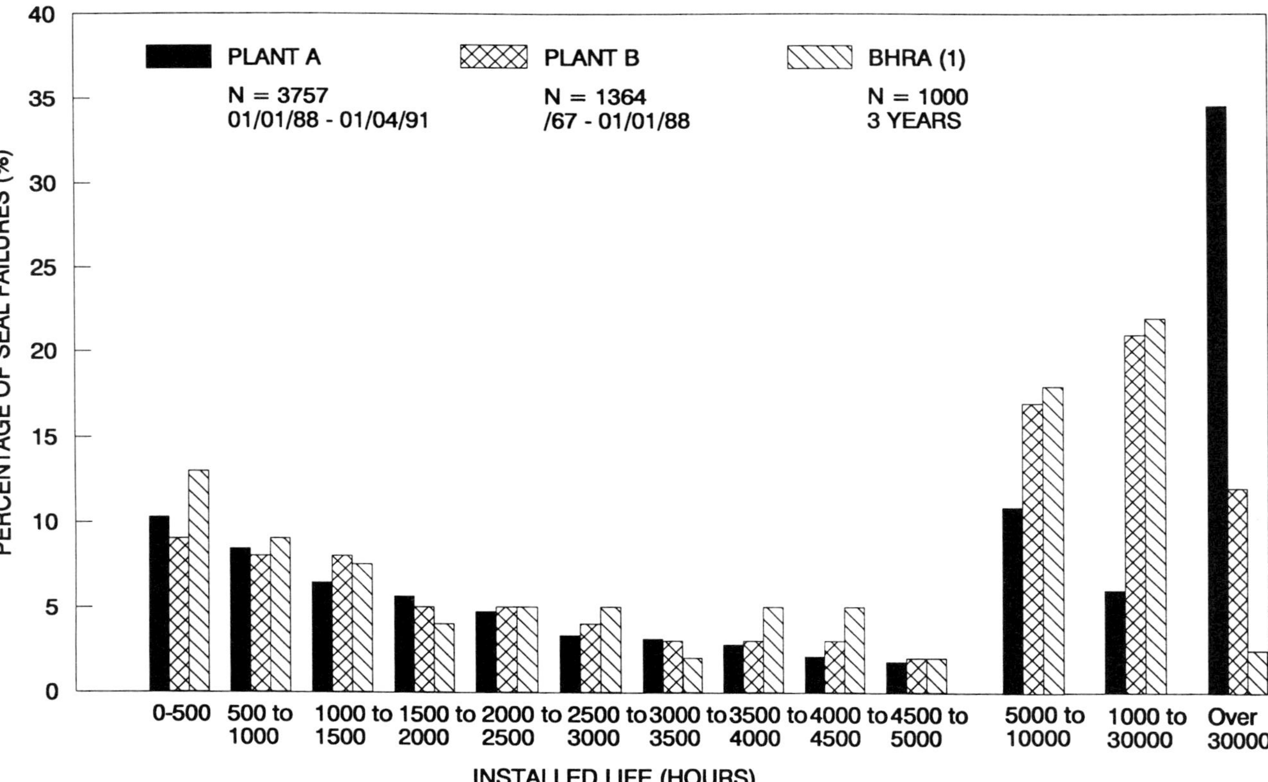

Figure 7: Comparing Mechanical Seal Life Distributions From Three Sources(6)

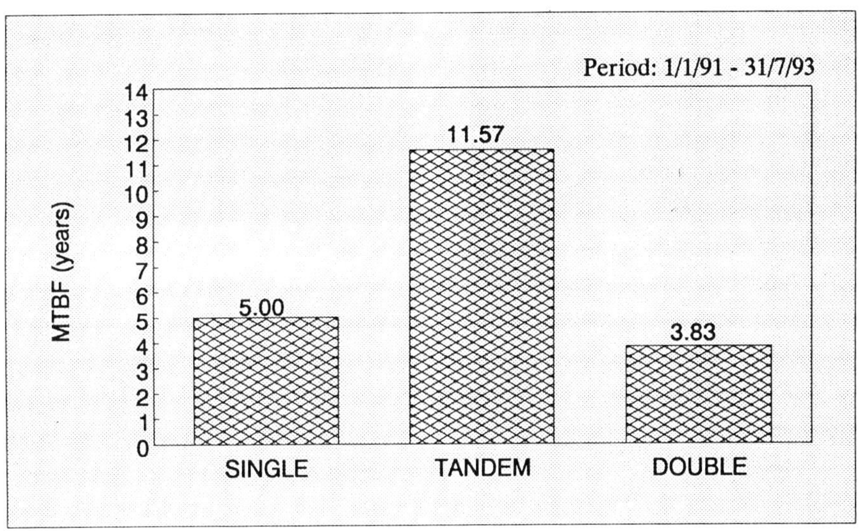

Fig. 8: MTBF By Seal Type at Plant C

Type	Number of Pumps With Seal Type		
	Black Oils	White Oils	OM&S
Single	123	258	184
Tandem	1	65	12
Double	38	38	0
Packed	16	14	48
Other	20	28	81

Table 6: Seal Types at Plant C

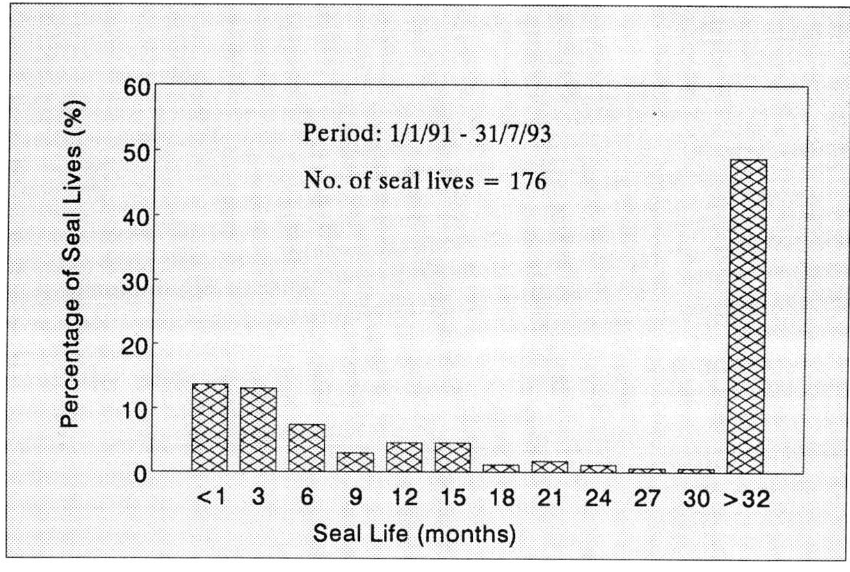

Fig.9: Black Oils Division Seal Life Distribution at Plant C

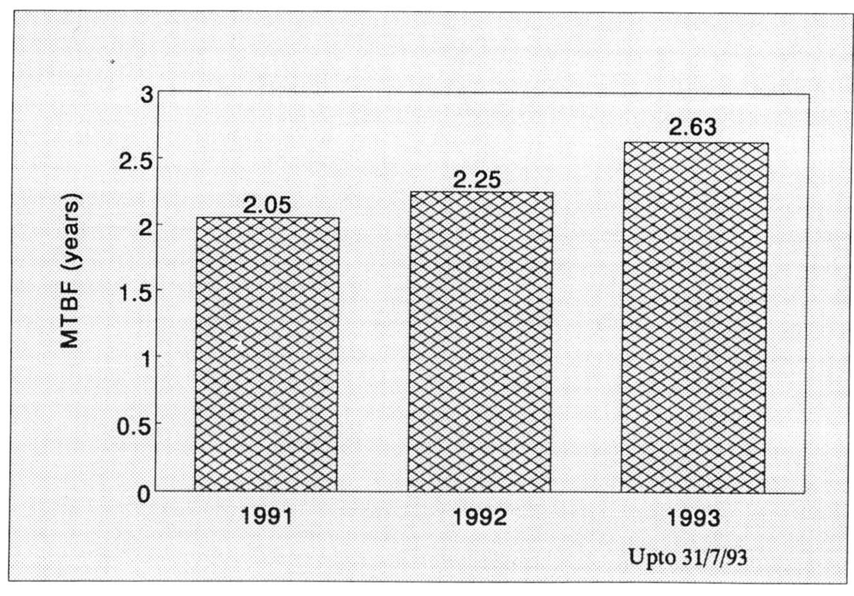

Fig.10: Black Oils Division Seal MTBF at Plant C

14th International Conference on Fluid Sealing, Firenze, Italy,
6-8 April 1994. Organised by BHR Group Limited, Cranfield,
Bedford, MK43 0AJ, UK; Tel: 0234 750422

Seal check systems check
which double seal is leaking

By Ir J.A.M. ten Houte de Lange

Introduction

Double mechanical seals have proven to be a reliable method to obtain negligible seal emission levels in process plant. It can therefore be expected that the environmental pressure will result in a further increase of the double seal population.

In a plant which has a number of double mechanical seals working, it may be considered advantageous to use a central barrier liquid system to provide pressurization and circulation of barrier liquid to each of the double seals.

In a central barrier liquid circulation system the circulation is invariably caused by a circulation pump, which feeds into a number of parallel circulation loops, each loop feeding a double mechanical seal. (Fig. 1)

Although the circulation rate for a double seal may depend upon the size and the duty, there is the problem of distribution of the flow between the different loops. In order to keep the distribution arrangement simple, it is usual to work with a relatively high safety margin on the circulation. On the average circulation rates are of the order of 5 - 10 l/min. per double seal. (Fig. 2)

When using such a central system there may come a time when one of the double seals develops an unacceptable level of barrier liquid leakage. This fact will become apparent by the barrier liquid consumption of the central system.

When this happens one has to determine which of the double seals is leaking. If the leakage is on the atmospheric side seal it can be visible or it can be easily detected by collection of the leakage.

The problem is, however, that there is also a chance that the leakage is on the product side seal and the seal shows no outward sign of leakage. In the central system, the barrier liquid leakage will be noticed by monitoring the barrier liquid level in the reservoir.

In a reservoir of say 1200 * 2000 * 1000 mm high a level difference of 10 mm will correspond with 24 litres barrier liquid. A level drop a day of 10 mm, which is just discernable, will represent a leakage rate of 1 l/h.

A detection level of the order of 0.5 - 1 l/h may therefore be sufficient in most cases. When lower

detection limits are required, for instance with the more expensive barrier liquids, it can be desirable to design the storage vessel accordingly.
For very low detection levels it may be better to consider individual barrier liquid systems.

Leakage monitoring

A rather obvious method to monitor the seal leakage would be to compare the outlet flow with the inlet flow to the loop. This method is, however, impractical because a leakage rate of 1 l/h represents only a fraction of a percent of the circulation flow. This is below the measuring accuracy of most flow meters. *)

It remains therefore necessary to isolate the double seal completely and to monitor the leakage by means of a volume sensitive device.
Whilst there is no problem to do this whilst the seal is stationary, it will be necessary to consider the influence of the frictional heat development and heat input from the pump when isolating a running double seal.

Temperature increase

During the isolation period, the nett heat input of the seal will heat up the barrier liquid and the seal components. The rate of temperature increase depends upon:

$$\frac{dT}{dt} = \frac{\Sigma Q}{\Sigma (m * sh)}$$

Where T = Temperature (°C)
       t = time (s)

$\dfrac{dT}{dt}$ = rate of temperature increase (°C/s)

ΣQ = nett heat input of seal = frictional heat + heat input - heat extraction (watt)

m = mass of resp. barrier liquid & seal components (kg)

sh = specific heat of resp. barrier liquid & seal components (J/kg.°C)

Thermal expansion

The barrier liquid present in the seal cavity will expand corresponding to the temperature increase during the isolation period.

The rate of thermal expansion of the barrier liquid is given by:

$$\frac{dV}{dt} = V * (Cl - Cm) * \frac{dT}{dt}$$

in which: V = Volume of the barrier liquid in the seal cavity (cm³)

$\dfrac{dV}{dt}$ = rate of thermal expansion of the barrier

liquid   dt   (cm^3/s)
        Cl   = coefficient of cubic thermal expansion
for the   barrier liquid
        Cm   = coefficient of cubic thermal expansion
for the seal   components

* Note: It has been attempted to use a hydraulical equivalent of the Weatstone bridge to compare the incoming and outgoing flow with accuracy.
This development has, however, never reached a stage to warrant the use of this principle in practice.

Example

    Pump   : 370°C
    Shaft   : size 80 mm diameter 3000 RPM
    Barrier liquid: temperature 100°C
                    pressure 20 Bar
                     spec. gravity 0,85
                    spec. heat 2000 J/kg.°C
                    coefficient of cubic thermal expansion
    700 * 10^{-6}

Barrier liquid volume in seal 1,5 litre

Estimated weight of the metal parts of the seal + seal housing 30 kg
Spec. heat metal parts 490 J/kg.°C
Coefficient of cubic thermal expansion 48 * 10^6

    Frictional heat seal   2260 Watt
    Heat input (from pump)   3970 Watt
    Heat extraction (box cooling)   - 1040 Watt

              ΣQ =   5190 Watt
                                    ============

    m * sh metal parts   = 30 * 490   = 14.700 J/°C
    m * sh barrier liquid = 1.5 * 0,85 * 2000 = 2.550 J/°C

                                               17.250 J/°C
                                              ==============

$$\frac{dT}{dt} = \frac{5190}{17.250} = 0,30 \ °C/s$$

$$\frac{dV}{dt} = 1500 * (700 - 48) * 10^{-6} * 0,3 = 0,29 \ cm^3/s$$

Temperature & volume

It has been established in practice that the isolation period does not need to be longer than 10 seconds to perform a viable leakage measurement. For the above example this would mean a temperature increase of only 3°C during the isolation period.

Since the example describes a duty at the higher end of the applicational range one can conclude that the effect of temperature increase is virtually negligible, even if the isolation period is being increased.

The volume increase during the isolation period of 10 seconds is of the order of 3 cm^3. Although this may seem a low figure it is possible that a relatively sudden expansion of the order of 1 cm^3 can lead to pressure increases which can lead to seal damage.

It will therefore be necessary to address this problem either by:
a) including a pressure relief device
b) including a volumetric expansion device

Negative leakage
The rate of thermal expansion during the isolation period constitutes of a change in volume in the opposite direction to the leakage. It could therefore be considered as a negative leakage. For the example this negative leakage represents

$$0,29 * \frac{3600}{1000} = 1.0 \ \text{l/h.}$$

Whilst this is a value at the top end of the range negative leakages of the order of 0.5 l/h are common.

Accuracy of the leakage indication
The nett result is that the leakage check will only indicate a loss of volume if in case that the leakage rate exceeds the negative leakage. It can therefore be concluded, that unless special measures are adopted, the leakage check will only detect relatively large leakage rates. In practice this has been experienced as a great advantage since smaller and less significant leakage rates remain excluded from detection. The accuracy is of a level which corresponds with the detection requirements for the central system.

Leakage measurement
Accurate leakage measurement is possible if the effect of thermal expansion is known. This is not usually the case.

For ambient temperature applications it is possible to make a leakage measurement whilst the seal is stationary. A seal may, however, leak considerably more whilst running.

The conclusion is that it remains of more interest to check a running double seal and to consider the seal check as an _indication_ of leakage rather than an accurate _measurement_.

Seal check systems

The final execution of the seal check system depends on practical and economical parameters. One can have a <u>hand operated</u> or a <u>semi automated system</u>. In either case one requires plug or ball valves which can be shut off very quickly.

Both two valve and one valve executions are possible. In a one valve system the second valve is being replaced by a non-return valve.

Hand operated systems

In a two valve system it will be expedient to operate both valves with one handle. Such a handle is then made spring loaded so that both valves open automatically as soon as the handle is being let loose. (Fig. 3)
This implies that the handle is to be kept in hand during the isolation period and the leakage indication device must therefore be positioned in a way to make this reading possible. The valves and the leakage indication device have in consequence been mounted on a separate panel, which adds to the cost.

Semi automated system

In a hand operated system it remains to the judgement of the operator for how long the isolation is to be prolonged. In a semi automated system the isolation time can be set with a timer. Another advantage is that the manual operation is eliminated and the operator can concentrate on watching the leakage detection device. A panel is no longer required.

The most convenient execution uses one ball valve on the inlet in combination to a non return valve on the outlet side. This non return valve acts at the same time as a pressure relief device. (Fig. 4)

It was decided to open the ball valve pneumatically against the action of a spring. A pneumatic timing device sets the time that the valve is being kept closed. After the blow off of the pneumatic pressure the valve opens automatically by spring force.

In order to eliminate any possibility of a hang up of the pneumatic feeding valve, it is recommended not to use a pneumatic supply but to apply instead a <u>moveable pneumatic feeder</u> of the type which is used for pressurizing car tires. The pneumatic feeder is only used to energize the closing action and cannot be left to continue activation of the valve.

Leakage detection device

The leakage detection device can be:
a) a volumetric device, such as a spring loaded plunger
b) a pressure gauge

Spring loaded plunger

A spring loaded piston device as per fig. 5 has been successfully in use for many years. The movement of the piston is being used to monitor the change in volume of the barrier liquid in the double seal. The indication sensitivity depends upon the diameter of the plunger.

The accuracy is being increased by applying a smaller diameter. Reducing the plunger diameter does however increase the friction to axial force ratio which introduces the risk of a hang up of the plunger specially when the system pressure is low.

The device must be set for the barrier liquid pressure of the central system. The spring loaded piston has sufficient expansion possibility for normal operation.

Pressure gauge

A normal pressure gauge can also be used as a leakage detection device. The indication sensitivity depends upon the volumetric displacement characteristic of the pressure gauge (graph 1).
It is important to realize that this displacement characteristic is also influenced by the amount of air trapped in the piping to the pressure gauge.

The effect of air presence has been explored by fitting the pressure gauge respectively in the normal way ('up') and in a reverse way ('down'). Graph 2 shows a typical leakage rate curve for a pressure gauge.

The accuracy of the leakage rate indication increases with the system pressure. It goes down very rapidly at lower pressures. It can be concluded from graph 2 that the accuracy is still acceptable down to barrier pressures of approximately 7,5 Bar absolute.

The difference between the 'up' and the 'down' curve demonstrates the reduction in the indication accuracy due to the presence of air. It can be concluded that it is worthwhile to reduce the non ventable air volume to a minimum.

Conclusion

Seal check systems have been successfully in use with the object of finding out which double seal, out of a given population of double seals, is leaking. Its sensitivity is adequate and prevents false alarms. Ongoing improvements have been introduced to improve the user friendliness, the safety and the cost of the system.

CENTRAL BARRIER LIQUID SYSTEM
SEAL
SEAL
SEAL
SEAL
* INDIVIDUAL SEAL CHECK SYSTEMS

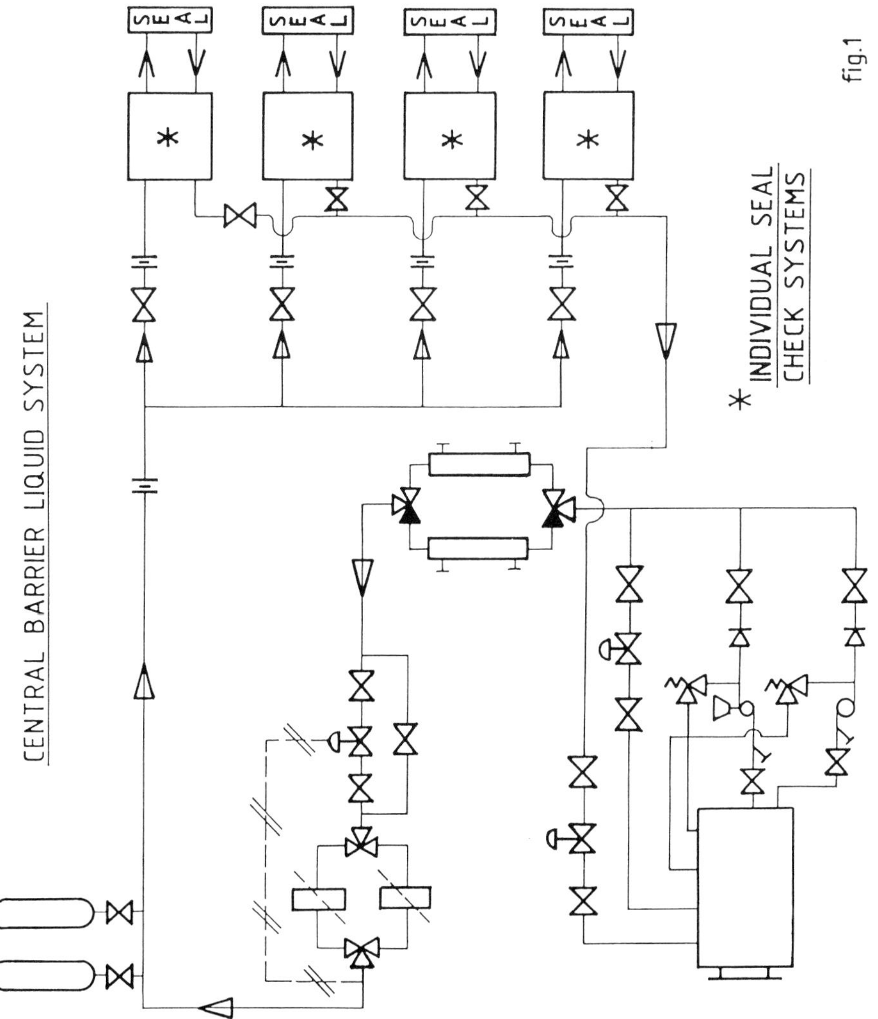

fig.1

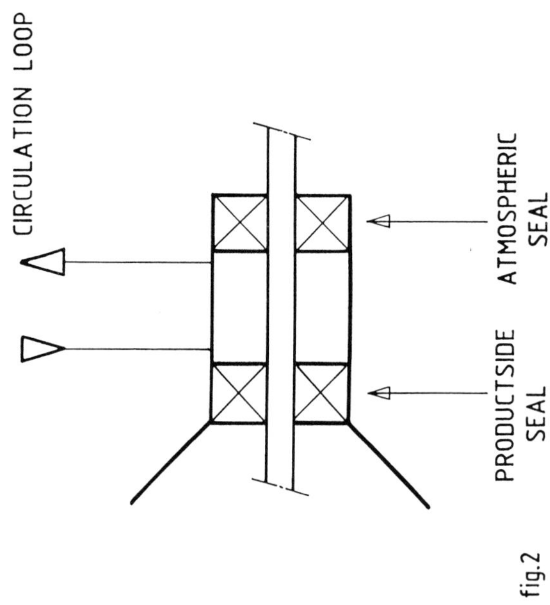

fig.2

PANEL.
L.C.D.
SEAL CHECK SYSTEM WITH
HAND OPERATED VALVES.
□ = PRESS. RELIEF VALVE (OPTIONAL)
L.C.D.=LEAKAGE CHECK DEVICE
O =SPRING LOADED HANDLE

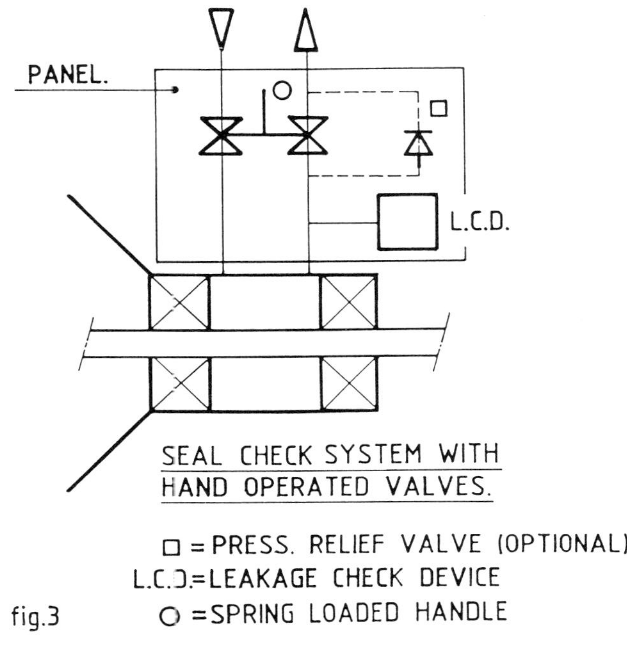

fig.3

L.C.D.
SEAL CHECK SYSTEM WITH
PNEUMATICALLY OPERATED VALVE
□ = NON RETURN VALVE
L.C.D. = LEAKAGE CHECK DEVICE
△ = PNEUMATIC TIMER
O = PNEUMATIC ACTUATOR (SPRING LOADED)

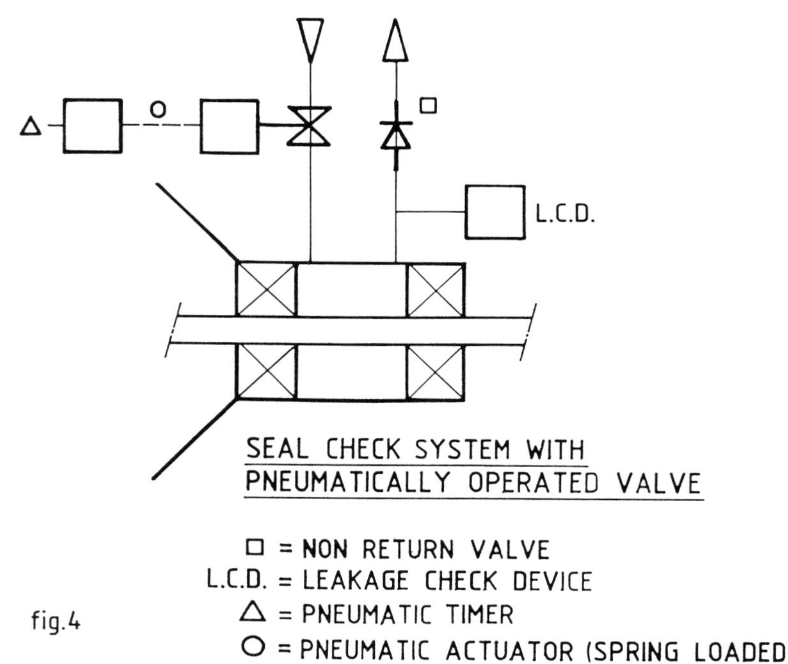

fig.4

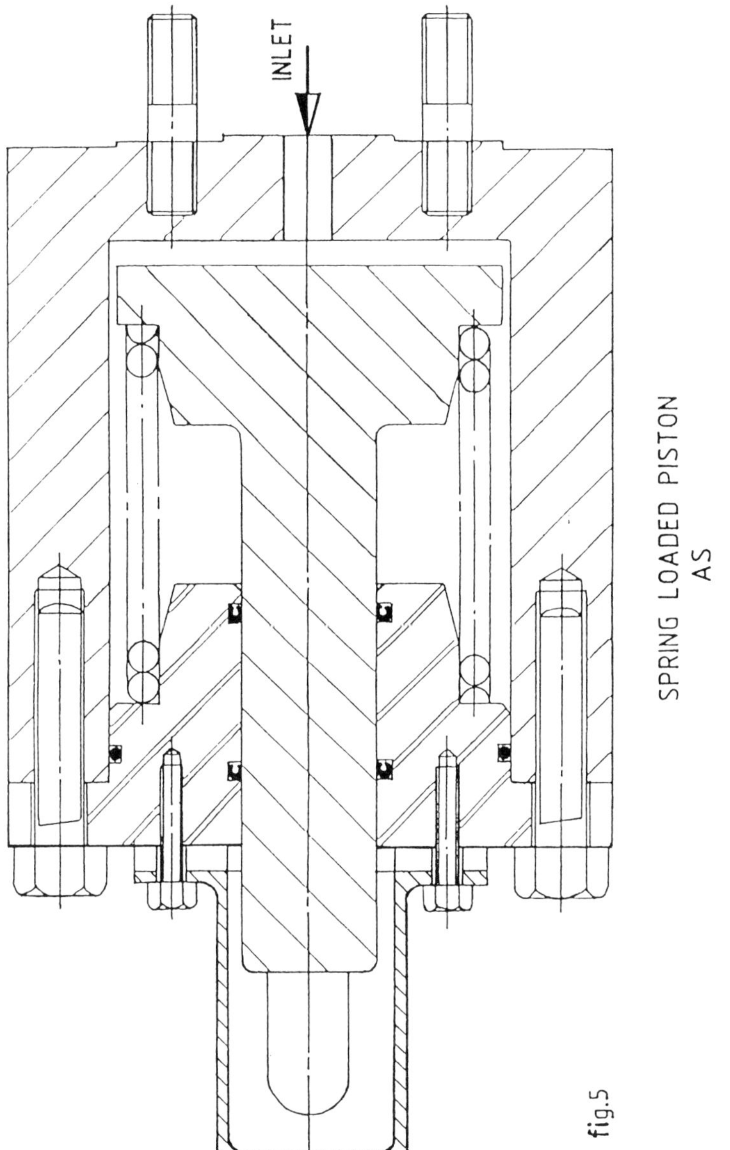

INLET
SPRING LOADED PISTON
AS
LEAKAGE CHECK DEVICE
fig.5

DISPLACEMENT VOLUME CHARACTERISTIC OF A PRESSURE GAUGE.

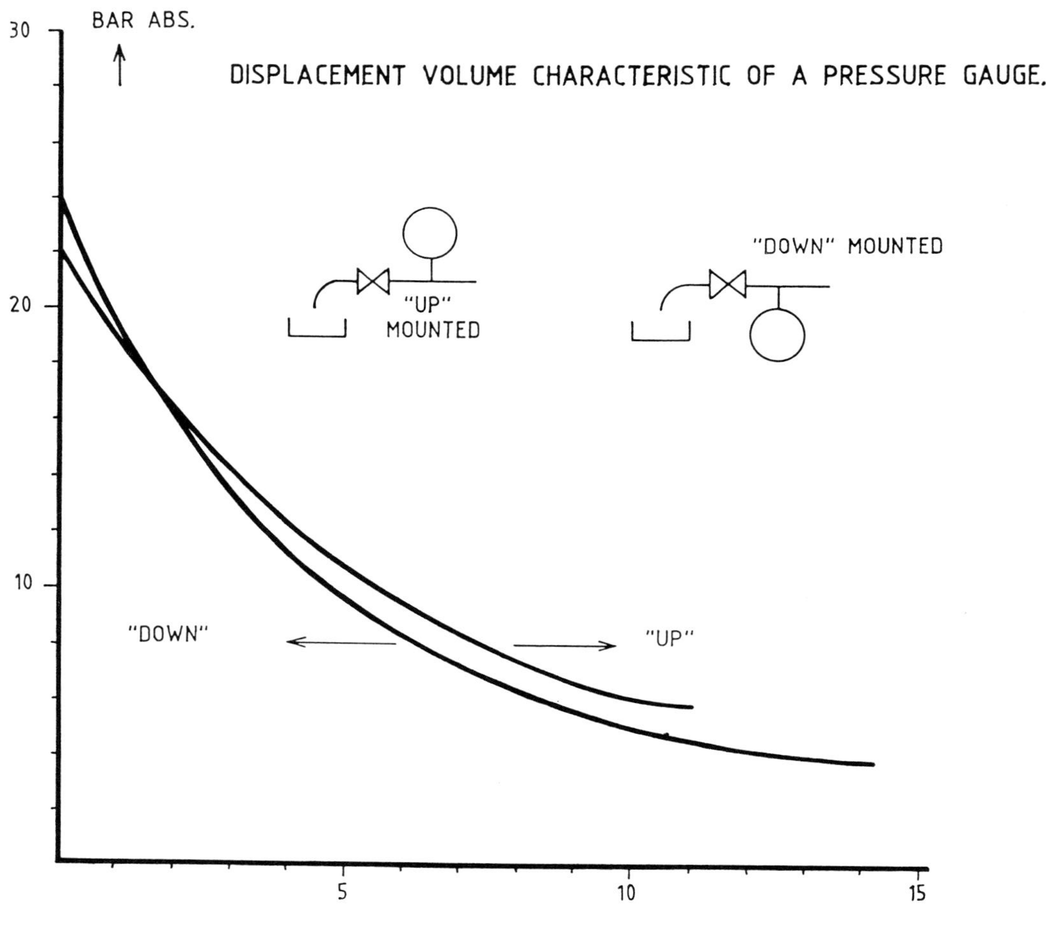

GRAPH 1.

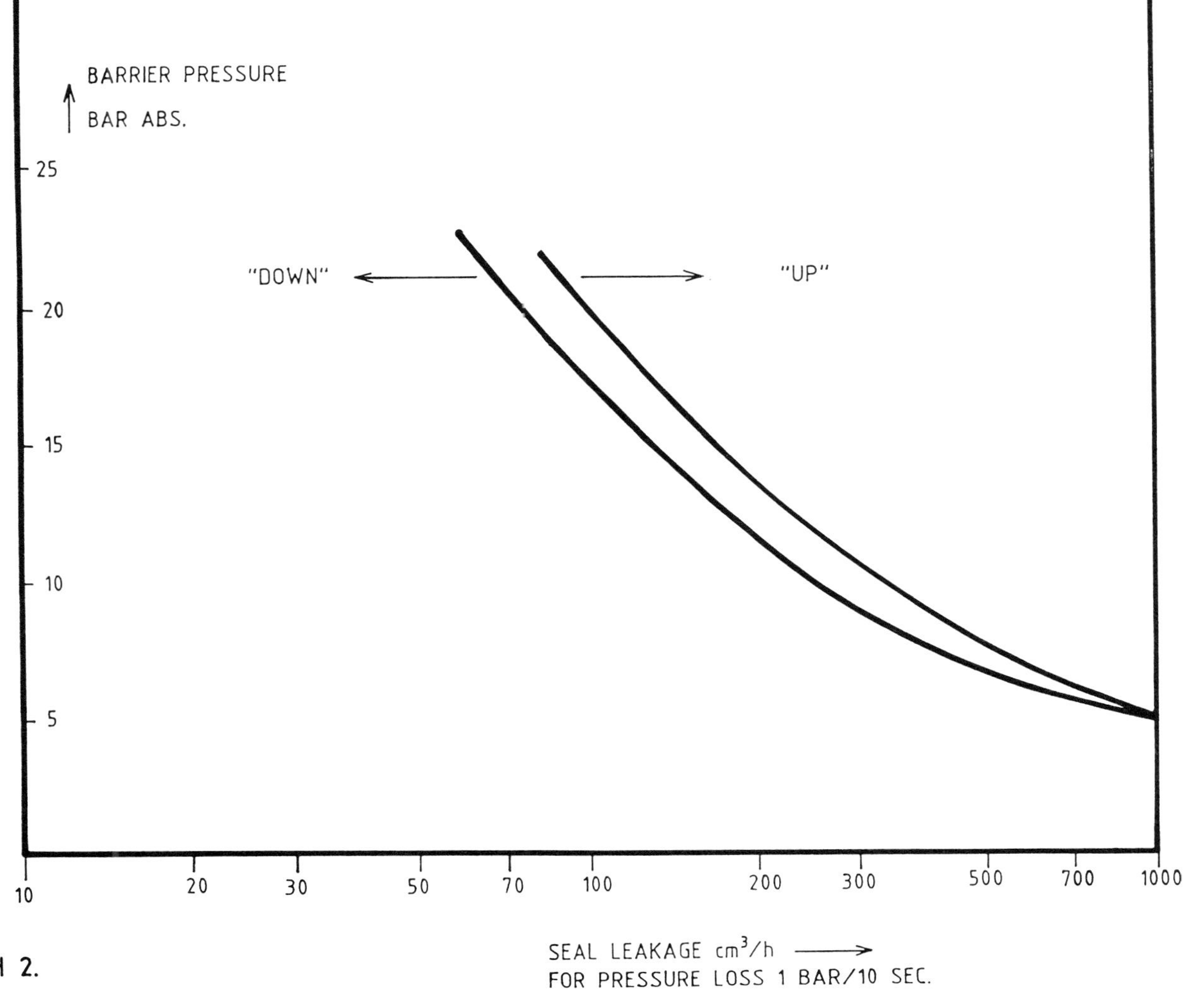

GRAPH 2.

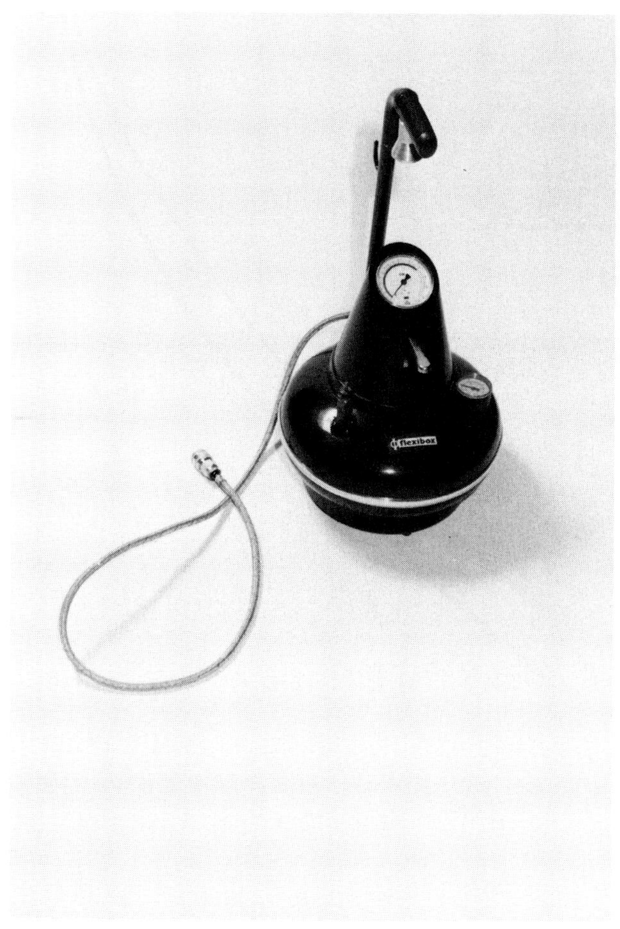

MOVEABLE PNEUMATIC FEEDER

SEMI AUTOMATIC SEAL CHECK SYSTEM

14th International Conference on Fluid Sealing, Firenze, Italy,
6-8 April 1994. Organised by BHR Group Limited, Cranfield,
Bedford, MK43 0AJ, UK; Tel: 0234 750422

KNIFE EDGE MECHANICAL SEAL FOR LOW EMISSION OF HIGH VISCOSITY FLUIDS

Yukio Goto, Kanji Ohba
Engineering Department, Nippon Pillar Packings, Co., Ltd.,
Nonakaminami 2-chome, Yodogawa-ku, Osaka, Japan

SUMMARY

Hitherto, when sealing a liquid which is solidified between seal faces due to sliding heat or shearing of liquid itself, such as latex and resin liquid of high viscosity, a conventional mechanical seal having a seal face width of 3 to 8 mm has not performed well. To solve the problem in regard to sealing of solidifiable fluid, a knife edge mechanical seal has been developed.

The knife edge mechanical seal is intended to achieve the purpose of sealing by designing the seal face width of about 0.5 mm in a form of knife edge, applying 10 to 60 times higher spring pressure than conventional mechanical seals', and cutting off and eliminating solid matter formed between the seal faces. The knife edge mechanical seal has radically changed the philosophy of the mechanical seal for high viscosity fluids, and is expected to contribute to emission-free sealing in the difficult-to-seal liquid fields including latex, electrode position paint, pulp, high viscosity resin, and food materials.

This paper introduces the sealing mechanism of knife edge mechanical seal, sealing performance, test results, design method, and typical application cases.

1. INTRODUCTION

The mechanical seal exhibits its function by balancing wear and leakage by interposing the fluid to be sealed between the seal faces as a lubricating film. Therefore, in the case of solidifiable liquid or high viscosity liquid, it is hard to form the lubricating film with a conventional mechanical seal, and hence difficult to seal. In particular, latex, which is a solidifiable liquid, is easily solidified by shearing or friction, and has been hitherto regarded as one of the most difficult-to-seal liquids. Even with a double seal having the sealing liquid at high pressure, the liquid film between the sealing faces is solidified by the counter flow into the seal faces (due to diffusion or pumping), which often led to massive leakage in a short time.

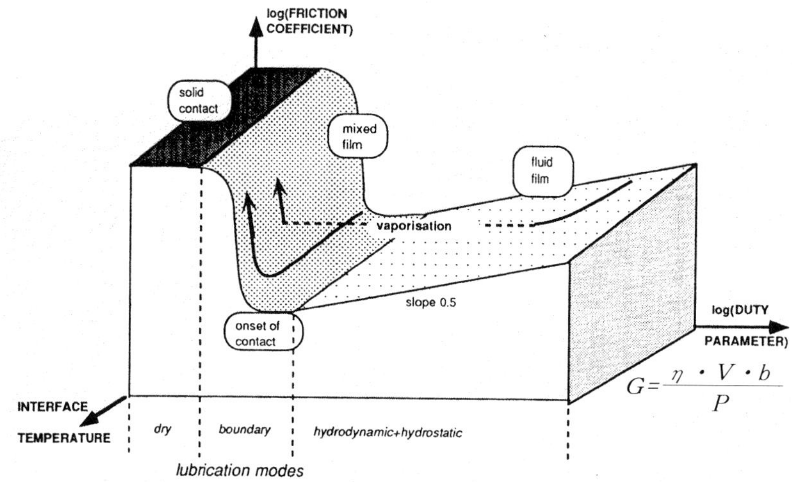

$$G = \frac{\eta \cdot V \cdot b}{P}$$

Fig. 1 f·G curve

As a parameter for expressing the lubricating state of seal face, the duty parameter curve as shown in Fig. 1 is known[1] .   In the diagram, f designates seal face friction coefficient, η designates viscosity index, V designates seal face velocity, b designates seal face width, P designates net closing force.   Generally, in the boundary lubrication condition, the mechanical seal exhibits a stable performance.   In the case of high viscosity liquid where η is far larger compared with ordinary water or oil, it tends to be in the hydrodynamic lubrication condition , and the leakage increases.   In this case, the leakage may be somewhat limited by decreasing b and increasing P, but it has its own limit, and the effect is not big in regard to the solidifying liquid. Accordingly, we have reached the concept of cutting down the leakage by decreasing b to the minimum limit for continuously cutting off and eliminating the liquid film solidified on the seal face.   In this case, the lubricating region is shifted from the boundary region to the dry region, but the function of the mechanical seal may be sufficiently fulfilled as far as the required abrasion life can be guaranteed by using wear resistant materials in the seal faces.

In this way, we set about development of a mechanical seal of a new type, i.e. a knife edge mechanical seal, to open a new dimension in the field difficult to seal with the conventional seals.

2. STRUCTURE AND FEATURES OF KNIFE EDGE MECHANICAL SEAL

The knife edge mechanical seal is a mechanical seal having the seal face as sharp as a knife edge.   The seal face width is about 0.2 to 0.6 mm, extremely narrower than the conventional mechanical seal, and about 1/10 of the so-called "narrow face seal" which is used widely in recent years. That is, the seal face has a line contact as against a conventional face contact, and it is characterized by using hard materials such as tungsten carbide as the materials for the both seal faces.   In addition, to inhibit formation of solid matter between the faces, and to eliminate the formed solid matter or sticky deposits, the spring pressure is set to measure about 10 to 60 times higher than in the conventional seal.

The structure of parts other than the seal faces is designed in various forms depending on the liquid property, pressure, velocity and other conditions. Fig.2 shows the structure of a typical knife edge mechanical seal using metal formed bellows.   The structure using the metal formed bellows is suited to a viscous liquid and solidifiable liquid, which may cause seal hang up, for the reason that the secondary packing is not provided.   Besides, designing in a wide range is possible because the balance ratio can be set so that the effect of the fluid pressure may be minimum.   Features of conventional mechanical seals and knife edge mechanical seals are comparatively shown in Tables 1 and 2.

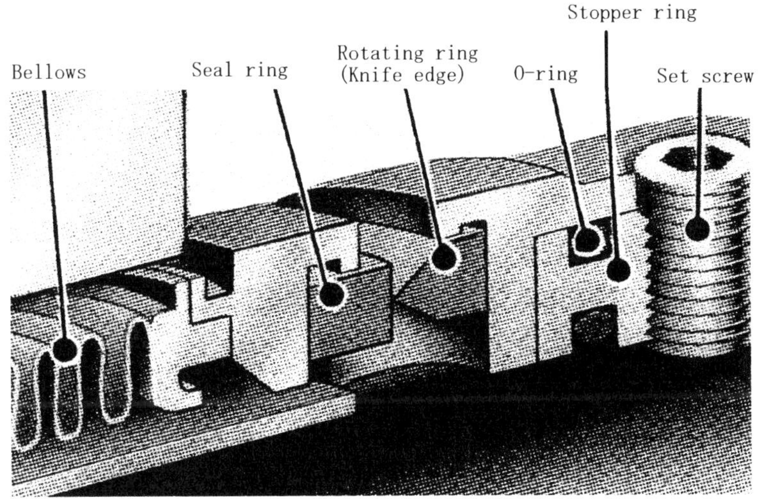

Fig.2 Typical knife edge mechanical Seal

Table 1. Seal specification

Specification	Conventional mechanical seals	Knife edge mechanical seals
Seal face width b mm	3 - 8	0. 2 - 0. 6
Balance ratio B	0. 75 - 1. 25	0 - 0. 5
Spring pressure ps MPa	0. 1-0. 3	1 - 6
Seal face materials	Soft material vs. hard material	Hard material vs. hard material

Table 2. Properties

Feature	Conventional mechanical seals	Knife edge mechanical seals
Resistance to solidifiable fluid Resistance to precipitating fluid Resistance to high viscosity fluid	Leaks due to formation of solidified matter or inclusion of solid matter on seal faces.	Formation of solidified matter is limited by narrow face width, and formed matter is cut off and removed by sharp blade edge.
Thermal strain Pressure strain	Leaks due to face distortion caused by radial strain.	Effects of radial distortion are eliminated.
Resistance to pressure fluctuations Resistance to reverse pressure Resistance to high pressure	Seizes or leaks due to contact with outer or inner circumference caused by radial strain of seal face.	Hardly influenced by pressure because the face width is minimum and B=0-0. 5
Cooling characteristic	The atmospheric side tends to be cooled poorly because of wide face, possibly leading to seizure or abnormal wear.	Cooling is excellent for same level of heating.

3. TESTS

Tests were conducted in two stages, basic test and practical test. The basic test was intended to evaluate the sealing and wearing performance at various spring pressures. The practical test was intended to determine the optimum design specifications and confirm applicability in regard to solidifiable fluids (latex, electrode position paint). The test seals were of stationary seal using formed bellows without secondary packing in order to prevent sticking. Type A shown in Fig. 3 is a standard type, and type B shown in Fig. 4 is a special type including a coil spring capable of applying a large spring pressure. These design specifications are as follows.

    Seal materials   : Tungsten carbide vs. Tungsten carbide
    Seal face width : 0.25 to 0.35 mm
    Balance ratio   : 0 to 0.5
    Spring pressure : 1 to 6 MPa

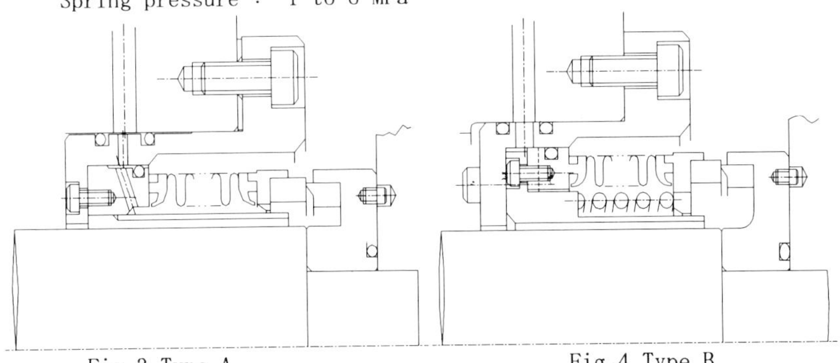

Fig. 3 Type A
Stationary seal using
formed metal bellows

Fig. 4 Type B
As Type A with a coil spring

3.1 Basic test
3.1.1 Apparatus and conditions
The outline of the test apparatus and test conditions are shown in Fig. 5.

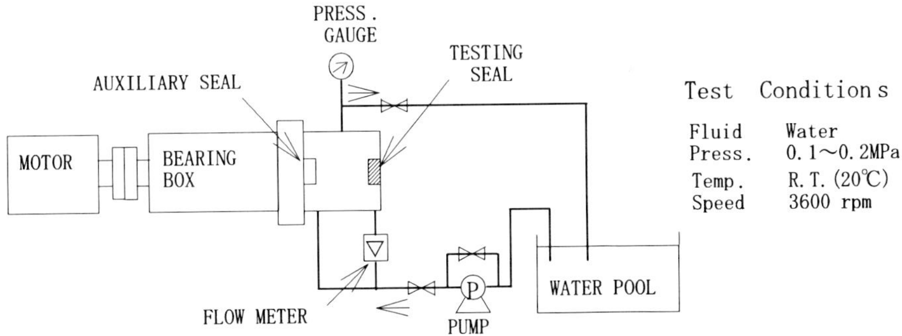

Fig. 5 Basic test apparatus & conditions

264

3.1.2 Results and deliberation

a) Spring pressure (ps)

Test results of leak amount and spring pressure are shown in Fig. 6. At $ps \geqq 2$ MPa, the leak was slight and stable. As shown in Fig. 7, meanwhile, at $ps = 5.5$ MPa, the wear was insignificant, being 2.7 μ m/90 hrs at edge side and 0.3 μ m/90 hrs at flat side, and crack and other abnormalities were not observed. In the mechanical seal, generally, as the closing pressure is increased, the sealing performance is enhanced, but the wear increases and the life is shortened. Accordingly, the spring pressure is usually set around 0.1 to 0.3 MPa. If the spring pressure is excessive, the friction of the seal faces increases, and the heat generation increases to cause abnormal wear or cracks.

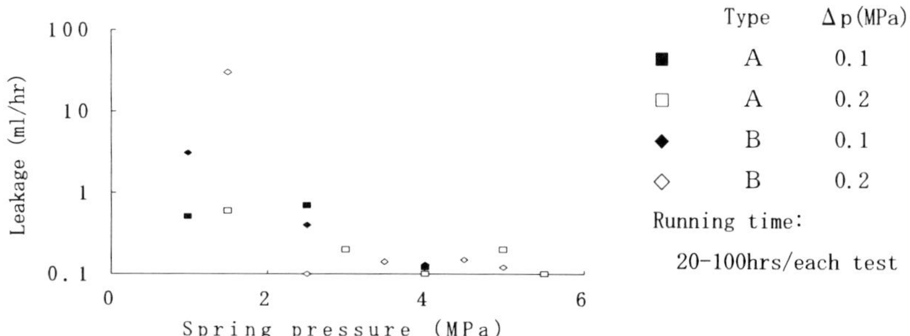

Fig. 6 Effect of spring pressure on leakage.

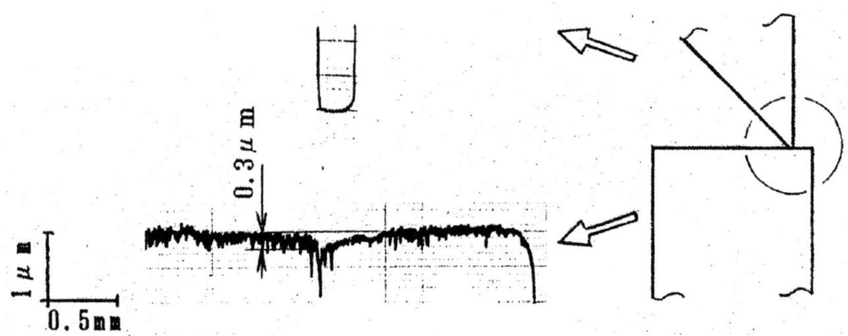

Fig. 7 Face profiles of the basic test
Seal type:A, ps:5.5MPa, Δp:0.1MPa, Running time:90hrs

Generally, the coefficient of friction of a material is determined by the properties of the material itself as expressed in Amonton's law and others, and is not related with the closing force as shown in formula (1).

$$f = \frac{F}{P} = \frac{\tau}{\sigma_c} \tag{1}$$

The heat generation H on the seal face is expressed as follows.

$$H = \frac{f \cdot P \cdot V}{J} \tag{2}$$

Where $\quad P = \pi \cdot d \cdot b \cdot p, \qquad p = (B - k)\Delta p + ps \tag{3}$

f : coefficient of friction of seal face, $\quad$ F: shearing force, $\quad \tau$: shearing stress, $\quad \sigma_c$: compressive stress of material, $\quad$ J:mechanical equivalent of heat, $\quad d$: mean diameter of seal face, $\quad p$: net closing pressure, $\quad K$: pressure drop coefficient, $\quad \Delta p$: differential pressure.

Depending on the combination of seal face materials, there is an empirical limit of withstanding the heat H. At this time, in the knife edge mechanical seal, since the face width is minimum, the opening force by liquid pressure between the faces can be ignored, and the balance ratio is nearly zero, so that the pushing force by liquid pressure can be also ignored. Therefore, formula (3) is rewritten as follows.

$$P \doteqdot (\pi \cdot d \cdot b) \cdot ps \tag{4}$$

That is, if the spring pressure ps is increased, the heat generation may be lowered below the level of a conventional seal, because the face width b is small. It suggests that the spring pressure may be taken at a sufficiently large value within a certain limit.

b) Sealing performance

From the test results shown in Fig. 6, the knife edge mechanical seal is confirmed to exhibit an excellent sealing performance equivalent to a conventional seal. Generally, the theoretical leak Q of mechanical seal in the hydrodynamic lubrication condition is expressed as follows, designating the sealing face gap as h.

$$Q \doteqdot \frac{\pi \cdot d \cdot h^3}{12\eta b} \cdot \Delta p \tag{5}$$

The knife edge mechanical seal is anticipated to leak much because the face width b is extremely small as compared with the conventional seal. Indeed, the face width of knife edge mechanical seal is 0.2 to 0.6 mm, which is 1/10 to 1/20 of the conventional seal. It, however, corresponds to about 1/2 to 1/3 of the rate of the sealing faces gap h influencing Q. With this value of b, therefore, it is possible to set a sufficiently small leak amount Q.

But, since h becomes small, a material possessing wear resistance is required, and the face materials are tungsten carbide vs. tungsten carbide.

266

c) Seal face strain

To maintain parallel flat faces by suppressing strain of seal faces is one of the most important technical subjects of mechanical seal. The seal faces are deformed to the radial direction by heat and pressure. As shown in the model in Fig. 8, the deformation can be approximated in the following equation, designating the rotational angle as θ.

$$\delta = \theta \cdot b \tag{6}$$

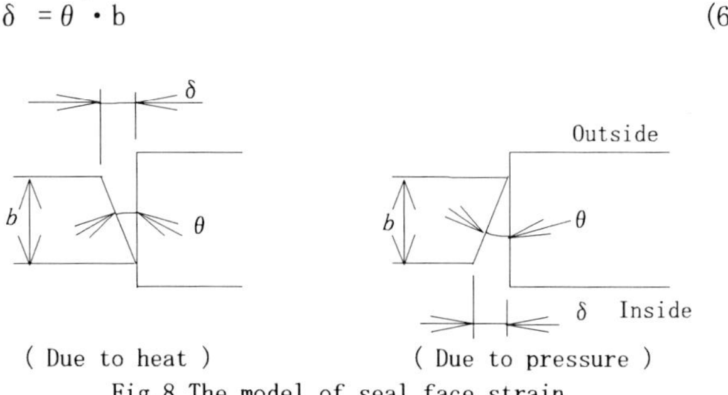

(Due to heat)   (Due to pressure)

Fig. 8 The model of seal face strain

If θ is the same, δ can be lowered to 1/10 to 1/20 in the knife edge mechanical seal as compared with the conventional seal.

d) Torque

Fig. 9 shows measurements of torque T in knife edge mechanical seal and conventional seal. The torque of the seal face is expressed as follows.

$$T = f \cdot d /2 \cdot P \tag{7}$$

Where P of the knife edge mechanical seal possesses a constant torque property not so influenced by the fluid pressure as shown in formula (4).

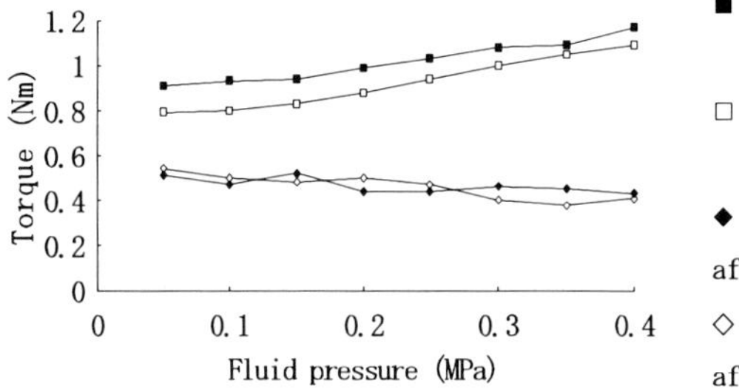

Fig. 9 Dynamic torque

3.1.3 Evaluation

The test results showed that the knife edge mechanical seal possesses the sealing performance equally comparable with that of a conventional seal, in a limited region, as far as the spring pressure and the balance ratio are set appropriately.   On the basis of these results, practical tests were conducted in regard to the solidifiable liquid and viscous liquid which were hard to seal by the conventional seals.

3.2 Test in latex
3.2.1 Apparatus and conditions

The apparatus and test seals were the same as used in the basic test.   The specified test conditions were that quenching water of 0.1 MPa was circulated on the inner side of the seal while latex of head of 1 m was collected outside.   As latex samples, the types reputed to be most difficult to seal among the manufacturers were selected.

3.2.2 Results

Test results are summarized in Table 3.

In test No.5, fresh liquid was supplied about every 12 hours because the latex is solidified in the seal box and the liquid concentration changes while observing the state of the seal box in a transparent case .   Fig.10 (a) shows the profiles of the seal face after operation of test No.4, and Fig.10 (b) shows the profiles after test No.5. As these results show, at ps = 2.5 MPa, solid matter was formed on the seal face, and the leak increased.   At ps = 4 MPa, the solid matter was completely cut off and a favorable sealing performance was shown.   The wear was 2 μm/100 hrs at edge side and 0.5 μm/100 hrs at flat side, and was satisfactorily slight.   Fig.11 shows the transition of leakage rate in test No.5.   A slightly large leak was measured right after starting, but it decreased within several hours, and remained low thereafter. Fig.12 and 13 show the solidified latex in the seal box and the seal face after test No.5.   These test results confirmed that the knife edge mechanical seal can be sufficiently used in the latex.

Table 3. Test results

Test No.	Seal	Spring pressure MPa	Running time hrs	Leakage ml/hr		Evaluation of sealing	
1	Type A	1.0	0.04	Large	Began to leak in 1 min after start, and blew out and leaked in 2 min.	Not cut off at all.	×
2		2.5	1.5	260	Gradual increasing tendency from 0.5 hr after start.	Not cut off sufficiently.	×
3		4.0	35	0.2	Almost stable.	Cut off sufficiently.	○
4	Type B	2.5	17	Large	Gradually increased from 1 hr after start, overflowing thereafter.	Not cut off sufficiently.	×
5		4.0	100	0.4	See Fig. 11.	Cut off sufficiently.	○

Note: ○ Good
     × Failure

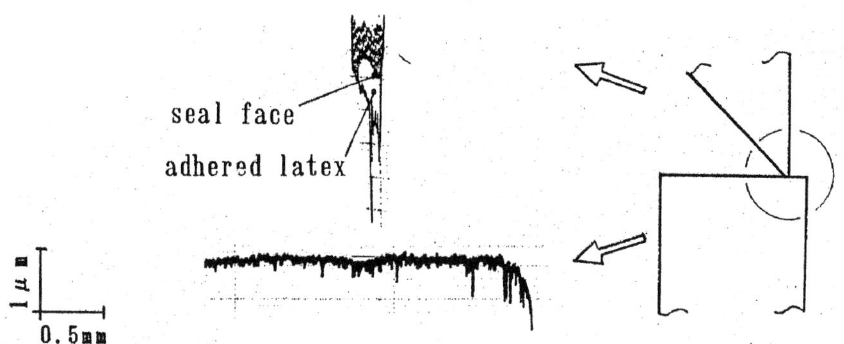

Fig. 10 (a) Face profiles of test No. 4

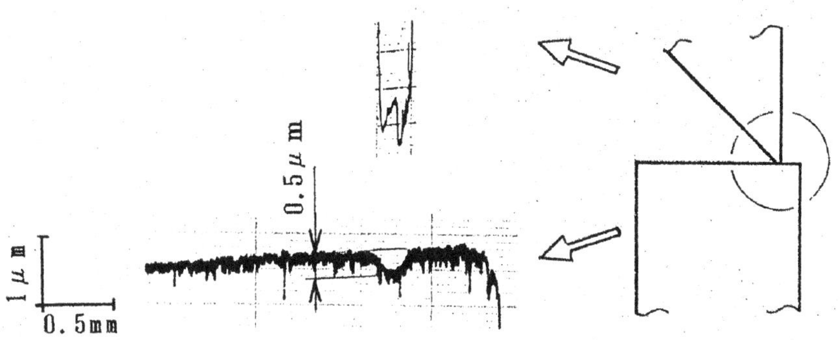

Fig. 10(b) Face profiles of test No. 5

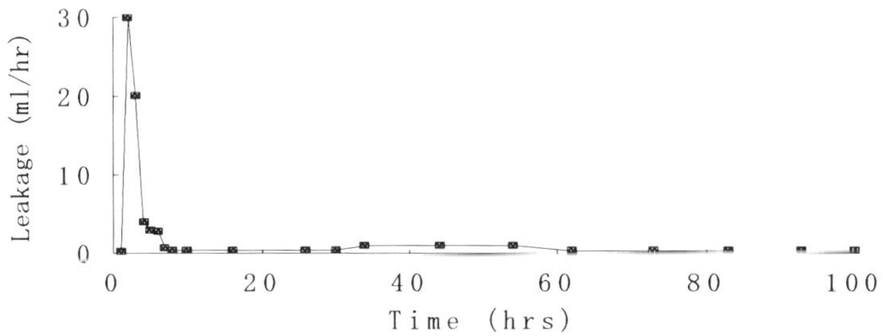

Fig. 11 Test NO. 5 at latex

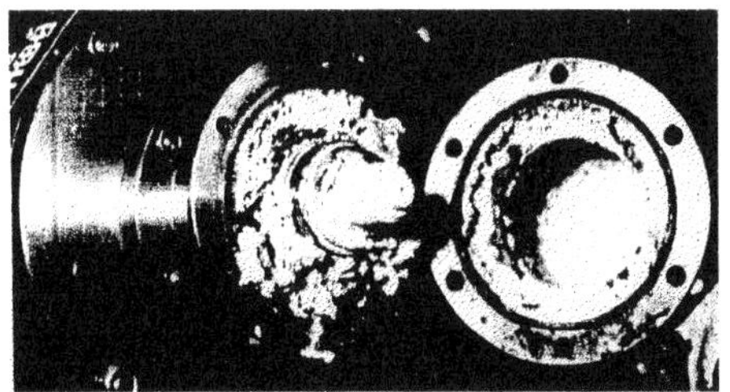

Fig. 12 Sollidified latex in the seal box

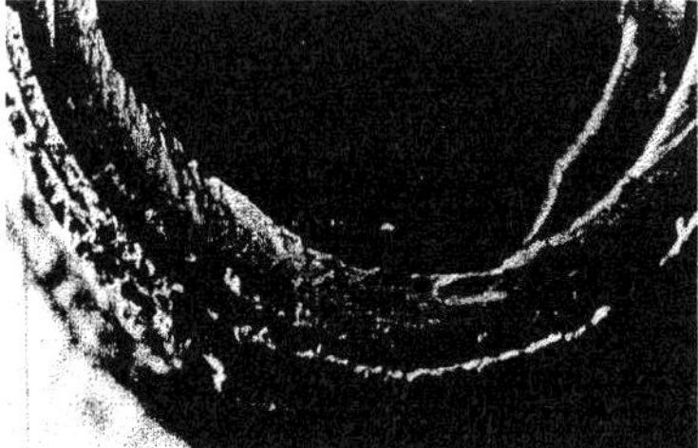

Fig. 13 Seal face after test No. 5

3.3 Test in electrode position paint

The electrode position paint is widely used for automobile industry and others, and is classified into anionic type and cationic type.   The cationic paint is a mixed liquid of resin, solvent, pigment and others, and is a very difficult-to-seal liquid, like latex, for the mechanical seal.

3.3.1 Apparatus and conditions

The test seal was of type B, and the cationic liquid pressure was set about 0.1 MPa higher than the quenching liquid. Varying the seal specifications, tests were conducted 17 times in a total of 3,800 hours, and the optimum design value was determined and the reliability was confirmed.

3.3.2 Results

The results of 1632 hours durability test conducted in accordance with final specification are shown below. The leak changes in the whole period of the durability test are shown in Fig. 14. Although less stable, the leaks settled within an allowable range. Fig. 15 shows the profiles of the seal face after the durability test. The wear was 23 μ m (0.014 μ m/hr) at edge side and max. 62 μ m (0.038 μ m /hr) at flat side, which were satisfactory values. Thus, the knife edge mechanical seal was confirmed to exhibit also a sufficient function in cationic paints.

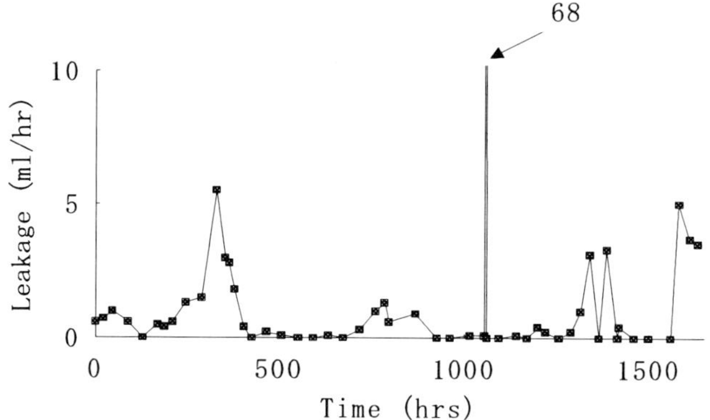

FIg. 14 Durability test at cationic paint

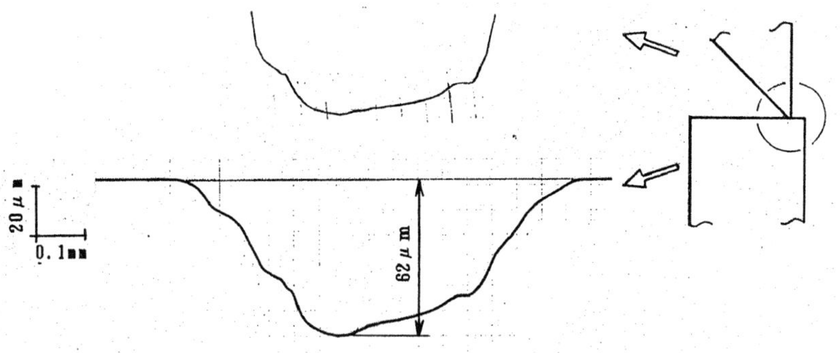

Fig. 15 Profiles of seal face after the durability test

4. APPLICATION EXAMPLES

Principal preferable design plans for application of the Knife edge mechanical seal are listed below. Basically, the design structure except the edge is same as conventional mechanical seals.

a) Quenching water

It is recommended to flow the quenching water at a small flow rate (1-3 ℓ/min) because the liquid as handled is often poor in lubricity and it is necessary to discharge solid matter cut off and eliminated by the edge. At this time, it is advised to form a quenching water passage by installing a baffle for making cleaning easy. For sealing of quenching water, a lip type seal or a simple mechanical seal is provided.

b) Prevention of sticking

It is important to avoid sticking of the secondary packing or clogging of the spring caused by solidifiable fluid. For this purpose, it is preferred to use the metal formed bellows with a large pitch. If there is a sufficient allowance in the seal box size, a monocoil spring may be installed on the quenching side, so that a long-term stability of load could be obtained.

c) Precision of seal and equipment

It is better to provide the edge side on the rotary side in order to minimize the effects of shaft runout, case inclination or centering precision.

d) Shape of edge

To produce the edge by processing a hard material such as tungsten carbide, the edge is preferred to have an obtuse angle rather than an acute angle, considering the strength. The angle and shape should be determined further in consideration of wear life or repairability.

Principal applicable liquids include latex (ABR, SBR), tacky paint (anionic, cationic), various paints, monomer, nylon, other high viscosity resins, white liquor, green liquor, black liquor, other pulp liquors, molasses, chocolate, animal and vegetable oil, corn starch, milk, other foods, tar, asphalt, and other high viscosity hydrocarbons, among others.

5. CONCLUSIONS

The knife edge mechanical seal developed for the purpose of sealing latex presented satisfactory results as a low emission mechanical seal in the demonstration experiment in the market. Henceforth, its applications are expected to expand widely into other fields than latex. In such cases, since the difficult-to-seal liquids are always main objects, it is important to study the field specification sufficiently.

6. REFERENCE

1) R. K. Flitney, B. S. Nau, Performance testing of Mechanical seals, 13th International Conference on Fluid Sealing, BHR Group, April 1992

7. APPENDIX

The symbols in the paper are as follows.

b	seal face width
B	Balance ratio
d	Mean diameter of seal face
f	Coefficient of friction of seal face
F	Shearing force
h	Sealing face gap
H	Heat generation
J	Mechanical equivalent of heat
K	Pressure drop coefficient
p	Net closing pressure
ps	Spring pressure
Δp	Differential pressure
P	Net closing force
Q	Leakage rate
V	Seal face velocity
η	Viscosity
τ	Shearing stress
σc	Compressive stress of material
θ	Deformation angle
δ	Deformatio of seal face

14th International Conference on Fluid Sealing, Firenze, Italy,
6-8 April 1994. Organised by BHR Group Limited, Cranfield,
Bedford, MK43 0AJ, UK; Tel: 0234 750422

TITLE: **NEW DEVELOPMENTS IN BELLOW SEALS FOR
IMPROVED PERFORMANCE & RELIABILITY**

J.G. EVANS - Flexibox Ltd, Manchester, UK

1.0 INTRODUCTION

Whilst being a popular sealing solution, the welded metal bellow seal has not advanced significantly over the past 20 years and is still viewed in some quarters as a fragile arrangement, intolerant to process upsets.

Flexibox have developed the Saflex range of twin ply formed metal bellow seals which retain the advantages of edge welded bellow seals whilst offering the robustness, pressure capability and versatility of a pusher seal.

This paper discusses the concept behind this development, details some testing and development work, and finally highlights some practical in-field applications.

2.0 METAL BELLOW SEALS - A BRIEF HISTORY

Metal bellow seals have been in existence for over 50 years. In fact the first seal patented by the founder of Flexibox in 1938 was a metal bellow seal, where the bellows were formed by mechanically deforming a metal tube.

The familiar welded metal bellow seal, owes its development to the aerospace programmes of the 1940s, 50s and 60s from where it has evolved into the general purpose seal which it is today.

It is produced by welding (usually TIG or plasma), alternately on the outside and inside diameter, a number of pre-formed metal diaphragms approximately 0.1 - 0.15 mm thick. In this way a flexible bellow core, typically 10mm long is produced. This core is then welded at each end to a drive ring and end ring which holds one of the seal faces as shown in Fig. 1. The bellow acts as both spring to pre-load the seal faces and secondary seal, replacing the sliding packing of the pusher seal.

Over the past 20 years, there have been few developments in welded bellows seal technology, although, there has been an increasing use of more exotic materials for the bellow such as Inconel and Hastelloy extending the pressure and temperature limits and improving the corrosion resistance.

Undoubtedly, whilst the welded metal bellows seal has gained popularity in several areas, most notably the USA, it is still viewed by many operators as being insufficiently robust for general usage.   Indeed, there are areas such as Light Hydrocarbon service, where specifications specifically recommend against using such seals.

3.0   **EDGE WELDED BELLOW SEALS VERSUS PUSHER SEALS**

3.1   WELDED BELLOW SEALS - **Advantages**

3.1.1   No Sliding Packing

A significant number of seal failures occur as result of seal hang up. This is where the surface on which the sliding 'O' ring or packing locates, becomes contaminated by leaked product. The increased resistance offered by the leakage may exceed the force offered by the spring resulting in the seal losing its flexibility and eventually leading to heavy seal leakage.

Operators used to handling both bellows and pusher seals would probably cite the absence of the sliding 'O' ring as the most important feature of the bellow seal, as the risk of hang-up is virtually eliminated.

Additionally, by replacing the sliding packing, the metal bellow seal is capable of operating at far higher temperatures than the pusher seal.

3.1.2   INHERENTLY BALANCED

To permit mechanical seals to operate at pressures above 10 bar g, a degree of hydraulic balance is designed into the seal to ensure that the sealing pressure at the face is less than the pressure to be sealed. In this way a fluid film between the seal faces is maintained.

To achieve this hydraulic balance, a stepped sleeve must be provided or alternatively a step must be machined into the pump shaft.

A bellow seal is inherently balanced and thus requires only a plain sleeve, or alternatively can be mounted directly onto the pump shaft, thus providing a simple and compact balanced seal.

3.2   WELDED BELLOW SEALS - **Disadvantages**

3.2.1   Lack of Robustness

Probably the single biggest dislike of any operator using bellows seals is their general lack of robustness in comparison with a pusher seal. The diaphragms are generally 0.1 - 0.15 mm thick, resulting in weld beads of approximately 0.3mm - 0.5mm diameter. A typical 3" seal will have in the region of 5 metres of weld length, any damage to which could result in catastrophic failure.

3.2.2   Susceptibility to Fatigue

One of the most common reasons for bellow seals failing is fatigue. Fatigue is a consequence of excessive cyclic stresses due to misalignment within the seal assembly. Misalignment may result from incorrect seal or pump assembly or in some cases be due to incorrect operation of the sealing arrangement. For instance, it is common for a steam quench to be applied to single high temperature bellow seals. It is possible, particularly on vacuum service for the seat to become dislodged by the steam pressure, resulting in misalignment of the seat. The constant cyclic action of the bellows required to 'follow' the misaligned seat can cause rapid fatigue as shown in fig. 2.

3.2.3   Lack of Axial Flexibility

Although, less of a problem nowadays, with increasing usage of cartridge seal arrangements, bellow seals generally permit only a small amount of compression to achieve the required face pre-load - typically 2-3mm. On a hot pump, this can cause problems due to the growth in the shaft causing either severe over or underloading of the seal faces. In the extreme, seal face contact can be completely lost, resulting in massive leakage.

3.2.4   Low Pressure Capability

The majority of welded metal bellows seals available have a pressure limit of only 20 bar g. This is a function of both the material strength and construction, plus the influence of a phenomenon known as balance line shift.

It is widely known that under increasing pressure, the diaphragms deform, causing the effective or balance diameter of the bellows to reduce. It is the relative position of that balance diameter to the seal face which determines the seal balance ratio. Most seals operate with balance ratios in the range of 65 - 80%.

Thus a welded bellows seal which at low pressure may be designed to operate with a balance ratio of 75%, may at a pressure as low as 10 bar be operating with a balance ratio approaching that of an unbalanced seal. At higher pressures, therefore by virtue of it becoming more unbalanced, a bellow seal will generate more frictional heat than an equivalent pusher seal.

3.2.5 Clogging due to Solids

When installed, the gap between adjacent diaphragms is typically less than 1mm. Thus on services where there is a high degree of entrained solids, clogging of the bellows can occur resulting in 'hang-up' of the bellows. In such services it is often recommended to rotate the bellows element to centrifuge solids from between the convolutions. This can help, but rotating the bellow increases the possibility of it failing due to fatigue, as noted in the draft API standard 682.

4.0 A NEW CONCEPT IN BELLOWS SEALS - THE TWIN PLY FORMED BELLOWS

4.1 PHILOSOPHY

It became clear during the mid 1980's that increasing pressures on operators to improve seal reliability and reduce emissions would require an improvement in metal bellows seal technology.

A new design of seal incorporating the advantages of the welded bellows seal whilst maintaining the benefits of pusher seals would be required - in particular their robustness and higher pressure capability.

It was clear that to continue with the welded diaphragm construction would limit advances. High pressure versions of welded bellows already existed but were generally regarded as being too stiff and specialised for general purpose use.

The twin-ply formed bellows was identified as the way forward and a four year development programme was set in place.

4.2 GENERAL CONSTRUCTION

The twin ply formed bellow is made by an entirely different method than the edge welded seal.

The start point is two tubes of Inconel, which are produced by cutting and folding sheet material extremely accurately and joining with a single longitudinal butt weld. The tubes differ in diameter by twice the thickness of the material, and the smaller of the tubes is inserted in the larger tube, with the butt weld 180° apart.

Then using a series of rollers and mandrels, the composite tube is formed into a bellow and in the process both tubes are brought into intimate contact. (See Fig. 3).

With the initial prototype seals the ends of the bellow core was a short cylindrical portion which was attached to the drive and end rings using various mechanical techniques such as swaging, gluing and shrinking (See Fig. 4). However, in some cases the location proved to be unreliable and an alternative method for terminating the end of the bellow core and attaching to the end and drive rings was developed.

With production units, the end convolution is flattened and neatly cut at the ID to create the ideal shape for welding to the end and drive rings. The sturdy construction of the end flange which is now four thicknesses of material is ideally robust for storage, handling and produces a weld bead at each end of the core which is four times larger and hence stronger than the normal welded bellows seal.

Thus not only is the weld substantially stronger than the edge welded seal, the weld length is also reduced by approximately 90%!

4.3   <u>WHY TWIN PLY?</u>

The reason for using a multi-ply bellow seal is very simple. The pressure capability of any bellows seal is dependent on the material thickness, the material strength and the basic profile. The axial stiffness is dependent on the cube of the material thickness.

Thus for a typical edge welded bellows of material thickness equal to 0.16 mm, the axial stiffness is proportional to 0.16^3. To achieve the same pressure rating using a twin ply bellows, the equivalent thickness of each ply need only be 0.08mm, as the forming process ensures that both plies are in intimate contact. The pressure rating is proportional to the number of plies and the material thickness. In the case of the twin ply bellows, as with metallic membrane couplings very slight sliding between the two tubes, means that the stiffness is proportional to 2×0.08^3. The ratio of stiffness is thus:-

$$\frac{0.16^3}{2 \times 0.08^3} = 4 \text{ to } 1.$$

The twin ply construction therefore allows a much better axial stiffness/pressure capability and offers a high margin of safety in the design. This is particularly important for the DIN L1K versions where the seal must have a reasonable amount of working compression within a very short axial length.

Theoretically increasing the number of plies lead to even greater stiffness/pressure capability. However, the availability of material in particular thicknesses and the ability to form into butt welded tubes sensibly limits the number of plies to two.

4.4 MATERIALS OF CONSTRUCTION

Inconel 625 was selected for the bellows material due to its superior corrosion resistance and strength in the formed condition. Unlike other grades of Inconel it requires no heat treatment for strength or high temperature operation.

With the High Temperature Version of the seal which operates up to 450°C all metal components are manufactured in Inconel 625 and the carbon seal face is uniquely mounted by a technique which squeezes the carbon between Inconel. This gives the face exceptionally stability and pressure vs temperature characteristics - a design feature first incorporated in high duty seals.

In the General Purpose and DIN L1K versions the drive and end rings are manufactured in duplex stainless steel, and can operate up to 250°C.

All versions of the seal offer carbon vs SiC or 2 hard face options.

5.0 **TESTING & DEVELOPMENT**

Prior to any field trials taking place, over 12,000 hours of testing were carried out in the laboratory on a variety of seal sizes ranging from 25 to 105 mm.

Testing was designed to investigate:-

i)   Pressure limits
ii)   Resistance to fatigue
iii) Torsional stiffness
iv)   Temperature limits
v)   Balance shift

5.1 Dynamic and Pressure Testing

Dynamic testing was carried out up to 60 bar, but generally at 30 bar on water at 50°C and 3000 rpm and 0.25 degrees misalignment (significantly more than a welded metal bellows will tolerate). All tests were conducted for a minimum of 10^8 cycles (550 hours) to confirm the fatigue resistance. To determine the limit of misalignment capability, misalignment was gradually increased until failure was induced. This occurred when running a 100mm seal with 1mm TIR misalignment at 30 bar, a remarkably high level of misalignment.

Additionally static testing was carried out to 140 bar (forward) and 300 bar reverse).

On one occasion, a 60mm seal was reverse pressurised to failure at 320 bar g.

A number of long term tests at elevated temperature were conducted to confirm the creep resistance at 500°C.

5.2 Torque Testing

One area where the formed bellows was predicted to benefit over the edge welded seal was under torque. Torque testing was carried out to determine the torsional stiffness of the bellows and the ultimate maximum torque capability. Fig. 7 shows the results for both formed and welded seals and demonstrates the significantly higher torque capability of the Saflex seal before yielding when compared with the welded seal.

This graph also demonstrates the relatively small angular deflection which occurs in a welded bellows seal before yielding. Some seal designs incorporate overload 'dogs' to prevent the bellows yielding - particularly common on bitumen type service where cold product can bond the seal faces together at start-up. With such a small angular deflection allowance it can be very difficult to obtain the correct degree of clearance in the overload dogs, using the welded arrangement.

With the formed seal the increased deflection before yielding permits more easy installation of a dog drive but more importantly has demonstrated that the dog drive becomes redundant in most cases, as the bellow can tolerate such an increase in torque.

Additionally, due to higher torsional stiffness of the formed seals, there is less need for vibration dampers sometimes found on welded bellow seals.

5.3 Balance Line Shift

Testing was carried out to determine the degree to which the balance line or effective diameter altered with increasing pressure. Special test capsules were constructed to allow both edge welded and formed seals to be pressurised and the resultant load measured.

As expected, the "open" design of the formed convolution resulted in very low levels of balance line shift as can be seen from   Fig. 8.

This aspect of the design allows the 30 bar pressure capability of the seal to be fully exploited.

6.0 **FIELD EXPERIENCE**

The first seals were supplied for field trials in 1991 in several locations around the world and in a number of different markets including:-

i)   Chemical industry
ii)   Refinery industry
iii) Pulp and paper
iv)   Mining and slurries

The longest running seal has run for over 2½ years and is fitted in Australia on an application operating at:-

Seal size	:	40mm
Speed	:	2950rpm
Product	:	30% Caustic + H_2S
Temperature	:	100°C
Pressure	:	12 bar g

To-date over 520 seals have been supplied around the world. Over 60 double seals have been supplied in sizes ranging from 50 - 95mm.

One such example has been operating successfully at a UK refinery for 10 months, where previously the double pusher seal operated for 2 months maximum. The 80mm seal shown in Fig. 9 is installed on a critical, unspared coker charge pump operating at 420°C and the seal operates with a barrier pressure of 14 bar g.

There is a trend in continental Europe for pressurised tandem arrangements to be used. One advantage is that any solids are centrifuged away from the seal faces. The seal shown in Fig.10 installed at a French refinery and has been operating for over 12 months. The 80 mm seal is fitted on a pump operating at 370°C and a barrier oil pressure of 13 bar g.

The use of such bellows are not limited to refinery applications. A Swedish customer, who has used Flexibox pusher seals for many years in marine service identified the formed bellow seal as an ideal solution for preventing marine growth on the sliding diameter. This growth has caused hang-up problems with some pusher seals. The 145 mm seal shown in Fig. 11 offers a simple design, with split, replaceable seal faces, operating on sea water at 4 bar g and 1450 rpm. The excellent misalignment capability is more than sufficient to cope with that encountered on-board the vessel.

The versatility of the formed bellow seal though is perhaps demonstrated best by Fig. 12 which is a tandem version being supplied on a new refinery in Thailand. This is the first refinery to be built requiring seals to the new API 682 specification, which provides limited axial space for mechanical seals. Due to the versatility of the Saflex range, a suitable arrangement has been supplied complying with the specification and providing an elegant sealing solution .

7.0   <u>CONCLUSIONS</u>

A requirement existed for a step improvement in metal bellows seal technology which had not really significantly advanced since the 1970's.

By using a twin ply formed bellow, the Saflex range of High Temperature General Purpose and DIN L1K seals answer that need.

This design marries the benefits of simplicity associated with edge welded seal to those of robustness, versatility and pressure capability of a pusher seal.

Such a seal will be a significant aid to improved seal reliability and reduced emissions.

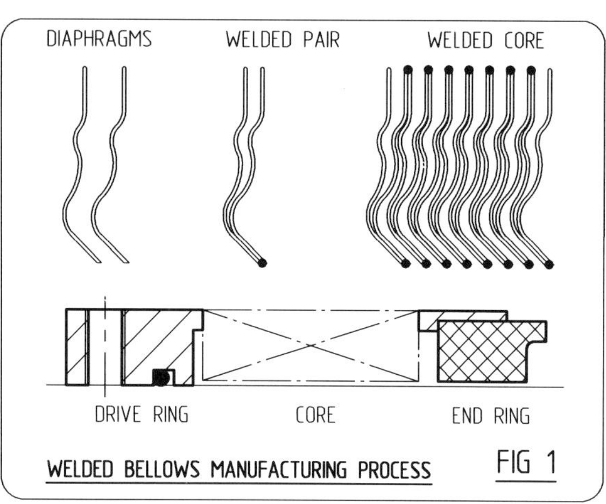

flexibox
DIAPHRAGMS
WELDED PAIR
WELDED CORE
DRIVE RING
CORE
END RING
WELDED BELLOWS MANUFACTURING PROCESS
FIG 1

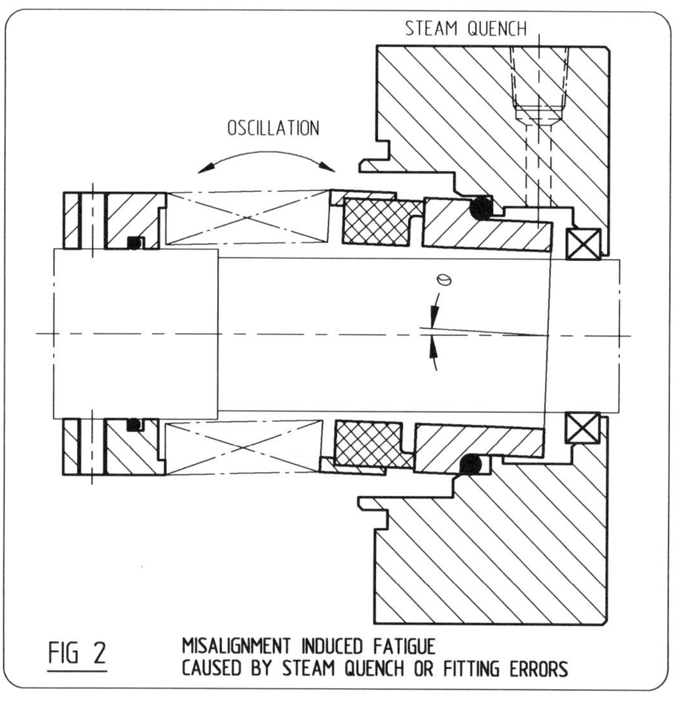

STEAM QUENCH
OSCILLATION
FIG 2
MISALIGNMENT INDUCED FATIGUE
CAUSED BY STEAM QUENCH OR FITTING ERRORS

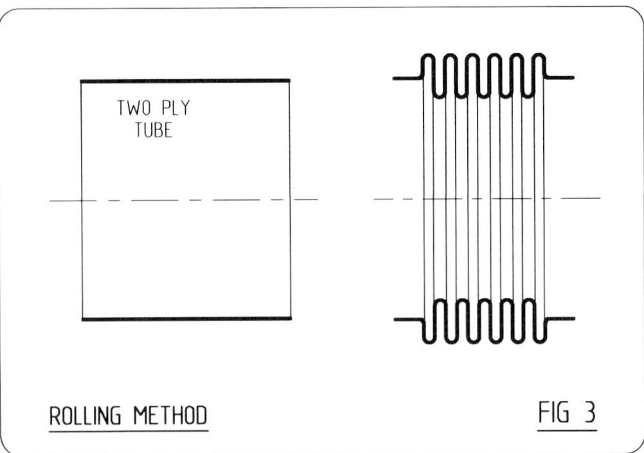
⊕flexibox
TWO PLY
TUBE
ROLLING METHOD
FIG 3

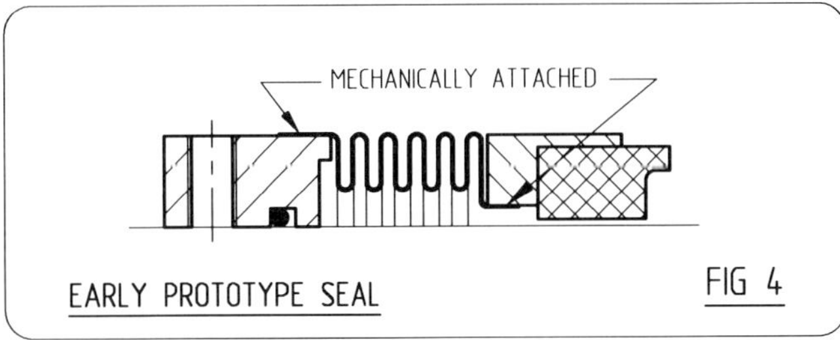
MECHANICALLY ATTACHED
EARLY PROTOTYPE SEAL
FIG 4

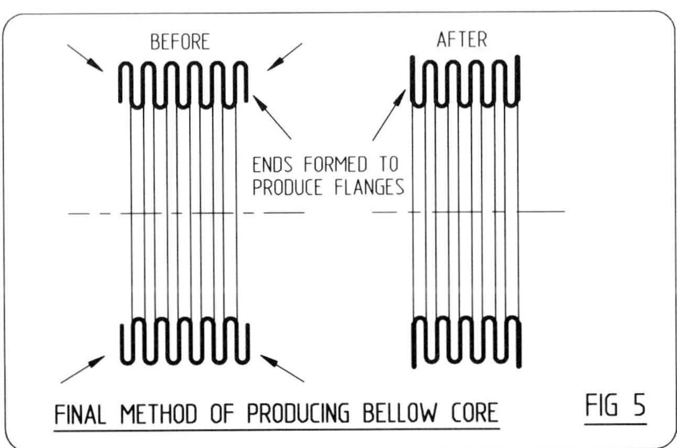
BEFORE
AFTER
ENDS FORMED TO
PRODUCE FLANGES
FINAL METHOD OF PRODUCING BELLOW CORE
FIG 5

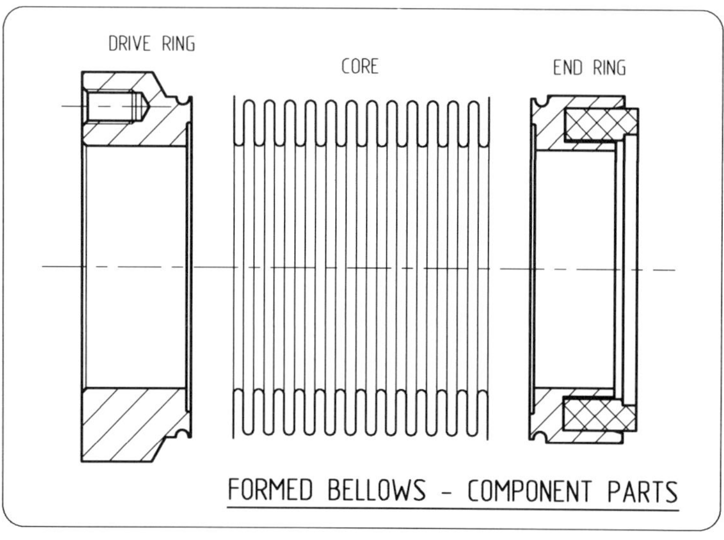

FORMED BELLOWS - COMPONENT PARTS

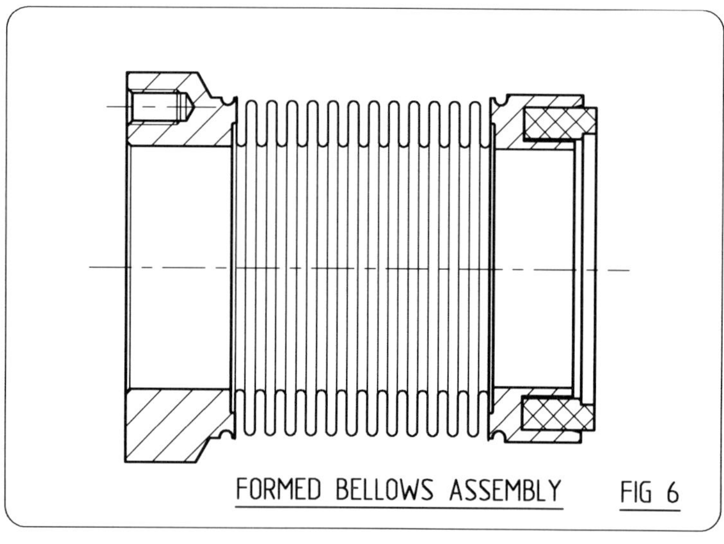

FORMED BELLOWS ASSEMBLY   FIG 6

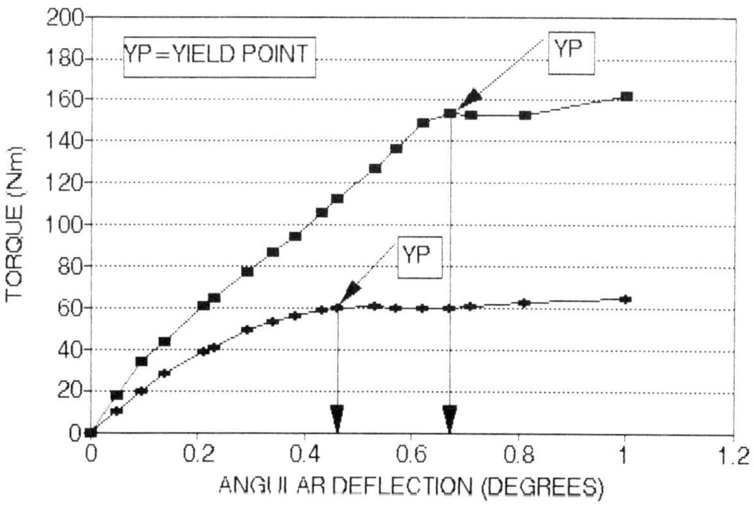

COMPARISON OF WELDED & FORMED SEAL
UNDER TORQUE TEST

FIG 7

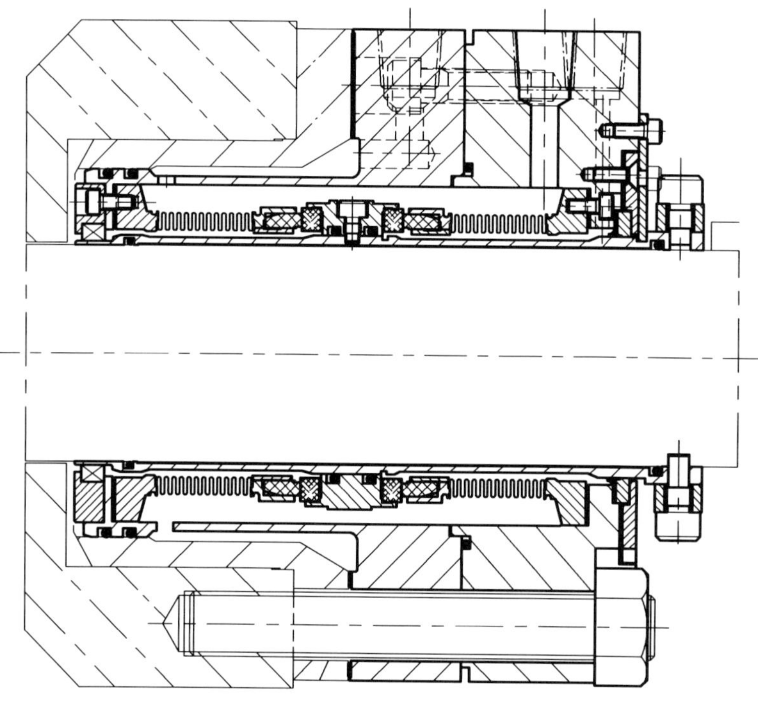

FLEXIBOX MECHANICAL DOUBLE
BELLOWS SEAL ON
HIGH TEMPERATURE SERVICE

FIG 9

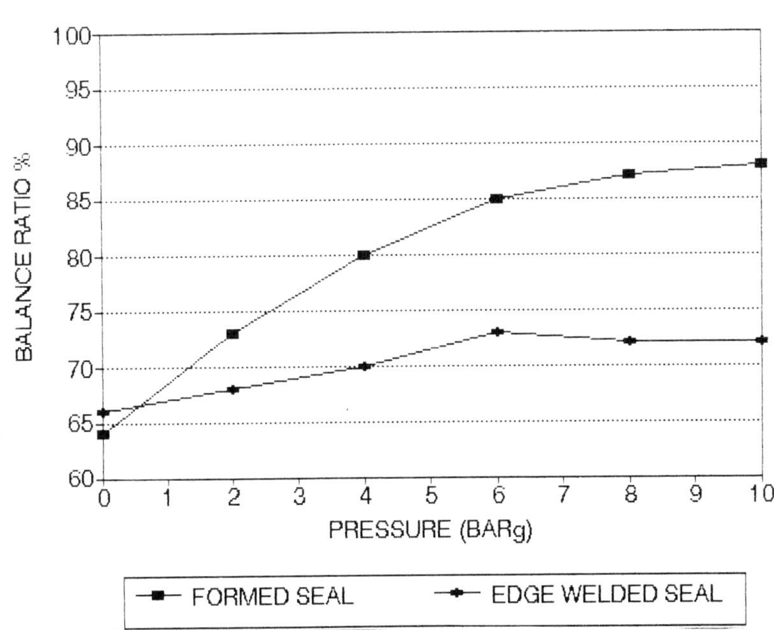

VARIATION OF BALANCE LINE WITH PRESSURE

FIG 8

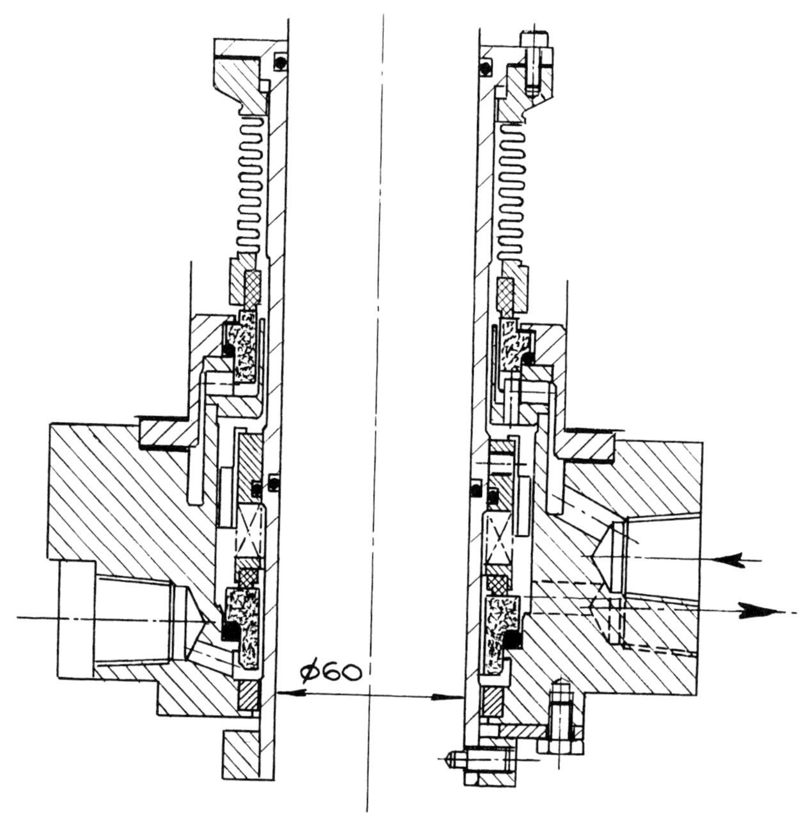

PRESSURIZED TANDEM (DOUBLE) SEAL
FOR HIGH TEMPERATURE SERVICE

FIG 10

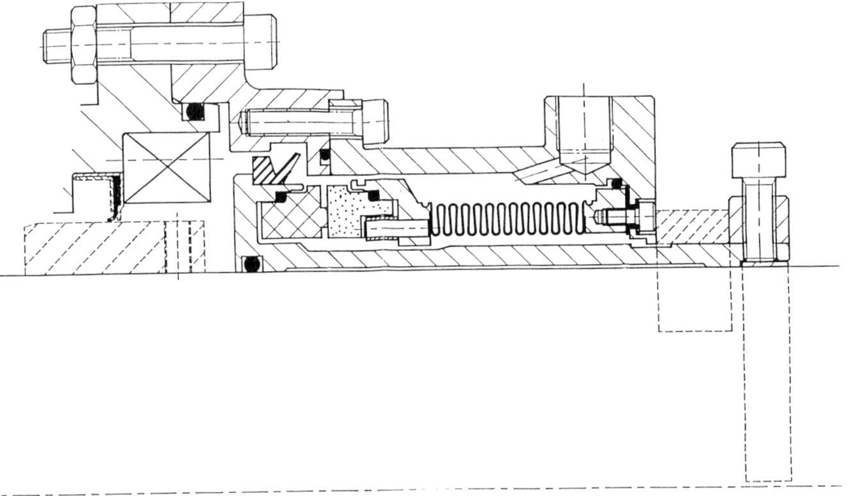

flexibox
150mm WATER JET SEAL WITH
SPLIT FACES - SiC Vs SiC
FIG 11

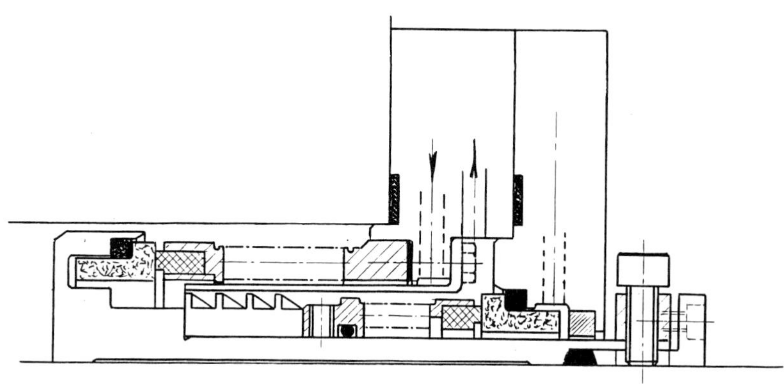

TANDEM ARRANGEMENT FOR REFINERY
IN THAILAND

FIG 12

14th International Conference on Fluid Sealing, Firenze, Italy,
6-8 April 1994. Organised by BHR Group Limited, Cranfield,
Bedford, MK43 0AJ, UK; Tel: 0234 750422

A SHAFT SEALING SYSTEM WITH GAS BARRIER

A.I.Golubev, N.V.Degterev.
Gidromash, Moscow, Russia

Summary

When pumping liquids hazardous for environment (toxic,
explosive and so on) the usual practice is to use double
mechanical or tandem seals for pump shaft, supplying a
neutral liquid (water, oil) in chamber between the sealing
stages. The use of sealing fluid sometimes is impossible
due to contamination of pumping fluid and spoiling the
process technology. A gas barrier sealing system is
discussed, which consists of an internal mechanical seal
and external radial gap seal to which a neutral gas
(usually nitrogen) is supplied. The seal was successfully
used in the process pumps, pumping dichlorethane in PVC
production lines. A similar sealing system was designed for
a submersible tank pump.

Centrifugal pumps handling chemically agressive, toxic,
flammable and explosive liquids are usually provided with
double mechanical seals of the pump shaft. A tandem seal
which is less practicable may be also considered as a
double seal. These seals are complicated in design, are not
reliable and do not exclude the pumping fluid contamination
with the sealing fluid (oil, water). There are cases when
such contamination must be avoided because it spoils the
process technology. External circulation, cooling and
cleaning systems must be used for double seals. These
systems are usually not hermetic and lead to contamination
of environment. These drawbacks of double mechanical seals
led to the development of a new sealing system in
Gidromash, which uses neutral gas (nitrogen) for a sealing
fluid. Some sealing systems using a neutral gas for a
sealing fluid are known in practice. They consist of
internal mechanical seal and external rigid radial gap
seal. A neutral gas with pressure lower than sealing
pressure is supplied to the chamber between the sealing

stages. This sealing system also has drawbacks. The leakage
of the pumping media is mixed with neutral gas. The vapor
concentration of the pumping fluid is lessened, but the
contamination of the environment is not excluded. The bush
of the radial gap seal is rigidly connected with the pump
casing, so the radial gap must be of sufficient size. (It
must be more than the sum of the shafts run out and radial
misalignment of the bush axis). It leads to excessive gas
leakage.

A sealing system with gas barrier developed in Gidromash
consists of a mechanical seal and a radial gap gas-static
seal (Fig.1).

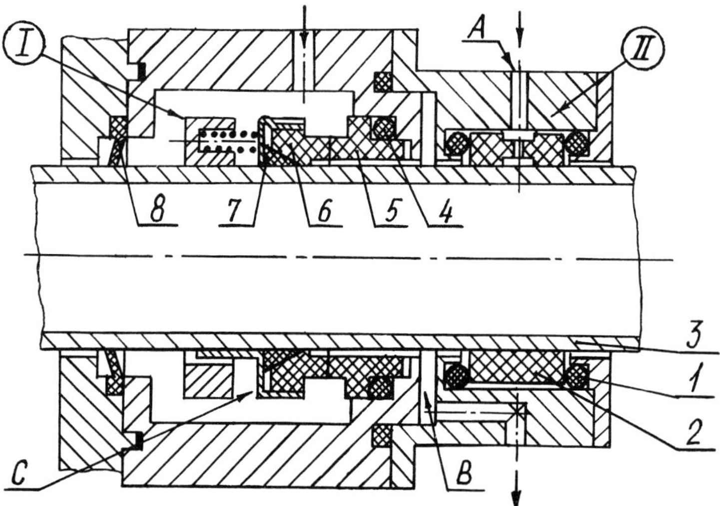

Fig.1 Process pump seal for PVC production technologies.

A Fig.1 shows a gas-static seal with throttle adjustment of
a radial gap. The acting principle of a gas seal is similar
to a gas-static bearing. The sealing gas is supplied
through the port "A" to a circular chamber "B" and flows
through the cylindrical throttles. At the part of the flow
channel, where the gap is minimal, the pressure nears to a
gas supply pressure. This leads to developing a force,
centering the sealing bush 2 relative to shaft acting
against the counteraction of elastic ring friction between
the rings 1 and bush 2. The self-adjustment of bush 2
relative to shaft excludes the contact between the same and
so excludes the wear. The pressure of gas supply should be
above the pressure in chamber "B". Part of the barrier gas

flows into this chamber and then mixing with vapors of pumping liquid flows out by the drainage pipe. Other part of the sealing neutral gas flows into atmosphere. The self-adjustment of the sealing bush lets the manufacturer to make the seal with small gaps and provides small gas leakage. At the same time the gas supply in central part of the seal excludes leakage of vapors of pumping media into atmosphere.

A seal unit shown on Fig.1 was developed for centrifugal pumps used in Kaloush "Chlorvinyl" syndicate (Ukraine) for pumping the following media:
- dichlorethane-64%;
- vinylchloride-32%;
- hydrogen chloride-4%.
- the media contains up to 0.1% of hard particles.

The pumps were supplied with double mechanical seals. Oil was used in the seals for the sealing fluid. Experience has shown that there were frequent cases of oil leakage into the pumping media. The oil carbonized at the friction couple of the internal mechanical seal, which led to seal failure. Contamination of pumping media by oil led to clogging of rectifying columns and to stopping of the process technology. The seals of centrifugal pumps mentioned above work at liquid pressure equal to 2.5 MPa, temperature - +80 C, rotation velocity - 1500 rpm. The seal shown on fig.1 was used for these conditions. The main mechanical seal is placed on the 75 mm shaft sleeve. Its overall dimensions are in accordance with standard ISO 3069-74. Friction couple of the seal consists of non-rotating ring 5 and rotating ring 6 : both are made of GAKK 55/40 siliconized graphite. The secondary 0-ring and wedge spherical ring 7 seals are made of Teflon. Sealing bush 2 of radial gap seal is made of lead-impregnated carbon graphite. The central part of the bush is perforated with 24 ports, the diameters of which are calculated. Radial movement of bush 2 is ensured by its installation on the rubber 0-rings which are the secondary seals for sealing gas - nitrogen, which is available in the PVC production technology. As there are hard particle impurities in the pumping media, a supply of pure dichlorethane is provided in chamber "C" between the mechanical seal and Teflon lip seal 8. Consumption of pure dichlorethane is valve regulated. In case, when dichlorethane supply is stopped, the lip seal 8 prevents pumping media entrance to the mechanical seal. Dichlorethane leakage through the mechanical seal, usually in vaporized state, flow to chamber between the mechanical and the radial gap seals, where it mixes with the nitrogen in-flow and the mixture is drained through the drain-pipe to the sanitary column.

Calculations of the radial-gap seal are based on the condition of maximal centering gas pressure force.

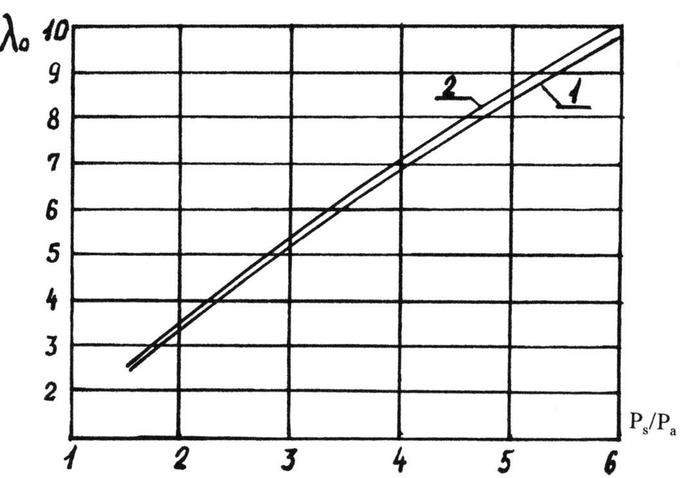

Fig.2 Relation between optimal factors $\lambda 0$ and gas pressure P_s: 1 - L/D=0.2; 2 - L/D=0.5.

The graphs on Fig.2 show the calculated relations of gas-dynamic factors $\lambda 0$ depending on input gas pressure

$$\lambda_0 = \frac{12\mu d_0 n_0}{c^2 \sqrt{P_a \rho_a}} \cdot \frac{L}{D}$$

where
- μ - dynamic viscosity of gas;
- d_O, n_O - diameter and numbers of input throttle ports;
- L ,D - length and internal diameter of sealing ring;
- c - radial gap in seal;
- P_a, ρ_a - absolute pressure and density of gas after the seal.

Curves on Fig.2 are calculated in presumptions of laminar and isothermal flows through the gap; eccentricity of ring 2 relative to shaft was considered small compared to radial gap in seal. Besides, the flow of gas in throttle ports was considered isoenthropic (isoenthropic coefficient is equal to 1.4, the flow rate coefficient - 0.8). After calculating the optimal quantity of factor λ_0 the number and diameter of input throttle ports are evaluated. Finally, the following dimensions of gas seal were selected: D = 85 mm; L = 25 mm; n = 24; do= 1.6 mm; c = 0.04 mm. Mass flow rate of sealing gas (nitrogen) G_O is evaluated at input pressure 0.3 MPa:

$$G_0 = \frac{\pi c^3 P_a \rho_a}{12\mu} \bullet \frac{D}{L} \bullet \left[\left(P_m / P_a \right)^2 - 1 \right]$$

where
- P_m = P_a+K(P_s-P_a) - maximal pressure of gas at central part of bush 2;
- K = 0.6 - 0.61 - coefficient.

The calculated flow rate of gas through the seal is G_O = $1.8 \bullet 10^{-3}$ kg/s, which is equal to 5 m^3/h at normal conditions. The industrial two years experience with pumps equipped by described seals has shown its high efficiency and reliability.

The submersible tank pump is designed for pumping the phtalic and maleic anhydrides in dye production technologies. The vapor mixtures of said anhydrides with air are explosive. To prevent explosions the tanks are pressurized with nitrogen above the liquid level. At the sealing part of the shaft there is nitrogen saturated with liquid vapors. The temperature of the pumped phtalic anhydride is +135 to +240 °C, the maleic anhydride - +55 to +150 °C. The nitrogen pressure is near to atmospheric (±2 KPa). The pump seal was developed instead of the used before double mechanical seal. It consists of two seals (Fig. 3) - the first of the radial gap gas-static type I and the second - mechanical stop seal II.
Design of the radial gap gas-static seal is similar to the one described above, Fig. 1. Nitrogen is supplied to the seal. The main difference between the seal on Fig.1 is in the use of Teflon lip seals 5 as secondary seals instead of rubber O-rings due to rubber indurability under described conditions. The stop seal is intended for preventing leakage at non-running pump. At the pump stop the sealing contact of rings 1 and 2 is obtained by the force of springs 3. At pump start centrifugal forces of weights 4 act upon the springs to discontact the rings 1 and 2. To provide safety of pump operation the following systems are installed:
- local evacuation system in the zone of nitrogen outflow;
- blocking and protection systems to prevent the pump start at non acting evacuation system, pressure drop of

nitrogen supply behind the prescribed limit, the
temperature drop of sealing casing lower than the product
melting point (+5 ^{O}C);
* sound and light alarm systems at emergency pump shut-off;
* the nitrogen supply line is provided with non-return
valve.

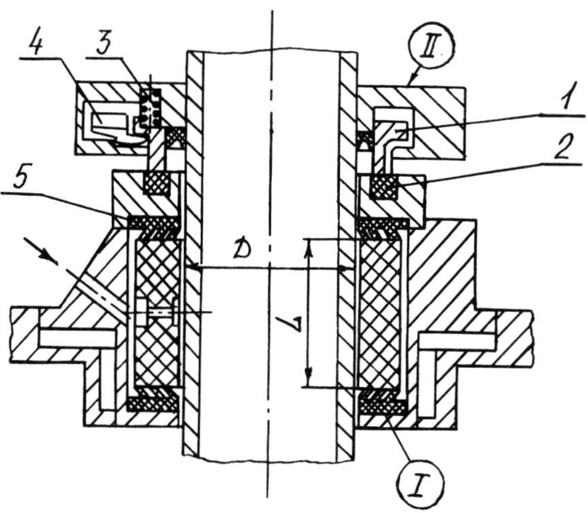

Fig.3 The seal for submersable tank pump.

The calculations of radial gap seal were done by the
described procedure. The following dimensions of the seal
were obtained: L = 40 mm, D = 50 mm, d_O = 0.3 mm, n_O = 12.
The input pressure of nitrogen P is 0.2 MPa. Mass flow of
nitrogen G_O is equal to $42.9 \bullet 10^{-6}$ kg/s (0.2 m^3/h at normal
conditions). The seal was tested and delivered to the
customer.

Conclusions

The neutral gas-barrier sealing systems are effective for
pumps handling toxic, explosive and other agressive
liquids. They have advantages over the double mechanical
seals in reliability and better protection of environment
from contamination.

Literature
Degterev N.V. and Furmanov B.A. "New seals constructions
for pumps transporting toxic, flammable and explosive
liquids". Chimicheskoe i nephtyanoye mashinostroyeniye, 4,
April 1985, p.p. 10 - 11, in Russian.

14th International Conference on Fluid Sealing, Firenze, Italy,
6-8 April 1994. Organised by BHR Group Limited, Cranfield,
Bedford, MK43 0AJ, UK; Tel: 0234 750422

ECCENTRIC SEALS FOR NUCLEAR PUMPS

R. METCALFE, T.A. GRAHAM and W.C. WONG
Fluid Sealing Technology Unit, AECL Research
Chalk River Laboratories, Chalk River, Ontario, Canada K0J 1J0

ABSTRACT

End face seals can be mounted eccentrically to promote lubrication, cooling
and lifetime in high performance applications. Such seals have proven
useful for high PV (pressure x velocity) or vaporizing conditions—low
pressure, volatile liquids—where conventional balanced seals that rely on
hydrostatic or boundary lubrication experience problems.

Eccentricity intermittently exposes the heated seal face to ambient
temperature and carries fluid across the seal in both directions, thus
cooling the interface. There can in theory be more fluid carried upstream to
the higher pressure side than leaks back to the lower pressure side of the
seal in fully-flooded liquid conditions. These cooling and lubricating
effects are described and explored. The alternating heating-cooling of the
rotating seal face as it is covered and uncovered every revolution is shown
to be the predominant cooling mode in typical conditions.

A tandem seal cartridge containing two identical, but 180° offset, eccentric
seals of 95 mm balance diameter was developed for the main heavy-water
pumps of the NRU research reactor at Chalk River Laboratories. The
requirements were low pressure (0.3 MPa), low leakage (0.2 mL/min) and
long, reliable lifetime (seven years). The seals were arranged to put full
pressure across the outboard seal initially, with the inboard seal providing
complete back-up in the event of outboard seal failure.

Laboratory testing of the eccentric seal cartridge is described. This
comprised basic characteristic testing, short-term running and long-term
running (1500 h, 1700 rpm). There was very low leakage and wear. The
first seal cartridge was installed in the reactor in November 1992 and has
since been operating without problem.

Nomenclature

A	- boundary layer shape factor
A_u	- area of rotating face half-heated, half-cooled in "under-eccentric" seals $[m^2]$
A_o	- area of rotating face fractionally heated and cooled in "over-eccentric" seals $[m^2]$
B_h	- hydraulic balance ratio
B_n	- hydraulic "inbalance"
C_r	- rotor material heat capacity $[J/(kg.K)]$
C_w	- sealed fluid heat capacity $[J/(kg.K)]$
e	- eccentricity $[m]$
e_{crit}	- critical eccentricity just to cause static tilted opening between seal faces $[m]$
f	- frequency $[s^{-1}]$
F_t	- total seal closing force $[N]$
h	- seal face uniform separation $[m]$
H_{cond}	- overall seal face heat transfer coefficient for conduction $[W/K]$
H_{alt}	- heat transfer coefficient for rotating face area alternately heated-cooled $[W/K]$
k_r	- thermal conductivity of the rotor $[W/(m.K)]$
p_i, p_o	- gauge pressure inside and outside the seal face $[Pa]$
Δp	- pressure drop across the seal face $[Pa]$
q	- heat flux $[W/m^2]$
Q	- concentric, flat-faced seal leakage $[m^3/s]$
Q_e	- extra flow through seal faces due to eccentricity $[m^3/s]$
r	- radial co-ordinate $[m]$
r_i, r_o	- raised seal face inner and outer radii $[m]$
Δr	- raised seal face width, $r_o - r_i$ $[m]$
R	- seal aspect ratio, r_i / r_o
$T, \Delta T$	- temperature, temperature above ambient $[K$ or $°C]$
T_a	- ambient temperature in the seal surroundings $[K$ or $°C]$
T_{sf}	- average temperature of the stationary seal face and between seal faces $[K$ or $°C]$
V_r	- radial component of velocity $[m/s]$
W_{cond}	- total heat conducted away from seal face through both seal rings $[W]$
W_g	- total viscous heating between seal faces $[W]$
x, y, z	- co-ordinate directions
ρ_w	- sealed fluid density $[kg/m^3]$
μ	- dynamic viscosity $[Pa.s]$
θ	- angle with respect to eccentricity axis position $[rad]$
ω	- angular velocity $[rad/s]$

BACKGROUND

End face seals can be deliberately mounted eccentrically to improve their performance. Such seals have proven useful for high PV (pressure x velocity) or vaporizing conditions—low pressure, volatile liquids—where conventional balanced seals that rely on hydrostatic or boundary lubrication experience problems.

For nuclear primary pumps, eccentricity was first used in the late sixties (1). This worked effectively in the Caorso (Italy) plant. The principle was borrowed from previously developed elliptical and wavy seals (2), where all of the rotating seal face was arranged to alternate between being covered then uncovered by the opposing stationary, non-circular, raised seal face. This had been found to promote lubrication, cooling and lifetime in high performance applications (3).

One problem with early designs was non-uniform loading and the resulting deflection of the seal faces, which became excessive at the greater pressures and sizes of subsequent nuclear primary pumps. This always occurs with elliptical or wavy seals, but can be minimized in an eccentric seal by placing the raised seal face on the flexibly-mounted seal ring, preferably the stationary one, then offsetting it to be eccentric (4). This ring can then maintain its axisymmetric shape and loading (Figure 1).

Without this "special" arrangement (4), eccentricity in conventional seals continued to cause problems of "consequential tilting." Accurate centering of shafts in large nuclear pumps was difficult, and too much eccentricity was found to cause tilted contact or separation unless the balance ratio (closing force) on the seal was boosted to compensate (5).

Through the eighties, leakage of conventional seals became more troublesome because of tighter regulations against the release of toxic or other hazardous fluids. Concepts for minimizing or eliminating leakage were sought. Rather than relying on auxiliary pumping devices or external support systems, passive means were explored for using the inherent rotation of a seal to pump upstream as much or more than might otherwise leak. Various complex face geometries incorporating hydrodynamic features were pursued (6, 7). It was realized that "deliberate tilting" of an appropriately arranged eccentric seal is a simple solution (8).

OBJECTIVE

The overall objective of the work described here was to improve the understanding of eccentricity effects and to take advantage of them in high performance nuclear applications.

The principles, theoretical tilt limits and heat transfer advantages of the eccentric arrangement are shown. Development of a tandem seal cartridge containing two identical, but 180° offset, eccentric seals of 95 mm balance diameter for the main heavy-water coolant pumps of the National Research Universal reactor (NRU) at Chalk River Laboratories is described. This was tested for more than 3000 hours before being installed.

ECCENTRIC SEAL THEORY

General Principles

Eccentricity creates hydrodynamic lift at a seal's edges, which act as Rayleigh steps. It also carries fluid across the seal in both directions, thus rapidly replacing the fluid that is heated between the seal faces with cool fluid. For further cooling effect, it alternately covers and then uncovers one seal face during every revolution, intermittently exposing this heated seal face to ambient temperature.

Eccentricity may cause the seal faces to tilt relative to each other. This consequential tilting is generally unwanted, since it increases leakage and may cause rubbing contact and complete tilted separation. However, if the seal is arranged to prevent consequential tilting, the closing force can be offset to cause deliberate tilting about a different axis, and in theory carry more fluid upstream to the higher pressure side than leaks back to the lower pressure side. For this, the seal must be flooded both sides with liquid and the tilt must cause a thicker film around the half of the seal where eccentricity carries liquid upstream than at the half where it is carried downstream (8). Pressure, speed, viscosity, separation and other factors determine the net result, but upstream pumping may be accompanied by full lubrication of the seal faces, giving long, reliable life.

Tilting due to Eccentricity

For analysis, seal faces are assumed essentially flat and rigid, with one of them flexibly-mounted to move axially or tilt about a diameter. The sealed fluid is assumed to be incompressible liquid, flowing under laminar, isoviscous conditions between seal faces with curvature slight enough to be neglected ("narrow seal" approximation).

For seals that do not have the "special" arrangement (4) and are therefore sensitive to eccentricity, there is a critical amount that will cause problems of rubbing or excessive leakage. To derive these for inside- or outside-pressurized seals, the "inbalance," B_n, of the seal has been defined as,

$$B_n = \frac{\pi (p_i + p_o)(r_o^2 - r_i^2) - 2F_t}{2\pi (p_i - p_o)(r_o^2 - r_i^2)}$$

This has been corrected from Ref. (5). The critical amount of eccentricity, e_{crit}, just to cause the seal faces to tilt completely open under static conditions is then,

$$e_{crit} = \frac{2r_o B_n (1 - R^2)}{(1 + R^2) - 2B_n (1 - R^2)}$$

This is plotted as a function of balance ratio and aspect ratio of the seal in Figure 2.

A lesser amount of eccentricity is sufficient to cause first tilted contact if the seal is fully lubricated by the hydrostatic effects of face coning, as shown in Ref. (5). This is also plotted in Figure 2, which shows both static effects for inside- and for outside-pressurized seals.

Extra Cooling due to Eccentricity

<u>Introduction to Cooling</u>

Eccentric seals that are designed with the "special" arrangement shown in Figure 1 are not susceptible to tilting problems, because there is no such moment on the flexibly-mounted seal ring, only on the fixed seal ring—the rotating one in this case. This can be made essentially rigid (carbide) and fixed to the shaft.

With full film liquid lubrication between the faces of a conventional concentric seal, temperature rise due to viscous heating can readily be calculated as a function of separation, speed, pressure, and the surrounding thermal conditions (9). Liquid enters between the seal faces at the upstream ambient temperature and becomes hotter as it passes through. The temperature distribution throughout the seal is steady-state, corresponding to the particular running conditions. Viscous heat escapes from the sealing interface either by conduction through the seal rings or by heating of the leakage.

The eccentric seal, by contrast, has two additional mechanisms for transfer of heat away from the sealing interface. One is the extra flow of liquid being dragged radially in and out of the seal due to the eccentricity. The other is the extra heat given up to ambient by portions of the rotating seal face that are alternately uncovered (cooled) then covered (heated)—akin to the "pumping" of brakes to prevent overheating.

<u>Extra Cooling due to Extra Flow in Eccentric Seals</u>

To assess the cooling effect of the extra flow, consider first the leakage from a concentric flat-faced seal with uniform separation, h, pressurized from the outside,

$$Q = \frac{-\pi h^3 (p_o - p_i)}{6\mu \ln(r_i / r_o)}$$

When the stator is mounted eccentrically, there is extra flow dragged into the sealing interface around half the outside and half the inside circumference by the rotating face's velocity component, V_r, (see Figure 1) normal to the edge of the raised seal face on the stator (10). This gives a total extra inflow of,

$$Q_e = - A \int_0^\pi \int_0^h r_i\, V_r\, dz\, d\theta + A \int_\pi^{2\pi} \int_0^h r_o\, V_r\, dz\, d\theta$$

where $V_r = - \omega\, e \sin\theta$. This gives inflow (i.e., extra through-flow) as,

$$Q_e = 2A\, \omega\, e\, h\, (r_o + r_i)$$

The value of the boundary layer shape factor, A, depends on the velocity profile of liquid near the rotating face as it approaches the edge of the seal face. This will vary slightly with radial position and rotational speed, but its lower bound is $A = 0.5$ (i.e., velocity profile the same outside as between the seal faces), and its upper bound is $A = 1$ (i.e., uniform fluid velocity equal to that of the rotating face). The latter will be used subsequently as the better approximation, since the seal face separation is always extremely small compared with other clearances.

The conventional (concentric), pressure-driven flow is compared in Figure 3 with the extra eccentric flow over a range of seal gaps, pressures and eccentricities for the NRU seal, as specified in Table 1. Eccentric flows are seen to be relatively larger at smaller gaps and lower pressures, showing where major improvements in convective cooling due to the extra flow should be anticipated. This also indicates the potential for extra lubrication of the seal faces.

Table 1: Material and Geometric Details of NRU Seal for Water

DETAILS OF NRU SEAL									
	Seal Face Material				Sealing Face Details				
Part	Type	Density (kg/m^3)	Thermal Conductivity (W/m.K)	Specific Heat (J/kg.K)	Outer dia. (mm)	Inner dia. (mm)	Balance dia. (mm)	Stator Eccentricity (mm)	Spring Force (N)
Rotor	Tungsten Carbide	14900	90	180	102	93	95.3	2.4	89
Stator	Carbon Graphite	1400	20	750					

Viscous heating in the liquid film between the seal faces becomes a limiting factor in low leakage, high speed seals because it leads to vaporization and loss of lubrication. For uniform separation total heat generation for a concentric seal is given by (9),

$$W_g = \pi \, \mu \, \omega^2 \, (r_o^4 - r_i^4) \, / \, 2h$$

The corresponding steady-state temperature rise at the seal face can then be calculated, since the seal faces are both essentially at the same temperature, with heat being removed by flow across the seal faces and by conduction through them (9).

The heat conducted through both seal rings for a concentric seal, W_{cond}, is determined using an axisymmetric finite-element code and applying a nominal temperature to the seal faces (11). This gives an overall seal face heat transfer coefficient, H_{cond}, where,

$$W_{cond} = H_{cond} \, (T_{sf} - T_a)$$

There is also heat convected away by the leakage, Q, and by the extra inflow of the eccentric seal, Q_e. Considering just these three components of cooling, the average seal face temperature (for the stationary face in this case, since the rotating face varies according to whether it is covered) would be,

$$T_{sf} = T_a + W_g \, / \, [\rho_w \, C_w \, (Q + Q_e) + H_{cond}]$$

Temperatures are in fact reduced by the further cooling effect of eccentricity described next.

Extra Cooling due to Alternating Heating-Cooling of Eccentric Seal Rotor

A fourth component of cooling, the extra due to alternating heating-cooling, was ignored in the previous section. However, once every revolution of an eccentric seal such as shown in Figure 1, some portions of the rotor surface alternate between being uncovered and cooled by exposure to ambient liquid outside or inside the extent of the stationary seal face, and being covered and heated by friction against the stationary seal face. When uncovered, the rotor surface can be assumed to be effectively quenched to ambient temperature (infinite convective heat transfer coefficient), because the speed is normally very high, with an extremely steep velocity profile at the surface at exit from the seal face. When covered, it can be assumed to be raised immediately to the average temperature of the stationary seal face.

Assuming one-dimensional conduction in the rotor, the instantaneous heat flux entering or leaving a point anywhere on the rotating seal face is given by conditions at the surface,

$$q = -k_r \, [dT/dz]_{z=0}$$

To find the total entering or leaving during each revolution, this must be integrated over the covered or uncovered time, respectively, for the whole portion swept by the stationary seal face.

Since $[dT/dz]_{z=0}$ varies with time according to the transient heat conduction equation,

$$\partial^2 T/\partial z^2 = (\rho_r C_r / k_r)\partial T/\partial t,$$

the total heat can most readily be found by combining all portions of the rotating seal face that are subject to alternating heating-cooling and applying the average heating-cooling cycle for this area.

Two cases must be considered. The seal is "under-eccentric" if some radial locations on the rotating face always remain covered, and it is "over-eccentric" if all radial locations are uncovered at some time during each revolution (Figure 4). Note that the NRU seal is approximately neutrally-eccentric (i.e., borderline between these two). The average conditions that apply to each portion of the rotating seal face are shown in Table 2.

Table 2:   Fractions of Time Heated or Cooled during each Revolution

Radial Portion of Rotating Seal Face	Under-Eccentric $e < \Delta r / 2$	Over-Eccentric $e > \Delta r / 2$	
$< (r_i - e)$	Always cooled	Always cooled	
From $(r_i - e)$ to $(r_i + e)$	Half-heated; Half-cooled	Heated	Cooled
From $(r_i + e)$ to $(r_o - e)$	Always heated	$\Delta r / (\Delta r + 2e)$	$2e / (\Delta r + 2e)$
From $(r_o - e)$ to $(r_o + e)$	Half-heated; Half-cooled	of each rev.	of each rev.
$> (r_o + e)$	Always cooled	Always cooled	

For an under-eccentric seal, the area of the alternating portions of the rotating face that provide extra cooling over that of a non-eccentric seal is,

304

$$A_u = \pi \, (r_i + e)^2 - \pi \, (r_i - e)^2 + \pi \, (r_o + e)^2 - \pi \, (r_o - e)^2$$

Extra cooling is then calculated as the heat absorbed (then rejected) each revolution by this area, multiplied by frequency, given that this area has the average stationary seal face temperature, T_{sf}, applied during half the revolution, and the ambient temperature, T_a, applied during the other half.

For an over-eccentric seal, the alternating area is,

$$A_o = \pi \, (r_o + e)^2 - \pi \, (r_i - e)^2$$

Calculation of the extra cooling contributed by eccentricity in this case must take account of the fact that, on average, the stationary seal face temperature, T_{sf}, is applied for less than half each revolution, and the ambient temperature, T_a, is applied for more than half.

One-dimensional time transient conduction analysis for these cases was performed using a commercial finite-element program (12). A unit alternating face temperature, half-heating, half-cooling, was applied to a unit area of the NRU seal rotor, whose axial thickness is 19.0 mm and whose rear face was assumed to be at ambient temperature, T_a— an assumption known to be reasonable from steady-state axisymmetric thermal analysis (9). Heat-cool cycles were applied at the rotational frequency, starting with the rotor wholly at ambient temperature and proceeding until little further change in axial distribution occurred from one cycle to the next.

The result for 30 cycles is shown in Figure 5, where the cumulative heat transfer per unit temperature and area of the rotating face is seen generally to increase with time because more heat enters the face per cycle than leaves. A quasi-steady equilibrium condition is reached when the increase becomes linear; i.e., when the difference between the "heat-in" and the "heat-out" is the same from one cycle to the next, this difference being the heat conducted to the rear. This is almost reached after the 30 cycles shown, which represents only about one second of total time.

For the NRU seal, the quasi-steady "heat-in" amount defines the extra cooling, due to the alternating heating-cooling of this neutrally-eccentric seal rotor. The eventual variation in axial temperature distribution during one revolution (after 30 cycles) is shown in Figure 6 for the input pertaining to this seal. It illustrates that the effect is truly only skin deep.

The effects of rotational speed and over-eccentricity were explored by varying the revolution time and the fractions for heating and cooling. By multiplying the resulting time-transient values for "heat-in" per revolution by the appropriate areas and frequencies for each case, the effects on seal face cooling were calculated in terms of H_{alt}, the heat transfer coefficient for this component of cooling.

<u>Effect of Total Extra Cooling due to Eccentricity on Seal Face Temperature</u>

To see the full effects of extra cooling due to eccentricity (from both extra flow and alternating heating-cooling), the stationary seal face temperature equation can now be rewritten to include all four components of cooling,

$$T_{sf} = T_a + W_g / [\, \rho_w \, C_w \, (Q + Q_e) + H_{alt} + H_{cond} \,]$$

The effects of these four components on stationary seal face temperature, separately and combined, are shown in Figure 7 for the NRU seal at its nominal eccentricity (e = 2.4 mm = 0.53 Δr), speed (1700 rpm) and water pressure (200 kPa).

Figure 8 shows the total effect on stationary seal face temperature of various pressures for the NRU seal at nominal eccentricity and speed, whereas Figures 9 and 10 similarly show the effects of various eccentricities and speeds.

DESIGN AND TESTING OF THE LOW PRESSURE NRU SEAL

A tandem seal cartridge containing two identical, but 180° offset, eccentric seals of 95 mm balance diameter was developed for the main heavy-water coolant pumps of the NRU reactor at Chalk River Laboratories. These five pumps operate at low pressures, between almost zero to 0.2 MPa, with frequent stops and starts. Normal and maximum rotational speed is 1700 rpm. Requirements for the seals were less than 0.2 mL/min leakage and a reliable lifetime of seven years.

The previous seal arrangement was difficult to install, often exceeded the desired leakage, and its lack of a back-up seal compromised its integrity.

The eccentric seals were designed into a single-unit cartridge to facilitate installation (Figure 11). They were arranged to expose the outboard seal to full pressure initially, with the inboard seal providing complete back-up in the event of outboard seal failure. This outboard "main seal," worked as the primary seal, while the inboard "backup seal" remained flooded with water at full pressure on both sides.

Material and geometric details of both seals in the cartridge are given in Table 1. Each stationary seal ring (stator) was axisymmetric and eccentric. It was shaped to give a geometric balance ratio of 0.75. Adding the spring load, the effective balance ratio exceeded unity for sealing pressures below 0.26 MPa. Without eccentricity, such seals in NRU would operate under boundary lubrication conditions with the danger of vaporization between the faces. With eccentricity, the faces became hydrodynamically lubricated and cooled under all operating conditions.

The seal cartridge was tested as shown in Figure 11. Tests were performed in two stages: short-term tests to evaluate the characteristics and suitability of the seal, and then a long-term test to verify the reliability of the final design.

The short-term tests included sealing pressures from 0.04 to 0.4 MPa; rotational speeds of 500, 1000 and 1700 rpm; temperatures between 24 and 70°C; and spring load between 85 and 135 N. Leakage was less than 0.1 mL/min for all cases, and power requirement variations were less than 0.2 kW over and above the drive power for fully lubricated conditions (~0.5 kW, for two seals and bearings). The sealing faces after tests were observed in excellent shape, with little wear—the least for least spring load.

The long-term test that followed was a 2300-hour test, as summarized in Figure 12, including stop-starts and intermittent axial movements. Mean leak rate was less than 0.05 mL/min, and power remained low, as before. Post-test examination showed the faces of both seals to be in good condition. Carbon stator thicknesses were measured for overall wear—0.025 mm for the main seal and 0.013 mm for the back-up (Figure 12).

Profilometer traces (Figure 13) showed some irregular wear on the main seal faces in the first 730 hours, but there was much less wear and no significant change in face profile on either seal during the remainder of the 2300 hours. The wear pattern for the main seal (Figure 13) shows a groove about 3 μm deep in the centre region of the rotating face, with mirror-image "peaks" on the stationary face. The backup seal is barely worn.

The overall wear rate for both seals was sufficiently low to meet the requirement for seven years of extrapolated seal lifetime. The measured trend demonstrates a decelerating wear mode, suggesting even longer lifetime. The first eccentric seal cartridge has been operating sucessfully in an NRU pump since Nov. 1992. Further installations are planned for 1994.

DISCUSSION

Seals not having the "special" arrangement shown in Figure 1 are very sensitive to eccentricity (Figure 2). For example, pre-supposing a typical outside-pressurized seal is fully lubricated by the conical-convergence between its seal faces, with 70% or 75% balance ratio and 0.9 aspect ratio (say, 50 mm outside radius and 5 mm face width), it will tilt into contact if eccentricity exceeds about 1.6% x 50 mm = 0.8 mm. Furthermore, Figure 2 shows it will tilt completely open if eccentricity exceeds about 2 mm.

However, eccentricity is shown to be extremely effective for seal cooling. This is not primarily because of the extra flow (lubrication) dragged between the seal faces by the eccentricity, which is often insignificant compared with the "normal" pressure-driven leakage (Figure 3). It is because of the alternating heating-cooling.

The transient conduction calculations show the effectiveness of this component of cooling. Quasi-steady conditions are reached very quickly after heating-cooling is applied. Just one second after start-up of the neutrally-eccentric NRU seal, about 90-95% of the heat absorbed during the heating phase is already being removed during the cooling phase, based on half heating and half cooling (Figure 5). Apart from the eccentricity, it is the rotor conductivity and heat capacity that govern the cooling effectiveness.

The alternating temperature "wave" penetrates very slightly into the rotating surface (Figure 6), and the heat transfer effect is therefore essentially independent of rotor thickness, or the boundary conditions elsewhere. For the NRU seal, the effect is mainly in the 1 mm layer at the rotating seal face, rather than the rest of the 19 mm total thickness.

All four components of cooling are shown separately and together in Figure 7: the conventional two of conduction through the seal rings and heating of the leakage, and the extra two peculiar to eccentric seals, i.e., transfer by the extra flow of liquid being dragged radially in and out of the seal, and the alternating cooling. The alternating cooling due to eccentricity is by far the most effective cooling component for the NRU seal at any face separation. It reduces stationary seal face temperature by about a factor of five over the other three effects combined. Note that seal face temperature will be reduced by this factor regardless of the means of heat generation, e.g., seals operating with boundary lubrication.

The effects of pressure on the NRU seal at otherwise nominal conditions (Figure 8) show that there is no significant difference in stationary seal face temperature at normal separations below about 4 μm. Figure 3 shows that leakage would become excessive for separations greater than this.

The NRU seal has a nominal eccentricity equal to half the face width. The effects of more or less than this (Figure 9) show that there are diminishing returns as eccentricity is increased further. A limit is quickly approached whereby the heat absorbed is the maximum possible, and there is more than enough time during the cooling fraction of each revolution to conduct this heat back out of the rotating face. However, the cooling effectiveness of the NRU seal is shown in Figure 9 to drop off rapidly as the eccentricity is decreased from the neutral amount nominally built in.

Finally, the effects of double or half speed on the NRU seal at its nominal eccentricity and operating pressure are shown in Figure 10 to cause more than proportional changes in stationary seal face temperature rise. This is because viscous heating depends on speed squared, whereas the predominant alternating cooling effect is less than proportional to speed and approaches a limit with further increase.

The calculated results for the NRU seal have assumed constant viscosity water based on 50°C. The eccentric effects also assumed a boundary layer shape factor of unity. The estimated errors in these and other approximations are less than 50%. The assumption that the seal remains back-flooded is an important one, and the eccentricity effects shown in Figures 3 and 7-10 would be reduced by about half if water were drained away from where it exits the seal face. In practice, centrifugal effects tend to hold any leaking liquid in place at this exit in the usual case of pressure-driven leakage from outside to inside.

Testing of the NRU seal confirmed that its torque remained as low as for fully lubricated conditions throughout, though the measured leakage was very small. Wear of the back-flooded "backup seal" was almost zero, while the minor wear of the "main seal" may in fact have been much less if greater efforts had been made also to back-flood this seal.

CONCLUSIONS

Eccentricity in the "wrong" arrangement of an end face seal can cause problems of tilted contact and opening between the faces, with excessive friction and leakage. However, in the "right" arrangement, eccentricity is a simple and effective means of inducing extra lubricating and cooling effects that can extend seal operating range or lifetime.

Cooling calculations for an eccentric seal designed for the NRU research reactor attribute its cooling primarily to alternating heating-cooling of the rotating seal face as it is covered and uncovered every revolution. Calculations show this cooling to be several times more effective than all other conductive and convective cooling combined. The rotor conductivity and heat capacity govern its effectiveness. The alternating temperature "wave" penetrates very slightly into the rotating surface, and the cooling is therefore essentially independent of rotor thickness, or the boundary conditions elsewhere. Further eccentricity, beyond the neutrally-eccentric amount designed into the seal, gives diminishing returns.

The NRU seal cartridge was tested over a wide range of conditions, including 2300 hours at low pressure nominal operating conditions. Wear of the "main seal" met its seven-year lifetime target, while the similar "backup seal" showed no measurable damage and confirmed its long-term integrity. It was installed in November 1992 and continues to operate successfully.

ACKNOWLEDGEMENT

This work was performed at the Chalk River Laboratories of AECL Research. Support provided by the CANDU Owners Group is gratefully acknowledged.

REFERENCES

1. Webster, G.R., "Primary Circulators—Water Reactor Circulating Pumps," Nuclear Engineering International, <u>16</u>, No. 187, December 1971.

2. Metcalfe, R., "End Face Seals in High Pressure Water—Learning from those Failures," Lubrication Engineering, <u>32</u>, 12, December 1976.

3. Vilim, P. and Zeller, L.A., "Shaft Seal for a Pump," <u>Canadian Patent 716.850</u>, August 1965.

4. Metcalfe, R., "Rotary Shaft Face Seal," <u>US Patent 4,026,564</u>, May 1977.

5. Metcalfe, R. and Brown, G.W., "Eccentricity of Balanced End Face Seals," <u>Proc. 10th Int. Conf. on Fluid Sealing</u>, Paper L3, BHR Group, Cranfield, April 1984.

6. Etsion, I., "Zero-Leakage Non-Contacting Mechanical Face Seal for Rotary Machines," <u>US Patent 4,407,509</u>, October 1983.

7. Buck, G.S. and Volden, D., "Upstream Pumping: A New Concept in Mechanical Sealing Technology," Lubrication Engineering, <u>46</u>, 4, April 1990.

8. Metcalfe, R., "Eccentric Face Seal With Asymmetric Closing Force," <u>US Patent 5,180,172</u>, January 1993.

9. Metcalfe, R., "Performance Analysis of Axisymmetric Flat Face Mechanical Seals," <u>Proc. 6th Int. Conf. on Fluid Sealing</u>, Paper D1, BHR Group, Cranfield, February 1973.

10. Sneck, H.J., "The Misaligned, Eccentric Face Seal," <u>Proc. 4th Int. Conf. on Fluid Sealing</u>, Paper 15A, BHR Group, Cranfield, May 1969.

11. Metcalfe, R., "The Use of Finite Element Deflection Analysis in Performance Prediction for End Face Seals," <u>Proc. 3rd Symp. on Engineering Applications of Solid Mechanics</u>. University of Toronto, June 1976. (Also available as AECL Report, AECL-6525, Atomic Energy of Canada Limited, Chalk River, Ontario).

12. "ABAQUS Version 5.2 User Manuals," Hibbitt, Karlsson and Sorenson, Inc., 1992.

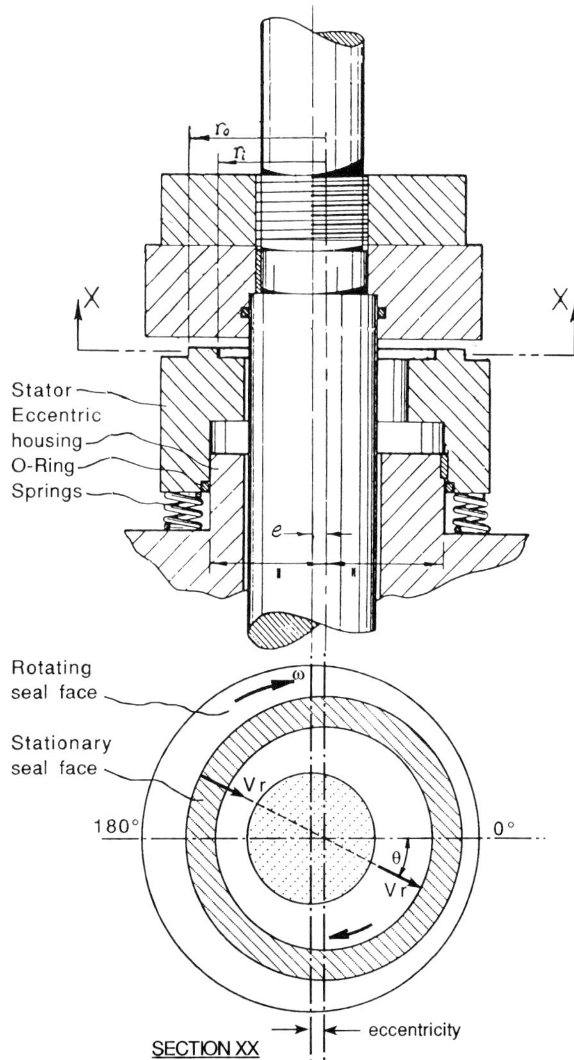

Figure 1: Tilt-Resistant Eccentric Seal Arrangement. The stator is mounted eccentrically, but because its shape and loading are axisymmetric, it has no tendency to tilt relative to the rotor, which is fixed rigidly to the shaft. The rotating seal face has a velocity component, V_r, normal to the edge of the stationary seal face.

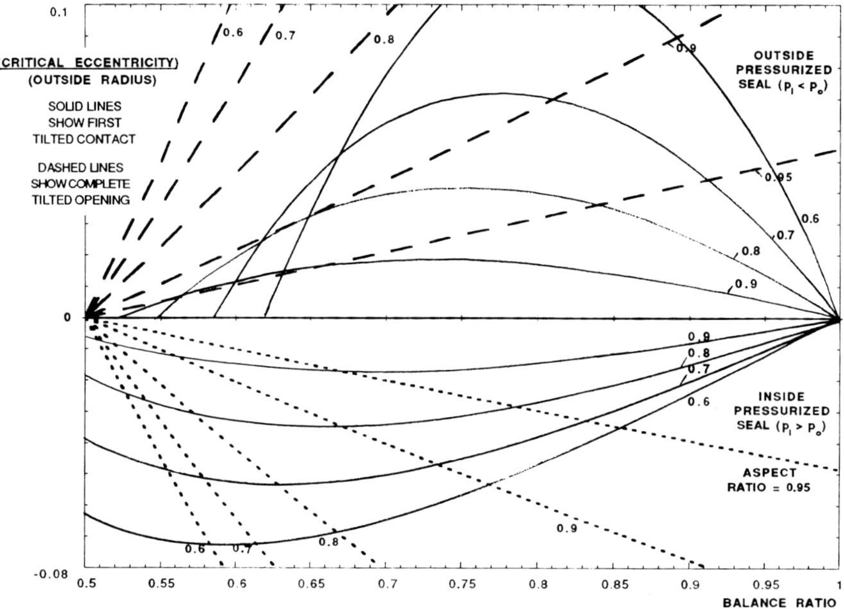

Figure 2: Critical Eccentricities for Tilt-Sensitive Seal. If the raised seal face is not on the flexibly mounted seal ring, there are two critical eccentricities for static conditions: one causing tilted contact of a fully lubricated conical seal; another causing complete tilted opening of any seal.

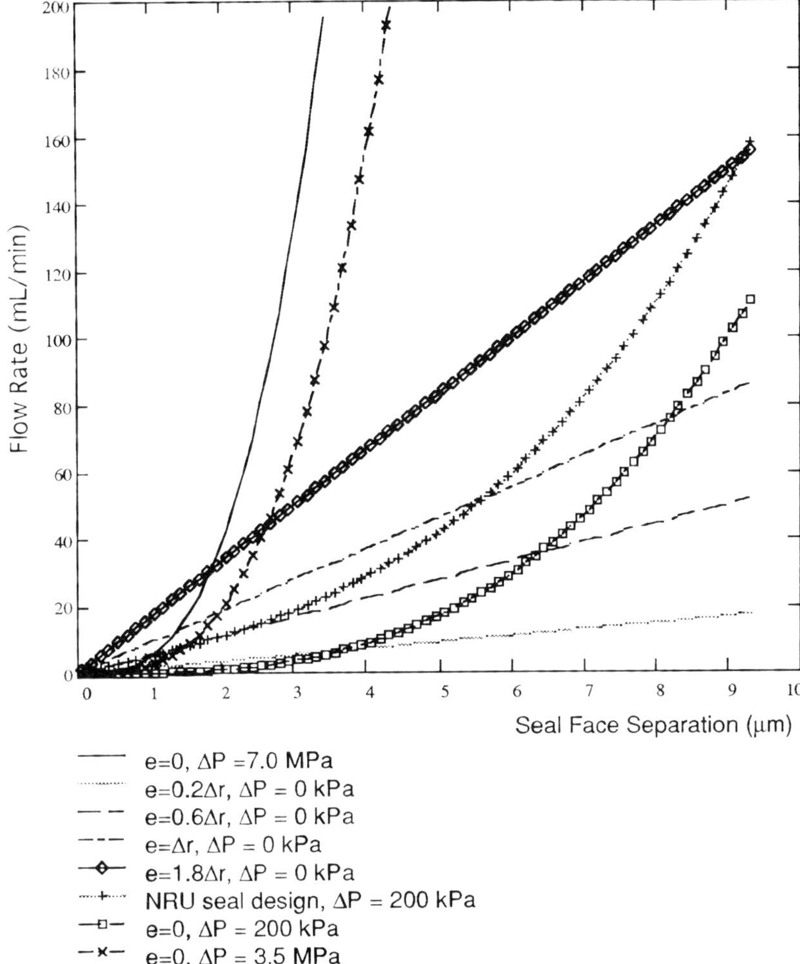

Figure 3: Comparison of Conventional and Eccentric Seal Face Flows.
Pressure-driven leakage of a conventional concentric seal at various face separations is compared with the extra flow dragged into and out of the seal face by eccentricity for the NRU seal size.

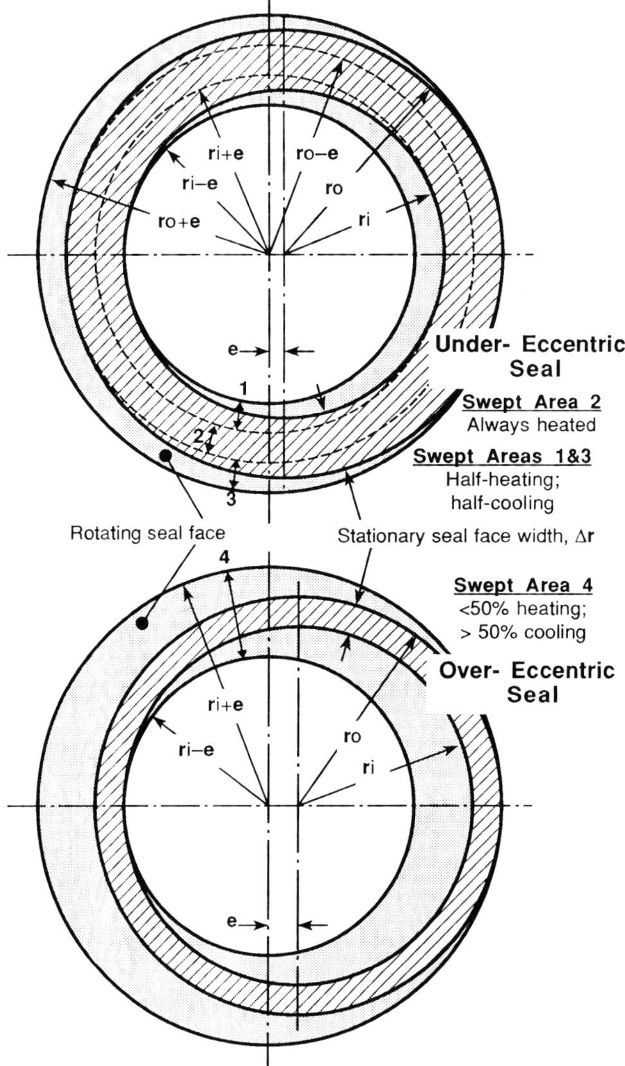

Figure 4: Heating and Cooling Areas for Eccentric Seals. The seal is "under-eccentric" if some radial locations on the rotating seal face always remain covered; "over-eccentric" if all locations are uncovered at some time. It is "neutrally-eccentric" if borderline between these two.

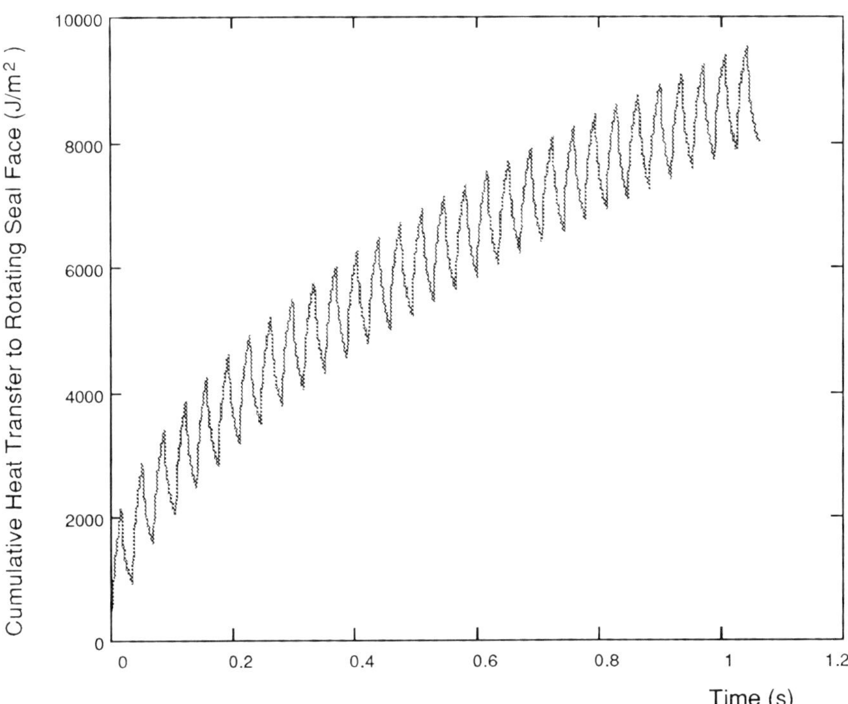

Figure 5: Cumulative Heat Transfer into NRU Seal Rotor. This is the unit temperature and area result of alternating (1700 rpm) face temperature, from ambient to 1 K above, giving half-heating, half-cooling at the face of the 19 mm thick tungsten carbide rotating face of the NRU Seal, cooled to ambient at the rear.

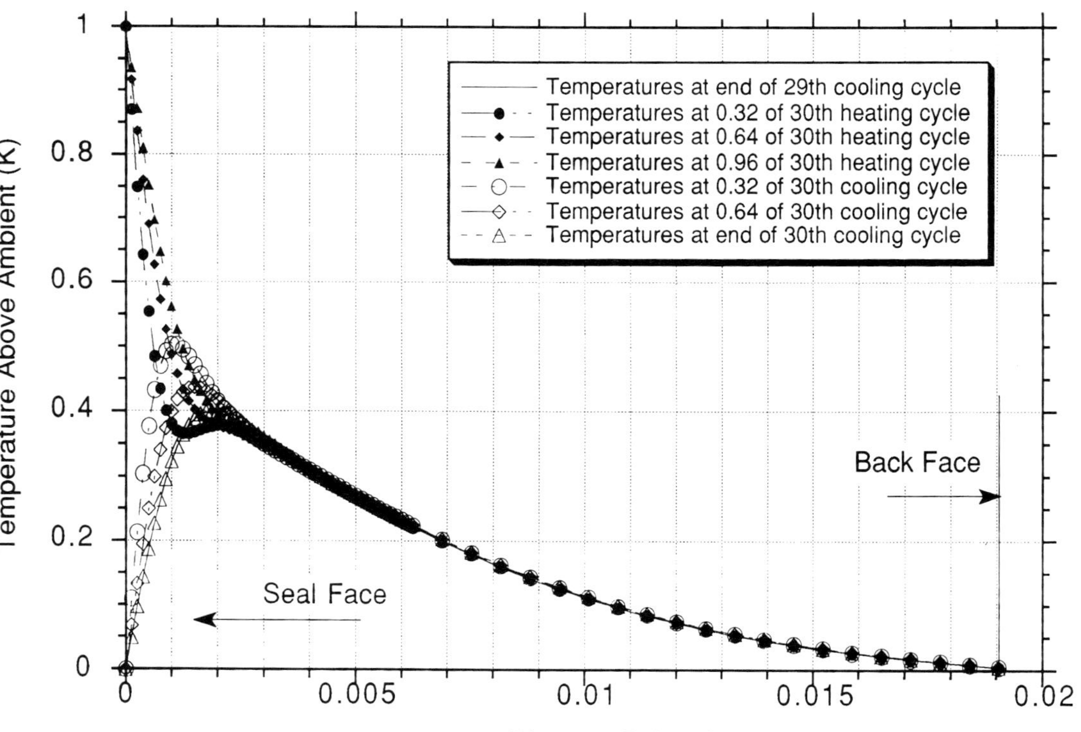

Figure 6: Quasi-Steady Cyclic Temperature Profile for NRU Seal Rotor. The alternating temperature applied to the surface is shown after 30 cycles, when there is little further change. Only a thin surface layer responds to the cyclic variation.

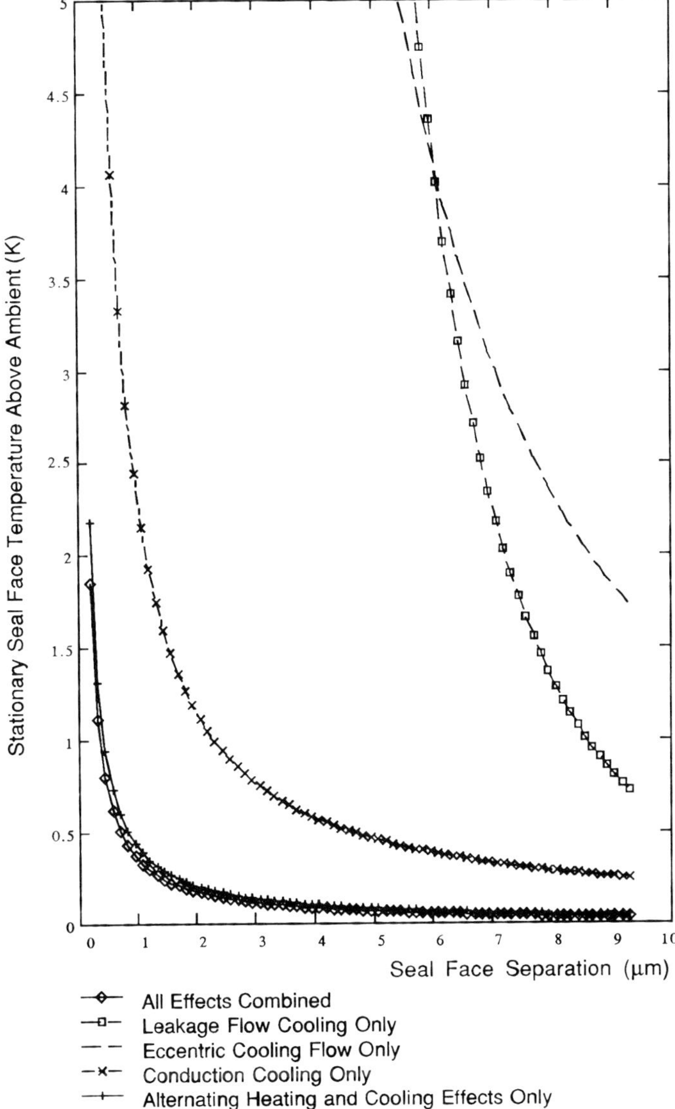

Figure 7: Effects of the Four Components of Stationary Seal Face Cooling.
The separate and combined effects are shown at various face separations for the NRU seal
(-"neutrally-eccentric," 1700 rpm, 200 kPa, sealing water at 50°C).

316

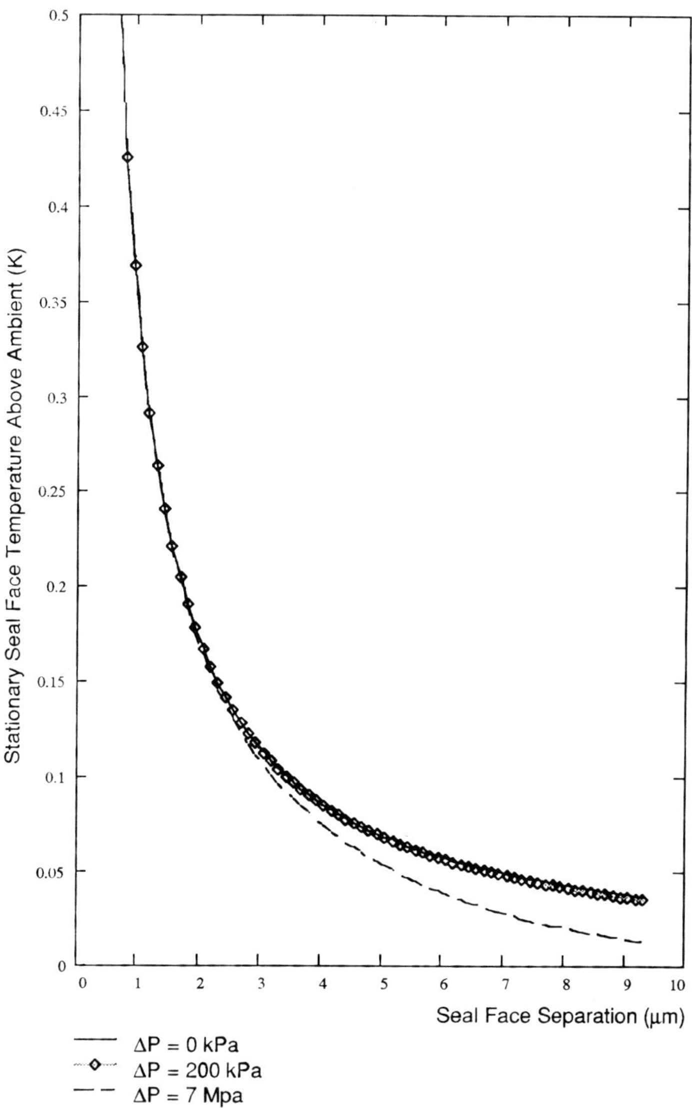

Figure 8: Effects of Pressure on NRU Seal Stationary Face Cooling. The total effect of all cooling is shown at various face separations for the NRU seal (~"neutrally-eccentric," 1700 rpm, sealing water at 50°C).

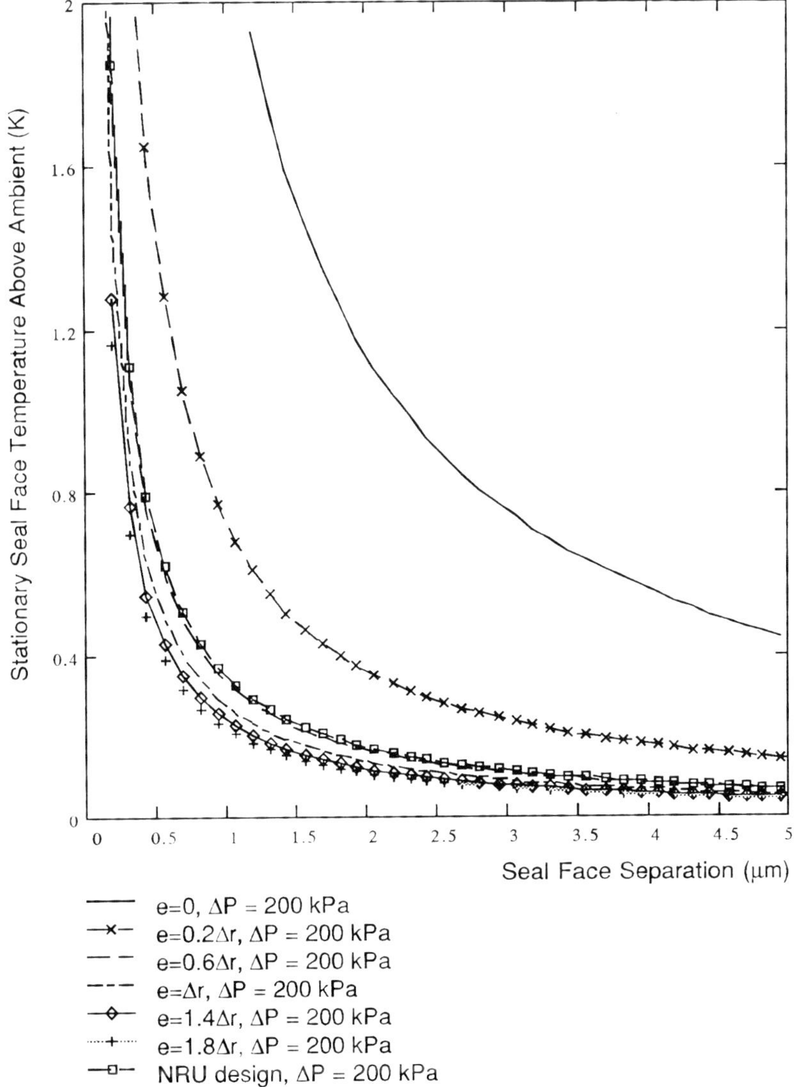

Figure 9: Effects of Eccentricity on NRU Seal Stationary Face Cooling.
The total effect of all cooling is shown at various face separations for the NRU seal (1700 rpm, 200 kPa, sealing water at 50°C).

318

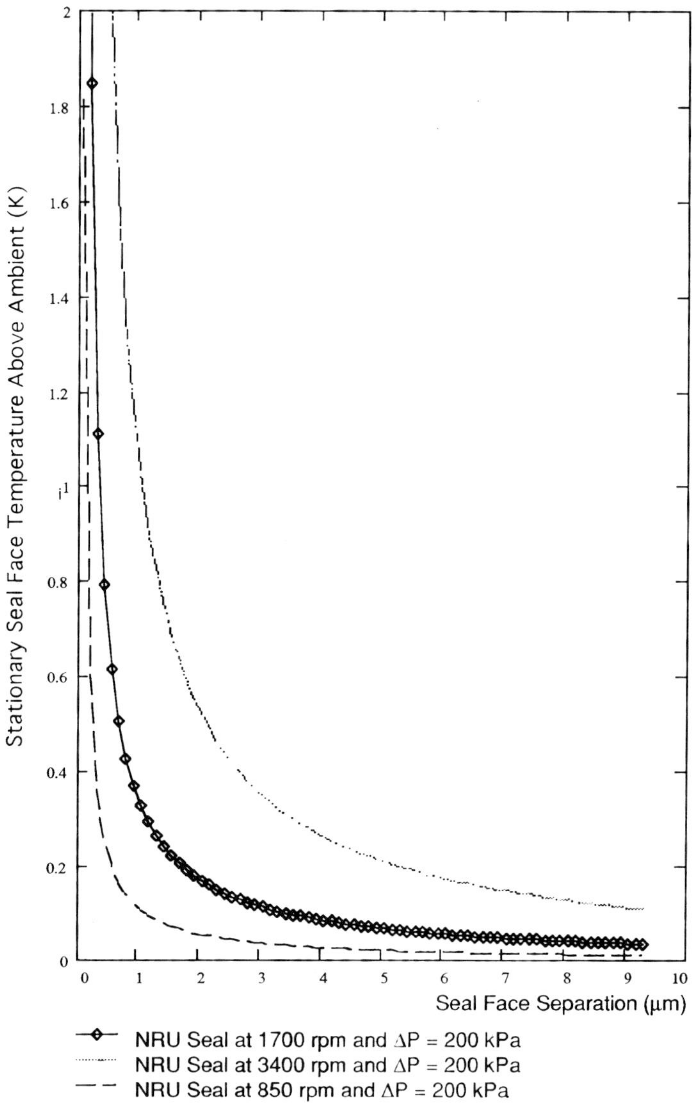

Figure 10: Effects of Speed on NRU Seal Stationary Face Cooling. The total effect of all cooling is shown at various speeds for the NRU seal (~"neutrally-eccentric," 200 kPa, sealing water at 50°C).

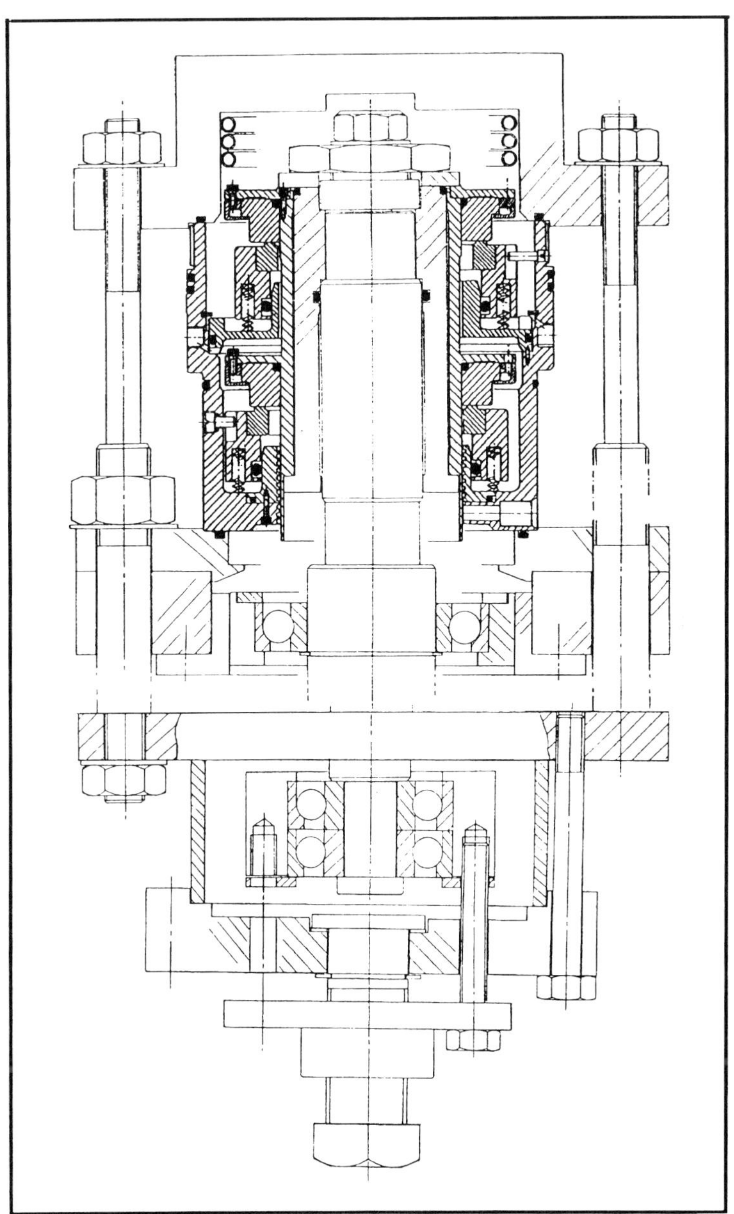

Figure 11: NRU Main Heavy-Water Pump Seal Cartridge in Seal Tester.

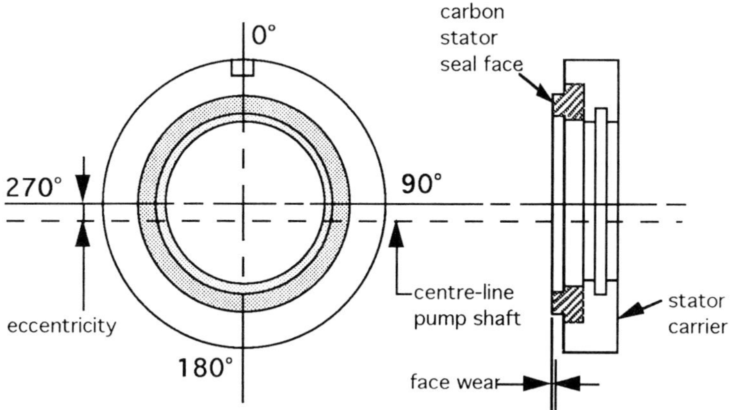

Seal	Test Conditions (Water)				Cumulative Test Time (hours)	Cumulative Average Wear (mm)	Seal Face Profile
	Pressure (MPa)	Temp. (°C)	rpm				
Main	0.17	20	1700		0	0	See Fig. 13
Seal	0.17	to			730	0.013	
	0.17	30		Stops and starts	1300	0.019	Not shown
	0.17			2.5 mm axial cycles	1500	Not measured	None
	0.34				2300	0.025	See Fig. 13
Backup	None	20	1700		0	0	See Fig. 13
Seal	(both	to			730	0.010	
	sides	30		Stops and starts	1300	0.010	Not shown
	of seal			2.5 mm axial cycles	1500	Not measured	None
	flooded)				2300	0.013	See Fig. 13

Figure 12: Long-Term Test Conditions and Face Wear of Low Pressure Seal

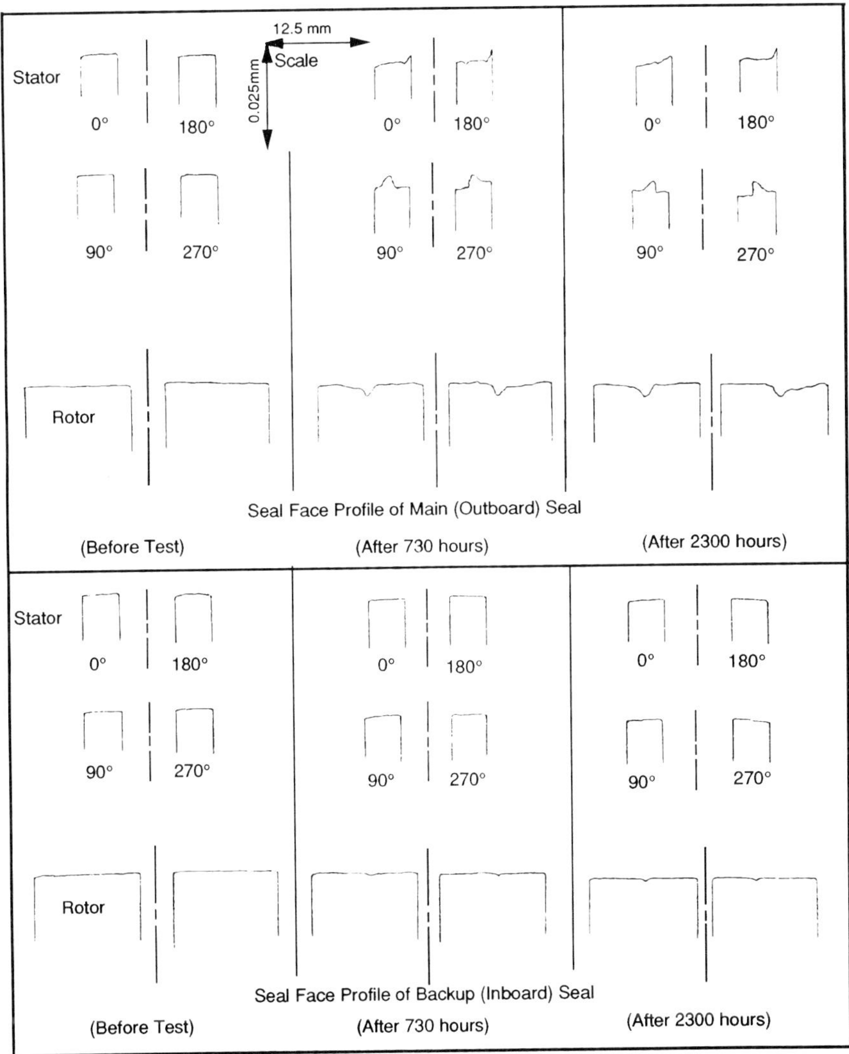

Figure 13:   Profilometer Traces of NRU Seal Faces After Testing.

14th International Conference on Fluid Sealing, Firenze, Italy,
6-8 April 1994. Organised by BHR Group Limited, Cranfield,
Bedford, MK43 0AJ, UK; Tel: 0234 750422

sealing of volatile liquids

by Ir J.A.M. ten Houte de Lange
Flexibox B.V., Vlaardingen - the Netherlands

Synopsis
This paper deals with advanced concepts for mechancial seals for sealing
volatile liquids such as light hydro carbons and water under conditions
near theyr boiling point.

Introduction
Like pumps, Mechanical seals cannot handle boiling liquid. For pumps, it is
common to arrange for the pumped liquid to have a net positive suction head
(NPSH) in order to prevent evaporative problems at the suction side of
the pump.

Fig. 1a shows the principle of the net positive suction head and the
resulting envelope of stable operation for a pump.

Similarly Mechanical seals will require a temperature margin (Δ-T) in order
to maintain a stable liquid film between the sealfaces.

Fig. 1b shows the envelope for stable operation for a Mechanical seal.
The principle of the Δ-T or temperature margin was first published in 1969
by A. Lymer [1].

This temperature margin can be converted into an equivalent pressure margin
and is therefore in principle fully comparable with the NPSH requirement of
the pump. This is shown in fig. 1c.

Unlike the pump, the Mechanical seal is generating heat and for this reason
its required pressure margin can be considerably higher than the NPSH
requirement for the pump.

This fact can lead to special measures being necessary to increase the
available temperature or pressure margin at the Mechanical seal.

As fig. 2 shows, the methods to make the seal application enter the stable
area are: [1] [18]

a) Increasing operating pressure
b) Reducing the operating temperature

Ad a) Increasing the operating pressure is in practice rather difficult.
      Furthermore as can be seen in fig. 2 this principle may not be
      effective under all possible conditions.

Ad b) Reducing the temperature involves some way of cooling which is an
      extra cost. When the operating temperature is already very near to
      the coolant temperature, cooling looses its effectivity.

Such measures invariably involve a certain complication and for this reason
a project was started in 1977 with the object to produce 'pump NPSH seals',
that is to say Mechanical seals for which the pump NPSH would be sufficient.

<u>The seal film under boiling conditions</u>:
The temperature limit for the Mechanical seal is related to the boiling phenomena occurring between the sealfaces.
A low sealface temperature will tend to move the boiling location towards the atmospheric end of the sealface.

The result of this will be that a larger part of the sealface area is liquid lubricated. Fig. 3 shows the temperature and the pressure diagram and the temperature margin of such a boiling seal film.

For a variety of reasons it is common practice to pressurise seals on its outside diameter. The separation between a liquid and a vapour phase area within the sealface is then being accentuated by the centrifugal field.

The described model for a boiling seal film has actually been photographed by Orcutt [2].

The equations for the boiling film equilibrium have been investigated by Hughes [3]. This investigation showed that the boiling film equilibrium has a high hydrostatic pressure. It is however, likely to collapse as soon as the temperature rises above the temperature dictated by the temperature margin. This collapse is accompanied by dry running and an increase in running temperature. This fact in turn leads to temperature deformation of the faces in the converging direction which can then give cause to hydrostatic opening of the seal. This results in a massive leakage which coincides with a cooling of the seal rings.

The seal with insufficient temperature margin will thus pass through a cyclic opening and closing movement which can happen with enough force to damage the sealfaces. This effect is in practice known under various names, face vaporization, popping or rattling. [4] [5] [6]

<u>The pump NPSH seal</u>:
A way to counter act this mechanism of seal destruction and seal leakage is to increase the closing pressure on the sealfaces to a point to suppress the possibility of hydrostatic opening of the seal. [5] [6]
This method of stabilizing the seal was at the time against the current opinion which regarded the 'Δt' purely as a mechanism caused by frictional heat. An increase of the face load would therefore only increase the 'Δt' requirement.

Furthermore one has to consider that the balance ratio for mechanical seals were still based to give acceptable seal life with the then current face materials such as stellite versus carbon.
The introduction of hard face material as tungsten carbide and silicon carbide reduced the face wear dramatically and rendered it possible to increase the face load in order to obtain a greater stability for the seal.

The method of <u>multi point injection</u> (see fig. 5) has been developed by our company to provide for sufficient cooling for a seal which has been stabilized by increasing its face load [15] [20]. The high velocity jets which hit the seal ring surface at right angles create an intense turbulence, which increases the heat transfer coefficient considerably. Multi point injection also offers the possibility to penetrate through any vapour blanket which may occur around the seal rings.

Being the originator of this principle of <u>multi point</u> (or <u>multi port</u>) injection [20] as an effective means of cooling a Mechanical seal, the author finds now some satisfaction that this principle is now gaining general recognition. According to the third draft for API 682 a circumferential multi port arrangement will become mandatory for single seals with rotating flexible elements [19].

In the course of the experimental stage it was found that the seals with a stationary tungsten carbide or silicon carbide face running against a rotating carbon ring gave the better results.
This fact looks like being in contradiction with heat transfer logic.
The normal speed distribution is bound to make the liquid heat transfer coefficient for the rotary seal ring to be much higher than that for the stationary seal ring. This would lead to a higher heat transfer for a rotating ring in the high conducting material (tungsten carbide or silicon carbide). A possible explanation is however that operation near the evaporation limits endangers to insulate rotating seal rings with a blanket of vapour [7].

This effect would render it more attractive to concentrate on cooling a stationary seal ring in high conducting material with a forceful shower of multi point injection. Such a hard ring would also be more resistant to the possible erosive influence of the rather intense seal flush.
The resulting pump NPSH seal is shown in fig. 6.

<u>Face technology</u>
A relevant question is whether the seal works permanently in the collapse mode. This would imply that the faces are running exclusively with a vapour film.

The writer has attempted to look for a possible answer to this question by studying the faces of successfully used 'NPSH seals'.
An interesting aspect was that all studied rings showed a worn-in waviness in their sealface surfaces: [5] [6] [8] [9] [10] [11] [12]

The carbon rotary seal rings showed a two tops - two valleys saddle type wear (see fig. 7). The total amplitude of these waves varied from 1,6 - 5,2 microns.

The tungsten carbide and silicon carbide stationary seal rings showed a low amplitude multi wave wear pattern of 0,1 - 0,3 microns.
The number of waves appear to coincide with the number of multi point injection holes (see fig. 8).
That multi point injection cooling can create waves on the face of a stationary seal ring has also been demonstrated by using a rough lapped stationary seal ring surface. On the photograph (see fig. 12) the wave crests have become visible by the polishing action of the counter face.

The fact that the seals in question were tight is indicative to the fact that the sealfaces kept minimum film thickness at all times.
The waves occur originally due to some local change in heat generation, which result in an increased axial average temperature at such points.
The corresponding increase in axial length causes a wave, which is however being fully flattened by the axial force.

The result is a closed seal gap with considerable fluctuations in the net closing pressure along side the seal surface.

The natural number of worn-in waves is two. The amplitude is a function of the coefficient of thermal expansion and the stiffness of the seal ring. It is therefore not surprising that the amplitudes are much larger for the carbon ring, which has the lower axial stiffness.

In the case of the stationary seal ring it is obvious that the principle of multi point injection is creating cooler locations due to improved heat transfer, which can explain the multi wave pattern on the stationary seal ring (fig. 11).

Whilst the collapse or dry running mode will prevail at the locations with high closing pressure, it is likely that the seal can develop a boiling film equilibrium at the locations with a low closing pressure.

The entry of liquid between the faces is being helped by the impact of the injection jets, which coincide with the low closing pressure locations on the stationary seal ring.

In this case there will be some liquid lubrication available between the faces, which can be an explanation for the long and trouble free service which has been obtained with this type of seals in a great number of LPG and ethylene seals running under pump NPSH conditions.

Pressure deformation
Pressure deformations in the running-in stage are a potential cause of early seal failure of NPSH-seals. The low Youngs modulus of carbon puts the emphasis on using low deformation design carbon seal rings.

In the design used the axial and the radial component of the pressure deformation are in opposing directions. Under these conditions it is a matter of dimensioning whether these deformations cancel each other out. [13] [14]

It has been demonstrated that the required low levels of pressure deformation can be obtained in this way.

A new development: The multi point cleaning arrangement
Solids in the circulating liquid can potentially damage the seal rings or the seal faces. The solid content in the circulation flow should therefore be eliminated or reduced as much as possible.
As a part of an ongoing development a cleaning arrangement for the circulation flow has been tried successfully.
In this multi point cleaning arrangement (see fig. 13 a) the circulation flow to the seal faces show a considerably reduced solid content. The majority of the solids are being concentrated in part of the flow and are ejected axially at the back of the seal in to the stuffingbox.

Emission levels
A rather unexpected phenomena is that pump NPSH seals have lower emission levels than liquid seals. The leakage rate is to a large extend being governed by the kinematic viscosity, which is very much higher for a vapour than for a liquid.

The vapour cushion on the atmospheric side of the sealface (see fig. 3) offers therefore a comparatively high resistance to the leakage flow. This effect helps to reduce the leakage rate.

Seals handling volatile organic liquids are nevertheless often fitted with a secondary seal both for safety as well as for environmental protection.

Such secondary seals can be liquid lubricated seals or dry running contacting or non contacting standby seals.

Liquid lubricated secondary seals or tandem seals require a seal system to contain the barrier liquid.

The development of dry running contacting and non contacting standby seals has made it possible to meet stringent emission regulations without having to use expensive seal systems (fig. 9 & 10). [15]

Conclusion
Pump NPSH seals are successfully used on water or hydrocarbons, notably volatile hydrocarbons [21] on which they are used as single or as tandem seals.
Also they are now finding extensive use in conjunction with contacting or non contacting standby seals.

They are now working successfully in all corners of the world under conditions totally outside the temperature margin envelop for stable liquid seals.

They are the Mechanical seal equivalent to the supersonic plane flying at the other side of the sound barrier.

As an addition to the successful multi point (or multi port) injection a multi point cleaning arrangement has been developed.

Ir J.A.M. ten Houte de Lange

Literature

[1]   - <u>A. Lymer (Former Technical Manager Flexibox Ltd.)</u>
        An engineering approach to the selection and application
        of Mechanical seals, Fourth Int. Conference on Fluid Sealing,
        held in conjunction with the 24th ASLE Annual meeting in
        Philadelphia May 5 - 9 1969.

[2]   - <u>F.K. Orcutt</u>
        An investigation of the operating and failure of Mechanical sals.
        Journal of lubrication technology, October 1969.

[3]   - <u>W.F. Hughes, N.S. Winowich, M.J. Birchak and W.C. Kennedy</u>
        Phase change in liquid face seals.
        Translation of the ASME 74, vol. 100, January 1978.

[4]   - <u>N.M. Wallace</u>
        Basic concepts of seal function and design.
        Mechanical seal Practice for improved performance,
        ImechE guides for the Process Industries.

[5]   - <u>J.A.M. ten Houte de Lange</u>
        Seal film dynamics and controlled film sealing.
        Geneva International Technical Conference 17 & 18 May 1984.

[6]   - <u>J.A.M. ten Houte de Lange</u>
        Further developments in High Duty sealing.
        BHRA 11th Int. Conference on fluid sealing 8 - 10 April 1987,
        Cannes, France.

[7]   - <u>N.B. Barnes, R.K. Flitney and B.S. Nau</u>
        Mechanical seal chamber design for improved performance,
        9th International Pump Users Symposium Texas A & M University,
        Houston, Texas, March 3 - 5 1992.

[8]   - <u>D. Summer - Smith (Imperial Chemical Industries Ltd. U.K.)</u>
        Performance of Mechanical seals in centrifugal process pumps.
        Paper H1 - 9th International Conference on fluid sealing
        sponsered by B.H.R.A. Fluid Engineering.

[9]   - <u>A.O. Lebeck</u>
        Design of an optimum moving wave ant tilt mechanical seal face
        seal for liquid applications.
        Paper E2 - 9th International Conference on fluid sealing
        sponsered by B.H.R.A. Fluid Engineering, Cranfield.

        <u>A.O. Lebeck, J.L. Teale, R.E. Pierce</u>
        Hydrodynamic lubrication and wear in wavy contacting face seals.
        Journal of lubrication technology January 1978, vol. 100 181.

        <u>A.O. Lebeck</u>
        Face seal waviness prediction, measurements, causes and effects.
        BHRA, 10th Int. conf. on Fluid sealing - Paper G1.

[10] - <u>I. Taylor and D.N. Jones</u>
        High duty seal tests: some observations on lubrication.
        Paper 12 - 8th Technical Conference of the British Pump
        Manufacturers Association, 29 - 31 March 1983, Cambridge.

[11] - <u>B.S. Nau</u>
        Rotary Mechanical seals in Process Duties: an assessment
        of the state of the art.
        Mechanical seal Practice for improved performance,
        ImechE guides for the Process Industries.

[12] - <u>J.G. Pape</u>
        Fundamental aspects of radial face seals.
        Dissertation dated 22nd October 1969 at the Technical University
        in Delft, Netherlands.

[13] - <u>B.S. Nau</u>
        Calculation of distortions in Mechanical seals, B.H.R.A. report
        TN 954, May 1968.

[14] - <u>T.N. Cleaver</u>
        Materials of construction.
        Mechanical seal Practice for improved Performance,
        ImechE guides for the Process Industries.

[15] - <u>N.M. Wallace and J.A.M. ten Houte de Lange</u>
        Zero Emission solutions for Mechanical seals on light hydrocarbons.
        9th International Pump Users Symposium, Houston, Texas,
        March 3 - 5,1992

      - <u>N.M. Wallace and J. David</u>
        "The design and application of seals for light hydro carbon
        service"
        9th Technical conference of the British pumpmanufacturers
        association, April 1985 Warwick University

[16] - <u>H.K. Müller</u>
        Gleitringdichtungen - Vorgänge im Dichtspalt.
        Konstruktion 26 (1974) H.6.

[17] - <u>E. Mayer</u>
        Axiale Gleitring Dichtungen.

[18] - <u>H.K. Müller</u>
        Abdichtung Bewegter Machineteile, Verlag Ursula Müller,
        Waiblingen, Germany

[19] - <u>API stand 682, third draft October 1992</u>

[20] - <u>Ir J.A.M. ten Houte de Lange</u>
        "Mechanical seals for light hydro carbons"
        10 February 1978 publication of Flexibox B.V.

[21] - <u>Flexibox International reference list</u>
        Selected application: "Seals for light hydro carbon service"

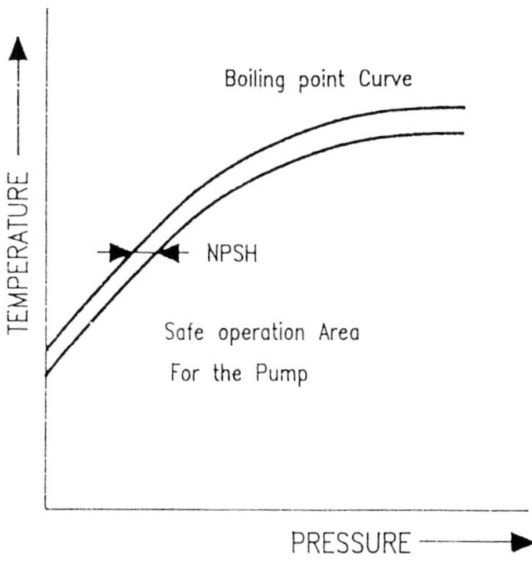

Fig. 1a

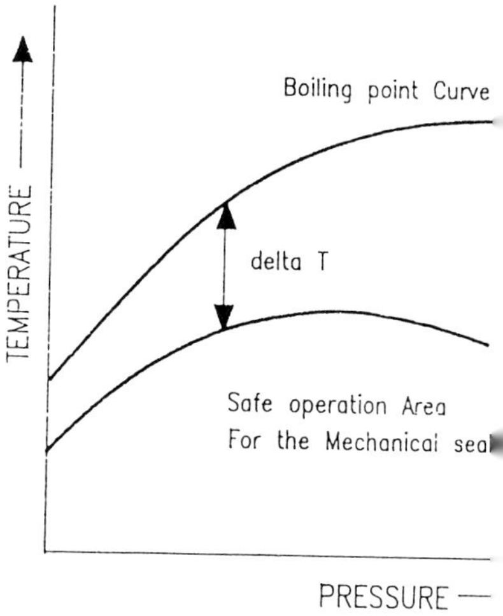

Fig. 1b

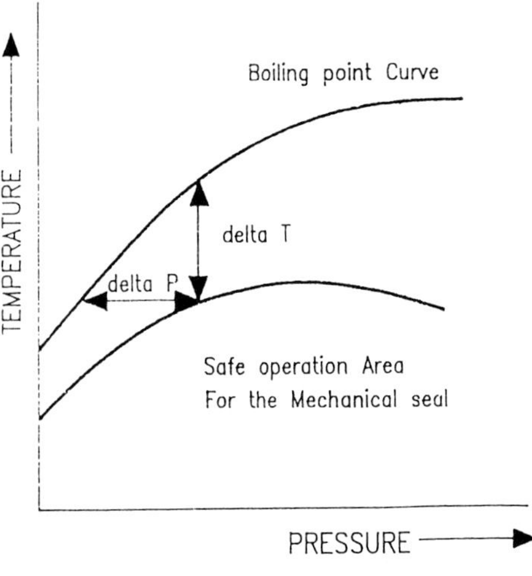

Fig. 1c

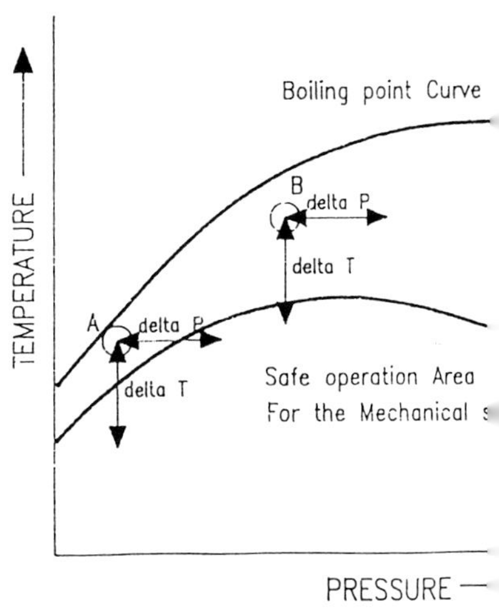

Fig. 2

Real boiling
Temperature
Boiling point at
System Pressure
TEMPERATURE
delta T
Product
Temperature
Atmospheric
Boiling Point
Product
Pressure
PRESSURE
Real Boiling
Pressure
Atmospheric
Pressure
LIQUID
VAPOUR
SEALING GAP WIDTH

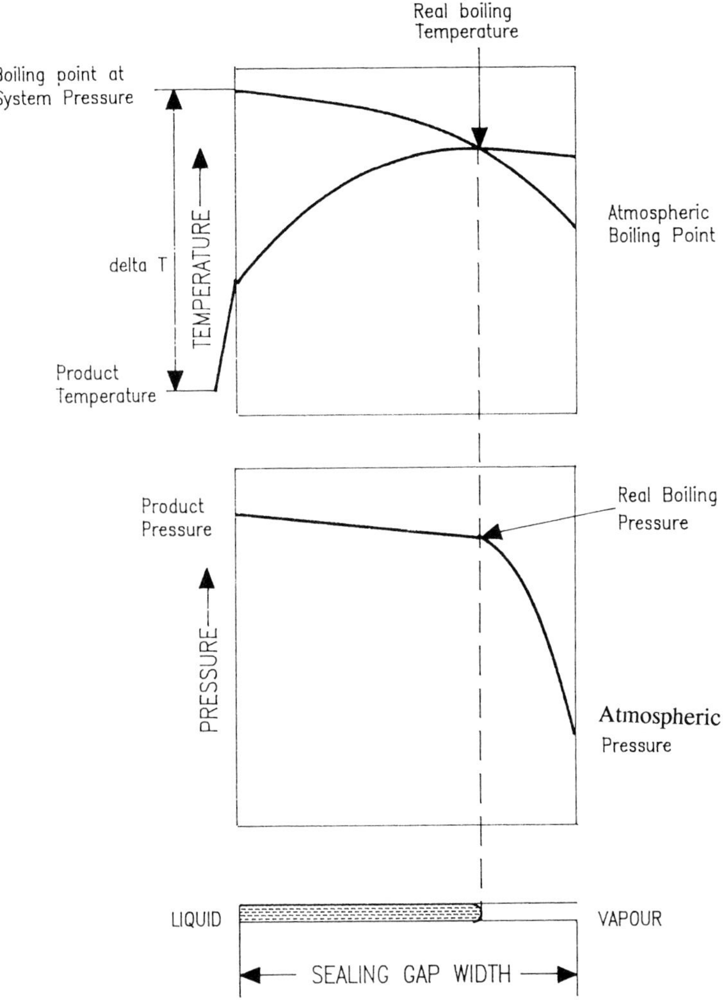

Fig. 3

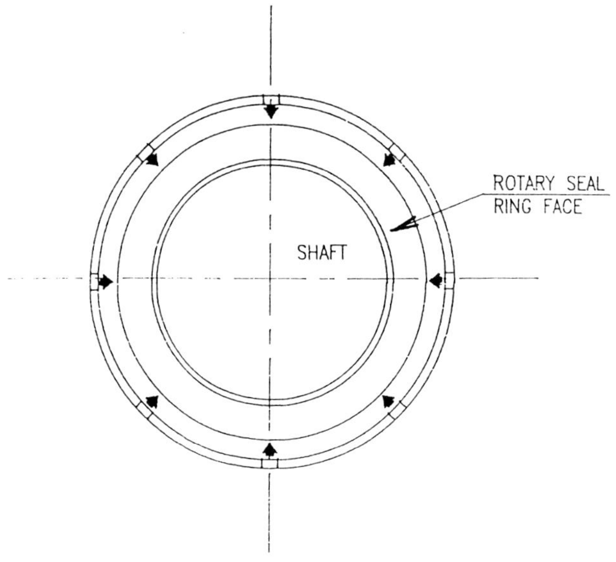

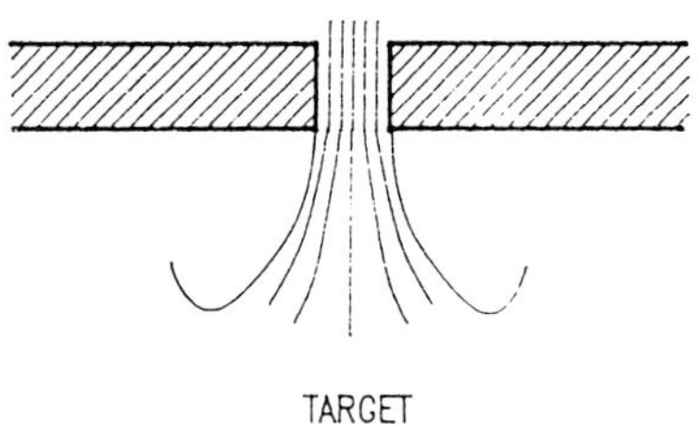

Fig. 5 Multi Point Injection

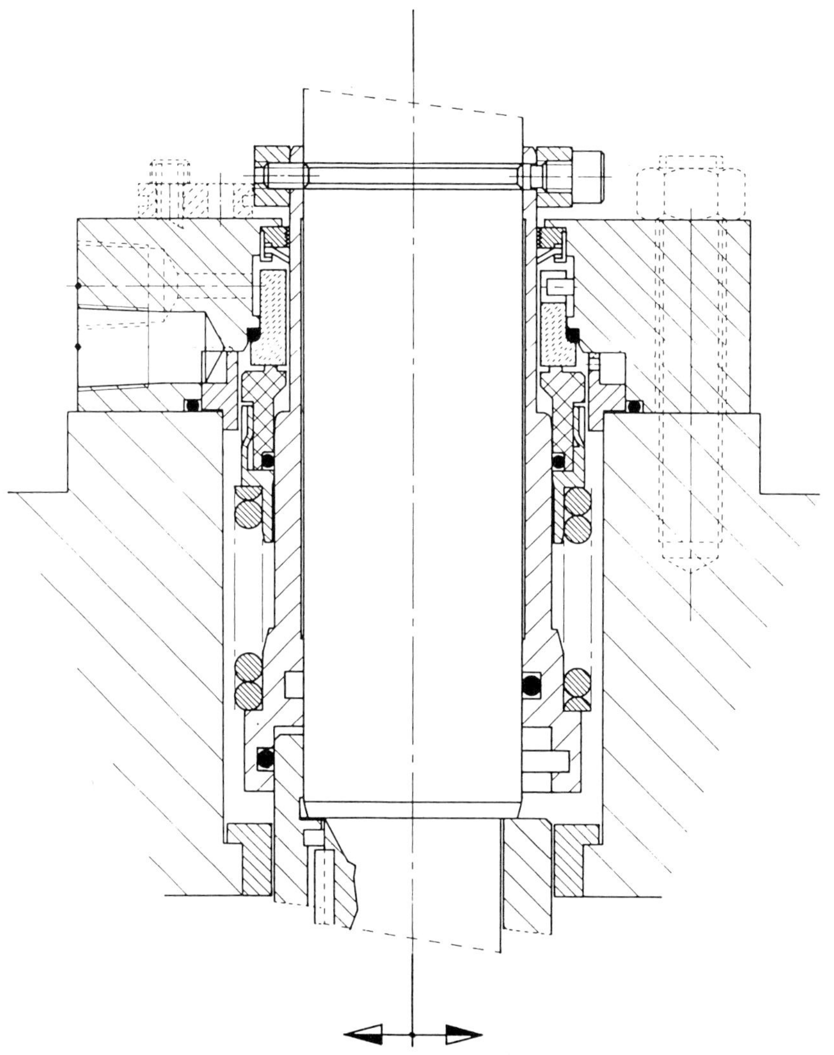

Fig. 6 Pump NPSH seal

334

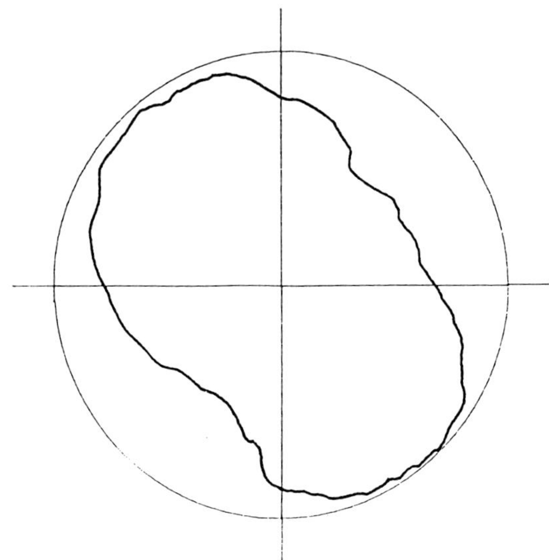

Fig. 7 Worn in saddle type waves
on Rotary Seal Ring

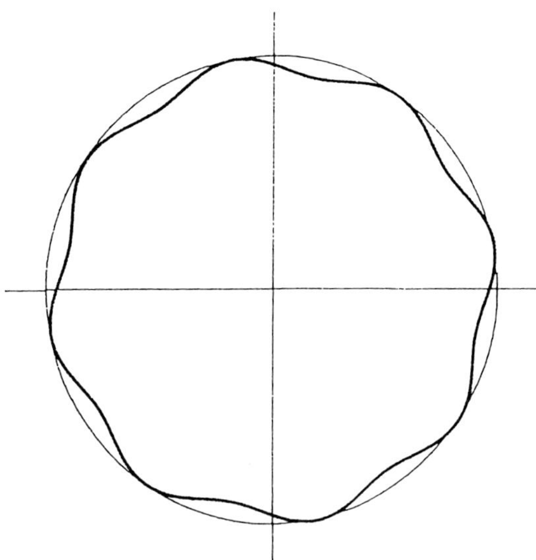

Fig. 8 Worn in Multi waves (Idealised)
on Stationary Seal Ring

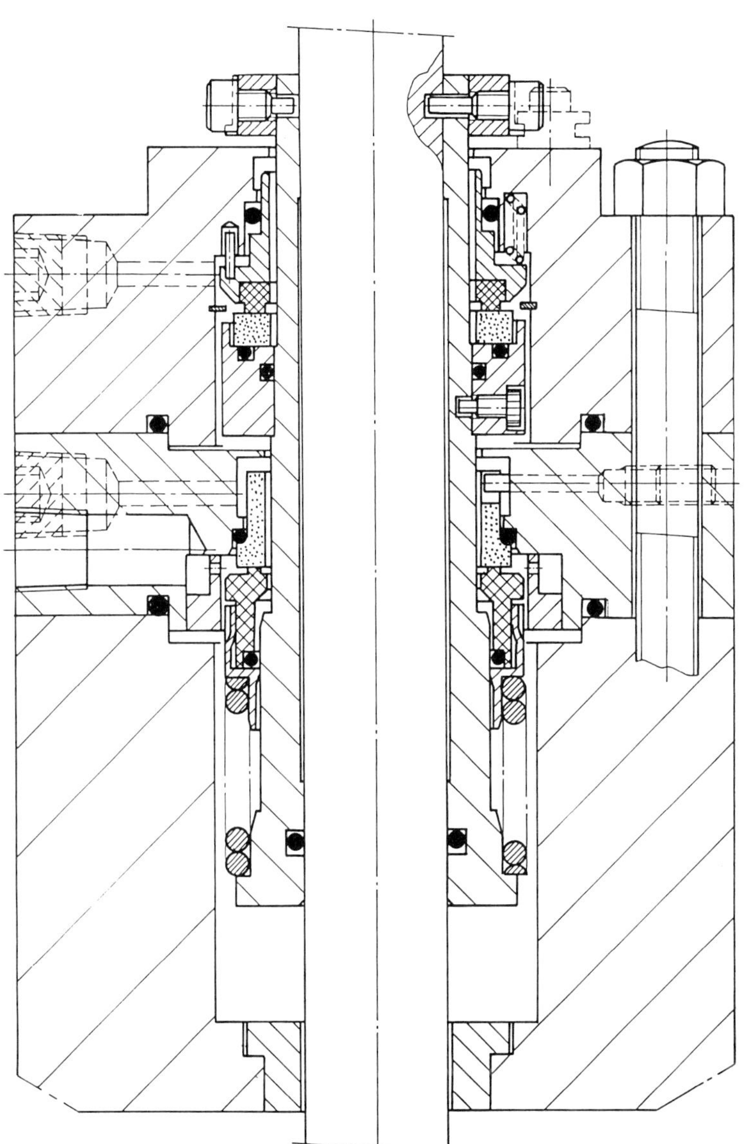

Fig. 9 PUmp NPSH seal with Contacting stand-by seal

336

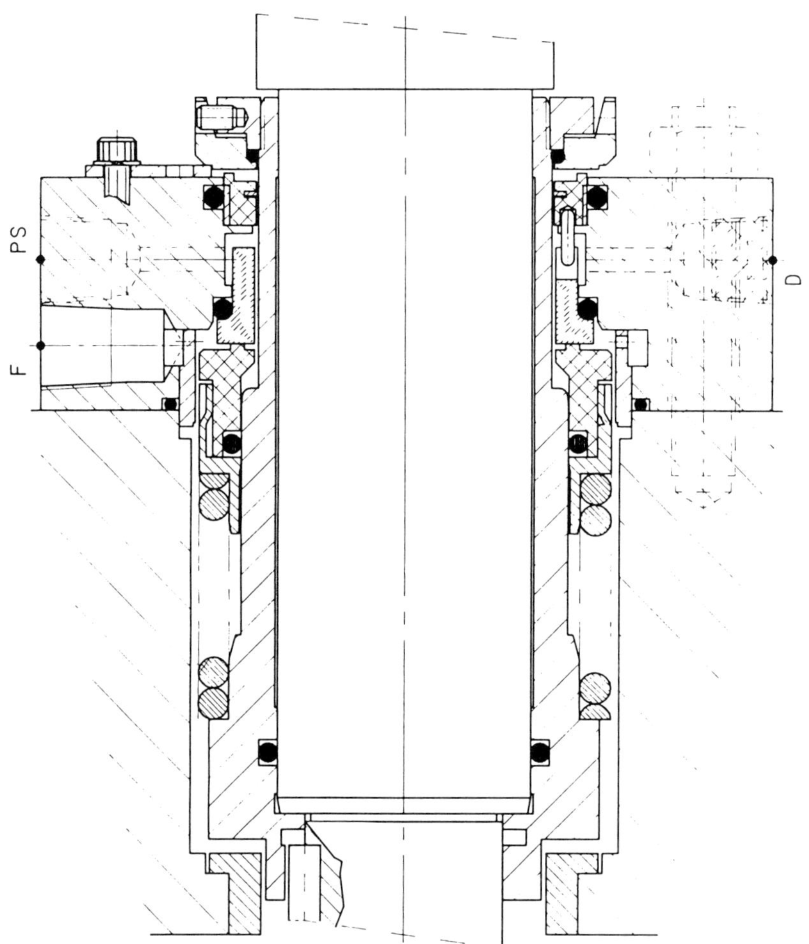

Fig. 10 Pump NPSH seal with Non-contacting stand-by seal

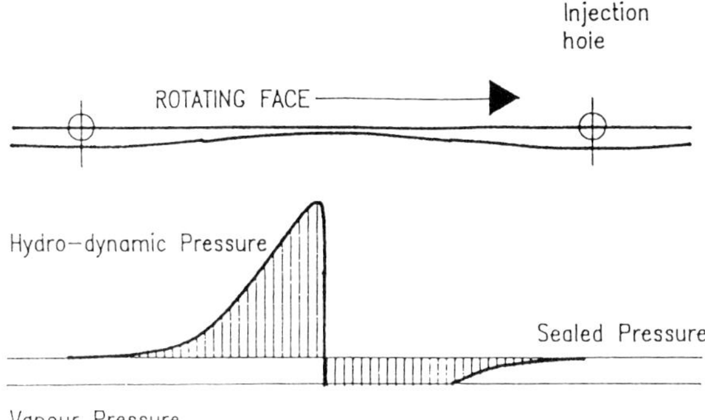

Fig. 11   Multi Point Injection & Lubrication

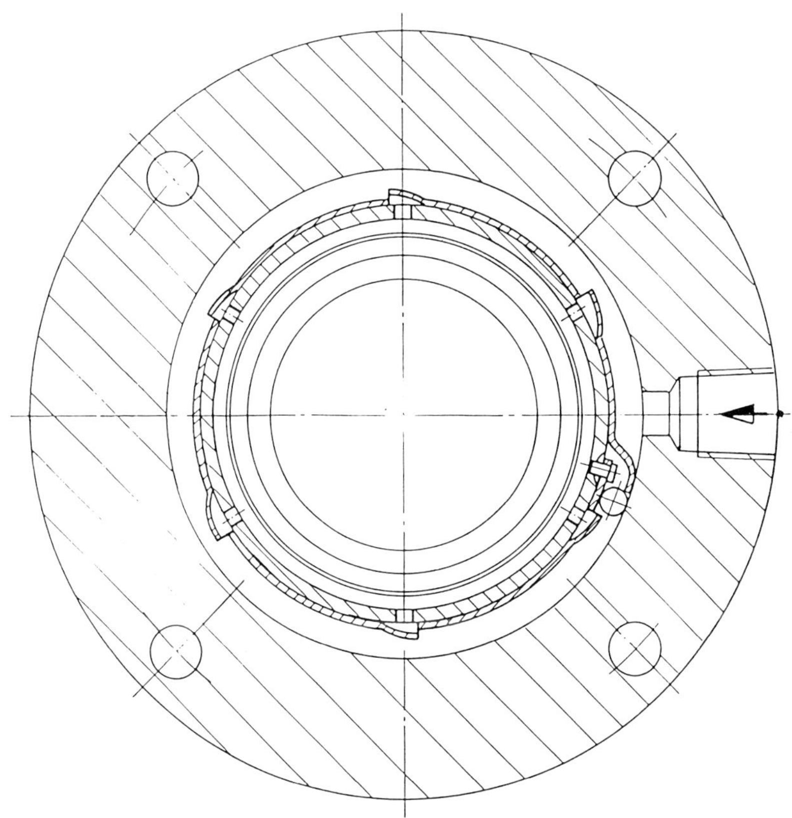

Fig. 13a   Multi Point Injection cleaning arrangement

MECHANICAL SEALS II - analysis/modelling

14th International Conference on Fluid Sealing, Firenze, Italy,
6-8 April 1994. Organised by BHR Group Limited, Cranfield,
Bedford, MK43 0AJ, UK; Tel: 0234 750422

THE ACCURACY OF ANALYTICAL SOLUTIONS FOR THE TEMPERATURE DISTRIBUTION IN MECHANICAL FACE SEALS

I. Etsion and M. Groper
Dept. of Mechanical Engineering
Technion, Haifa 32000, Israel

Abstract

The accuracy of two analytical solutions to the complex thermo-hydrodynamic problem of a mechanical face seal is evaluated. This is done by comparing the results of the seal's face temperature with the results obtained from an "exact" numerical solution. It is shown that even with relatively simple approximations fair accuracy can be achieved with errors of less than 33 percent.

Nomenclature

D	=	dissipation number, $\mu_f \omega^2 r_o^2 / kT_f$
G	=	geometry parameter, $r_o/h_i \sin 2\varphi$
H	=	convection coefficient
h	=	film thickness
$\bar{h}$	=	dimensionless film thickness, h/h_i
K	=	conductivity ratio, k/k^*
k	=	fluid thermal conductivity
k^*	=	rotor thermal conductivity
Nu	=	Nusselt number, $2r_o H/k$
r	=	radial coordinate
$\bar{r}$	=	dimensionless radial coordinate, r/r_o
T	=	temperature
T_m	=	mean local temperature
$\bar{T}$	=	dimensionless temperature, T/T_f
u	=	tangential velocity of rotor surface
y	=	axial coordinate
β	=	temperature-viscosity coefficient
$\bar{\beta}$	=	dimensionless coefficient, βT_f
μ	=	fluid viscosity
ω	=	rotor angular velocity

Subscripts

f	=	sealed fluid

342

i   =   inner radius
o   =   outer radius
1   =   at stator face
2   =   at rotor face
3   =   at O.D. rotor surface

Introduction

The face temperature distribution in mechanical face seals is an important information in cases of possible phase change due to boiling of sealed liquids or when face thermal distortion can occur. This temperature distribution may be obtained from a thermohydrodynamic (THD) analysis which is usually complex and often requires cumbersome numerical solutions by finite element analyses (FEA). Several FEA solutions of the THD problems in mechanical face seals can be found in Refs. (1) to (3), for example.

In order to save computing time and also to enable more elegant parametric study of the THD problem analytical solutions, rather than numerical ones, are required. Due to the complexity of the THD problem in seals only very few attempts of such analytical solutions were made so far.

In Ref. 4 a series solution is presented which neglects seal curvature and, hence, may be appropriate for very narrow faces. Two more analytical solutions can be bound in Refs. (5) and (6). The solution in Ref. (6) is based on the assumption suggested in Ref. (5) that the heat flow path in the rotating seal ring can be approximated by a set of straight lines at a fixed angle φ to the rotor face as shown in Fig. 1. This approach resulted in a set of implicit equations for the temperature distribution of the seal mating faces and the fluid film between them. The accuracy of the solutions presented in Refs. (4) and (6) was evaluated in Ref. (7) and an attempt was made to find improved, more accurate additional analytical solutions. Some of the results obtained in Ref. (7) are presented in this work.

Model and Approximate Analytical Solution

A cross section of the seal is shown in Fig. 1. A face to face double seal arrangement is considered, which is symmetrical about a midplan of the rotor. Hence, due to this symmetry, only one stator has to be analyzed. The following assumptions are made:

1. The mating faces are aligned so that axisymmetry prevails.
2. The fluid is a Newtonian liquid of constant viscosity and density across the height of the sealing gap (these properties, however, vary in the radial direction).
3. The temperature, T_f of the sealed fluid surrounding the surfaces at the outer diameters of the rotor and stator is known.
4. The surfaces exposed to the air are insulated.
5. The convective heat transfer between the stator and the sealed fluid is much smaller compared to that of the rotor and hence, can be neglected.

Assumptions 4 and 5 actually mean that the stator can be considered as being perfectly insulated.

6. A steady-state condition is assumed.

The model to be analyzed is shown in Fig. 2. Surfaces 1 and 2 are the mating faces of the stator and rotor, respectively. Surface 3 is the rotor surface in contact with the sealed fluid. The sealing gap is of thickness h(r) where h(r) can be any arbitrary function of r.

The local average temperature, T_m, at a radius r, is defined as

$$T_m = \frac{1}{h} \int_0^h T dy \tag{1}$$

and the temperature T is obtained from a solution of the energy equation

$$k \frac{\partial^2 T}{\partial y^2} + \mu \left(\frac{\partial u}{\partial y} \right)^2 = 0 \tag{2}$$

Equation (2) is based on the assumption that most of the heat generated in the lubricating film is rejected by conduction. This is true especially in mechanical seals where it is desired to eliminate leakage.

Using the boundary condition $\partial T/\partial y = 0$ at surface 1, and performing a heat balance between the conduction in the rotor from surface 2 to surface 3, and the convection form surface 3 to the sealed fluid the solution for T_1, T_2 and T_m can be obtained in the form (see Ref. (6))

$$\left(\overline{T}_m - 1 \right) e^{\overline{\beta}(\overline{T}_m - 1)} = Df(\overline{r}) \tag{3}$$

$$\overline{T}_2 = \overline{T}_m - \frac{D}{3} \overline{r}^2 e^{-\overline{\beta}(\overline{T}_m - 1)} \tag{4}$$

$$\overline{T}_1 = \overline{T}_m + \frac{D}{6} \overline{r}^2 e^{-\overline{\beta}(\overline{T}_m - 1)} \tag{5}$$

The function $f(\overline{r})$ in Eq. (3) for the case of approximated straight lines for the heat flow path as shown in Fig. 1b was presented in Ref. (6) in the form

$$f(\overline{r}) = 2 \left(\frac{2\cos^2 \phi}{Nu} - K \ln \overline{r} \right) G \frac{\overline{r}^3}{h} + \frac{\overline{r}^2}{3} \tag{6}$$

In Eqs. (3) through (6) the temperature is normalized by the sealed fluid temperature T_f. The parameter D is the dissipation number in the form $\mu_f \omega^2 r_o/kT_f$, K is a conductivity ratio k/k^*, Nu is the Nusselt Number having the form $2r_o H/k$ and G is a geometric parameter in the form

$$G = \frac{r_o}{h_i \sin 2\phi} \tag{7}$$

344

where h_i is the film thickness at $r = r_i$.

As noted in Ref. (6) the weakest point of the above analytical solution is the lack of information for a preferred selection of the angle ϕ approximating the direction of the heat flow path. This point was resolved in Ref. (7) where the results of Ref. (6) were compared with an "exact" numerical solution of the THD problem. It was found that the analytical solution of Ref. (6) always overestimates the temperature distribution. It was further found that the best selection of the angle ϕ which minimizes the difference between the approximate and "exact" solutions, is such that the resistance to heat flow from the sealing gap through the rotor and into the surrounding sealed fluid is minimized. This gives the expression

$$\tan^2\phi = 1 + \frac{k^*}{r_o H \ell n \dfrac{r_o}{r_m}} \tag{8}$$

where $r_m = (r_o + r_i)/2$.

Results and Discussion

A typical example of a water seal was selected for evaluating and comparing the accuracy of the analytical solutions. The seal has an outer radius $r_o = 45$ mm and a constant sealing gap $h = h_i = 1\mu m$. The rotation speed is $\omega = 300$ rad/s and the rotor is made of stainless steel having a thermal conductivity $k^* = 45$ W/m °C. The ambient temperature of the sealed water is $T_f = 20$ °C. The various properties and corresponding dimensionless parameters are summarized in Table 1.

The optimum values for the direction of heat flow lines according to Eq. (8) are $\phi = 52.6$ degree at $r_i/r_o = 0.85$ and $\phi = 55.3$ degree at $r_i/r_o = 0.9$.

Figs. 3 and 4 present the results of the temperature distribution, T_1, along the stator face for the two cases of $r_i/r_o = 0.85$ and $r_i/r_o = 0.9$, respectively. These results were obtained from the "exact" numerical solution and from the various approximate analytical solutions. As can be seen the linear heat flow lines approximation (6) gives the poorer accuracy at both $r_i/r_o = 0.85$ and $r_i/r_o = 0.9$. The maximum errors obtained from this approximation are about 33 percent at both cases of r_i/r_o values. The accuracy of the series solution by Buck (4) is much better and is only about 1 percent.

Conclusion

The accuracy of two analytical solutions to the thermo-hydrodynamic problem of a mechanical face seal was evaluated. This was done by comparing the results of the face temperature distribution obtained from these solutions with similar results that were obtained from an "exact" numerical solution. The analytical solutions that were evaluated are a series solution offered by Buck (4) and a solution based on approximating the heat flow lines in the seal rotor as linear lines (6). The two solutions overestimate the face temperature. The maximum error obtained with the linear heat flow line approximation is 33 percent compared to only about 1 percent error with the series solution. The linear heat flow line approximation is however much simpler to use and may be helpful in cases where fast and easy evaluation of face temperature is needed.

References

1) Li, C.H., "Thermal Deformation in a Mechanical Face Seal", ASLE Trans. Vol. 19, No. 2, 1976, pp. 146-152.

2) Doust, T.G., and Parmar, A., "Transient Thermoelastic Effects in a Mechanical Face Seal", Proc. 11th International Conference on Fluid Sealing, BHRA, 1987, pp. 407 - 422.

3) Zeus, D. "Viscous Friction in Small Gaps Calculations for Non-Contacting Liquid or Gas Lubricated End Face Seals". Tribology Transactions, Vol. 33, No. 3, 1990, pp. 454 - 462.

4) Buck, G.S. "Heat Transfer in Mechanical Seals" 6th International Pump Users Symposium, 1989, pp. 9 - 15.

5) Dumbrava, M.A., and Morariu, Z. "Thermo-Hydrodynamic Aspects of the Double Mechanical Seals", Proc. 11th International Conference on Fluid Sealing, BHRA, 1987, pp. 394 - 406.

6) Pascovici, M.D. and Etsion, I. "A Thermo-Hydrodynamic Analysis of a Mechanical Face Seal", J. Tribology, Trans. ASME, Vol. 114, No. 4, 1992, pp. 639 - 645.

7) Groper, M. "An Improved Solution for the Temperature Distribution in a Mechanical Seal", M.Sc. Thesis, Technion, 1994

Table 1 Various properties and corresponding dimensionless parameters for a water seal

Property/parameter		Water seal
μ	Pa Sec	10^{-3}
k	w/m °C	0.65
H	w/m^2 °C	18000
β		0.0175
$\bar{\beta}$		0.35
D		0.014
K		0.0144
Nu		2490

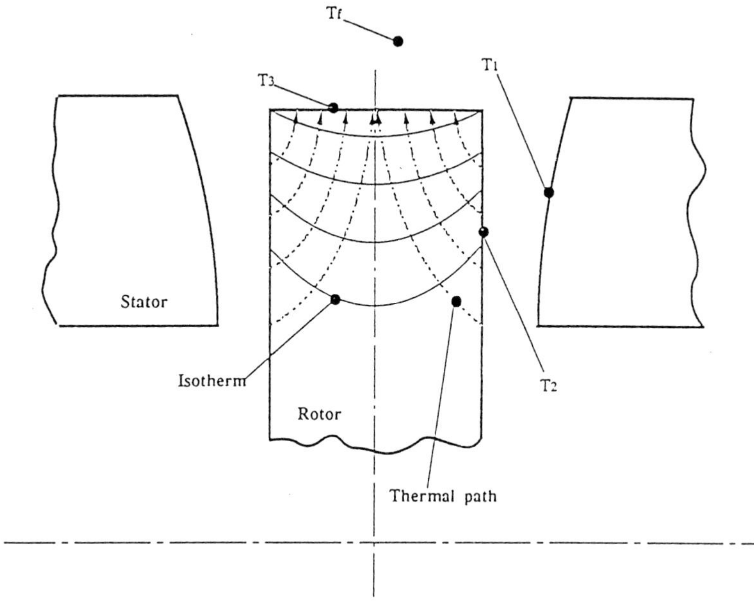

Fig. 1a   Isotherms and heat flow lines in a symmetric face to face double seal.

$$\eta_p \;=\; \frac{h^3}{12 \int_0^h \frac{z}{\eta} \left(z - \frac{\int_0^h \frac{z}{\eta} dz}{\int_0^h \frac{1}{\eta} dz} \right) dz}$$

Thermal deformations in sealing rings are calculated by the principle of virtual displacements: Here, the work of inner forces is equal to the work of outer forces. Deformations caused by mechanical loads are not considered.

4 Boundary conditions

4.1 Thermal boundary conditions

The following boundary conditions are implemented in the program system

- Neumann boundary condition (adiabatic case)

- Dirichlet boundary condition (known temperatures)

- Robbins boundary conditions (heat convection laws)

4.2 Hydrodynamic boundary conditions

At the inner and outer radius of the lubricant film only Dirichlet boundary conditions are necessary. To consider boundary conditions between pressure area and cavitation area a mass conserving cavitation algorithm developed by Kumar/Booker [1] is applied.

5 Calculation of power loss

Power loss has a great influence on the temperature distribution in the sealing system (i.e. the viscosity distribution in the lubricant film and the thermal deformations of the sealing rings) There are two possibilities to consider this effect:

- two-dimensional calculation

$$P \;=\; M\omega$$
$$M = \int_\Omega r\,\tau\,d\Omega$$

Under the assumption of a Newtonian fluid this yields

$$P = \int_\Omega r \left[\eta \frac{\partial v}{\partial z} \right]_{z=0} d\Omega$$

- three-dimensional calculation

The dissipation term of the energy equation can be used to calculate the power loss. Here the three-dimensional viscosity distribution is used.

$$Q = \int_\Omega \eta \left[\left(\frac{\partial u}{\partial z} \right)^2 + \left(\frac{\partial v}{\partial z} \right)^2 \right] d\Omega$$

6 Calculation of the leakage rate

The leakage rate can be evaluated by establishing a balance of inflow and outflow at the inner or outer radius of the gap. Considering the Reynolds boundary condition or Gümbel boundary condition, the mass conservation law is violated because the continuity equation is ignored within the cavitation area. A dynamic mass conserving cavitation algorithm developed by Kumar/Booker however allows to calculate accurate leakage rates.

One method to calculate the leakage rate is to integrate the radial velocities over the inner or outer boundary of lubricant film:

$$\dot{Q} = \int_\varphi \int_z u\, r\, dz d\varphi \tag{8}$$

This method however contains two numerical integrations (pressure calculation and velocity calculation) using the method of finite elements. In order to reduce the calculation error, a more suitable approach is used [2] by evaluation of the unmodified (no Dirichlet boundary conditions considered) system of equations of the finite element formulation of the Reynolds equation (9)

$$a(i,j)\,\bar{p}_j - Q_i^U - Q_i^{\dot{p}} - q_i = 0 \tag{9}$$

Here, variable a presents the stiffness matrix of the finite element system, q_i the flow of lubricant at node i, Q_i^U the shear flow and $Q_i^{\dot{p}}$ the expansion flow.

After having solved the pressure system equation, it is now possible to evaluate the leakage rate at these nodes by using equation (9).

7 Coupling of mechanical and thermal equations

Fig. 2 prescribes the iterative solution scheme to consider the dependence of temperature and viscosity. To speed up the iteration, best results have been obtained by starting the calculation with a thermal deformed gap geometry. Therefore the solution of the heat conduction equation in the sealing rings - assuming a start temperature T_{gap} in the lubricant film - is necessary.

Minimal film thickness h_{min} is determined by finding the load support equilibrium of the seal. Here, h_{min} is varied in Reynolds equation (secant method) as long as the equilibrium point of outer forces and the calculated pressure is found.

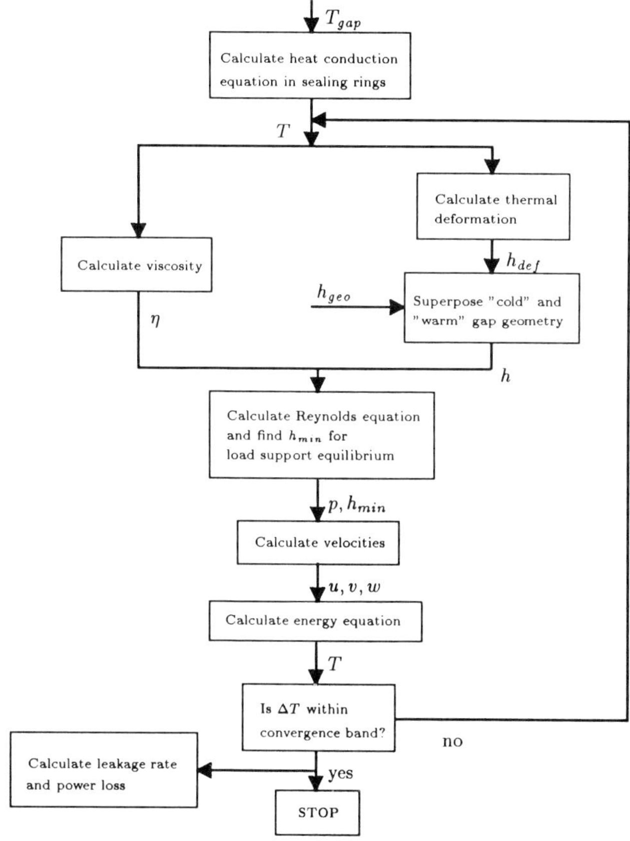

Figure 2: Flow chart for calculation

8   Evaluation

8.1   System values and material data

The basis of this analysis is a double mechanical seal with a rotating and a stationary ring. Table 1 presents the system values. The rotating ring consists of silicon carbide, the stationary ones consist of carbon loaded silicon carbide. Buffer medium is mineral oil. Fluid data and material data are shown in Table 2. The FEM-structure consists of 9660 nodes and 7630 linear brick elements (Fig.3).

Inner Radius Lubricant Film	$67.5\,mm$
Outer Radius Lubricant Film	$70.5\,mm$
Defined Waviness A	$0.4\,\mu m$
A-Gap	$0.5\,\mu m$
k-Factor	0.622
Revolution	$14\,800\,min^{-1}$
pf-Value	$41.06\,N/cm^2$
Pressure Product Side	$10\,bar$
Pressure Atmosphere Side	$0\,bar$
Cavitation Pressure	$0\,bar$

Table 1: System values of the calculated end face seal

Fluid		Sealing Rings	
Mineral Oil	ISO VG 46	Heat Conductivity $\lambda_{st/rot}$	$130\,W/mK$
Heat Capacity c_F	$2.4\,KJ/kgK$	Poisson's Ratio $\nu_{st/rot}$	0.17
Heat Conductivity λ_F	$0.13\,W/mK$	Coeff. of Therm. Exp. $\alpha_{st/rot}$	$3.4{\cdot}10^{-6}\,1/K$
Density ρ_F	$875\,kg/m^3$	Density ρ_{rot}	$3210\,kg/m^3$
Temp. Buffer Oil T_F	$50\,C^o$	Heat Capacity c_{rot}	$1.1\,KJ/kgK$

Table 2: Fluid data and material data

8.2   Thermal boundary conditions

Due to thermal symmetry on the rotating counter ring and the assumed constant temperature of buffer oil, only half the face seal system has to be considered, requiring Neumann boundary conditions at the line of symmetry (Fig. 4). Transmission of

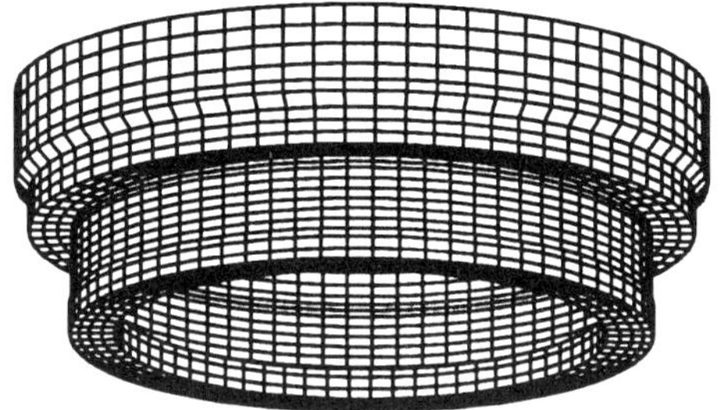

Figure 3: FEM-Structur (9660 nodes/ 7630 elements)

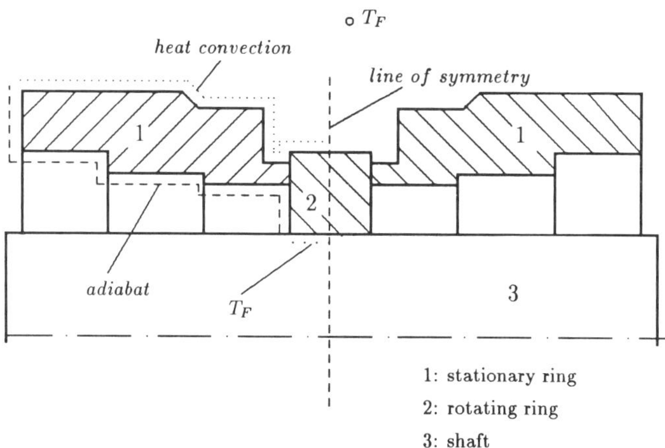

Figure 4: Thermal boundary conditions

heat takes place at the outer surfaces of both rings being in contact with buffer oil. Here, heat convection is described by heat convection coefficients α and ambient temperature T_F. Heat convection coefficients α are determined from heat convection laws of rotating cylinders, disks and cones [3,4,5]. Heat convection laws at the inner surface at the atmosphere side lead to very small α-coefficient values so that they may be neglected.

Assuming a constant temperature in the shaft and ideal heat convection between shaft and rotating counter ring requires Dirichlet boundary conditions $T = T_F$ at their contact area.

8.3   Results of calculation

The gap between the wavy stationary ring (measured waviness: A) and smooth rotating ring is approximated by the formula $h(\varphi) = \frac{A}{2} \cos(2\,\varphi)$ considering an A-type gap of $0.5\,\mu m$ (Fig. 5). Here, measured waviness is due to the machining operations. Bedding-in processes were not considered. The three-dimensional adjusting tem-

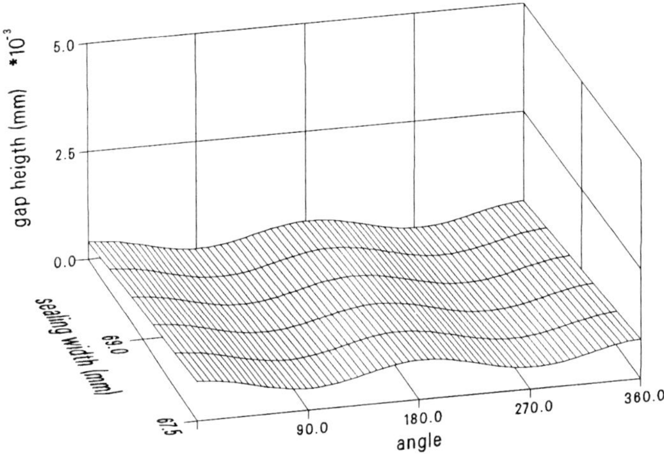

Figure 5: Resulting gap of unworn surfaces

perature distribution of the sealing system, later mentioned, yields locally different thermal deformations in the sealing rings which are responsible for the gap geometry prescribed in Fig. 7. Thermal distortions supply a V-type gap of about $1,5\,\mu m$ (Fig. 6). Hydrodynamic calculations require the gap geometry as a superposition of unworn surface and thermal deformations, as presented in Fig. 8.

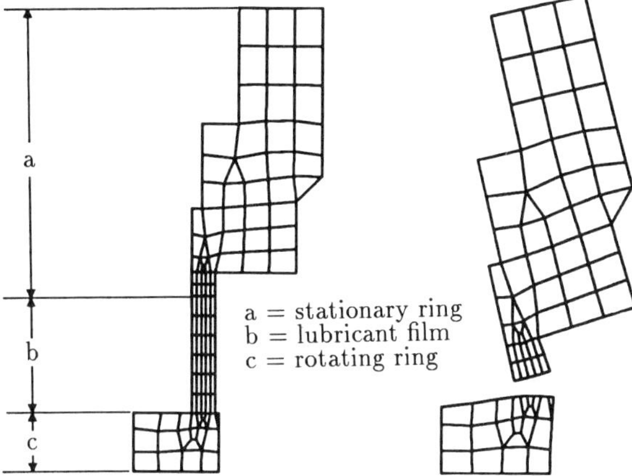

Figure 6: Effect of thermal deformation enlarged by a factor of 1000

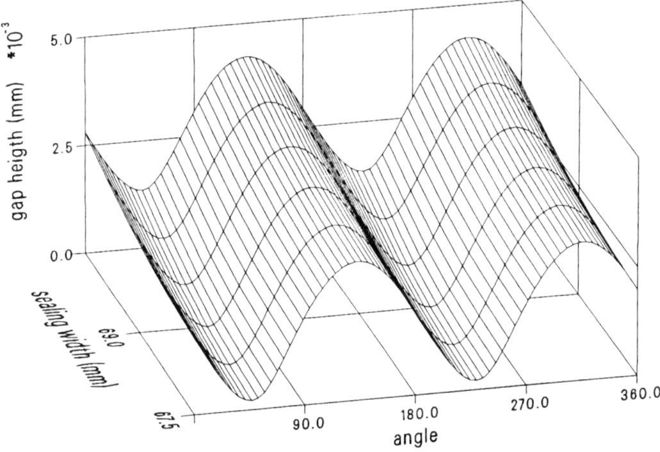

Figure 7: Gap distribution by thermal deformation

This calculation yields two symmetric so-called pressure hills (Fig. 9). Note that the pressure rise starts not at the beginning of the maximal gap which results from the principle of mass conservation as has been formulated in the *Kumar/Booker*

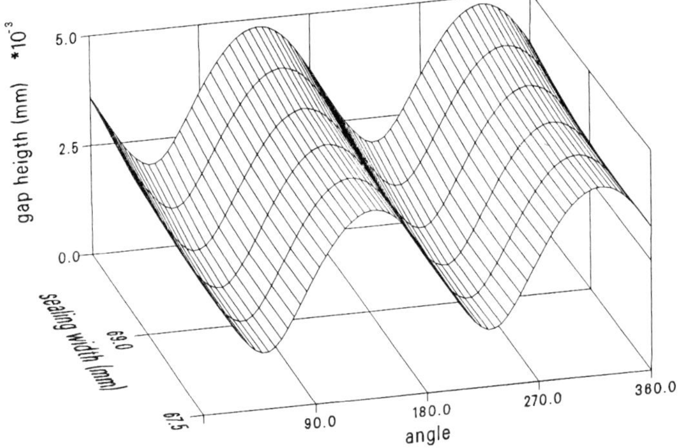

Figure 8: Gap geometry in working condition

algorithm. As a consequence of its low fractional film content (Fig. 10), the gap has to be filled before any positive pressure can be developed.

In addition, the filling-up process is supported by a radial Poiseuille flow having its origin in the high pressure gradient from the outer boundary to the inner radius of the lubricant film. This leads to a higher pressure rise in the outer areas.

The cavitation area is located beyond the smallest gap where ambient pressure p_a is assumed. Here again, the positive effect of the Poiseuille flow becomes apparent in the outer area: Hence, cavitation effects get restricted towards the inner area. The minimal fractional film content is about $\bar{\rho} = 0.31$.

The temperature distributions on the sliding face of the stationary ring, in the middle of the lubricant film and on the sliding face of the rotating ring are presented in Fig. 11, Fig. 12 and Fig. 13. Changes in power loss yield huge temperature gradients in circumferential direction.

The prescription of a given temperature $T = T_F$ at the inner surface of the rotating ring has great influence on the radial temperature decrease as shown in Fig. 12 and 13. The importance of three-dimensional calculations become apparent in Fig. 14: The sectional view in axial direction in the region of minimal and maximal gap shows the temperature distributions of the two sealing rings and the lubricant film. The difference between the maximal temperatures is about $30\,C^o$. Results of computer simulations and experimental investigations are presented in Tab. 3. Here T_{meas} is a

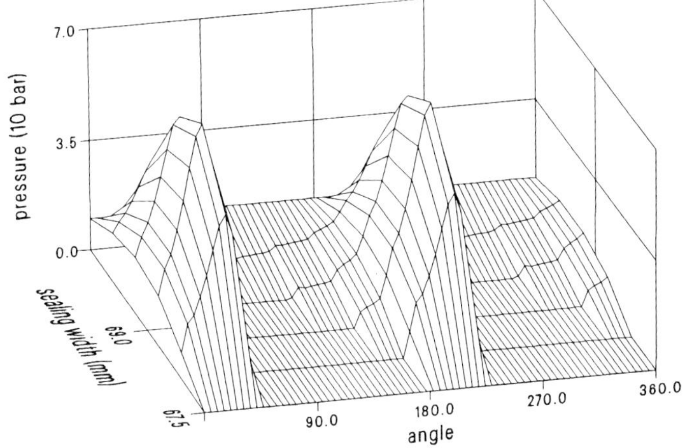

Figure 9: Pressure distribution

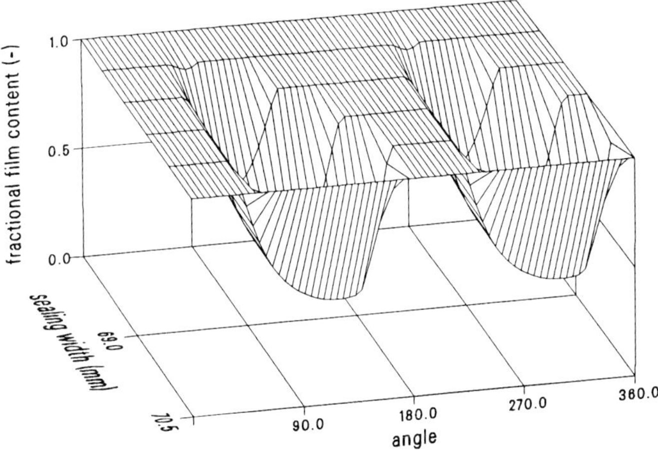

Figure 10: Distribution of fractional film content

temperature measured at the inner radius of the stationary ring $5\,mm$ above the gap
to verify the computer algorithms. Note that the calculation of turbulence power loss

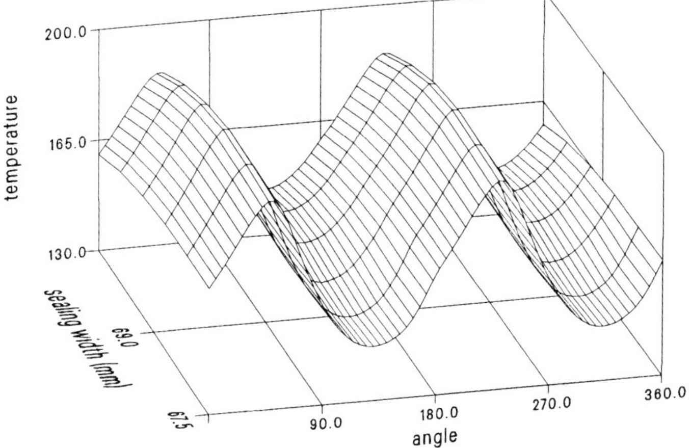

Figure 11: Temperature distribution on sliding face of stationary ring

		Calculation	test result
Minimal Gap Width	h_{min}	$0.41\,\mu m$	-
Maximal Pressure	p_{max}	$68.1\,bar$	-
Maximal Temperature	T_{max}	$201\,C^o$	-
Measured Temperature	T_{meas}	$120\,C^o$	$120 - 140\,C^o$
Power Loss in the Gap	P_g	$15.7\,KW$	-
Turbulence Loss	P_t	$9.9\,KW$	-
Total Power Loss	P_T	$25.6\,KW$	$30.95\,KW$
Leackage Rate	$\dot{Q}_l$	$1.88\,ltr./h$	$0.43\,ltr./h$

Table 3: Results calculation/bench test

is based on an empirical formula [6] considering smooth outer surfaces, a constant gap
between rotating ring and housing and a constant viscosity of the buffer oil. This
fact explains the difference between the calculated and measured total power loss.
Further explanations for differences between calculation and bench test are:

- an error rate of about $\pm 20\%$ in the applied heat convection laws

- idealized surface shape of the stationary ring

- the assumption of a constant buffer medium temperature $T_F = \frac{T_{in} + T_{out}}{2}$

- running-in processes and processes concerning the o-rings were not considered

- neglection of surface roughness effects

- assumption of a rotating ring without waviness; this is the most probable reason for differences between calculated and measured leakage rate

Future tasks to extend our computer simulations are:

- considering waviness on the sliding surface of the rotating ring,

- validation and implementation of new heat convection laws and determination of more accurate approximated ambient temperatures T_F in the buffer medium; The solution of three-dimensional Navier-Stokes equations in the buffer medium is intended,

- considering roughness on sliding surfaces.

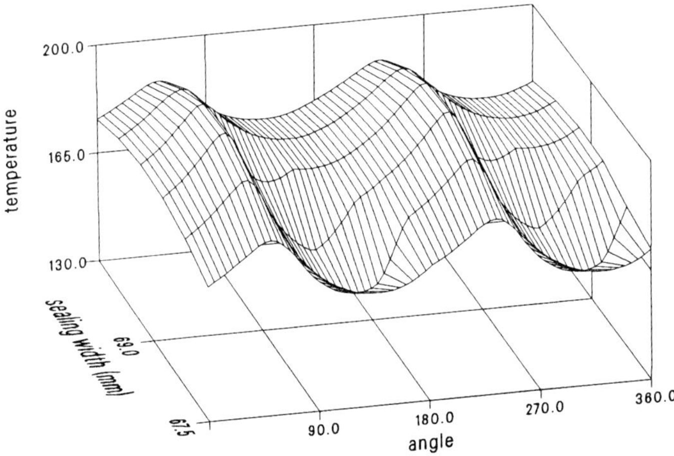

Figure 12: Temperature distribution in the middle of the lubricant film

382

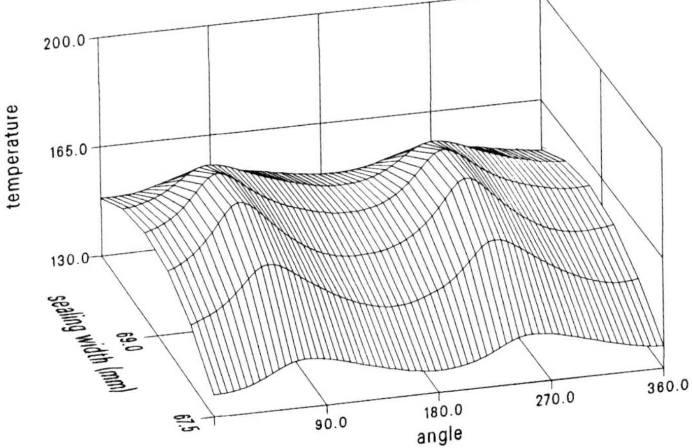

Figure 13: Temperature distribution on sliding face of the rotating ring

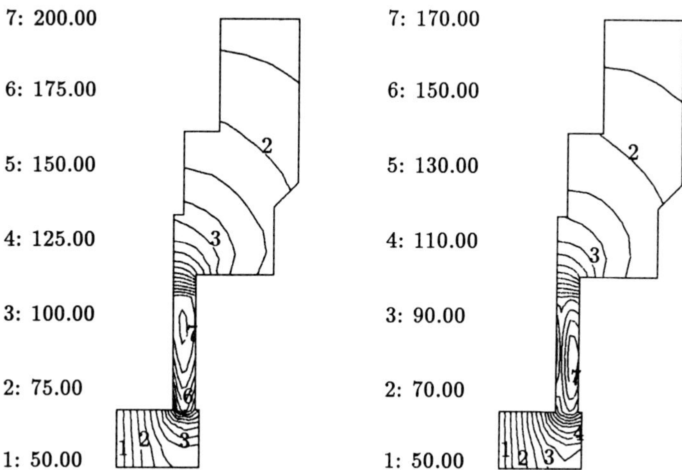

Figure 14: Temperature distribution of minimal (left) and maximal (right) gap

Acknowledgements

The simulation program has been developed in a joint research project (Bundesministerium für Forschung und Technik). The authors are grateful to D. Zeus and A. Eiletz (Feodor Burgmann Dichtungswerke) for their close cooperation and the test results, which enabled the comparison with the calculated results.

References

1. A. Kumar, Mass Conservation in Cavitating Bearings: A Finite Element Approach. PHD-thesis, Cornell University, 1991

2. J.F. Booker, K.H. Huebner, Application of Finite Element Element Methods to Lubrication: An Engeneering Approach, Journal of Lubrication Technology, October 1972, pp. 313-323

3. W.M. Kays, I.S. Bjorklund, Heat Transfer from a Rotating Cylinder With and Without Crossflow, Transactions of the ASME, January 1958, pp. 70-78.

4. H.T. Lin, L.K. Lin, Heat Transfer From a Rotating Cone or Disk to Fluids of Any Prandtl Number, Int. Comm. Heat Mass Transfer, Vol. 14, 1987, pp. 323-332

5. F. Kreith, Convection Heat Transfer in Rotating Systems, Advances in HEAT TRANSFER , Vol.5, 1968, pp. 129-251

6. H. Linneken, Der Radreibungsverlust insbesondere bei Turbomaschinen, AEG-Mitteilungen, Jg. 47, 1957, Heft 1/2

14th International Conference on Fluid Sealing, Firenze, Italy,
6-8 April 1994. Organised by BHR Group Limited, Cranfield,
Bedford, MK43 0AJ, UK; Tel: 0234 750422

ANALYSIS OF A HYDROSTATIC GAS SEAL WITH A COMPLIANT FACE

Richard F. Salant
School of Mechanical Engineering
Georgia Institute of Technology
Atlanta, Georgia 30332, USA

ABSTRACT

The potential advantages of a hydrostatic mechanical gas seal with a compliant face, analogous to a foil bearing, are investigated through a numerical analysis. Solutions to the isothermal compressible Reynolds equation are obtained, assuming the surface of one face is compliant and supported by an elastic foundation. The pressure distribution in the gas film, the opening force, the axial film stiffness, and the leakage rate are determined as functions of the design parameters. It is found that a properly designed compliant seal can provide stiffness characteristics superior to those of a conventional hydrostatic seal.

NOMENCLATURE

F	dimensionless opening force, $F^*/2\pi p_i^* r_i^{*2}$
F^*	opening force
h	dimensionless film thickness, h^*/h_i^*
h^*	film thickness
k	dimensionless axial film stiffness, $(h_i^*/2\pi p_i^* r_i^{*2})k^*$
k^*	axial film stiffness
m	dimensionless leakage rate, $(12\mu RT/\pi h_i^{*3} p_i^{*2})m^*$
m^*	leakage rate
p	dimensionless pressure, p^*/p_i^*
p^*	pressure
R	gas constant
r	dimensionless radial coordinate, r^*/r_i^*

r^* radial coordinate
s dimensionless foundation stiffness, $s^* h_i^* / p_i^*$
s^* foundation stiffness
T temperature
Δh^* coning of foundation, $h_o^* - h_i^*$
μ viscosity
$(\ )_i$ at ID
$(\ )_o$ at OD

INTRODUCTION

Since the 1970's, foil bearings with compliant surfaces have been developed [1]. In such bearings, the compliant surface deforms in response to the local film pressure. This allows one to design the bearing to deform in a favorable manner for a given application, by selecting a suitable elastic foundation design to produce the desired characteristics. Thus, for example, foil bearings can be designed to give increased load capacity and/or increased stability, as compared with conventional bearings [2]. It is believed that similar advantages could be obtained with a "foil seal," viz., a mechanical seal with a compliant face surface. An especially suitable application is the hydrostatic gas seal.

For dry gas sealing (without the use of a liquid buffer fluid), the flat faced hydrostatic mechanical seal is the simplest device to manufacture and use. However, such a seal has undesireable stiffness characteristics: the axial film stiffness is strongly dependent on the coning of the seal faces. The stiffness is positive only if the coning is positive (seal gap convergent in direction of leakage flow), and decreases very steeply as the coning is reduced. If there is zero coning, the stiffness is zero, and the seal is metastable. If the coning is negative (divergent gap), the stiffness is negative, and the seal is unstable. Therefore a hydrostatic seal must be designed to operate with positive coning [3]. However, the coning (defined as the difference between the film thicknesses at OD and ID) must be very small to obtain a thin enough film for acceptable leakage, typically on the order of a micron. Since mechanical and thermal deformations are of the same order, it is very difficult to engineer a seal with just the right amount of coning to give a desired stiffness. Furthermore, if operating conditions change, the coning will change significantly, and so will the stiffness. Therefore conventional hydrostatic gas seals run the risk of operating with very low or negative stiffness, resulting in instability and collapse of the gas film.

To improve the stiffness characteristics of a hydrostatic gas seal, a seal with a compliant face (a "foil seal") is proposed. A schematic diagram of such a seal is shown in figure 1. It is similar to a conventional seal, except the surface of one face is constructed of a foil skin, supported by an elastic foundation. The foundation could be constructed of metal subfoils or springs, or of an elastomeric material such as an open cell elastomeric foam. The elastic properties of the foundation could be constant, or could vary spatially. For constant properties, the foil will deform most in regions where the film pressure is highest.

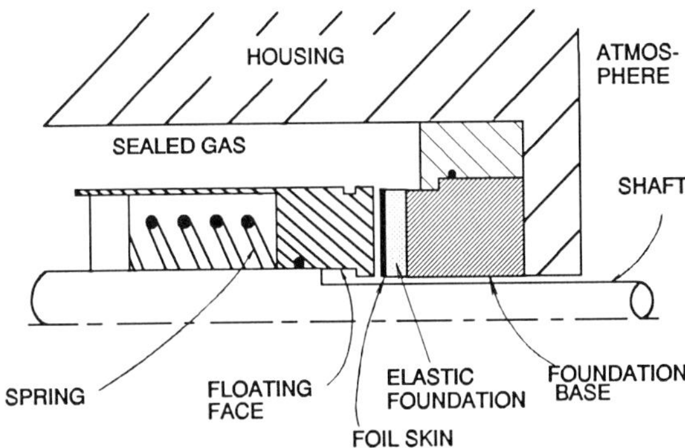

Figure 1 Schematic of compliant seal.

Consider the operation of the above compliant seal (with constant elastic properties) when the base of the foundation experiences no coning. This would correspond to the case of the conventional seal with no coning and zero film stiffness. Now, however, the pressure distribution in the film (which is highest at the OD and drops as the ID is approached) causes the foil to deform in such a way as to produce positive coning of the face surface. This results in positive stiffness. Even for cases when the foundation base experiences negative coning (corresponding to negative stiffness and instability for a conventional seal), the foundation elasticity can be chosen such that the deformation of the foil produces enough positive coning to compensate for the negative coning of the base and insure stability.

ANALYSIS

To assess the performance of the hydrostatic gas seal with a compliant face, a numerical analysis is performed. The pressure distribution in the gas film, between the seal faces, is governed by the isothermal Reynolds Equation. For an axisymmetric film, the equation takes the dimensionless form,

388

$$\frac{d}{dr}(rh^3\frac{dp^2}{dr}) \ = \ 0 \tag{1}$$

For a seal with a compliant face, the dimensionless film thickness can be expressed as,

$$h \ = \ h_o - (h_o-1)\frac{(r_o-r)}{(r_o-1)} + \frac{p-1}{s} \tag{2}$$

where h_o is the *undeformed* film thickness at the OD, normalized by the film thickness at the ID. h_o is related to the coning of the foundation, Δh^*, through,

$$h_o \ = \ 1 + \frac{\Delta h^*}{h_i^*} \tag{3}$$

The dimensionless leakage rate (mass flow rate) is given by,

$$m \ = \ rh^3\frac{dp^2}{dr} \tag{4}$$

The governing equations are solved with an iterative finite difference scheme. Equation (1) is discretized using the micro-control volume technique [4], and the resulting algebraic equations are solved with the Tridiagonal Matrix Algorithm. The grid consists of 800 nodes. h is evaluated from equation (2), using the pressure p from the previous iteration. The computation is completed when the maximum deviation in dimensionless pressure between succeeding iterations is less than 10^{-6}. Once the pressure distribution is obtained, it is integrated over the seal face to yield the opening force, F. The film stiffness is obtained by recomputing the pressure distribution and opening force at a slightly larger film thickness, h + δh, and noting,

$$k \ \approx \ -(1/\delta h)[F(h+\delta h)-F(h)] \tag{5}$$

and the leakage rate is obtained from the finite difference form of equation (4).

RESULTS AND DISCUSSION

Figure 2 shows the film stiffness of a seal, with $r_o = 1.2$ and $p_o = 10$, as a function of h_o (a measure of the coning of the foundation), for various values of the foundation stiffness s. Also shown is the film stiffness for a conventional seal with rigid faces. First consider the conventional seal. From the figure it is seen that the stiffness drops very rapidly when h_o is less than 2, to a value of 0 at $h_o = 1$ (zero coning), and to negative values when $h_o < 1$ (negative coning). This is a very undesireable stiffness curve, as discussed in the Introduction. Next, consider the case $s = 5$. The film stiffness curve is extremely flat, even for values of h_o as low as .5 (negative coning). While the magnitude of the film stiffness is less than the maximum stiffness for a conventional seal, since it is flat and positive the seal should always be stable. As the foundation stiffness is increased to 10, and then to 15, it is seen that the curve becomes less flat in the region $h_o < 2$, but the magnitude of the film stiffness in the region $h_o > 2$ increases.

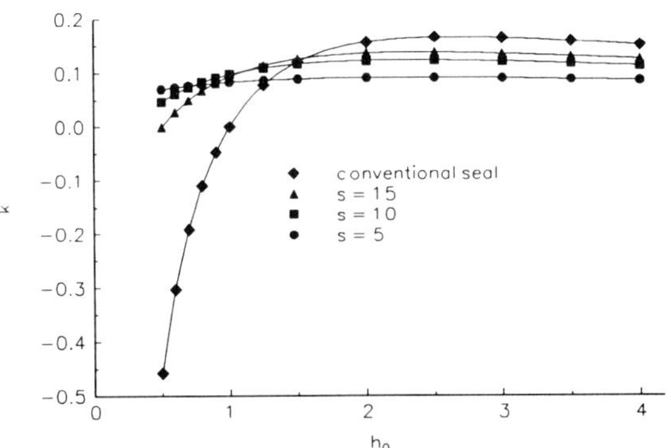

Figure 2 Film stiffness, $r_o = 1.2$, $p_o = 10$.

The corresponding opening force, F, is shown in figure 3. It is seen that in the compliant seal the opening force is less dependent on h_o than in the conventional seal. The lower the foundation stiffness, the less the dependence.

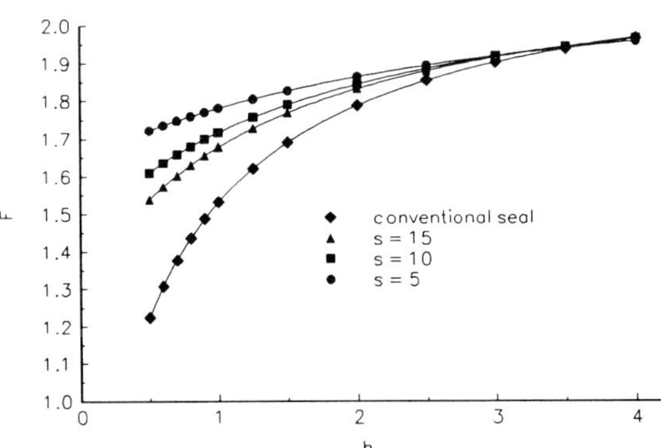

Figure 3   Opening force, $r_o = 1.2$, $p_o = 10$.

390

Figure 4 shows the dimensionless leakage rates for the cases considered in figure 2 and 3. For a given value of h_o, the conventional seal has the lowest leakage rate. For the compliant seal, the lower the foundation stiffness, the higher the leakage rate. Thus, it is seen that the choice of foundation stiffness involves a tradeoff between flatness of the film stiffness curve, magnitude of film stiffness, and leakage rate.

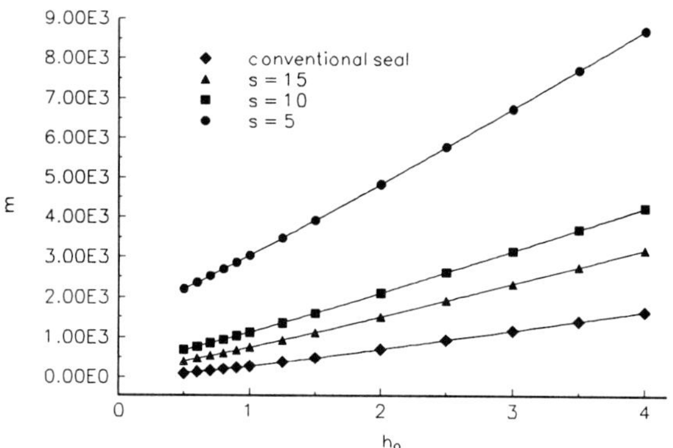

Figure 4 Leakage rate, $r_o = 1.2$, $p_o = 10$.

Better performance characteristics of the compliant seal can be obtained by constructing a foundation whose stiffness varies with radius. Figure 5 shows the film stiffness for a compliant seal whose foundation stiffness is given by $s = 5 + 25(r_o-r)/(r_o-1)$. It is seen that the film stiffness curve is very flat, and only slightly below the maximum stiffness of the conventional seal. The corresponding opening force curve is in figure 6. These curves are superior to all of the curves for the constant foundation stiffness cases (figures 2 and 3).

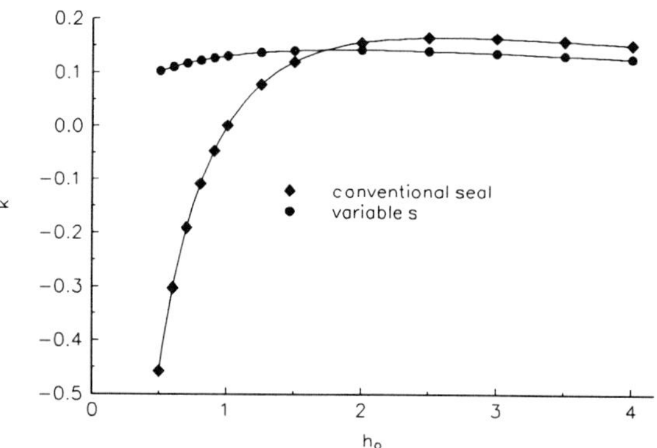

Figure 5 Film stiffness, $r_o = 1.2$, $p_o = 10$, variable s.

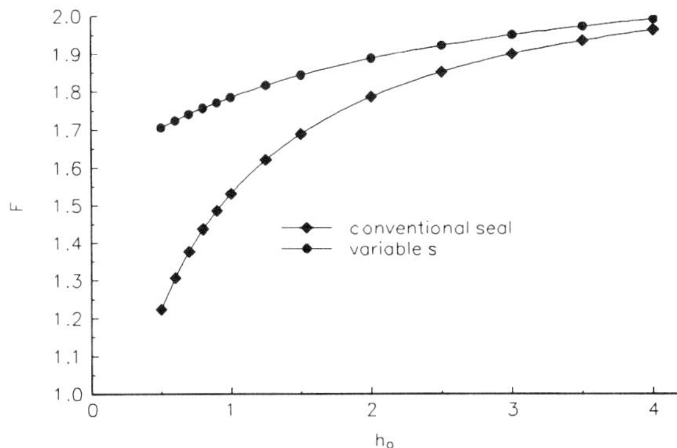

Figure 6 Opening force, $r_o = 1.2$, $p_o = 10$, variable s.

Figure 7 shows the corresponding leakage rate. While the leakage rate is still greater than that of a conventional seal for a given h_o, it is much lower than corresponding values with a constant foundation stiffness (figure 4). Furthermore, in considering the curve for the conventional seal, it should be noted that such a seal cannot be used, practically, with values of $h_o < 2$, so the low leakage rates predicted for $h_o < 2$ are just hypothetical, with no real meaning. In contrast, the compliant seal can be used at the lower values of h_o with the corresponding lower leakage rates.

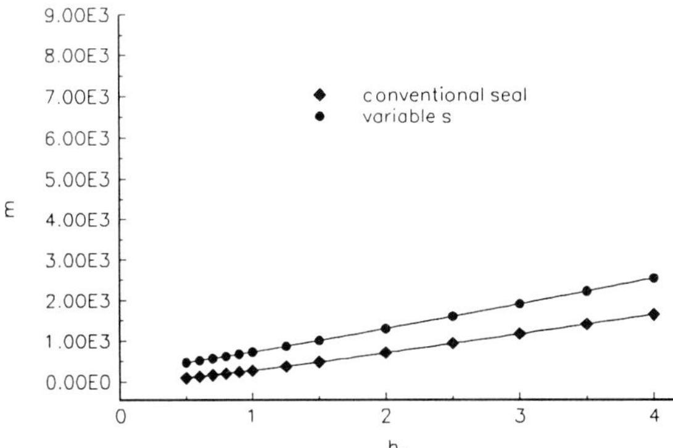

Figure 7 Leakage rate, $r_o = 1.2$, $p_o = 10$, variable s.

The substantial improvement in leakage rate that is achieved with a variable foundation elasticity, can be understood by considering the film thickness distribution.

Figure 8 shows the film thickness distribution for both constant and variable foundation stiffness cases. Note that for the constant s case, going from the OD to the ID, the film remains quite thick until just before the ID is reached, at which point the film necks down drastically. This is because the pressure distribution is nonlinear, as shown in figure 9. Since high pressure persists throughout most of the radial extent, deformation occurs in the same region, resulting in a thick film throughout that region. Such a film distribution results in a high leakage rate. This is remedied by utilizing a foundation with a variable stiffness, in which the stiffness is highest near the ID. From figure 8 it is seen that this produces a film that is much thinner than the constant foundation stiffness case over most of the radial extent, resulting in a lower leakage rate.

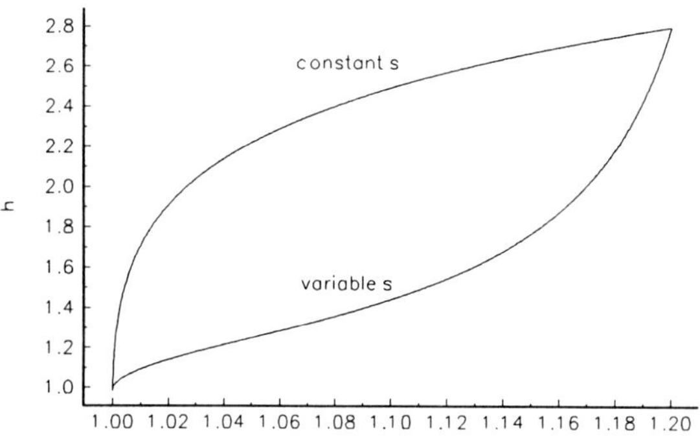

Figure 8 Film thickness distributions.

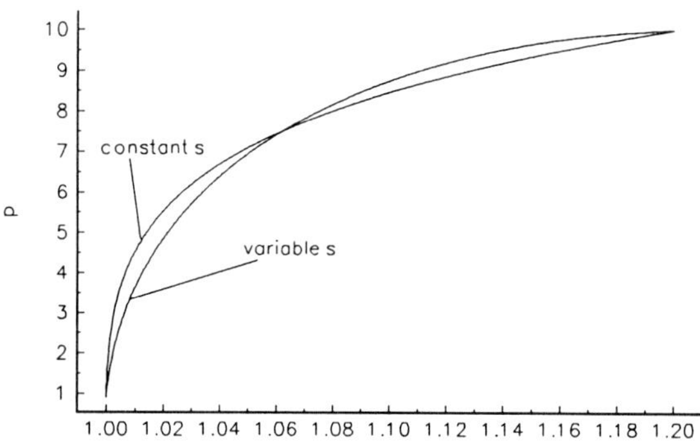

Figure 9 Pressure distributions.

All of the results discussed above are for a moderate sealed pressure, $p_o = 10$. Now consider a high pressure case, $p_o = 100$. For higher pressures, the foundation stiffness must be increased. Therefore for this case the foundation stiffnesses are all scaled up by a factor of 10. The results are shown in figures 10 to 12 for the variable foundation stiffness. Comparison with figures 5 to 7 shows that the high pressure results are very similar to the moderate pressure results, and the discussion above is equally applicable to the high pressure case.

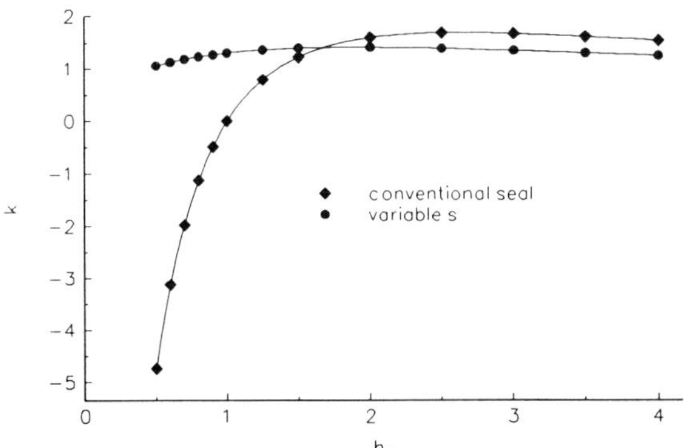

Figure 10 Film stiffness, $r_o = 1.2$, $p_o = 100$, variable s.

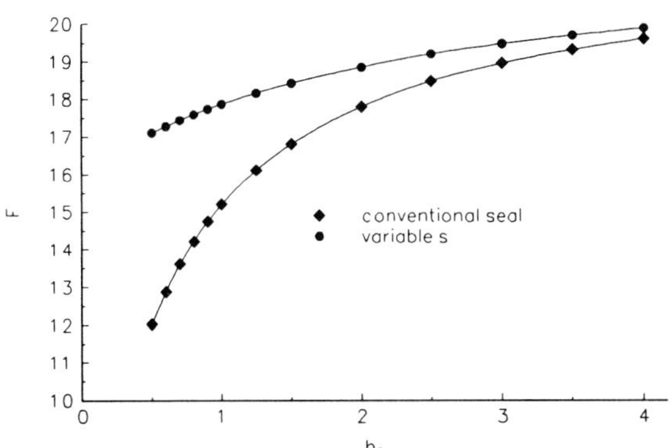

Figure 11 Opening force, $r_o = 1.2$, $p_o = 100$, variable s.

394

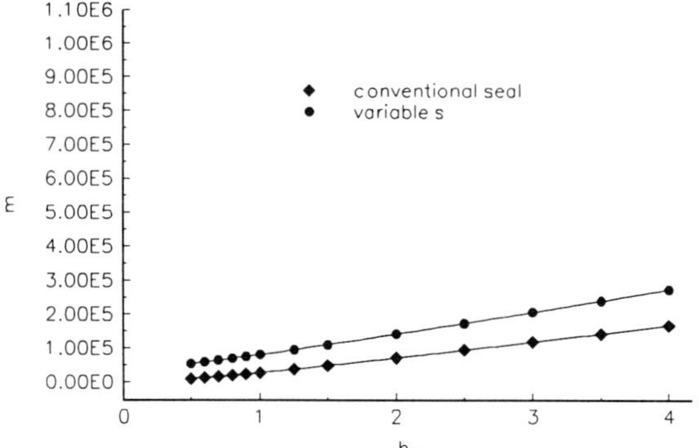

Figure 12 Leakage rate, $r_o = 1.2$, $p_o = 100$, variable s.

CONCLUSIONS

The results of this study indicate that a hydrostatic mechanical gas seal with a compliant face has axial stiffness characteristics superior to those of a conventional seal, and will be stable to axial disturbances even when the foundation base experiences negative coning. Furthermore, with the appropriate choice of a radially varying foundation stiffness, the leakage characteristics of the seal are comparable to those of a conventional seal.

There are, however, a number of questions, both conceptual and practical, that still must be addressed. In the former category fall questions regarding the angular stiffness of the seal, its dynamic response, and its tracking capabilities. In the latter fall questions regarding the skin material, and the design of the elastic foundation. These are suitable areas for future research.

REFERENCES

1.   Heshmat, H., Shapiro, W. and Gray, S., "Development of Foil Journal Bearings for High Load Capacity and High Speed Whirl Stability," ASME Journal of

Lubrication Technology, **104**, pp. 149-156, 1982.

2.	Peng, J.P. and Carpino, M., "Calculation of Stiffness and Damping Coefficients for Elastically Supported Gas Foil Bearings," ASME Journal of Tribology, **115**, pp. 20-27, 1993.

3.	Salant, R.F. and Key, W.E., "Improved Mechanical Seal Design Through Mathematical Modelling," *Proc. First International Pump Symposium*, pp. 37-46, Texas A&M University, 1984.

4.	Patankar, S.V., *Numerical Heat Transfer*, McGraw-Hill, New York, 1980.

14th International Conference on Fluid Sealing, Firenze, Italy,
6-8 April 1994. Organised by BHR Group Limited, Cranfield,
Bedford, MK43 0AJ, UK; Tel: 0234 750422

MODELLING OF PLAIN FACE GAS SEAL DYNAMICS

Simon Leefe

BHR Group Limited
Cranfield, Bedford, MK43 OAJ, UK

ABSTRACT

A mathematical model is presented for the prediction of the behaviour, under various dynamic excitation conditions, of plain radial face gas seals. This model offers an integrated treatment of seal ring structural distortions, thin film lubrication and vibrational dynamics of the mechanically damped, bellows-mounted floating stator ring in four degrees of freedom. Seal face waviness and coning (thermal and mechanical) on both rings are catered for as well as angular misalignment, eccentricity and axial oscillation of the clamped, rotating ring.

Algorithms for the numerical solution of the governing equations are outlined, and some sample problems solved and discussed.

NOMENCLATURE
(See also Figs. 1-3)

a_v	rotor axial vibration amplitude (zero-to-peak)
a_{w1}	stator wave amplitude (zero-to-peak)
a_{w2}	rotor wave amplitude (zero-to-peak)
e	eccentricity of rotor geometrical centre from axis of rotation
h	film thickness
h_1	displacement (in the z-direction) of a point on the stator face from the datum plane
h_2	displacement (in the z-direction) of a point on the rotor face from the datum plane
n (subscript)	time step index
n_1	number of waves on stator face
n_2	number of waves on rotor face
p	interfacial pressure
$<p>$	circumferentially-averaged pressure at radius r
r	cylindrical polar radial coordinate
r_i	interface internal radius
r_m	interface mean radius
r_o	interface external radius
t	time
u_r	film-thickness-averaged fluid radial velocity
u_ϕ	film-thickness-averaged fluid circumferential velocity
w	weighting function relating interfacial pressure distribution to face coning

x	Cartesian coordinate - direction in datum plane
$\{x\}$	stator rigid body displacement vector
$\{x'\}$	stator rigid body velocity vector
$\{x''\}$	stator rigid body acceleration vector
y	Cartesian coordinate - direction in datum plane perpendicular to x
z	Cartesian and cylindrical polar coordinate - normal to datum plane, positive direction points from stator towards mating ring
z_1	displacement of stator centroid in z-direction
$\{F\}$	stator total applied force vector
$\{F_{constant}\}$	contribution to $\{F\}$ arising from peripheral pressure distribution and bellows mean compression
$\{F_{flexible\ supports}\}$	contribution to $\{F\}$ arising from bellows forces and moments
$\{F_{fluid\ film}\}$	contribution to $\{F\}$ arising from distribution of interfacial pressure and shear stress
$\{F_{mechanical\ dampers}\}$	contribution to $\{F\}$ arising from bellows and face-carrier dampers
$\{F_{NL}\}$	contribution to $\{F\}$ defined in equation (20)
$[K_{bellows}]$	linear stiffness matrix arising from stator bellows forces and moments
M	Mach number
$[M]$	stator face carrier inertia matrix
Re	Reynolds number
$Re_{critical}$	Reynolds number for laminar/turbulent transition
α	angular displacement of stator about x-axis
β	angular displacement of stator about y-axis
δ	rotor misalignment angle (out-of-squareness with respect to z axis)
δt	time increment
$\varnothing$	cylindrical polar circumferential coordinate
γ	ratio of sealed gas specific heat capacities
η	sealed fluid dynamic viscosity
θ_v	rotor axial vibration initial phase angle
θ_{w1}	angular displacement of first stator wave peak
θ_{w2}	initial angular displacement of first rotor wave peak
θ_δ	initial orientation of rotor misalignment axis with respect to $\varnothing = 0$
τ_r	stator face radial shear stress - laminar isothermal model
$\tau_\varnothing$	stator face circumferential shear stress - laminar isothermal model
ω	angular velocity of rotor
ω_v	rotor axial vibrational angular frequency
ψ_i	face coning angle due to interfacial pressure distribution
ψ_1	total stator coning angle
ψ_2	total rotor axisymmetric coning angle

INTRODUCTION

Given the obvious effect of radial face seal dynamics on the fluid film and hence on seal performance, it is surprising that until comparatively recently, this has been a neglected area for detailed modelling. As well as some experimental work, for example examining the effect of different mechanical dampers, theoretical effort up until the early 1970s concentrated on predicting the effect of known seal ring vibrations on performance, rather than allowing an interaction between the fluid film and seal ring dynamics. Furthermore, what work does exist is usually concerned with liquid sealing. Nau (1) reviews the early work, whilst other publications worth examining for their relevance to the current application are Tournerie and Frene (2), which considers waviness, coning and misalignment in a liquid film dynamic analysis with cavitation; Shapiro & Colsher (3), which couples seal ring dynamics in three degrees of freedom with a transient gas film analysis; Shapiro (4), who includes face friction, eccentricity, misalignment and secondary seal friction in a five degree-of-freedom dynamics model, but whose interfacial dynamic analysis is a perturbation solution based on a parallel film; and the work of Hughes and associates on two-phase seal dynamics, referenced in the previous Sub-Section. The other principal authors active in this field are Etsion and Green, whose wealth of published material is adequately captured in two references (5,6). These deal with noncontacting liquid seals.

Given the complexity of adequate modelling of face seal behaviour, it is not surprising that vibrational phenomena are typically studied in isolation. In practice however, mechanical and thermal effects have a profound impact on the sealing interface. Therefore, much can be learned about seal performance from an improved understanding of the *interactions* between these phenomena. For example, face pressure distribution modifies film taper which, in turn affects face pressure distribution. The pressure distribution affects the instantaneous force on the floating stator, which results in its displacent, thus altering the film thickness distribution, hence pressure field and frictional heat generation. This then modifies the amplitude of face coning, and so on. Almost all the phenomena governing performance are inter-related. Thus, there is a great deal to be gained from the integration of existing modelling approaches to particular phenomena into a single computer program.

MATHEMATICAL MODEL

Scope

As suggested in the Introduction, this analysis synthesises the modelling of a variety of phenomena which have previously been studied in isolation or in limited combination, whilst adding some novel features. Thus, whilst the scope, in terms of complexity of analysis, is very broad indeed, its application is restricted here to one particular type of seal, namely a plain radial-face gas seal with bellows-mounted stator and rigidly clamped rotating mating ring (see Fig. 4). This arrangement is typical of that found in low- to medium-powered rocket engine fuel and oxidiser turbopumps (7), where such devices seal both the liquid cryogen at the pump end and the hot exhaust gases at the turbine end. This application provides the rationale for some of the features incorporated into the model, a full and detailed account of which may be found in ref. (8).

The geometry of both seal faces is governed by coning from a variety of sources (both steady-state and transient), detailed below, and waviness of arbitrary order and amplitude. The working fluid is assumed to behave as an ideal gas through the sealing interface, where its flow is assumed to be laminar and isothermal. Dynamic excitation of the seal is provided by rotor axial vibration of arbitrary amplitude and frequency, rotating mating ring face out-

400

of-squareness and eccentricity. The dynamic response of the floating stator in four degrees of freedom (axial, torsional and rotational about two perpendicular diameters) is governed by face-carrier inertia and bellows stiffness in each degree and by Coulomb dampers acting on both face-carrier and bellows. Each gives a constant force opposing the motion of the stator face relative to its housing.

In order to keep the model within reasonable bounds of complexity, ring mode distortions are considered to axisymmetric. Circumferential variation in ring mode distortion could, in principle, be treated by an additional array of influence coefficients (and this may indeed be incorporated at a later date), but this is considered by the author to be a less significant contributor to the accommodation of face waviness than EHL effects which, together with roughness and mechanical contact has very recently been added, with very promising initial results (to be reported at a later date).

Gas film

The relationship between the distributions of pressure and film thickness between the seal faces is given by the isothermal, compressible Reynolds equation for an ideal gas:

$$\partial/\partial r\{ph^3.\partial p/\partial r\} + \partial/r\partial\varnothing\{ph^3.\partial p/r\partial\varnothing\} - 6\eta r\,\omega.\partial/r\partial\varnothing(ph) = 12\eta\partial/\partial t(ph) \qquad - (1)$$

in which $h(r,\varnothing,t)$ is calculated as:

$$h = h_1 - h_2 \qquad - (2)$$

the difference in elevation of points on rotor and stator face, respectively, at location $(r,\varnothing)$ above an arbitrary datum plane perpendicular to the shaft axis. In this way, the geometry of each face can be specified in terms of taper, waviness, stator displacements and mating ring out-of-squareness, eccentricity and axial vibration. The stator height distribution, h_1, is given by:

$$h_1 = z_1 + r(\alpha.\sin\varnothing - \beta.\cos\varnothing) - (r-r_m).\psi_1 + a_{w1}.\cos\{n_1.(\theta_{w1}-\varnothing)\} \qquad - (3)$$

where the first two terms represent initial displacements, the third accounts for coning and the last for waviness. Rotor height distribution is given (to first order) by:

$$h_2 = \{r-r_m+e.\sin(\omega t-\varnothing)\}.\psi_2 + a_{w2}.\cos\{n_2.(\omega t+\theta_{w2}-\varnothing)\}$$
$$- \delta\{r.\sin(\omega t+\theta_\delta-\varnothing) + e.\cos\theta_\delta\} + a_v.\sin(\omega_v t+\theta_v) \qquad - (4)$$

where the first term accounts for the combined effect of rotor coning and eccentricity, the second for waviness, the third for the combined effect of out-of-squareness and eccentricity and the last term for axial vibration.

Radial gas velocity (averaged across the film thickness) and shear stress are derived from the radial derivative of the pressure distribution:

$$u_r = -(h^2/12\eta).\partial p/\partial r \qquad - (5)$$

$$\tau_r = -(h/2).\partial p/\partial r \qquad - (6)$$

whilst circumferential velocity and shear stress also incorporate drag terms arising from the sliding:

$$u_\emptyset = -(h^2/12\eta).\partial p/r\partial\emptyset + r\omega/2 \qquad - (7)$$

$$\tau_\emptyset = -(h/2).\partial p/r\partial\emptyset + r\omega\eta/h \qquad - (8)$$

The velocities are used to calculate the leakage rate along with Reynolds and Mach numbers to enable checks that flow remains laminar and isothermal, in line with the assumptions. The shear stress and pressure distributions are used to calculate fluid film forces and moments acting on the floating stator ring.

Coning

Equations (3) and (4) contain terms in stator and rotor coning angles, ψ_1 and ψ_2, respectively. These arise from a number of sources as detailed below:

- Taper of either seal-ring assembly due to depressed or elevated ambient temperature, arising from differential thermal contraction or expansion between parts of an assembly

- Taper of stator face due to bellows force

- Taper due to sealed pressure differential

- Taper due to mating ring centrifugal inertia

- Pre-start-up taper on either ring, e.g. that arising from lapping or clamping

- Taper due to interfacial pressure distribution.

With the exception of the last of these, each of the above may be regarded as constant over an analysis. All but the last two may be conveniently evaluated by means of an influence coefficient and an operating parameter, whilst the pre-start-up taper is simply a constant. Taper arising from interfacial pressure distribution is worthy of further comment, since its treatment (9) is novel. If the pressure at a given radius is averaged out, circumferentially, a radial pressure distribution results:

$$<p(r,t)> = (1/2\pi). \,_0\!\int^{2\pi} p(r,\emptyset,t). \; rd\emptyset \qquad - (9)$$

If pressure $<p(r,t)>$ acting over an annulus between r and r+dr on one of the seal faces results in a contribution to the coning of that face of $w(r).<p(r,t)>.dr$ where w is a weighting function, then the taper at time t of that face arising from the instantaneous interfacial pressure distribution is:

$$\psi_i(t) = \,_{r_i}\!\int^{r_o} w(r).<p(r,t)>.dr \qquad - (10)$$

This approach lends itself very well to discretisation of the solution domain for the fluid film, resulting in the use of a series of face influence coefficients (the discrete analogue of the weighting function, w).

402

Equations of motion

The equations of motion of the face-carrier sub-assembly, are summarised in matrix notation as follows:

$$\{F_{fluid\ film}\} + \{F_{mechanical\ dampers}\} + \{F_{flexible\ supports}\} + \{F_{constant}\}$$
$$= [M].\{x''\} \qquad - (11)$$

where the vector of fluid film forces and moments may be evaluated explicitly from the pressure and shear stress distributions, the vector of mechanical damper forces and moments may be evaluated from the stator velocity components and the vector of flexible support (bellows) forces is given by:

$$\{F_{flexible\ support}\} = [K_{bellows}].\{x\} \qquad - (12)$$

These forces arise from the displacement of the bellows about its mean compression. The constant force vector consists solely of an axial component, arising from the bellows mean compression and the peripheral distribution of sealed and drain pressures.

The treatment of these equations in the solution algorithm is somewhat unusual and is discussed more fully in the following section.

COMPUTER PROGRAM

The governing equations for the fluid film are solved by finite difference methods, with the solution domain - the sealing interface - discretised in space over a user-defined grid. A semi-implicit algorithm is employed for the solution of the Reynolds equation, in which the second order spatial derivatives employ central differences whilst the drag term employs a backward difference. This was found to give improved stability. Much attention was given to the centering of the time discretisation, with a view to enhancing accuracy. This leads to a scheme whereby the film thickness is calculated at mid-timestep, with mating ring face geometry calculated from an explicit time-dependent formula (equation (4)) and stator face geometry by means of Taylor series expansion:

$$(h_1)_{n+1/2} = (h_1)_{n-1/2} + (\partial h_1/\partial t)_n.\delta t \qquad - (13)$$

Here, $(\partial h_1/\partial t)_n$ is evaluated from equation (3), using current values of stator displacements. In this way, p_{n+1} may be evaluated from p_n using $h_{n+1/2}$.

Having solved for pressure, fluid velocity is calculated and checks made against criteria representing laminar-turbulent transition ($Re = Re_{critical}$) and departure from isothermal towards adiabatic compressible flow ($M = 1/\sqrt{\gamma}$). Assuming no violation, leakrate is calculated, along with the stator face shear stress distribution, hence interfacial viscous heat generation. The pressure distribution is also used to update the face coning, as outlined in the preceding section. This then modifies the film thickness distribution at the next timestep.

The equations of motion of the floating stator are solved by an unusual variant of Newmark's method, again in attempt to enhance accuracy by providing a well-centred time discretisation. In this scheme, the velocity vector at the end of the timestep is calculated by means of an average acceleration over the timestep:

$$\{x'\}_{n+1} = \{x'\}_n + \delta t.(\{x''\}_{n+1} + \{x''\}_n)/2 \qquad - (14)$$

so that:

$$\{x''\}_{n+1} = 2.(\{x'\}_{n+1} - \{x'\}_n)/\delta t - \{x''\}_n \qquad - (15)$$

Similarly:

$$\{x'\}_{n+1} = 2.(\{x\}_{n+1} - \{x\}_n)/\delta t - \{x'\}_n \qquad - (16)$$

Substituting equation (16) into equation (15) yields:

$$\{x''\}_{n+1} = 4.(\{x\}_{n+1} - \{x\}_n)/\delta t^2 - 4.\{x'\}_n/\delta t - \{x''\}_n \qquad - (17)$$

Substitution of Newton's second law into the left hand side of equation (17) gives:

$$[M]^{-1}.\{F\}_{n+1} = 4.(\{x\}_{n+1} - \{x\}_n)/\delta t^2 - 4.\{x'\}_n/\delta t - \{x''\}_n \qquad - (18)$$

As suggested by equations (11) and (12), we may write:

$$\{F\}_{n+1} = [K_{bellows}].\{x\}_{n+1} + \{F_{NL}\}_{n+1} \qquad - (19)$$

where $\{F_{NL}\}$ is the vector of (nonlinear) forces arising from sources other than bellows support (linear) stiffness:

$$\{F_{NL}\} = \{F_{fluid\ film}\} + \{F_{mechanical\ dampers}\} + \{F_{constant}\} \qquad - (20)$$

Substituting equation (19) into equation (18) and the result into equation 17, then re-arranging gives the final result:

$$\{x\}_{n+1} = \left[[I] + \delta t^2/4.[M]^{-1}[K_{bellows}]\right]^{-1}.$$
$$\{\delta t^2/4\left([M]^{-1}.\{F_{NL}\}_{n+1} + \{x''\}_n\right) + \{x'\}_n.\delta t + \{x\}_n\} \qquad - (21)$$

so that the new displacement vector is calculated explicitly in terms of the old displacement, velocity and acceleration vectors, the fluid film forces (from the pressure and shear stress distributions just calculated) and an estimate of the mechanical damper forces. These estimates are made from velocities extrapolated from previous values by first order Taylor expansion. Since bellows lateral stiffness is very much greater than the axial and diametral components, the bellows stiffness matrix is decoupled and the inverses are trivial.

The structure of the solution algorithm is illustrated in the flowchart of Fig. 5. The initial pressure distribution is calculated by solving the Reynolds equation in which the film thickness and its time derivative is calculated by means of specified initial stator displacements and velocities and shaft rotational speed. Indeed, this procedure permits the evaluation of steady-state performance for pre-specified seal geometries in the absence of dynamic excitation.

SAMPLE RESULTS AND DISCUSSION

Since the number of parameters which can be modelled using this program is somewhat great, it is not possible to offer here a comprehensive parametric study of excitation and design parameters. However, a reasonable overview of the analysis capabilities may be obtained by consideration of the following four examples, all of which are based on a bellows-mounted seal and mating ring pair with principal characteristics given in Table 1. Bellows torsional stiffness is assumed infinite.

Rotating face angular misalignment

The conditions for this simulation are as follows:

Working fluid:	Nitrogen gas at 20°C
Sealed pressure:	6 bar absolute
Sealed pressure ratio:	6
Shaft speed:	14,000 rpm
Mechanical damping:	0.33 N coulomb friction at bellows, 0.81 N at face-carrier
Pre-installation net coning:	1.2 µm open to o.d. (pressure side)
Angular misalignment:	28 µm p-t-p runout at o.d. of sealing interface (32 mm dia.)

An initial film thickness of about 10 µm was specified with zero stator velocity, and the analysis started, so that the seal finds its own equilibrium conditions. The results are summarised graphically in Figs. 6-8. The seal settles very rapidly (within half a shaft revolution) into a stable operating condition, with the stator accurately tracking the mating ring, with a minimum film thickness of 0.3 µm. This can be seen by referring to the "swash", which is the magnitude of the stator tilt (or the resultant of the tilt, α and β, about a pair of perpendicular diametral axes. This remains absolutely constant at the value of the mating ring misalignment. The in-service coning angle settles to a value of 1.3 µm. At these conditions, the predicted leakrate is 0.5 g/hr and the viscous heat generation 1.6 W.

Eccentricity with coning

The conditions for this simulation are as follows:

Working fluid:	Nitrogen gas at 63°C
Sealed pressure:	16 bar absolute
Sealed pressure ratio:	16
Shaft speed:	14,000 rpm
Mechanical damping:	0.64 N coulomb friction at bellows, none at face-carrier
Pre-installation net coning:	1.2 µm open to o.d. (pressure side)
Rotating face eccentricity:	0.115 mm

The specified initial conditions were such as to allow the stator to find its equilibrium limit cycle from rest with a starting film thickness of about 2.5 µm. The results are summarised graphically in Figs. 9 and 10. Again, the rapidity with which a stable limit cycle is reached is notable, (for this example, within the first quarter of a shaft revolution). Two features of Fig. 9 are worth particular comment. The combination of coning and eccentricity results in a film thickness distribution which is non-uniform with respect to circumferential coordinate. This means that at any given instant there is a higher mean pressure on one side of the sealing interface than that which pertains diametrically opposite, i.e. there is a net pressure-induced moment about a diametral axis whose instantaneous orientation depends on the location of the point of minimum film thickness, itself determined by the eccentricity vector. The result is the inducement of stator swash - in this case of about 0.25 µm amplitude, i.e. about 0.3 of minimum film thickness. The second point worth noting is the larger in-service net coning (about 1.6 µm, as compared to 1.3 µm for the same pre-installation coning of the previous example). This results primarily from the combination of higher ambient operating temperature and sealed pressure.

Opposing wavy faces

The conditions for this simulation are as follows:

Working fluid:	Nitrogen gas at 183°C
Sealed pressure:	16 bar absolute
Sealed pressure ratio:	3.33
Shaft speed:	60,000 rpm
Mechanical damping:	0.33 N coulomb friction at bellows, none at face-carrier
Pre-installation net coning:	1.2 µm open to o.d. (pressure side)
Stator waviness:	2 wave system of 0.4 µm peak-to-peak amplitude
Mating ring waviness:	2 wave system of 1.6 µm peak-to-peak amplitude

The specified initial conditions were such as to allow the stator to find its equilibrium limit cycle from rest with a starting film thickness of about 2.5 µm. The results are summarised graphically in Figs. 11 and 12. The large in-service net coning is due to high ambient temperature, moderate sealed pressure differential (about 10 bar) and also centrifugal inertia of the mating ring at the very high shaft speed of 60,000 rpm. What is of more interest from the point of view of the seal dynamics is the twice synchronous axial vibration of the stator set up by the system of opposing face waviness. This arises because the mating ring waves have to "ride over" the stator waves each time peak opposes peak. The effect on the minimum film thickness is also notable. At about 0.2 µm, it is substantially less than in the previous example, in which it was about 0.9 µm, for the same sealed pressure (albeit for a higher drain pressure). The practical implications of this case are two-fold. Firstly, it shows how waviness can induce seal vibration. Secondly, it illustrates how opposed waviness increases the likelihood of mechanical contact between seal faces.

Shaft synchronous axial vibration (with and without mechanical damping)

The conditions for this simulation are as follows:

Working fluid:	Nitrogen gas at 200°C
Sealed pressure:	6 bar absolute
Sealed pressure ratio:	6
Shaft speed:	14,000 rpm
Pre-installation net coning:	1.0 µm open to o.d. (pressure side)
Axial vibration:	0.02 mm peak-to-peak mating ring runout, synchronous
Mechanical damping:	a) no damping
	b) 0.33 N coulomb friction at bellows, 0.81 N at face-carrier

Fig. 13 shows stator response in terms of axial vibration, for both cases. Clearly, the stator tracks the axial excitation, since in both cases the response very closely approximates a sinusoid with the same peak-to-peak amplitude (in fact slightly larger, as expected). Again, it can be seen that the in-service coning is large. This is due to the high ambient temperature. What is particularly interesting, however, is the difference in the film thickness and performance waveforms (Figs. 14 and 15, respectively) which result from the introduction of the mechanical dampers. There is a clear "squaring" of the previously quasi-sinusoidal forms, which is the consequence of damper hysteresis - the response is more characteristic of a "slip-stick" phenomenon, whereby the film thickness remains nearly constant at its minimum value over the half cycle during which the mating ring is moving towards the stator then nearly constant at its maximum value over the remaining half-cycle during which mating ring moves away from stator. It is also interesting to note that the consequence of this "square-wave" effect is to slightly *increase* the mean leakrate over a cycle, since the cubic dependence of leakage on film thickness ensures that the increased flow whilst film thickness is near maximum more than compensates for the decreased flow when it is near minimum. However, given the large value of coning for this example (hence relatively large *mean* film thickness in both cases) the leakrate figures are very similar between the two cases, as are other performance parameters such as viscous heat generation. Consideration of the minimum film thickness vs. shaft revs data for the two cases shows that the introduction of mechanical damping significantly *increases* the variation in this parameter over a cycle, since it tends to impede stator tracking of the mating ring. This should not, however, be taken to imply that mechanical damping is detrimental. On the one hand, as just noted, the impact on the leakrate and frictional heat is not great. On the other hand, where potential instability emanates from the fluid film itself (such as might be expected with a boiling film, for example), rather than oscillation resulting from an external source (such as misalignment or axial shaft runout), the presence of damping can be of assistance in seal stabilisation.

Numerical values of performance parameters

Clearly, the stator dynamics discussed above are both qualitatively and quantitatively plausible. However, experience with rig tests (10) on corresponding seals (notwithstanding the difficulties associated with instrumentation under conditions similar to those used in the examples given here) generally indicate an under-prediction both of leakage and of friction. This is explained by the omission from the model, as laid out herein, of two potentially significant phenomena. It is very common to find poor correlation between measured and predicted performance for plain-faced seals. This is because, in contrast to self-acting designs, the operating thickness of the lubricating film is on the order of a micron, (which, of course, is positively advantageous in terms of sealing performance). In practice, most seals employ carbon-graphite for one of the running faces. This material has a comparatively low modulus of elasticity, so that face pressures on the order of 1 MPa, typical of mechanical seal practice, produce local face deformation on the order of 1 μm, with the effect that regions of thin lubricant film, where pressure peaks tend to occur, see a greater deformation of the compliant face. In other words elastohydrodynamic effects may well be significant in terms of their effect on film thickness and pressure distribution and hence on leakrate. Furthermore, it will be noted that minimum film thicknesses on the order of a few tenths of a micron were indicated for the examples discussed. This is approaching values typical of surface roughness amplitude and partial mechanical contact is likely to occur in practice, i.e. seals will tend to run under mixed friction conditions. Clearly, for cases with less extreme face coning this becomes even more likely. This explains the under-prediction of friction by the model presented herein, which assumes perfectly smooth faces. Work currently being undertaken by the author (11) is addressing both of these deficiencies and preliminary indications suggest improved accuracy.

CONCLUSIONS

A mathematical model has been developed for the prediction of the behaviour, under various dynamic excitation conditions, of plain radial face gas seals. This model offers an integrated treatment of thermally and mechanically-induced seal ring structural distortions, thin film lubrication and vibrational dynamics of the mechanically damped, bellows-mounted floating stator ring in four degrees of freedom. Seal face waviness and coning on both rings are catered for as well as angular misalignment, eccentricity and axial oscillation of the clamped, rotating ring. This permits the simulation of geometrical imperfections arising from manufacturing and assembly tolerances, as well as their detailed interactions with the sealing interface and hence on seal performance.

The results of the simulations discussed above are qualitatively plausible in all respects. The prediction of stator dynamics appears to be quantitatively valid: misalignment and runout are accurately tracked, film thicknesses are in line with expected values. It is demonstrated how opposing systems of face waviness can result in axial vibration and how the combination of coning and eccentricity results in swash. Mechanical damping of the floating ring is shown to have little effect on performance parameters for the case considered, although its effect on the film thickness and leakage periodic waveforms is marked.

Quantitative discrepancies between measured and observed leakrates and friction are explained in terms of the absence, in the current model, of elastohydrodynamic effects and of surface roughness and mixed friction considerations, both of which are being developed in the current phase of model development, with promising initial results.

REFERENCES

1   Nau, B.S.
Vibration and rotary mechanical seals: a review
Tribology International Vol. 14, No. 1, Feb. 1981 pp. 55 - 59

2   Tournerie, B. and Frene, J.
Computer modelling of the functioning modes of non-contacting face-seals
Tribology International Vol. 17, No. 5, Oct. 1984   pp. 269 - 276

3   Shapiro, W. and Colsher, R.
Steady-state and dynamic analysis of a jet-engine gas-lubricated shaft seal
ASLE Transactions Vol. 17, No. 3, July 1974,   pp. 190 - 200

4   Shapiro, W.
Computer codes for face seal dynamics - vol. 2 Program documentation manual
Mechanical Technology Inc. Report No. MTI-83-TR-31 April 1983

5   Etsion, I. and Green, I.
Mechanical face seal dynamics. Final report of work done under USAF Contract F49620-83-C-0057 by Technion Research and Development Foundation, Haifa, Israel (EEC 159)
(AFWAL/TR-85-2082) (AD/A164978-9) Dec 1985

6   Green, I. and Etsion, I.
        Stability threshold and steady-state response of noncontacting coned-face seals
        <u>ASLE Transactions</u> Vol. 28, No. 4. 1985   pp. 449 - 460

7   Nosaka M. *et al.*
        Study on sealing characteristics of high-speed face-contact bellows mechanical seals
        for liquid hydrogen
        <u>Proceedings, JSLE International Tribology Conference</u> Vol III   1985   Paper 5E-7

8   Leefe, S.E. and Nau, B.S.
        Turbopump face seal technology: final report
        <u>BHR Group Limited Report no. CR 6062</u> March 1992
        Prepared for the European Space Agency under contract no. 8124/89/NL/PH(SC)

9   Nau, B.S. and Leefe, S.E.
        A review of some aspects of of the prediction of mechanical seal coning
        <u>Tribology Transactions</u>, Vol. 34, No.4, October 1991, pp 611-617

1 0   Leefe, S.E. and Nau, B.S.
        An experimental study of high-speed mechanical seal stability
        <u>STLE Preprint</u> No. 93-AM-2D-5,   1993

1 1   Leefe, S.E.
        Vibrational dynamics of turbopump plain face seals: incorporation of mechanical
        contact forces
        <u>BHR Group Limited Report no. CR 6184</u> April 1993
        Prepared for European space Agency under contract no. 10006/92/NL/PP(SC)

<u>TABLE 1</u> - <u>Principal characteristics of seal</u>

Sealing face o.d.	32.0 mm
Sealing face i.d.	28.3 mm
Mass of face carrier assembly	0.023 kg
Diametral moment of inertia of face carrier assembly	0.66 $kg.mm^2$
Polar moment of inertia of face carrier assembly	1.32 $kg.mm^2$
Bellows diametral angular stiffness	1.62 $Nm.rad^{-1}$
Bellows axial stiffness	11.9 $N.mm^{-1}$
Balance ratio (at zero Δp)	0.764
Nominal bellows force	13.6 N

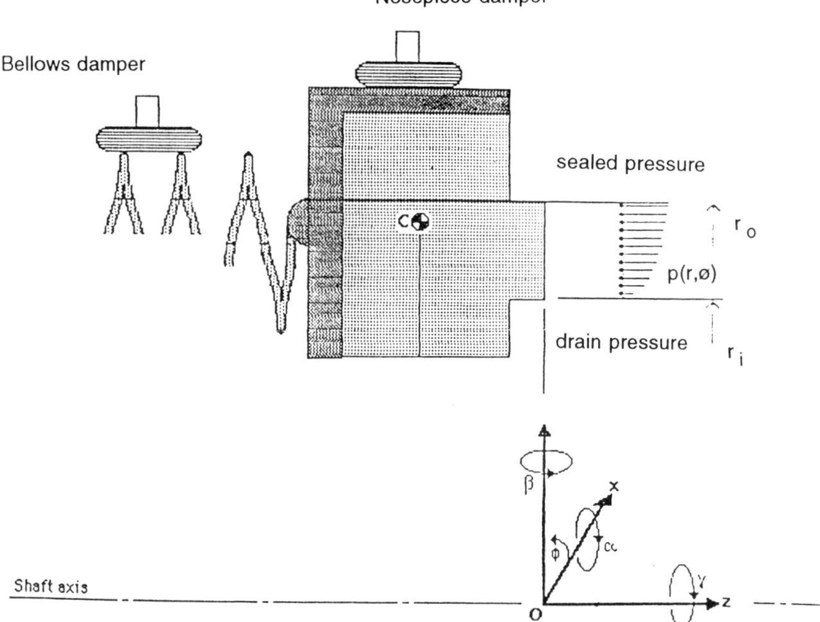

Fig. 1 - Coordinate system and displacements

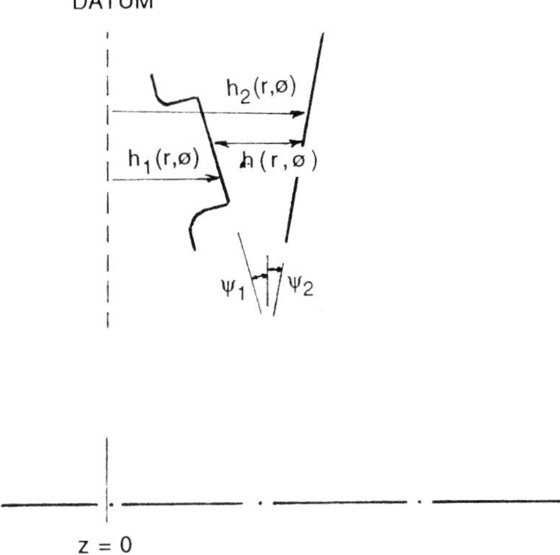

Fig. 2 - Definition of stator and rotor coning and face 'height' distributions

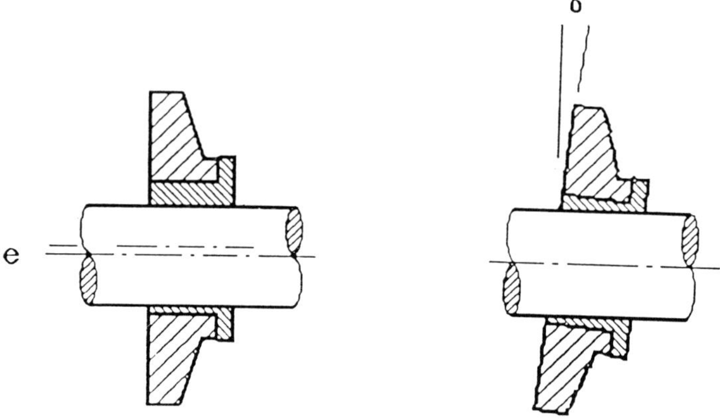

Fig. 3 - Rotor eccentricity and angular misalignment

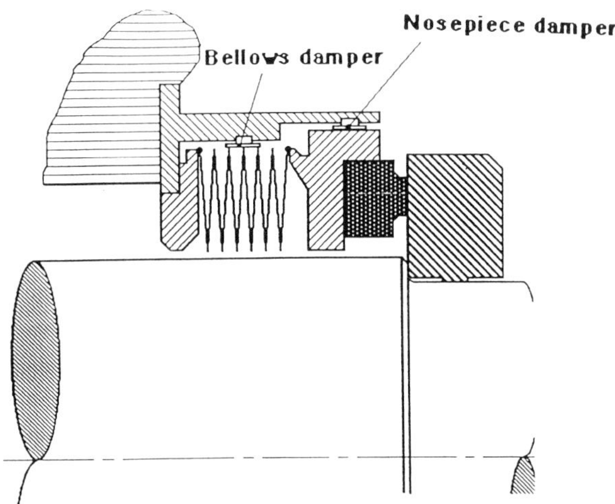

Fig. 4 - Generic seal type

414

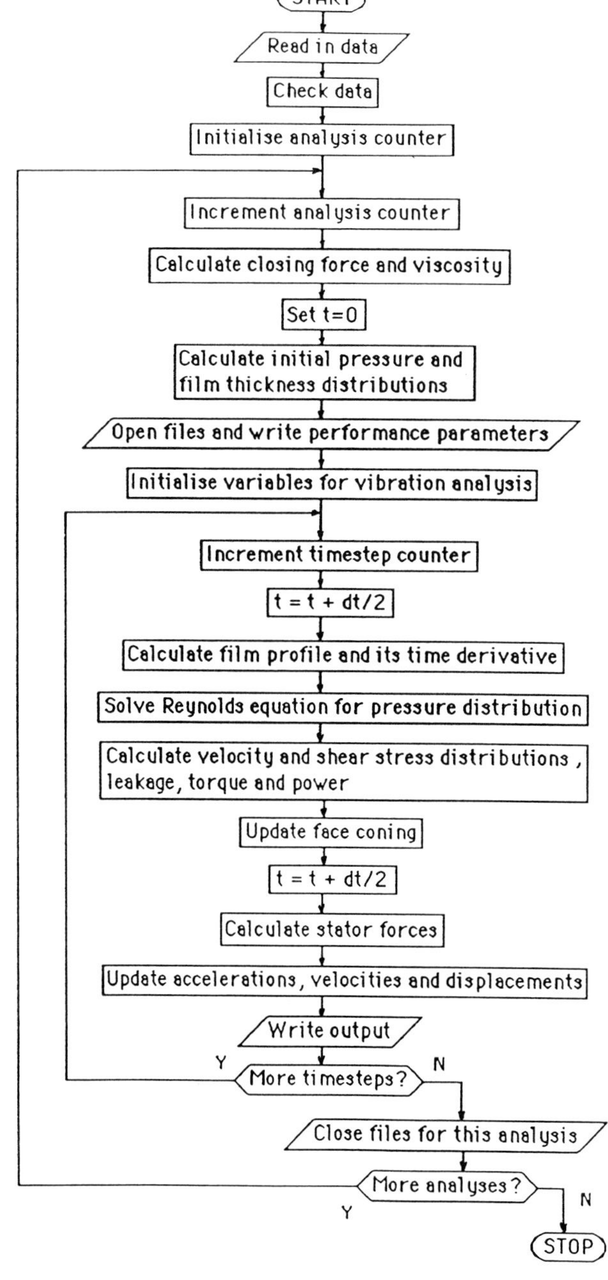

Fig. 5 - Program flowchart

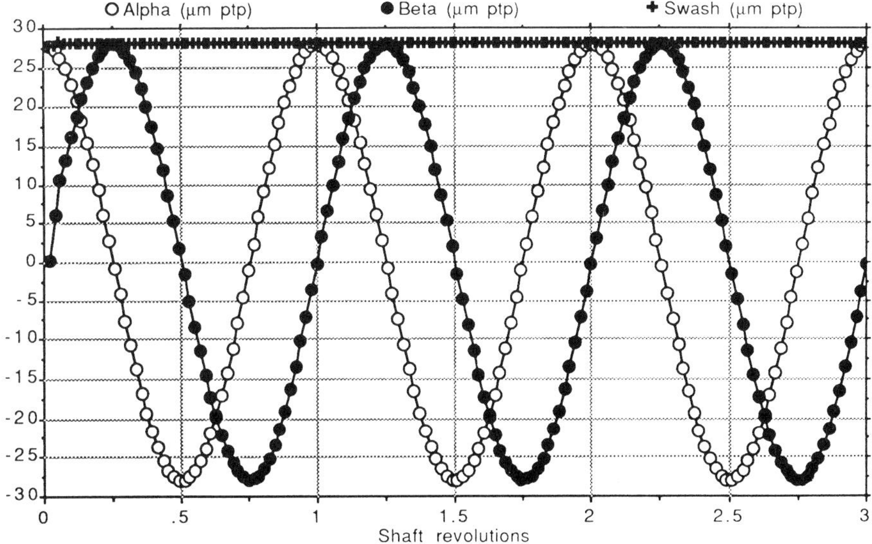

Fig. 6 - Stator tracking of angular misalignment

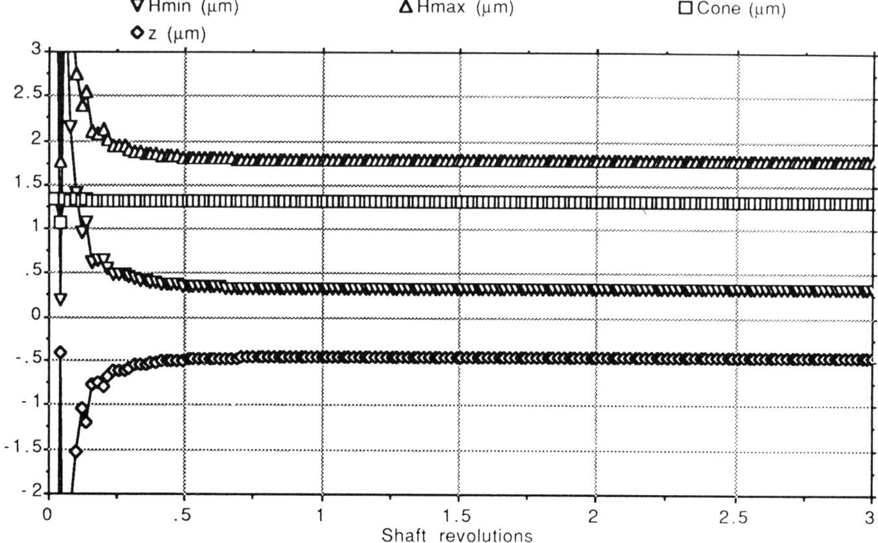

Fig. 7 - Response of maximum and minimum film thickness, nett coning and stator axial displacement to rotor angular misalignment

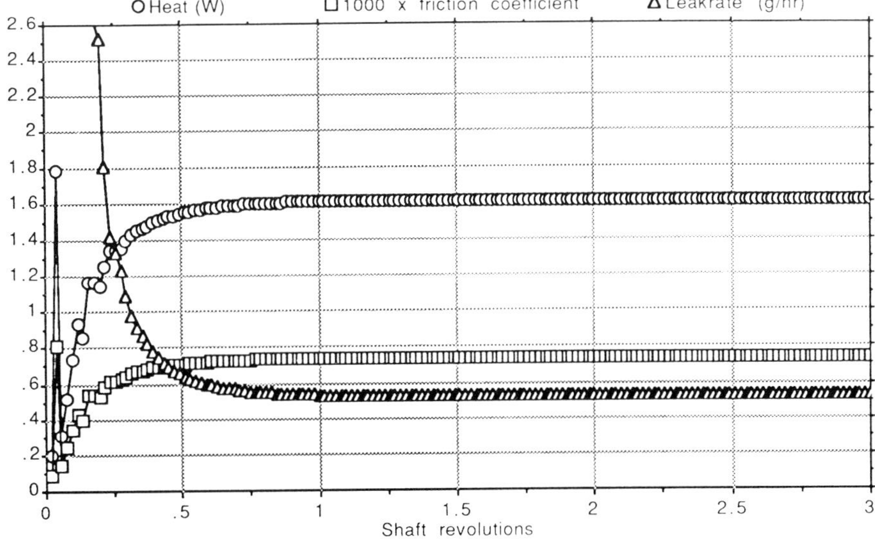

Fig. 8 - Performance parameters for angular misalignment excitation case

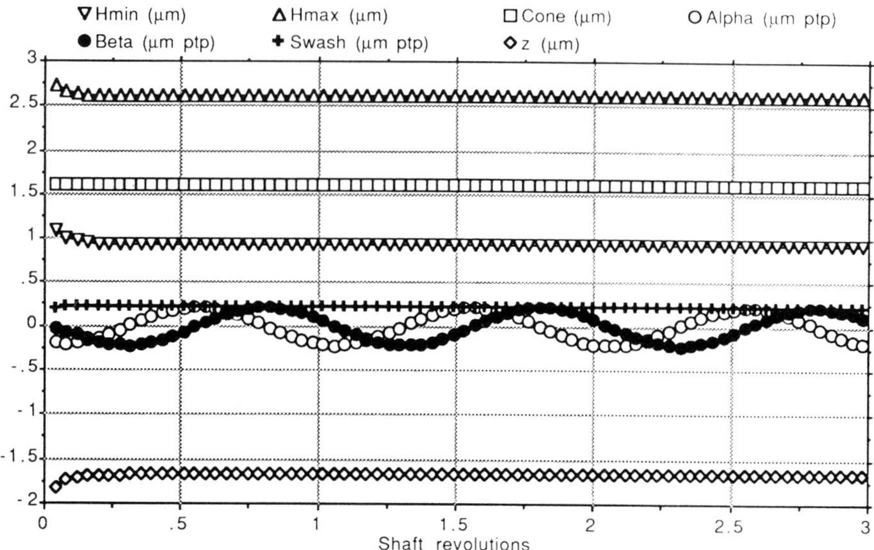

Fig. 9 - Stator response, nett coning and film thickness for combined coning with mating ring eccentricity

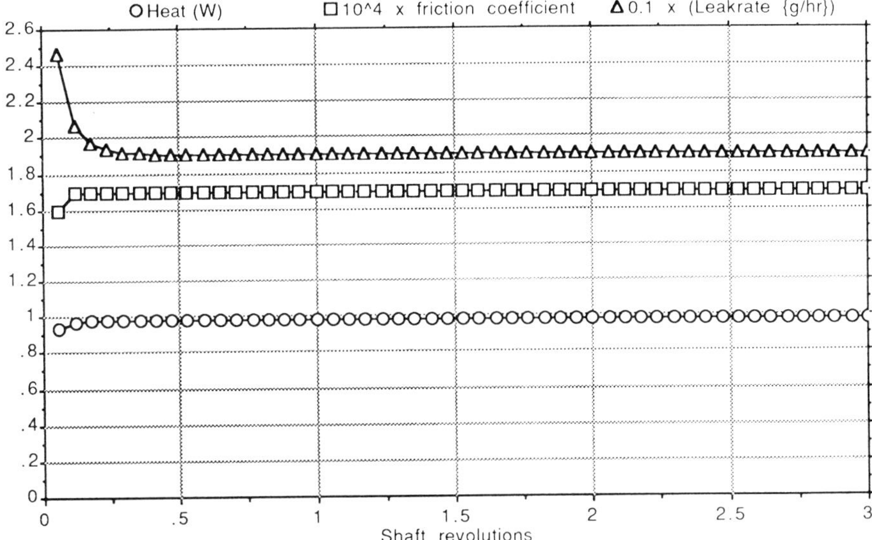

Fig. 10 - Performance parameters for combined coning and mating ring eccentricity excitation case

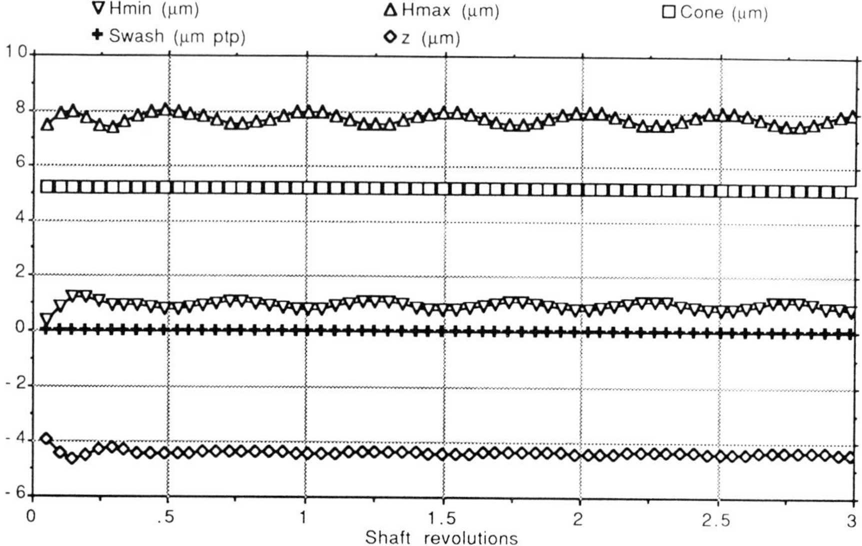

Fig. 11 - Stator response, nett coning and film thickness for opposed wavy faces case

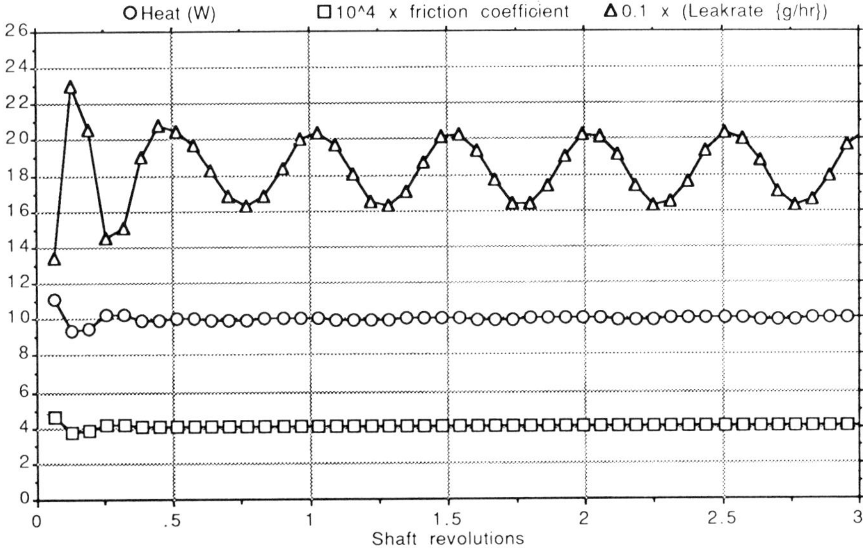

Fig. 12 - Performance parameters for opposed wavy faces case

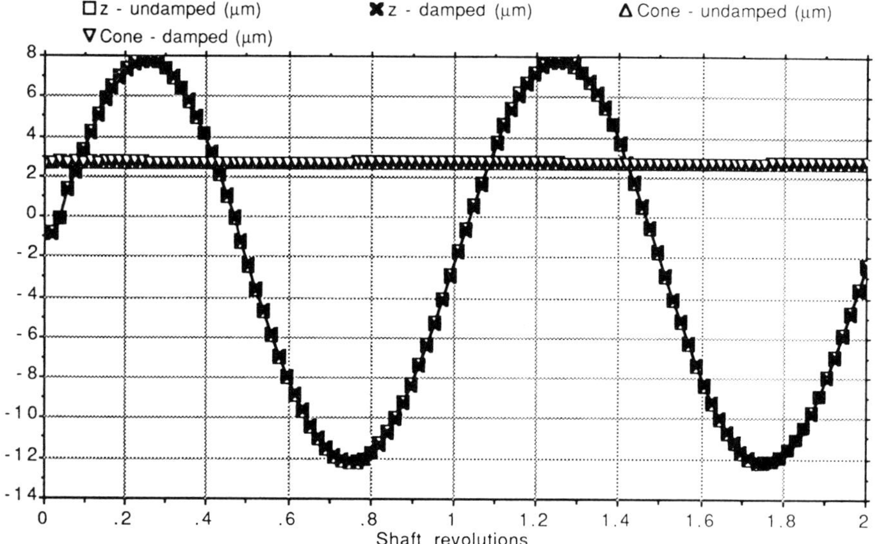

Fig. 13 - Stator response and coning for damped and undamped seals under shaft-borne axial vibration excitation

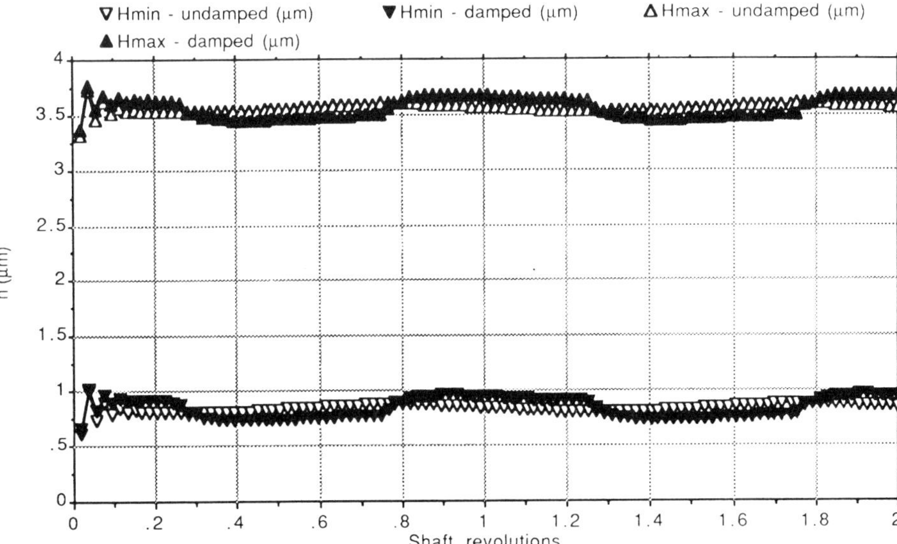

Fig. 14 - Maximum and minimum film thickness for damped and undamped seals under shaft-borne axial vibration excitation

424

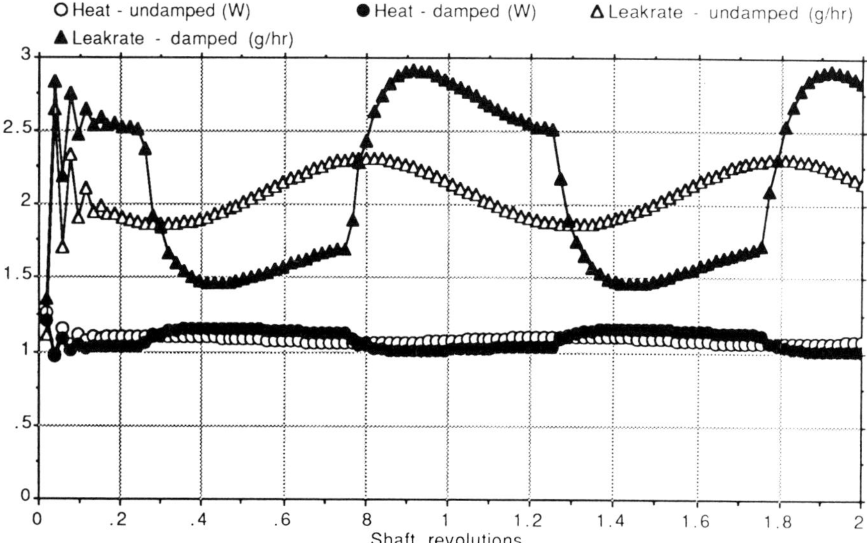

Fig. 15 - Performance parameters for damped and undamped seals under shaft-borne axial vibration excitation

MECHANCIAL SEALS III - techniques and experiments

14th International Conference on Fluid Sealing, Firenze, Italy,
6-8 April 1994. Organised by BHR Group Limited, Cranfield,
Bedford, MK43 0AJ, UK; Tel: 0234 750422

Long-term-tests of mechanical seals for hot water application

H.-J. Franke, R. Lachmayer
Institut für Konstruktionslehre, Maschinen- und Feinwerkelemente (IKMF)
Technische Universität Braunschweig, Germany

J. Nosowicz
Feodor Burgmann Dichtungswerke GmbH & Co.
Wolfratshausen, Germany

Abstract

Today mechanical seals with sliding material combinations of various "hard" carbides against "soft" carbon graphites are generally choosen for sealing of hot water which is used as heat transfer medium e.g. in building or machine heating systems. Typical conditions are temperatures between 90°C and 170°C and pressures between 0.4 and 1.6 MPa. Additionally the water is mixed with additives for general protection against corrosion and if necessary to prevent freezing. In spite of many applications in this field only very little testing of the tribological system - sealing gap / water at temperatures above 100°C - has been performed.

In the present study 100h and 300h - tests with hot water of 140°C were carried out, using balanced seals with seats made of silicon carbide, tungsten carbide or aluminia against carbon graphite seal faces. The influence of the different sliding material combinations on the seal operating behaviour is pointed out. A FEA of the sealing gap deformation confirm the test results. Various failure modes were discussed in view of mechanical and thermal properties, as well as tribochemical reaction.

Introduction

Although hot water sealing with mechanical seals is a very common subject, it actually makes high demands on the quality of sealing rings and elastomeric sealing components. Concerning performance and troubles of mechanical seals for hot water applications a number of theoretical and experimental investigations were reported discribing different aspects of the subject (Mayer [3], Müller [4], Waidner [7], Nosowicz [5]). Nevertheless there are still some questions open. For a better understanding of the whole system further tests are necessary.
The performance of hot water seals is influenced by a large group of parameters which can be concentrated into four groups:

<u>Thermo-mechanical behaviour</u>
- Temperature limits for the reliable operation of a seal in uncooled mode
  as a function of the sealing materials, especially the elastomeric secondary seals
- Thermal and mechanical deformations of the gap influenced by geometry, pressure, temperature and material [1,2,4]

<u>Lubrication of sealing gap and two-phase flow phenomena</u>
- Medium (pressure, temperature, vapour point, viscosity, content of solid matter)
- Mechanical seal (geometry, surface structure, spring force, sliding material combination)

<u>Effects of water chemistry and water treatment</u>
- Crystallization of soluble matters (silicates, phosphates, sulfites) [5] as a result of vaporizing in the sealing gap followed by an increase of wear rate and leakage
- Formation of a film on the sliding faces with subsequent increasing leakage
- Cementing of the sliding faces after long stand-still periods
- Preventing of the axial adjustment of the seal as a result of phosphate and silicate sedimentation on the atmospheric side of metal bellow seals or in the area of secondary seals [5]
- Corrosion by electrolytes or cavitation erosion influenced by the oxygen concentration [8]
- Deformation of the seal housing as a result of corrosion which can cause deformation of the seal faces [5]

<u>Instationary deformations and vibrational effects</u>
- Thermoelastic instability [6,7]
- Vibrations of the machinery

Main aspects for the present study are the influence of different sliding materials and of the water treatment on the lubricating condition in the sealing gap.

Experiments

Figure 1 shows the geometry of the tested seal and Table 1 the test parameters. As shown in Figure 2 for each test two identical seals were mounted to the shaft back-to-back to compensate the pressure related axial thrust. Thermocouples and piezo elements were arranged near the sealing gap for measuring temperature and pressure continously. In addition to these torque, sliding speed, circulation flow rate of the medium and leakage were recorded during the whole test.
An extraction of the test-programme is shown in table 2.
The mainly used ceramic materials - direct sintered silicon carbide and tungsten carbide - were tested in combination with seal faces made of antimony impregnated and resin impregnated carbon graphite. In addition to these tests aluminia against antimony impregnated carbon graphite was tested for comparision.

The mechanical and thermal properties of the tested materials are shown in Table 3.

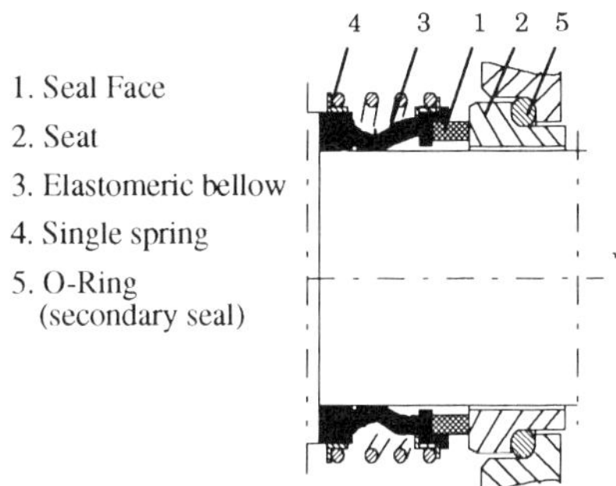

1. Seal Face
2. Seat
3. Elastomeric bellow
4. Single spring
5. O-Ring
   (secondary seal)

Seat material	Tungsten carbide (TC) Silicon carbide (SiC) Aluminia
Seal face	Antimony imp. carbon graph. Resin imp. carbon graph.
Medium	Demineralized water with Na_3PO_4 and $Na_2 SO_3$
Medium conduct. (μS/cm)	< 5
Sliding speed (rpm)	1000...3000
Temperature (°C)	140
Pressure (MPa)	0.4...1.6
Duration of tests (h)	100...300
Medium circul. (l/min)	1...1.2

Figure 1: Test seal (Burgmann MG 1, d_{Shaft} 53mm)

Table 1: Test conditions

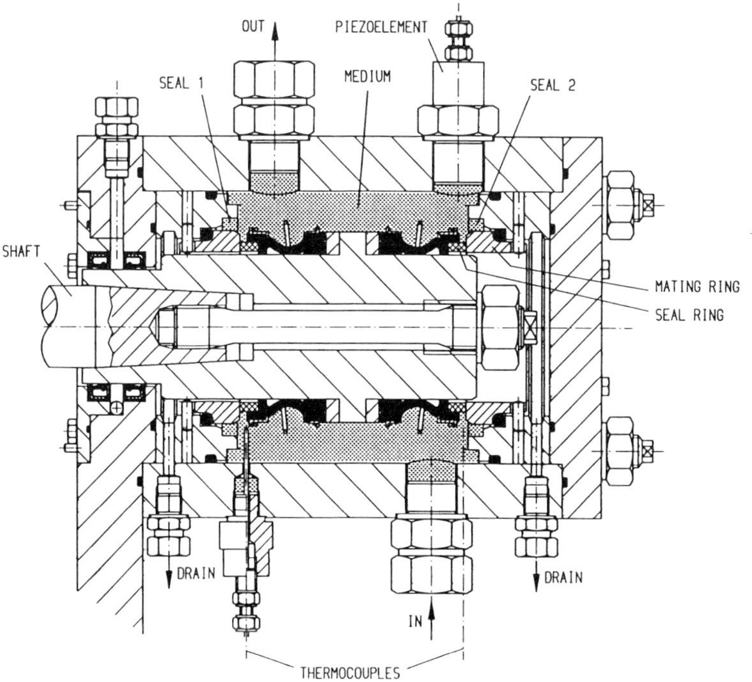

Figure 2: Test set-up

430

	N°	Sliding material seat	Sliding material seal face	T [°C]	p [MPa]	k b [mm]	t [h] v [m/s]	Water quality
Material Comb.	**2**	Silicon carbide	Carbon graphite resin imp.	140	0.5	0.826 3.8	100 9.75	demineralized
	3	Aluminia	Carbon graphite antimony imp.	140	0.5	0.826 3.8	100 9.75	demineralized
Additiv Influence	**7**	Silicon carbide	Carbon graphite antimony imp.	140	0.5	0.826 3.8	100 9.75	demineralized Na_3PO_4(15mg/l), Na_2SO_3 (5mg/l)
	8	Silicon carbide	Carbon graphite antimony imp.	140	0.5	0.826 3.8	100 9.75	demineralized Na_2SO_3(250mg/l)
Pressure Influence	**9**	Silicon carbide	Carbon graphite antimony imp.	140	1	0.826 3.8	300 9.75	demineralized
	10	Silicon carbide	Carbon graphite antimony imp.	140	0.5	0.826 3.8	300 9.75	demineralized
	11	Tungsten carbide with NiCrMo	Carbon graphite antimony imp.	140	0.5	0.826 3.8	300 9.75	demineralized
Variation of k	**12**	Silicon carbide	Carbon graphite antimony imp.	140	0.5	0.763 2.85	300 9.75	demineralized
	13	Silicon carbide	Carbon graphite antimony imp.	140	0.5	1.083 2.85	300 9.75	demineralized
Variation of carbon	**14**	Silicon carbide	Carbon graphite resin imp.	140	0.5	0.826 3.8	300 9.75	demineralized

Table 2: Extract of test-program

Material	Carbon graphite antimony imp.	Carbon graphite resin imp.	Silicon carbide	Tungsten carbide	Aluminia
Composition			SiC	Binder: NiCrMo	99,5% Al_2O_3
Density (g/cm^3)	2,15...2,4	1,7...1,83	3,0...3,16	14,3...14,6	3,8...3,9
Vickers hardness			2500..2800	1400...1700	1800.2000
Flexual strength (N/mm²)	80...120	55... 95	350...460	1300...3500	300...400
Tensile strength (N/mm²)	35...58	44...48	175...250	900	160...300
Compressive strength (N/mm²)	250...400	170...350	2100..3800	4000...4500	3000.3500
Young'-modulus 10^4 (N/mm²)	2,4...2,9	1,7...2,4	40...42	55...64	35...38
Thermal conductivity (W/mK)	7...20	8...14	90...100	80	29...30
Thermal expansion coefficient 10^{-6}(1/K)	3,7...5,2	4...5	3,5...4,0	5,0...5,6	6,6...8,5

Table 3: Mechanical and thermal properties of the tested sliding materials

In hot water applications the water treatment has to meet the following demands:

Protection against corrosion
      by forming layers
      e.g.: Sodiumphosphate Na_3PO_4, Hydrazin N_2H_4, Aquaris M 67, Bayrofilm T328
      by binding of oxygen
      e.g.: Sodiumsulfite Na_2SO_3, Aquaris R 66, Bayrotol T220, Nalco Surgard
Protection against freezing and corrosion
      Anti-freezing additives are based on glycol. They contain anti-corrosion additives
      because pure glycol-water mixtures are very corrosive
      e.g.: Varidos FSK, Antifrogen N or Preventol CI-2.

For the first serie of tests the following additives were chosen and used in different concentrations (5mg/l, 15mg/l, 250mg/l):

- Sodiumphosphate Na_3PO_4, which protect against corrosion by forming ironposphate-layers ($FePO_4$) and deposits of calcium carbonate (Ca_2CO_3) in heating systems.
- Sodiumsulfite (Na_2SO_3), which reduce oxygen concentration by forming (Na_2SO_4).

Further test results from Burgmann with the additives Helamin 906, Nafleet 9-11 and Nalco 43-19 are included in the discussion.

Test results

After each test the surfaces of seal faces and seats were analysed (Table 4) and the profile roughness was measured. Figure 3 shows for example the competition between test 9, 10 and 11:

Test No. 9	Seal face	Blistering and some small cracks, running grooves only along the atmospheric side of the sealing surface
	Seat	Good appearence with some deposits and small running grooves along the atmospheric side of the sealing surface
Test No. 10	Seal face	Dry running grooves and blistering over the whole width of the sealing surface
	Seat	Deep dry running grooves over the whole width of the sealing surface, high wear
Test No. 11	Seal face	Dry running grooves between the atmospheric side and the middle of the sealing surface, blistering all over the sealing surface
	Seat	Good appearence with small grooves at OD and ID of the sealing surface

Table 4: Appearance of the sliding surfaces

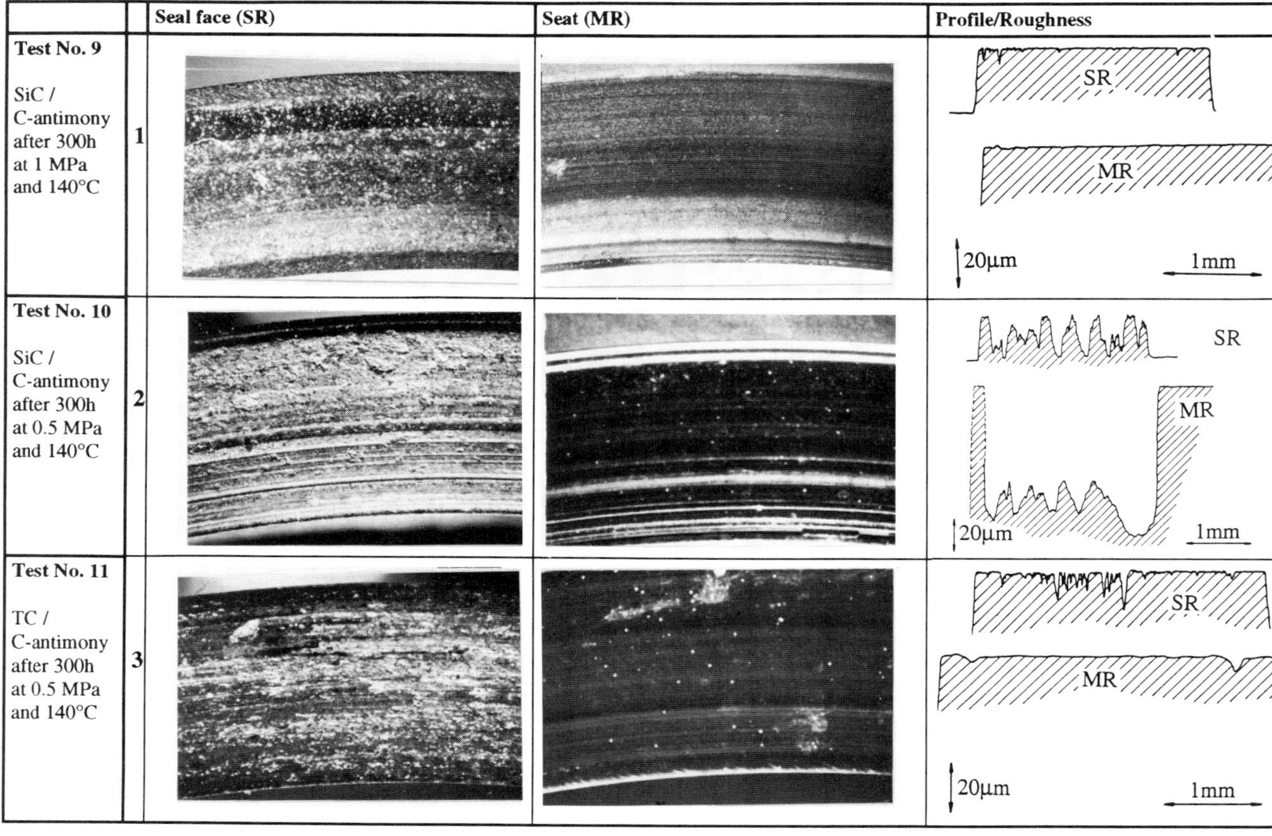

Figure 3: Appearence of wear

In addition to the appearence of sealing surfaces leakage rate was used to confirm the impression of the tribological situation. Figure 4 shows the leakagediagramms of the tests 9, 10 and 11. A surface demage of the sealrings as mentioned in test 10 goes together with high leakage and corresponding with the theory for the critical cases [9] the wear rate was increasing exponentially after a period of stable running.

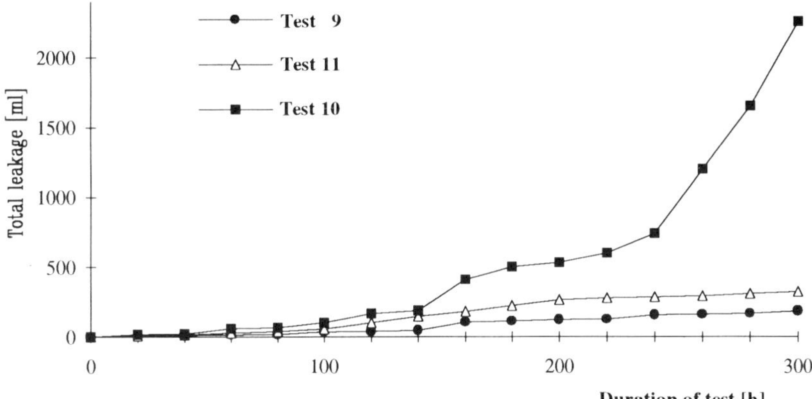

Figure 4: Leakagediagram of test 9, 10 and 11

To get a more exact idea of wear mechanism and tribochemical reaction deposits on the sealing rings and surface defects were analysed with REM. Chemical elements were identified with x-ray energy spectrum-analysis. The following examples are results of tests 10 and 11 and show four failure mechanism which illustrate the differences in the performance of silicon carbide and tungsten carbide:

Embedding of hard particles
Figure 5 and 6 show a carbon graphite sliding ring with embedded silicon carbide particles. The high number of such big embedded particles (e.g. about 30 p/mm² in test 10) leads to high abrasive wear and deap concentric running grooves in the SiC-seat. Some initial defects of the SiC-seat lead to an exponentially increasing wear. This mechanism was mainly observed with silicon carbide, which may have two reasons:

- the silicon carbide grains are much bigger than the tungsten carbide grains (10:1)
- the tested tungsten carbide is manufactured with a "soft" binder material (CrNiMo)

Formation of layers
An example for the tribochemical reaction of tungsten carbide is illustrated by Figure 7 and 8, showing a section of the seat (test 11). The deposits on the seat are shown in Figure 7 and, with reference to the spectrum-analysis (Figure 8), can be divided in:

- antimony layers resulting from wear of the sealing ring
- reaction products of metallic elements in the water and tungsten carbide respectively the binder

434

These deposits were accumulated in close vicinity to the OD and ID of the seat. Concentric running grooves at the seal face are the result.

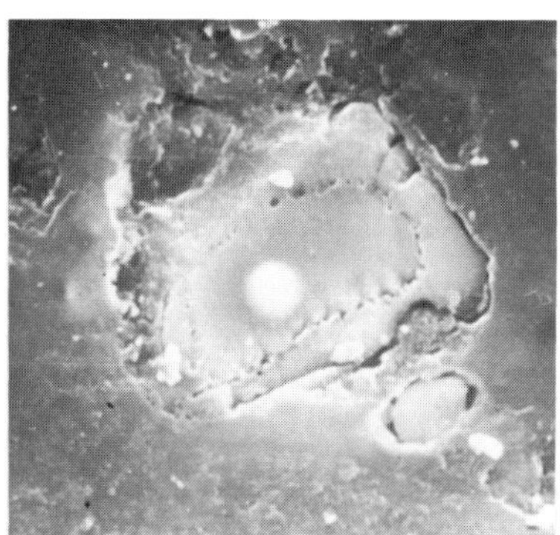

Figure 5: Carbon graphite seal face with an embedded SiC particle (Test 10, 6000 fold)

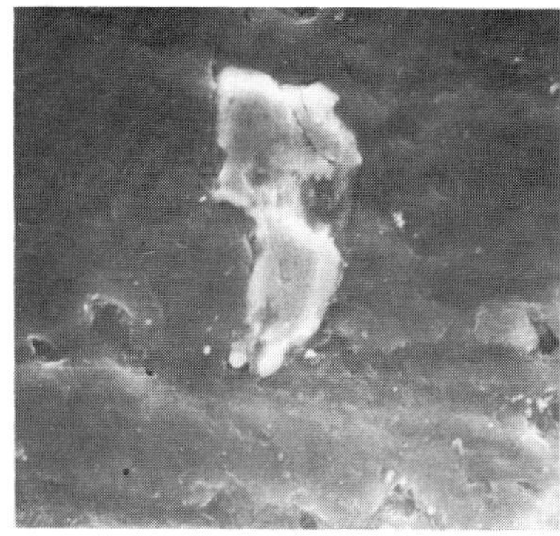

Figure 6: Carbon graphite seal face with an embedded SiC particle (Test 10, 2000 fold)

Figure 7: Deposits on the surface at the rim of a TC seat (Test 11, 120 fold)

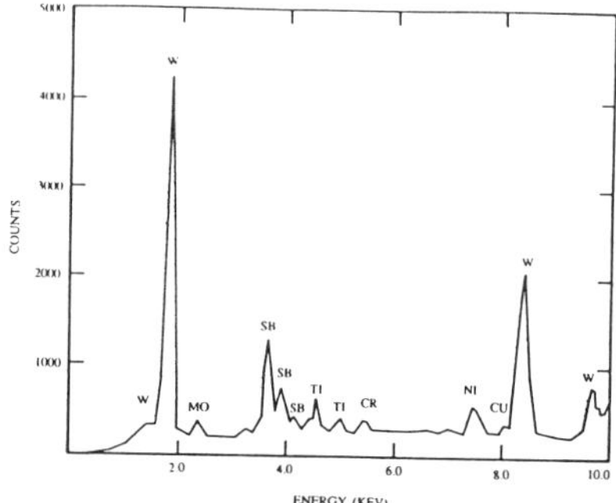

Figure 8: Spectrum energy analysis of deposits

Material pop-outs

Figure 9 shows the destroyed surface of an antimony impregnated carbon graphite seal face. At the bottom of the crater-like pop-out an accumulation of antimony can be recognized. It is assumed that the carbon graphite ring had some inhomogeneous areas with large pores filled with antimony. Due to the higher thermal expansion coefficient of antimony ($\alpha_{Sb} \approx 2.5\alpha_C$) carbon particles in the sealing surface will be forced out of the structure at high temperatures.

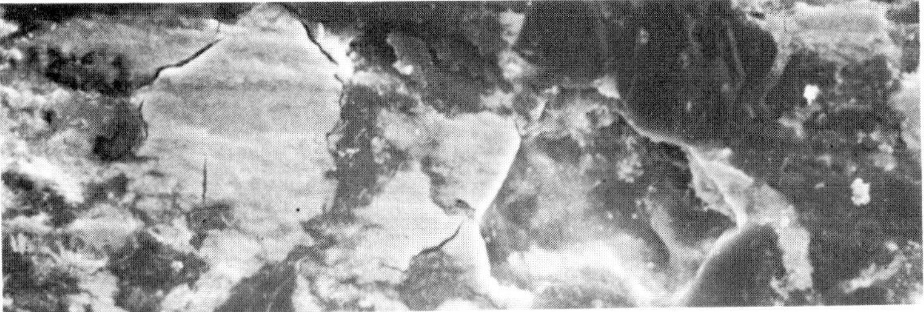

Figure 9: Destroyed section of an antimony impregnated carbon graphite seal face
(Test 10, 800 fold)

Build-up of deposits

In combination with tungsten carbide whitish deposits were accumulated in the asperities of the sealing surface of the antimony impregnated carbon graphite ring (figure 10). These deposits increased locally the coefficient of friction and therefore the energy input resp. the temperature. Blistering was the result of this increased temperature.

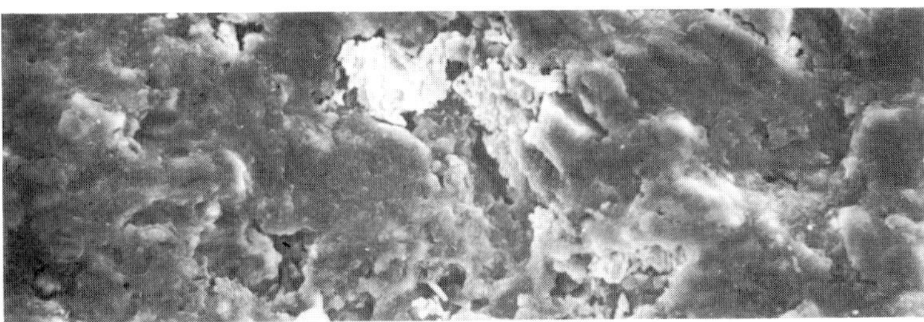

Figure 10: Antimony impregnated carbon graphite seal face with deposits and blistering
(Test 11, 1200 fold)

Discussion of the test results

1. Thermo-mechanical behaviour

To define the thermal and mechanical stress as well as the deformation of the sealing rings a FEA was carried out. Figure 11 shows the grid of the axis symmetric 3D seal modell. The mechanical stress in the sealing rings caused by low pressures is very small (max. 7 N/mm²). In the steady state condition the temperature-gradients and with them the thermal stress are low also. The temperature of the whole seal housing reached nearly the medium temperature of 140°C (after a running period of about 3...4h). The difference between the liquid temperature at the inlet and the outlet of the seal housing always was lower than 2°C and between medium- and atmospheric side of the seal lower than 1.5°C.

Nevertheless deformation took place. Figure 12 shows the qualitative deformations of seal faces and seats made of different materials. All the deformations are very small and for the current geometry and materials very similar, forming little divergent sealing gaps in the unused state of the seals.

With FEA it could be shown, that radial temperature gradients of only 0.2 K/mm cause deformations which turn the sealing gap from a divergent into a convergent state. The appearence of the sealing surface in Figure 6 can be the result of this effect. It is triggered by cyclic changes of the friction coefficient.

Figure 13 shows the deformation of the sealing gap at pressures of 0.5 and 1 MPa and for a modified seal geometry.

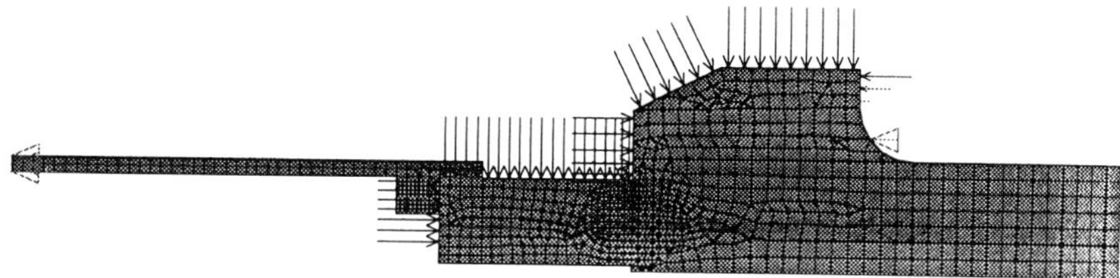

Figure 11: Modell for the calculation of sealgeomety with FEA

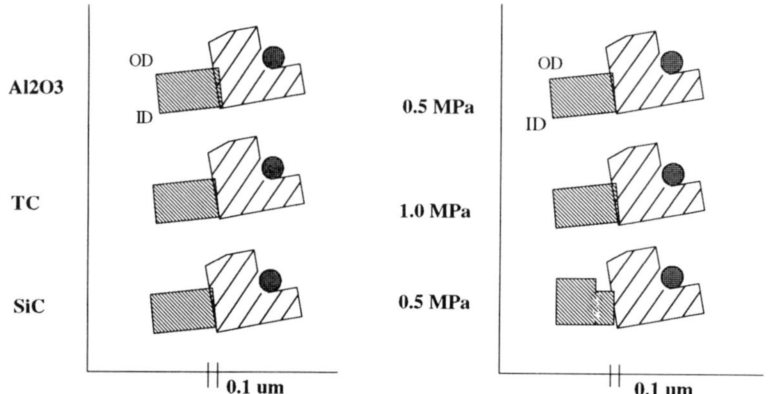

Figure 12: Deformations of seals made of differerent material (0.5MPa, 140°C)

Figure 13: Deformations of seals under different pressure (SiC, 140°C)

2. Lubrication of sliding surfaces and two-phase flow phenomena

Figure 14 shows different cases of the linear pressure drop in the sealing gap (A...E). Although the linear pressure fall is a simplified modell (compare e.g. Waidner [7]) two facts can be pointed out:

High pressure / Vaporization at ID:
>In the case of a new seal the vaporization point of the hot-water is situated in the area of ID (case A). High wear caused by partially dry-running shifts this position from ID to OD of the sealing gap.

Low pressure / Vaporization between ID and OD:
>The location of the vaporization point in the sealing gap depends on the pressure difference hot-water/atmosphere and the medium temperature. Case D and E illustrates in a simplified way the influence of the product pressure on the position of the vaporization point.

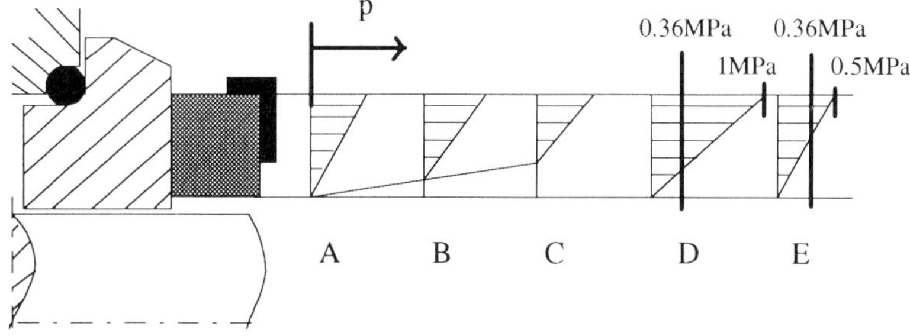

Figure 14: Pressure drop in the sealing gap
>A: New seal
>B: Seal with vaporization point between ID and OD
>C: Worsening of case B
>D: Pressure drop from 1MPa to 0.1MPa and temperature related vaporization point
>E: Pressure drop from 0.5 MPa to 0.1MPa and temperature related vaporization point

This interrelationship is universally valid for mechanical seals. The application range, however must be defined in relation to design, material combination and medium. Figure 15 shows for example the application range of the balanced H7-seal (Burgmann).

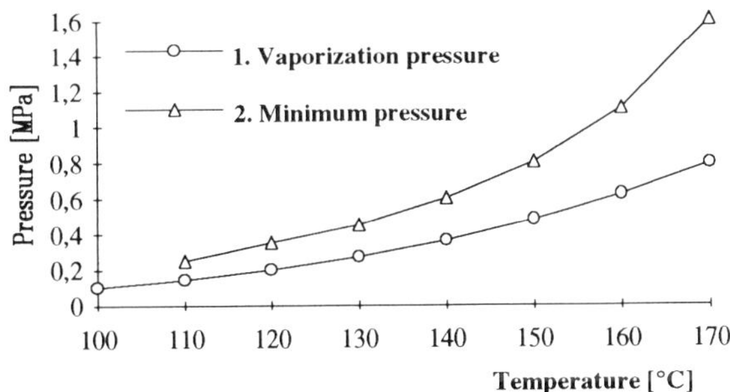

Figure 15: Application range for the balanced H7-seal, Nosowicz [5].
>Material combination: SiC / Antimony impregnated carbon graphite,

The minimum pressure for longtime operation is defined by line 2. At pressures in the area between line 1 (vapour pressure) and line 2, vaporization in the sealing gap is given, resulting in high wear, blistering and leakage.

Increasing the medium pressure helps to improve the operating behaviour of a hot-water seal. But the operating behaviour is also influenced by:

- sealing width
- balance ratio
- oxygen content.

This interrelationships have still to be investigated.

3. Effects of waterquality and water treatment [5]

Water quality

Natural waters contain undissolved and dissolved inorganic and organic solids and gases. Dissolved solids occur mostly in the form of inorganic salts (compounds of calcium and magnesium) and organic substances. Gases such as oxygen, nitrogen and carbon dioxide are often dissolved. In hot water application tribochemical reactions take place such as
- deposit build-up of calcium carbonate in the vaporization area of the sealing gap
- formation of irregular layers in combination with tungsten carbide as sliding material.
The result of these reactions is abrasive wear and increasing leakage.

Water treatment

The hot-water additives (Na_3PO_4, Na_2SO_3, Helamin, Nalco, Antifrogen N and Nafleet 911) were investigated in the Burgmann test lab for their effect on the running performance of the H7 seal in hot water. The 100h-tests were carried out with demineralized water at various pressures, temperatures and additive concentrations.

Non-vapour-volatile additives
    The dosis limits for non-vapour-volatile conditioning agents (e.g. trisodium phosphate + sodium sulphite, Nalco 43-19) must not be exeeded. Exessive dosing of these agents usually causes great difficulties for the sealing of hot-water pumps. Mechanical seals operated in heating water containing silicate are at particular risk: in addition to exerting a strong abrasive action, a high silicate content in circulating water with a residual hardness can produce firmly adhering coatings on seal components. Nalco 43-19, a mixture of silicates and phosphate, is one of these agents. With this water additive it is recommended, therefore, to aim at the lower dosing limit.

Vapour-volatile additives
    When using vapour-volatile water additives, e.g. ammonium, part of the conditioning agent can pass into the vapour phase under certain conditions of pressure and temperature. The emitted gas forms a ring around the sealing gap and leads to dry running of the seal. In these cases, the degassing of the water is intensified by incorporating cyclone seperators or throttles in the circulation line.

<u>Antifreeze and anticorrosion additives</u>

Antifreeze and anticorrosion agents form a special group of water additives. The antifreeze effect is based on glycols with a high boiling point. Freeze resistance is determined by the mixing ratio with water. Unless inhibitors are added, it is not possible to use glycol/water mixtures because of their corrosion-inducing properties, which are more extreme than water alone.

Among the constituents of anticorrosion inhibitors are silicates, whose abrasive action is a threat to the proper functioning of the mechanical seal. A further problem arises with the crystallized solids, which are smeared on the sliding face and result in higher leakage. A different category of water additives are the polyfatty amines and the fatty polyamines (e.g. Helamin 906). These agents form very thin hydrophobic, corrosion-inhibiting films on metallic surfaces. In this case the water is not polluted with abrasive solids. Laboratory investigations into the effect of adding Helamin revealed no negative repercussions for the mechanical seal.

Conclusions

Summarizing the test results and the discussion, the following conclusions can be derived concerning lubricating conditions, material selection and water chemistry:

- The most critical point for the application of hot water seals is the operation near the vapour pressure.

- Generall rules concerning the difference between medium temperature and vaporization point like $\Delta T > 10K$ are not sufficient. Figure 15 gives a more detailed application range.

- At poor lubricating conditions the operating behaviour of tungsten carbide is better than that of silicon carbide.

- Concerning the influence of the water treatment the tests showed that at Na_3PO_4 and Na_2SO_3 concentrations of $< 15mg/l$ the material combinations - antimony impregnated carbon graphite against tungsten graphite resp. silicon carbide - showed no significant change in operation.

- Further investigation reducing the area of poor lubrication will be carried out.

References

[1] D. Barnes, R. K. Flitney: Considerations affecting the design of pump
     S. Nau   chambers for improved
                             mechanical seal reliability
                             BPMA 12th Int. Pump Technical Conference, 1991

[2] H.-J. Franke,   Auslegung axialer GLRD mit einer rechner-
     R. Lachmayer: gestützten modularen Konstruktionsumgebung
                             VIII. Int. Dichtungskollquium, Köthen, 1993

[3] Mayer, E.:   Axiale Gleitringdichtungen, VDI-Verlag, Düsseldorf,
                             1977

[4] S. Müller, H. K. Müller: Wärmeübergang bei Gleitringdichtungen
                             Konstruktion 44 Seite 161...166, Springer-Verlag,
                             Berlin, 1992

[5] J. Nosowicz,   Mechanical seals in hot-water applications,
     W. Schöpplein: Special print series No. 44 Feodor Burgmann
                             Dichtungswerke GmbH & Co. Wolfratshausen,
                             Germany

[6] F. Salant, S. E. Hassan: Large-scale thermoelastic instability in hydrostatic
                             mechanical seals, 12th Int. Con. Fluid sealing,
                             Brighton, 1989

[7] P. Waidner: Vorgänge im Dichtspalt wasserabdichtender
                             Gleitringdichtungen, Diss. Universität Stuttgart, 1987

[8] W. Schöpplein,   Ursachen von Dichtungskavitation und Maßnahmen
     D. Zeus: zur Abhilfe, Maschinenmarkt, Würzburg 97,
                             1991, 24

[9] M. Woydt, K.-H. Habig: Technisch-physikalische Grundlagen zum
                             tribologischen Verhalten keramischer Werkstoffe
                             Forschungsbericht 133, BAM, Berlin, 1987

OPTIMISATION OF THE PUMPING RING IN A MECHANICAL SEAL WITH AN INTEGRATED COOLER FOR FEED-WATER PUMPS

Denis BUCHDAHL
E.D.F./Direction des Etudes et Recherches, Département Machines, 6 quai Watier, 78401 Chatou Cedex, France.

Roger MARTIN
E.D.F./Service d'Etudes et de Projets Thermiques et Nucléaires, Département Matériels, 12/14 Avenue Dutriévoz, 69628 Villeurbanne Cedex, France.

Gérard GUERET
Latty International, 1 rue Xavier Latty, 28160 Brou, France.

Michel BLANC
G.E.C. Alsthom Bergeron Rateau, 141 rue Rateau, 93126 La Courneuve, France.

<u>S</u>UMMARY :

To simplify maintenance, E.D.F. alongwith its collaborators* undertook the study of mechanical seal with integrated cooler used in feed-water pumps in the nuclear power plants.

The working conditions of the mechanical seal are :

- fluid :   water
- pressure :   2.7 MPa
- speed of rotation : 4576 rpm
- shaft diameter :   180 mm
- sliding speed :   45 m/s (approx.)

The cooler, integrated to the pump acts as a thermal barrier as well as a cooler of the mechanical seal.

The water circulation in the cooler is assumed by an integrated pumping ring in the rotary part of the mechanical seal, with a matching screw thread in the pumping case.

This assembly of mechanical seal/integrated cooler is tested in a test loop at the EDF/DER laboratory [1]. All working conditions are similar to that at site.

Tests with different configurations of the rotor/stator profiles are performed, i.e. different lengths and types of threading. Hydraulic performances and the global thermal balance of this assembly are studied.

Our basic aim during these tests is to optimise the hydraulic performance of the pumping ring so as to best cool the mechanical seal faces.

The different results obtained and the conclusions drawn during these tests are presented.

* -G.E.C. Alsthom Bergeron Rateau, France
  - Latty International, France

INTRODUCTION :

General layout of the feed-water pumps :

The feed-water pumps used in the french nuclear power plants (R.E.P. 1400 MW N4) are designed and manufactured by G.E.C Alsthom Bergeron Rateau.

These pumps are double ended. They are equipped with two mechanical seals, one on each side of the pump. The heat generated during the rotation of the mechanical seal (due to friction and turbulence) is evacuated by the integrated cooler. This cooler, integrated to the pump acts as a thermal barrier as well as a cooler of the mechanical seal.

This integrated cooler when compared to the previous design has simplified the operation of mechanical seals as it eliminates the use of all supplementary circuits such as external coolers, filters and valves.

The test loop at E.D.F./D.E.R. laboratory :

The entire assembly of the mechanical seal/integrated cooler and test loop are shown in Fig.1 & 2 respectively. The working conditions are as follows :

- speed of rotation   4576 rpm
- temperature of the water to be sealed   182°C
- shaft diameter   180 mm
- pressure of the water to be sealed (P)   2.7 MPa
- average tangential speed at the interface
  of the seal (V)   42 m/s
- P.V product   about 115 MPa m/s

The integrated cooler has two functions :

- it serves as a thermal barrier over the casing of the pump,
- it assures the cooling of the sweep water between the two faces of the mechanical seal.

Due to plant's cooling circuit limitations, the heat power evacuated by the integrated cooler is 35 kW maximum.

Consequently, for the fixed discharge rate of the cooler (45 l/min), the temperature rise of the coolant water is limited to 11°C.

The characteristics of the cooling circuit are :

- demineralized water, pH at 25°C   7
- maximum temperature of water at inlet   38°C
- maximum temperature of water at outlet   49°C
- maximum permitted difference between inlet
and outlet temperature   11°C
- discharge rate of the cooler   45 l/min

The circulation of the sweep water surrounding the faces is assured by an integrated pumping ring in the rotary part of the mechanical seal.

Pre-Trials on the test loop :

During initial trials on the test loop, with the first version of the Latty mechanical seal an abnormal level of vibration is observed thereby preventing us from proceeding further.

This first version of the mechanical seal consisted of a rotor with helical grooves and a plain stator. The radial clearance between the rotor and the stator is 0.8 mm.

The global level of the relative radial vibrations is too high (>50 µm peak to peak) and is highly fluctuating. At the fundamental frequency of 76 Hz the amplitude is weak. However, at about 20 Hz the amplitude of the vibration is very high and eminently variable in amplitude.

On analyzing this phenomenon, it is determined that the cause is hydrodynamic instability. This instability is a phenomenon which also occurs in the bearings at certain conditions [2].

Nevertheless, in the present case, we have shown that the hydrodynamic instability is due to the pumping ring of the mechanical seal. In fact, on modifying the geometry of the pumping ring, the hydrodynamic instability disappears :

- on providing the stator with circumferential grooves, the hydrodynamic instability decreases by about 50%,
- with axial grooves, this instability disappears completely.

Although this is satisfactory in the point of view of thermodynamic performance, the conception of the pumping ring is to be reviewed.

Therefore in collaboration with the seal manufacturer we undertook the study of rotor/stator profiles with different configurations.

First Trial

Firstly, as commonly admitted, the hydraulic performance of the pumping ring is to be increased. To achieve this, the plain stator is replaced by a stator with helical grooves having an inverse sense to that of the rotor. The same radial clearance (0.8 mm) between the stator and the rotor is maintained.

For these working conditions, with similar pumping ring (as also for different suppliers of mechanical seals) we obtain the following results :

- temperature of the sweep water at the inlet
of the mechanical seal faces :   about 55°C,
- raise in temperature of the cooler water : about 9°C.

Contrary to our expectations, the results of these trials are disappointing :

- temperature of the sweep water at the inlet
of the mechanical seal faces :   about 63°C,
- raise in temperature of the cooler
water :   about 13°C.

Mechanical seals faces seem to be badly cooled. Meanwhile, the heat exchanged is increased.

Consequently, as the trial conditions remain unchanged, the heat dissipated by the mechanical seal is the same.

The different possible reasons for this raise in the heat evacuated are:

- notable improvement in the cooling of the two mechanical seal faces,
- heating due to the pumping ring,
- extraction of calories on a level with the thermal barrier,
- mixing with the feed-water.

Considering the difficulty in the instrumentation of the integrated cooler, we could not give a global satisfactory reply to this phenomenon.

It is therefore impending on us to optimise the pumping ring. The objective is to obtain a low temperature of the sweep water.

The different trials undertaken

For a given profile of the groove and maintaining the same radial clearance (0.8 mm) between the stator and the rotor, different trials are undertaken by changing the length of both the threads.

The trials undertaken are :

- a rotor with helical grooves (figures 3a, 3b),
- a plain rotor (figure 4).

In all cases, the associated stator of different configurations is a rotor with helical grooves having an inverse sense The total length of the stator grooves is 6 mm more than that of the rotor as there is the possibility of an axial shaft displacement of $\pm$ 3 mm.

For the two different types of configurations evaluated, different types of tests are performed :

1) The parameter studied is the length of the groove of the stator and rotor. The different lengths of the rotor thread are:

- 43, 30, 20, 15, 12, 8 and 6 mm.

A complementary trial is undertaken by removing every alternate tooth on the 6 mm rotor length.
Decreasing the effective length of the pumping ring below 6 mm does not have any physical sense due to the helical profile of the grooves.

2) A plain rotor, with different associated effective lengths of the stator tried are :

- 12, 9 and 6 mm.

The different quantities measured on the test loop are:

- speed of rotation,
- pressure difference between the two ends of the pumping ring,
- temperature raise of the cooling water,
- inlet water temperature of the mechanical seal.

The results of the trials are shown in graph Nos. 5, 6, and 7.

For all these trials undertaken, the hydrodynamic instability disappears.

RESULTS :

As one would expect from the litterature [3] realised tests confirm that for a rotor and a stator having helicoïdal grooves :

- differential pressure of the pumping ring is high and feebly proportional to effective length,

- differential pressure of the pumping ring is proportional to (rotation speed)^2.

Main results of the tests with an integrated cooler are shown on figures 6 and 7 :

1) Different lengths of the groove of the stator and rotor :

- the raise in temperature of the coolant water decrease with the decrease in the effective length of the pumping ring. However the optimum is not obtained,

- the inlet temperature of the sweep water
between the mechanical seal faces decreases
with the decrease in the effective length of the
pumping ring (hydraulic performance),

- a complementary trial undertaken by removing
every alternate tooth on the rotor better
improved the global performance of the system.

2) Trials with the plain rotor :

- the raise in temperature of the coolant water
is low.

Due the high rotation speed, with a plain rotor, the
pressure difference between the two ends of the pumping
ring is quite high. It decreases feebly with the decrease of
the effective length of the stator.

Moreover, the trials undertaken at low speed, and at
full pressure show that contrary to our expectations, the
cooling of the entire mechanical seal assembly is always
assured satisfactorily.

The entire set of trials undertaken does not however
help us in obtaining an optimum cooling of the entire
mechanical seal assembly with the integrated cooler.

The best results, in term of low temperature of the
sweep water, are obtained by coupling a plain rotor with a
stator having a small groove length (6 mm). The clearance
between the rotor and stator is 0.8 mm. In this case the
performance registered is as follows :

- raise in temperature of the coolant
water :   7.6°C,
- temperature of the sweep water at the inlet
of the mechanical seal face :   50°C.

CONCLUSION :

The pumping ring with a plain rotor and grooved stator is an industrially lucrative solution. The advantages are:

- the axial position of the stator grooves does not need to be determined,
- decreased cost of machining and control,
- the **rotating parts** of the dead-end and motor-end does need not to be differentiated, thus decreasing the spare parts inventory and the risk of error.

Moreover this solution assures an excellent cooling of the mechanical seal faces. This better cooling further helps in :

- increase in the lifetime of the seal,
- relief in the margin with respect to the working conditions, thus a better tolerance as far as pertubations are concerned.

BIBLIOGRAPHY

1 - Mechanical seals qualification procedure of the main pumps of nuclear power plants in France.
D. Buchdahl, R. Martin, J.M. Girault.
Fluid Sealing, B.H.R.G. Brugge, 1992.

2 - Experiment of static and dynamic characteristics of spiral grooved seals.
T. Iwatsubo, B.C. Sheng, M. ONO.
Rotordynamic instability high-performance turbomachinery. N.A.S.A. Washington,1990.

3 - Studies on seals for rotating shaft of high pressure pumps.
A.I. Golubiev, Wear 8, 1965.

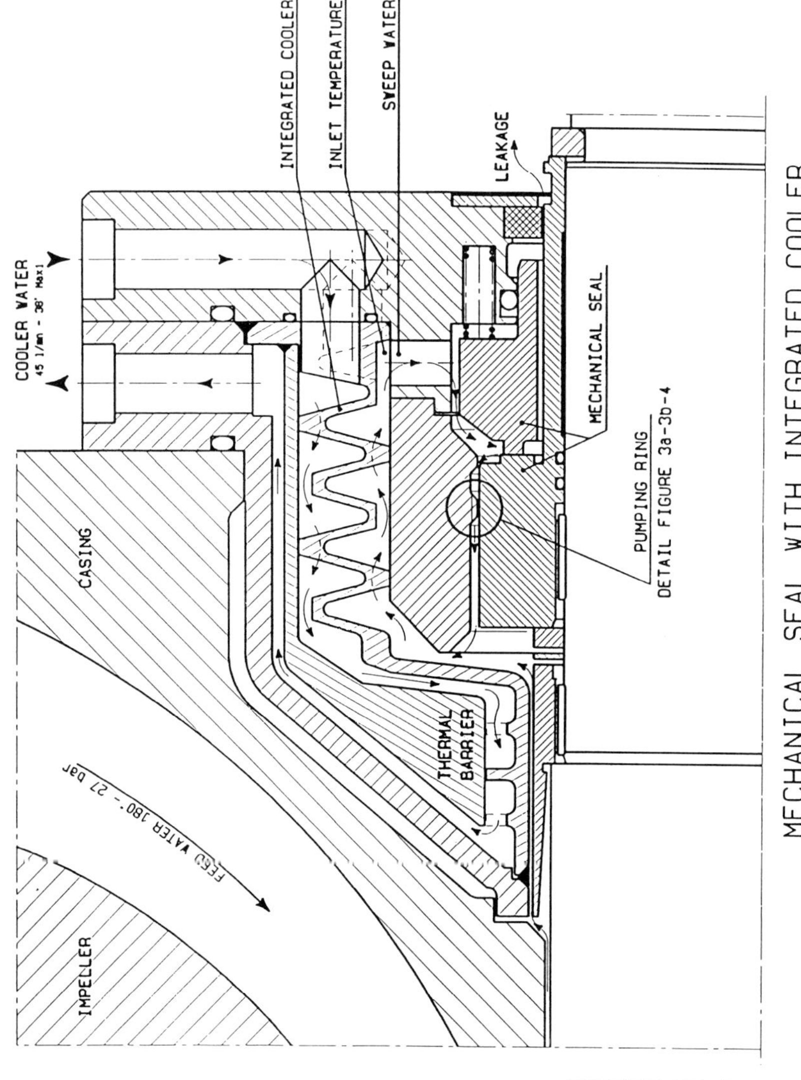

MECHANICAL SEAL WITH INTEGRATED COOLER

FIGURE N° 1

Test loop of the mechanical seal for feed-water pumps

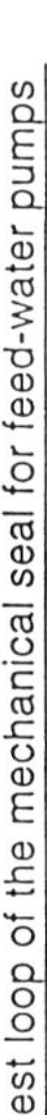
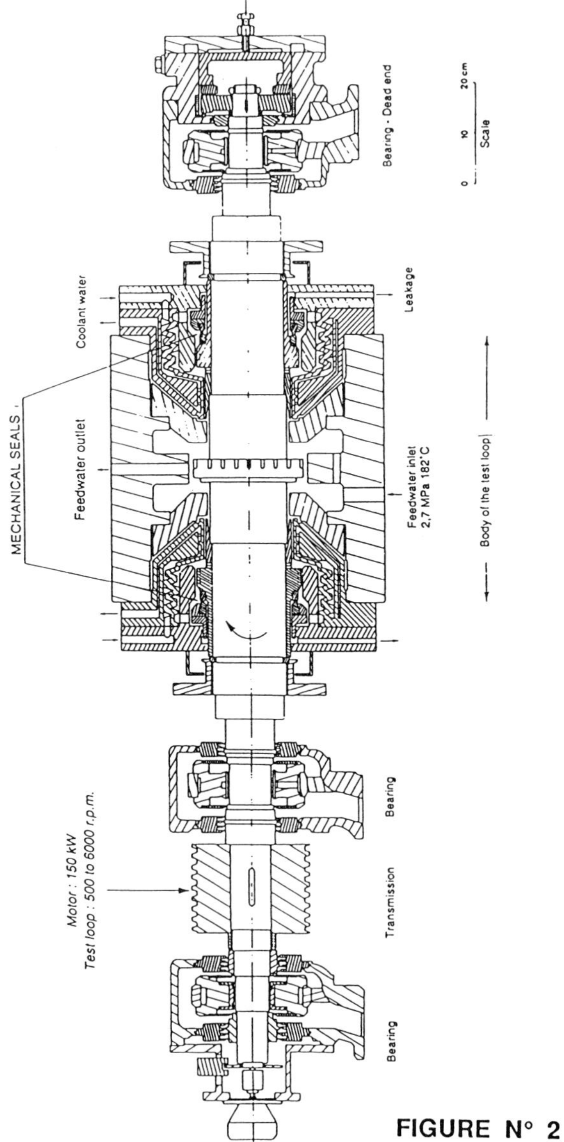

FIGURE N° 2

454

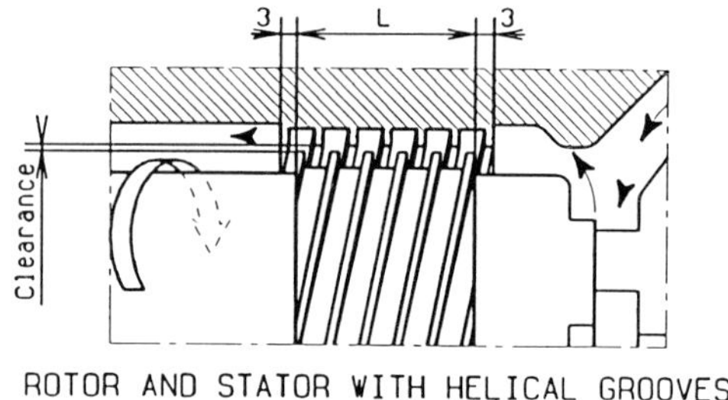

ROTOR AND STATOR WITH HELICAL GROOVES

FIGURE N° 3a

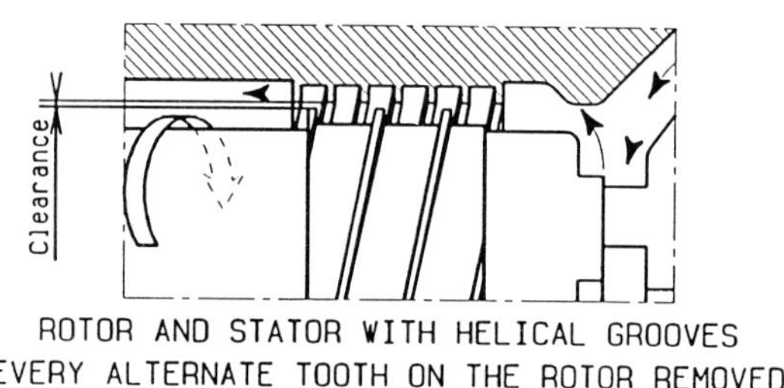

ROTOR AND STATOR WITH HELICAL GROOVES
EVERY ALTERNATE TOOTH ON THE ROTOR REMOVED

FIGURE N° 3b

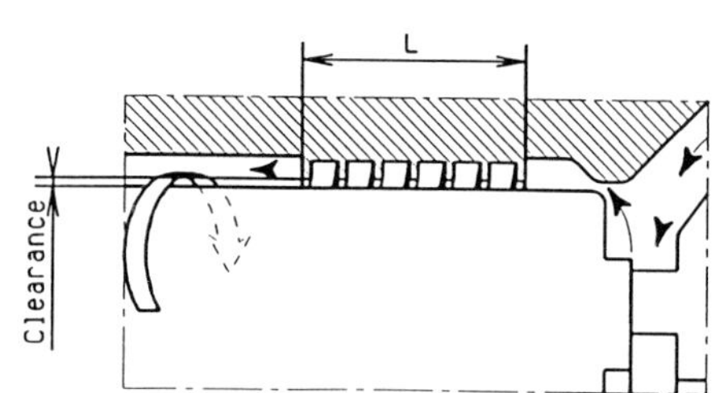

PLAIN ROTOR AND STATOR WITH HELICAL GROOVES

FIGURE N° 4

PRESSURE DIFFERENCE CREATED
BY THE PUMPING RING
AS A FUNCTION OF THE ROTATIONAL SPEED
FOR DIFFERENT CONFIGURATIONS
OF THE PUMPING RING

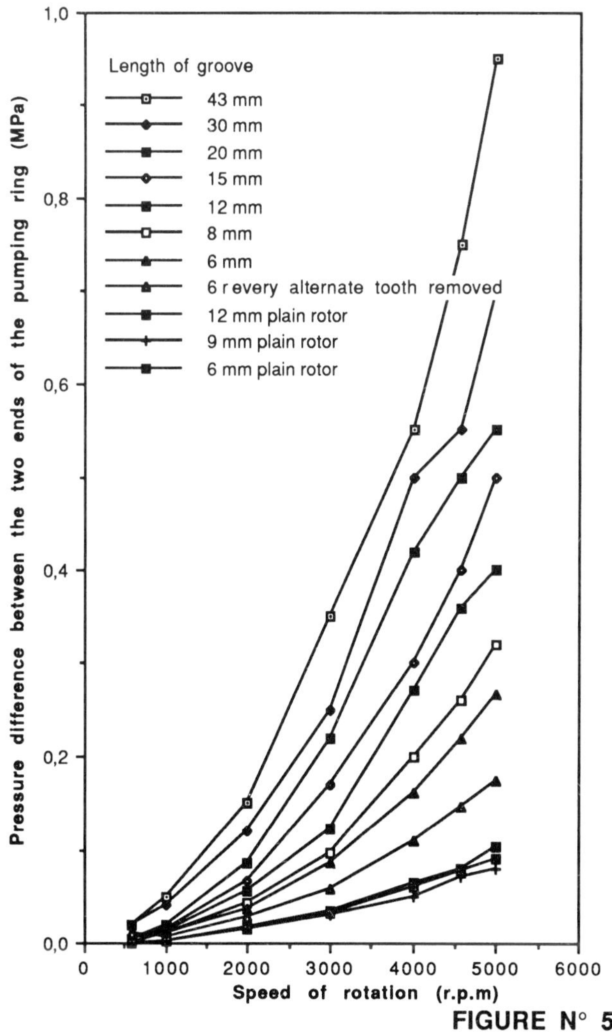

FIGURE N° 5

TEMPERATURE RISE OF THE COOLANT WATER
FOR DIFFERENT CONFIGURATIONS
OF THE PUMPING RING

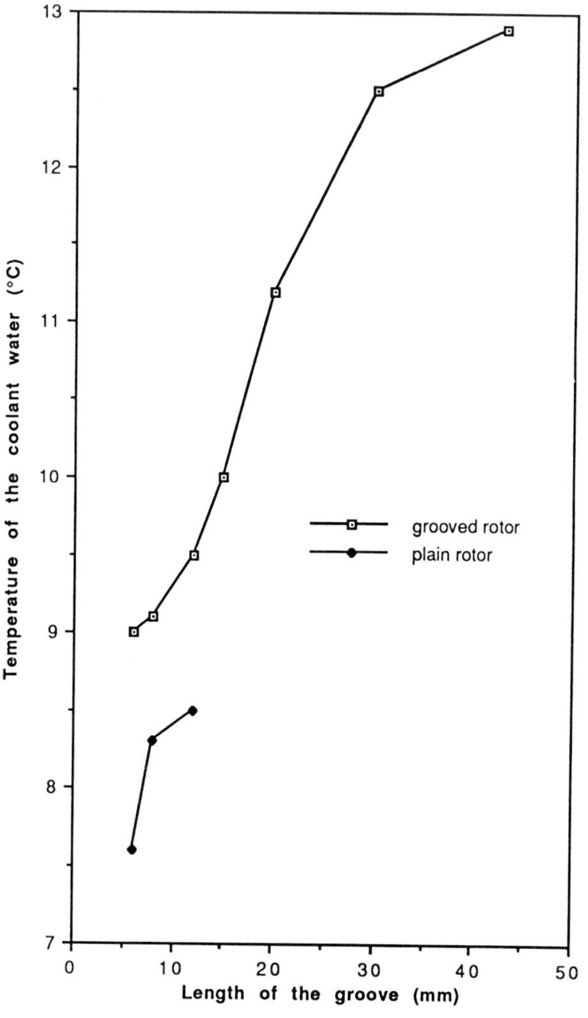

FIGURE N° 6

TEMPERATURE OF THE SWEEP WATER
AT THE INLET OF THE MECHANICAL SEAL
FOR DIFFERENT CONFIGURATIONS
OF THE PUMPING RING
70
Rotor with helicoïdal grooves
Temperature of the sweep water (°C)
60
Rotor with every alternate tooth removed
+
50
Plain rotor
40
0
10
20
30
40
50
Length of the groove (mm)

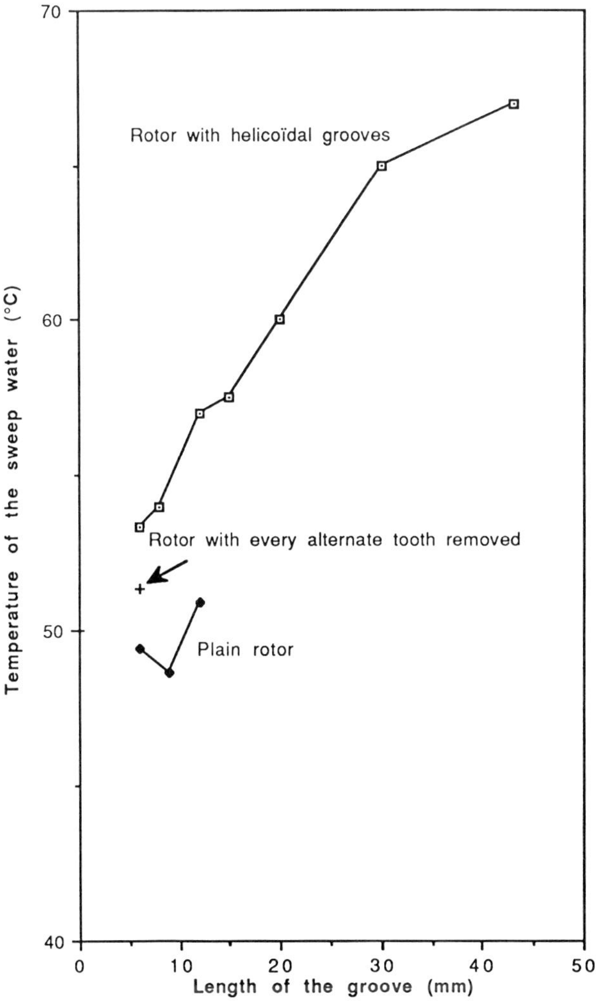

FIGURE N° 7

RECIPROCATING SEALS

14th International Conference on Fluid Sealing, Firenze, Italy,
6-8 April 1994. Organised by BHR Group Limited, Cranfield,
Bedford, MK43 0AJ, UK; Tel: 0234 750422

SEALING TECHNIQUES FOR SUBSEA CONTROL APPLICATIONS

M.C. THEOBALD
Research & Development Group
FSSL Ltd
Monarch House
Victoria Road
London W3 6UL
United Kingdom

ABSTRACT

This paper describes advances in hydraulic sealing performance for subsea control
systems used in the offshore oil industry. Particular reference is given to high
integrity dynamic metal seals working with both high pressure (1,725 bar proof) and
aggressive fluid media, such as injected chemicals to avoid corrosion, scale, wax
formation and other undesirable process conditions.

With the ever increasing demand to exploit more difficult subsea hydrocarbon
reserves, control system suppliers are faced with challenging sealing requirements.
The advent of High Pressure/High Temperature (HP/HT) Subsea Wells, together
with the on going need to develop new technology to reduce system costs, has
advanced sealing techniques in this high reliability sector of industry.

Typical pressures to control downhole hydraulic safety valves (DSV's) are now
commonly in the region of 1,035 bar working pressure for HP/HT wells. In order to
effectively seal such pressures, controls valves have been developed using metal-to-
metal sealing techniques, in conjunction with advanced surface treatments, to permit
drop tight leakages over many thousands of valve cycles.

Other developments in the control of injected chemicals have produced techniques
for sealing high pressure aggressive chemicals, where avoidance of elastomeric seals
is considered necessary.

INTRODUCTION

Hydraulic systems form an integral part of control techniques for offshore
operations involved in the recovery of hydrocarbons.

Subsea control systems designers are presented with a combination of requirements
in addition to high working pressures, such as high ambient external pressure and
extreme inaccessibility of subsea equipment. These factors have produced a range of
unique designs combining reliability with cost effectiveness in an industry dependant
on reducing overall costs.

461

Typical applications of subsea control systems include the operation of subsea wellhead actuators, control of safety isolation valves and most demanding of all, the actuation of downhole safety valves.

In order to achieve reliable and long life operations the full range of hydraulic sealing techniques are used to contain pressure up to 1,725 bar. These sealing methods include radial bore and face sealing for static fluid containment and positively energised metal shear seals for switching applications.

DESIGN CONSIDERATIONS - SEALING OPTIONS

In line with other hydraulic systems, subsea controls have the usual design requirements in order to reliably seal hydraulic fluids, which may be either water based or mineral oil fluids.

Particular considerations occur in relation to materials compatibility, with possible contamination from the hydrocarbon process fluid and the requirements to install and retrieve equipments underwater. Retrieval and disengagement of mating parts can create vacuum conditions, leading to water inrush and the displacement of 'o' ring seals.

Face Seals

Typical seal materials are viton or nitrile rubbers conforming to conventional 'o' ring compression standards. These are used alone up to medium working pressures (100 bar), but higher pressures require the use of PTFE back-up rings. These ensure sealing from both internal and external sources with modified containment profiles to suit.

Face seals offer a safety advantage in devices with one atmosphere chambers. These can gradually pressurise over long periods of exposure to well pressures (690 bar) and subsea pressure (310 bar). When dismantling, face seals quickly vent entrapped pressure prior to release of the final threads of holding bolts. Bore seals can create situations where internal trapped pressure is being held by the final threads of the last fixing bolt. This can results in the end cap etc, becoming a possible projectile and a serious safety hazard.

Bore Seals

This technique embraces sliding seals with typical materials being loaded PTFE, for internal small (0.025/mm) clearances, or 'o' rings with back up rings for external use with mating parts from typically, two halves of tubular hydraulic couplings.

Bore seals can be subdivided into categories allowing multiple make ups, or single make-ups which require replacement prior to subsequent engagement.

Materials for multiple make-up seals can be either elastomeric, as previously described, or spring energised metal part section 'o' rings, with gold plated coating for lubrication and in-fill properties.

These give up to 100 make-ups which is sufficient for most equipment, including considerable factory and integration testing.

Single make up seals, such as metal crushable conical joint rings, form high integrity seals with good resistance to process fluids. Designs tend to configure the retrievable part the of equipment with the conical seal, such as to facilitate replacement prior to deployment. However this type of seal has become less frequently used, due primarily to the added complexity of spares and repair considerations.

Shear Seals

The shear seal principal is a long established valve technique. It has become the dominant method for subsea hydraulic directional control valves. The single shear seal (figure 1) is shown to be pressure unbalanced towards the sealing plate. The shear seal spring is normally of just sufficient strength, to ensure that the sealing surfaces remain in contact when unpressurised.

When the shear seals and seal plate are lapped to better than 2 light bands of flatness, the faces will seal effectively leak tight at all the working pressures required. The shear seal can be slid to and fro over ports in the seal plate giving the basis of a directional flow valve.

There are many principles of use of shear seals, but discussion here will be restricted to the pair of back to back shear seals, which form a balanced pair when mounted in a spool between two seal plates.

In one method of use, one seal plate is the inlet and the other the outlet. Alternatively one seal plate may contain both inlet and outlet ports and the other may be blank (the slave seal plate).

A valve containing one inlet port and one outlet port gives 3 way, 2 position directional control. This is the most frequently used actuator control valve type used subsea.

Care is taken to ensure that the shear seal land is less than the seal plate port size, to ensure that the hydraulic actuator being controlled can never be totally blocked off by the shear seal land. This guarantees the actuator can always close by depressurisation of supply pressure in an emergency. This deliberate interflow which occurs when the land is in the mid position over the function port, is sometimes achieved by a fine slit into the sealing land. More usually the sealing land is relieved to a depth of 0.1mm to reduce the area in contact with the sealing plate. This reduces the hydraulic unbalanced force on the shear seal and hence the frictional resistance to motion. A low frictional deadband is desirable on valves which are required to drop out (switch off) when supply pressures are reduced.

Shear seals of this design seal against internal pressure. A shear seal design for external pressure control is shown in figure 2. For valves requiring control of both internal and external pressures ie switchover or commanded shuttle type valve, a partially balanced shear seal has been developed, in which the sealing force is totally obtained from the shear seal spring.

4 way, 2 position valves are obtained by the addition of an extra port to the selecting seal plate. These valves frequently do not require the interflow condition, with shear seals which overlap the seal plate ports being used. This can increase the land area and friction forces. Profiled shear seals have been introduced, with crescent shaped ports to return the friction levels to normal.

Materials used are stainless steels for compatibility with water based fluids and the stringencies of fluid cleanliness requirements. Shear seals and seal plates from hardened AISI 440C have given excellent cycling test lives of 20,000 to 40,000 cycles with media of NAS class 6 working cleanliness. Typically NAS class 12 is the design limit.

To restate the achievement of these electro-hydraulic control valves using shear seal technology.

- Working Pressure   70 to 1,035 bar

- Control (Pilot) Pressure   70 to 380 bar

- Leakage (mostly from pilot solenoid)   0.1cc/minute maximum

- Qualification cyclic test   10,000 cycles minimum

- Fluids either water based or mineral oil   NAS class 12

- Full range of hydraulic control functions

SUBSEA CONTROL APPLICATIONS

Hydraulics Couplings

Self sealing couplings are a common device in subsea hydraulic control systems. They are used at the ends of umbilicals and hoses to prevent loss of fluid when piping is disconnected. Couplings comprise male and female halves with metal/metal sealing poppets which are drop tight at all working pressures. When the pipelines are made up, the couplings open to permit free fluid flow in either direction.

The couplings are classified into two types:

1)   Stab type couplings, which can be used in multiples of up to 30, pulled in simultaneously on module bases, or side mounted stab plates.

2)   Individual (or diver mateable) single pair of couplings, frequently keyed to prevent incorrect connection. These are made up by divers or Remotely Operated Vehicles (ROV's).

In the simplest, but highly successful form, diver mateable couplings have a single elastomer sealed body joint with back up ring (figure 3). This gives leak free make and break at pressures to 1,035 bar. The primary body joint seal is a metal to metal sealing cone similar to the standard J.I.C cone.

The equivalent stab coupling uses dual 'o' rings with back-up rings, giving redundant sealing of the body joint (figure 4). This allows greater make up latitude.

Cone gasket seal versions of these couplings (figure 5) with metal to metal body joint sealing rings have had much use in subsea systems. These conical joint rings produce high sealing forces at the inner and outer edges when the seal ring is deformed into its sealing position.

This successful seal has the drawback, that when used subsea with an 'o' ring type seal in a coupling body joint, the hydraulic lock between the seals can cause the conoseal to deform incorrectly. When demated the seals remain in the female, retrievable, part of the coupling so they can be brought to the surface and replaced. This seal can permit a one or two make-ups without replacement.

Cup type metal seals, which are pressure and spring energised to seal (figure 6), avoid the hydraulic lock and have permitted 690 bar to be repeatedly made and broken under full pressure.

These pressure energised cup seals, made of inconel have gold plating to fill surface defects on the mating parts. Over 100 make and break cycles at maximum working pressure is achievable without leakage.

The ideal secondary body joint seal to be used with the metallic cup seal, is a plastic 'U' packing (figure 7), as this prevents the creation of vacuum during subsea extraction and hence avoids wash out of seals due to inrush of sea-water.

The metallic cup has two pressure energised sealing surfaces. From reliability considerations it was hypothecised that a seal with one trapped static sealing edge and only one pressure energised sealing edge could confer an advantage. The resulting 'J' profile seal also proved an effective metal to metal sealing device and has been qualified for subsea use.

The metallic seals were qualified when contaminated with 'Arizona test dust', a fine graded sand, and gave 100 pressurised make and break cycles before unacceptable leakage occurred.

Directional Control Valves

Although the earliest designs of subsea electro-hydraulic control valves used soft seated poppet valves, these were soon superseded by the higher pressure capability metal shear seal type of valve.

There are several types of control using shear seals, with a pair of back to back shear seals contained within a valve spool as shown in Figure 8.

Each shear seal is spring and pressure biassed against its sealing face. The spool is actuated to the open and closed position with the shear seals traversing the function port. The method of sealing is basically leak tight with permissible leakage of one or two drops per minute at pressures to 1,035 bar.

Surfaces have been developed to avoid wear over the life of the valve, particularly when using water based fluids. Hardened stainless steels of the AISI 440 type have recently been replaced by nitrided stainless steel of the 316L type. This has permitted plug welding of seal plates and gives prospect of satisfying the NACE (well fluid compatible) requirement. The nitriding is a plasma process giving hardness of approximately 1000 Vickers Diamond Hardness for 0.1mm deep, over the small area used by the shear seal.

The NACE design requirement precludes the use of hardened steels in applications which may be exposed to hydrogen sulphide. This is a constituent (sour gas) of some production crude fluids. These gases are known to leak back from the control line of downhole safety valves and can reach the control valves.

The spool and shear seals carry external seals which are elastomers with a PTFE profiled covering. The PTFE gives the low friction required for a valve which is spring reset at specific pressures and also forms the anti-extrusion ring for the 'o' seal.

This combination has passed cyclic test at 1,035 bar and proof tests at 1,550 bar on valves of the following types; which can be either solenoid piloted or direct pilot operated:

- Electro Hydraulic Directional Control Valves

- Choke Control Valves

- Dump Valves

- Safety Valves

- Switch Over (Commanded Shuttle) Valves

Material developments are in progress to extend the use of these principles into the control of methanol at 690 bar. The lack of lubricity of this fluid has required the use of tungsten carbide, tungsten nickel and stellite.

Coatings of PTFE impregnated nickel are currently being tested. Chemical injection fluids can be extremely aggressive towards elastomers and methods of avoiding these are being developed ie., the elastomer sealed shear seal pair being replaced by a single crushable shear seal (figure 9).

Chemical Injection

Another requirement for offshore operations is the control and monitoring of a range of injected chemicals. This prevents undesirable effects on the transportation of fluids both to and from the hydrocarbon field.

These types of phenomena are unique to the oil industry, and range from the control of corrosion in flowlines, requiring small dosing of treatment chemicals, to the large scale injection of methanol. This chemical is used to prevent hydrate formation as the process fluid moves through different pressure and temperature phase states.

Typical current control techniques involve using dedicated umbilical lines to individual wells with topside pumps providing supply pressure. However with the need to maintain flow rates with variable well back pressure, local distributed controls are able to monitor and automatically adjust to provide constant flows.

Particular problems occur when designing systems for offshore chemicals, in relation to the lack of lubricity with methanol, to aggressive chemicals designed to prevent build up of scale or the formation of emulsions.

The poor lubricity, combined with a searching reaction to any inbuilt lubricity, as is particularly the case with methanol, produce unique difficulties in shear seal valves. The close contact of surfaces flat to less than 2 lightbands, and of fine surface finish in presence of high interface pressures of over say 2,070 bar, makes them prone to galling under these conditions.

Increased hardness is a current technique for coping with these media. Plasma nitriding of stainless steel has been introduced and the plasma nitriding of stronger substrate's is being considered. In plasma hardening, the particular surface of the component required to be hardened is given a 0.1mm thickness of nitride at nominally 1,000 Vickers Diamond Hardness.

Tungsten Carbide inlays, and coatings often cooperating with solid items of tungsten carbide are in use. To obtain the advantages of using dissimilar metals, tungsten nickel is used with satellite or tungsten carbide.

Surface treatments such as PTFE impregnated electroless deposited nickel are showing advantages over other materials with inherent long operational life properties.

SYSTEM SUMMARY

Subsea control systems are used for the following applications:

        Wellhead Actuators
        Down Hole Safety Valves
        Seabed Safety Isolation Valves
        Chemical Injection Valves
        Hydraulic Distribution Systems
        Production Chokes

High reliability equipment with proven design history or extensive test qualification is required to cover the following typical performance requirements.

Working Pressure (internal)	0 to 1035 bar
Ambient Depth pressures	0 to 3,000 metres
Working Temperatures : generally	-10° + 70°C
occasionally	-10° + 150°C
Leakage external	zero
Leakage across valves	0 - 0.1 cc/min
Cyclic Test Life	10,000 cycles minimum
Working media	Water based fluids and mineral oils
Chemicals	Anti-Corrosion
	Anti-Scale
	Methanol
	Emulsion Inhibitors
External Environment	Sea water
	Transformer oil

CONCLUSIONS

Considerable advances have been made in recent years to address the increasing pressure requirements for more difficult hydrocarbon subsea developments. This has seen working pressures increased from 515 bar to 1035 bar for downhole safety valves and control fluid temperatures elevated to 150°C. The hydraulic fluids have been developed and tested, with particular requirements to avoid the tendency to solidify or throw down adhesive residues over prolonged periods at high temperatures.

Although significant development is currently being focused on the replacement of hydraulic systems with all-electric subsea actuators, the requirement for hydraulics is set to continue for many years to come.

Through the rigourous use of qualification testing and the careful selection of materials, it should be possible to advance even further the range and operating parameters of hydraulic systems, for subsea hydrocarbon production with particular emphasis on effective fluid sealing.

FIGURES

Figure 1   -   Single Shear Seal

Figure 2   -   Shear Seal - External Pressure Control

Figure 3   -   Diver Mateable Hydraulic Coupling

Figure 4   -   Manifold Hydraulic Coupling

Figure 5   -   Conical Joint Ring

Figure 6   -   Metal Cup Seals

Figure 7   -   Anti In-rush Seal

Figure 8   -   Directional Shear Seal Control

Figure 9   -   One Piece Shear Seal

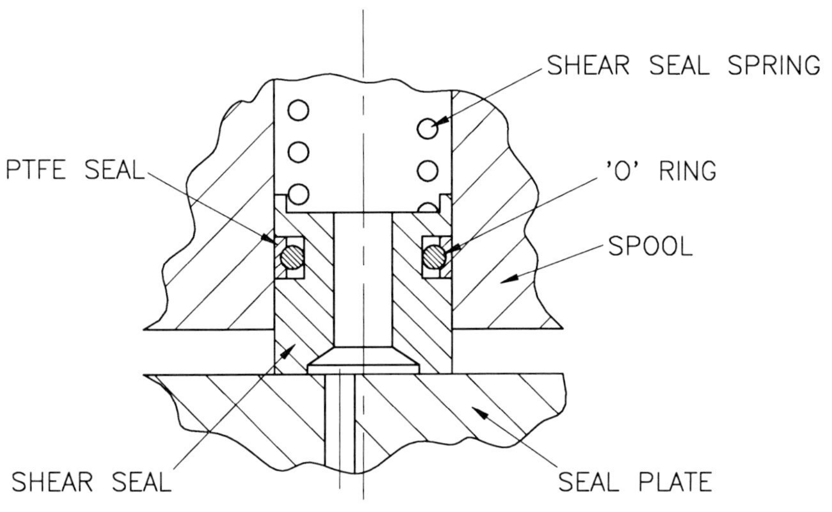

FIG 1 SINGLE SHEAR SEAL

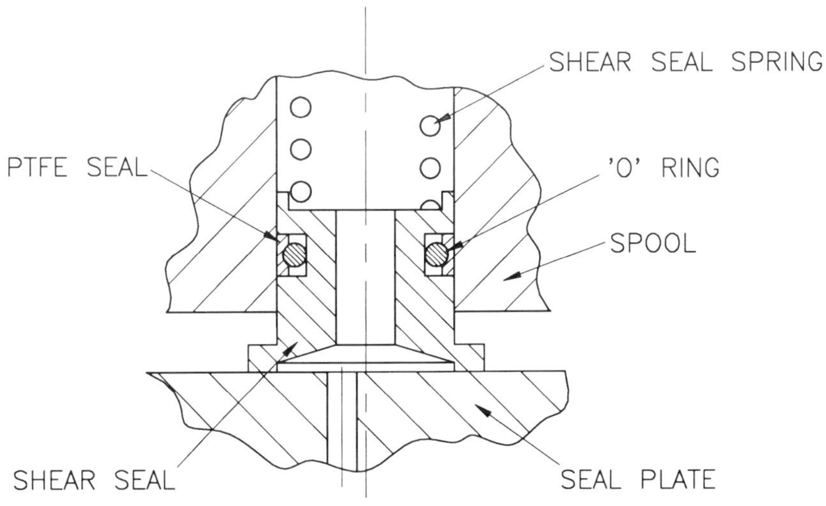

FIG 2 SHEAR SEAL, EXTERNAL PRESSURE CONTROL

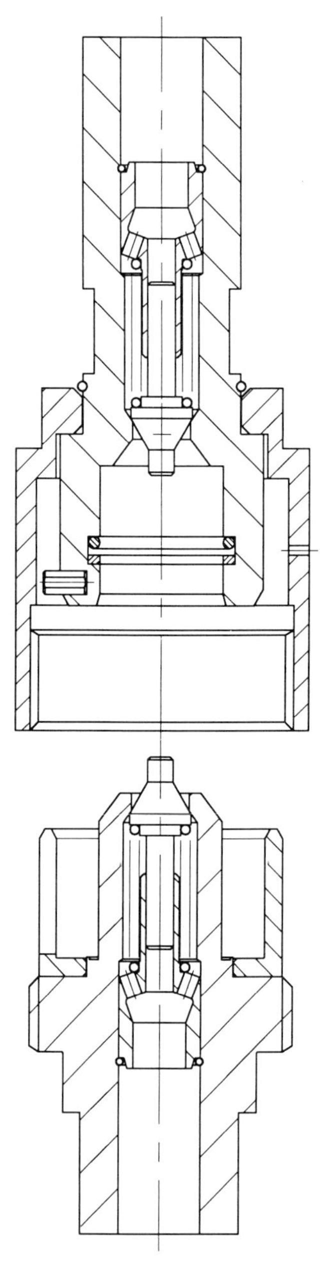

FIG 3   DIVER   MATEABLE   HYDRAULIC   COUPLING

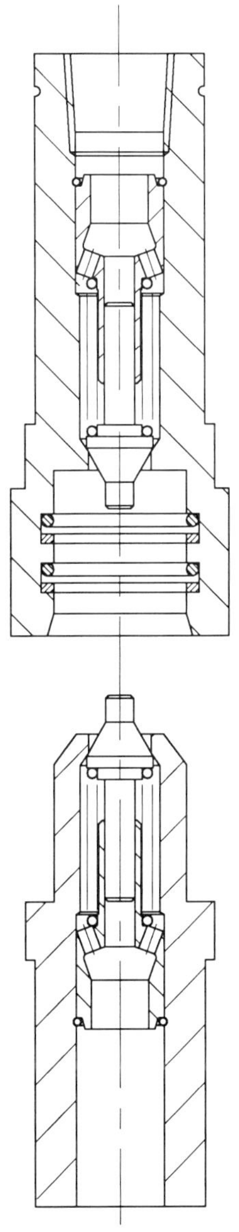

FIG 4 MANIFOLD HYDRAULIC COUPLING

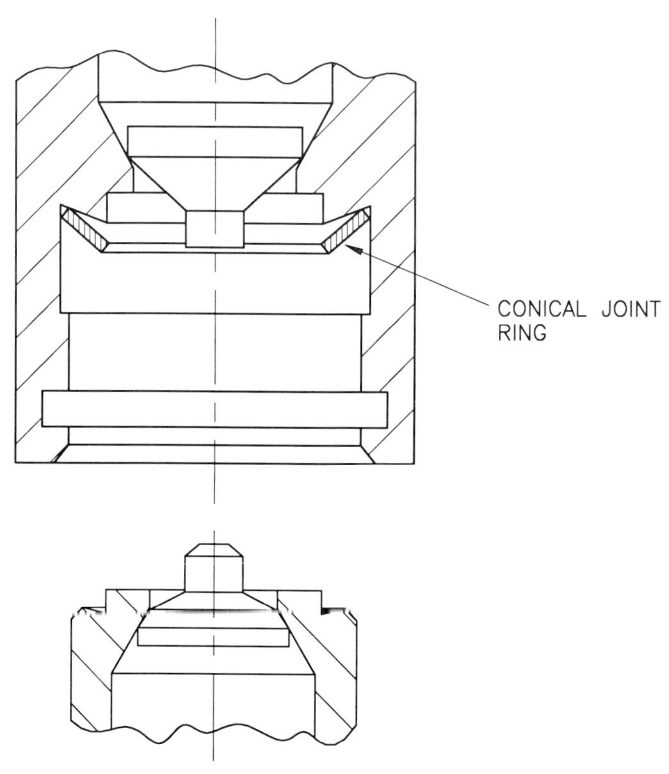

FIG 5 CONICAL JOINT RING COUPLING

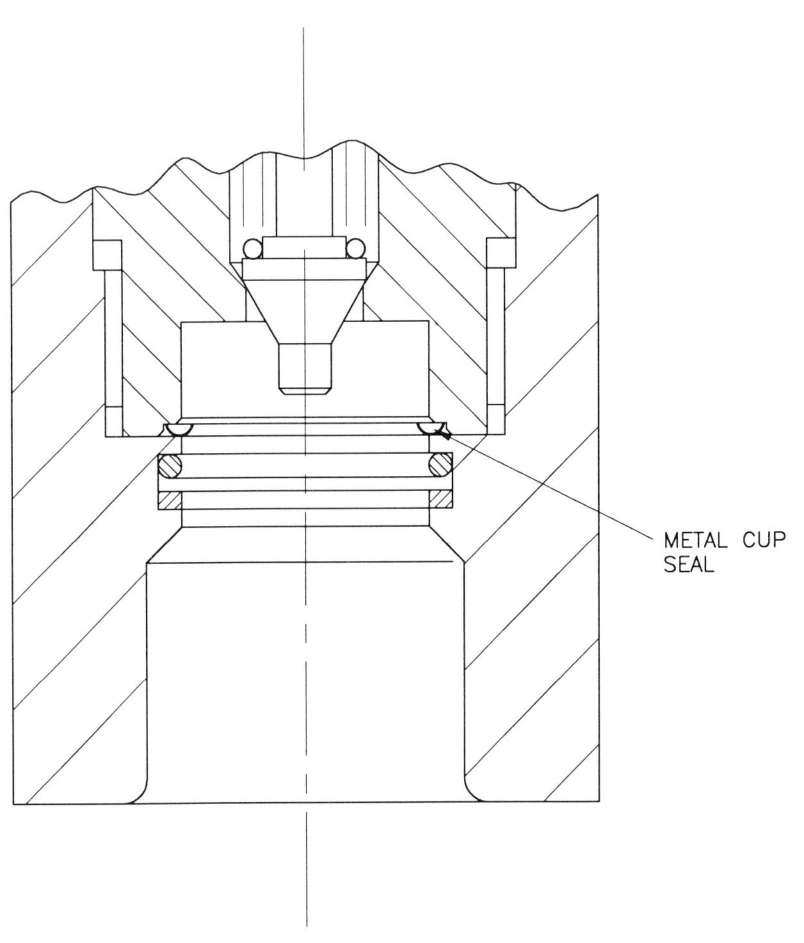

FIG 6 METAL CUP SEAL COUPLING

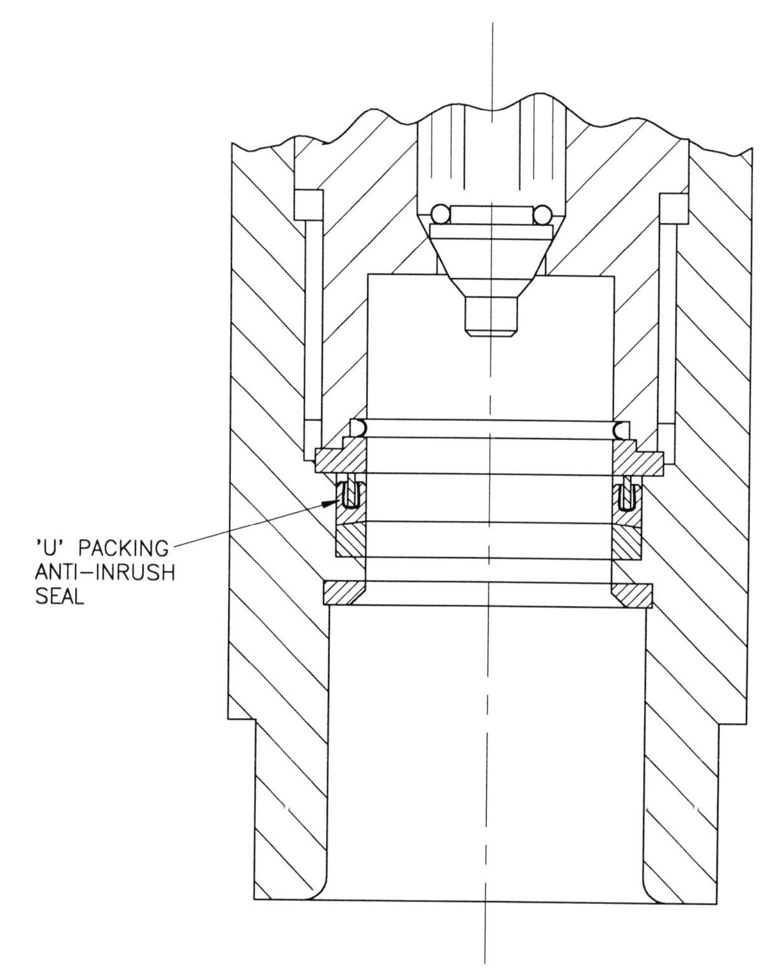

FIG 7 ANTI-INRUSH SEAL

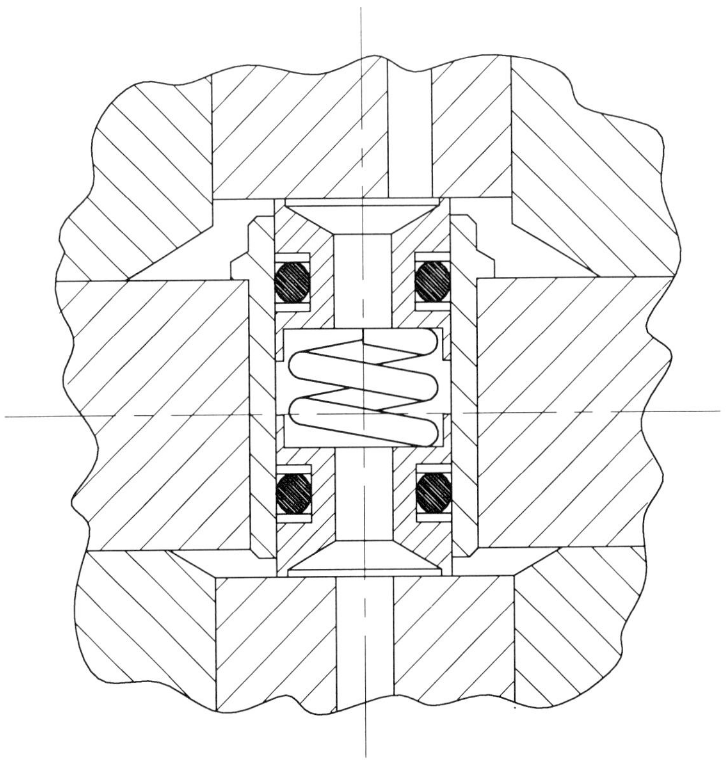

FIG 8 DIRECTIONAL SHEAR SEAL CONTROL

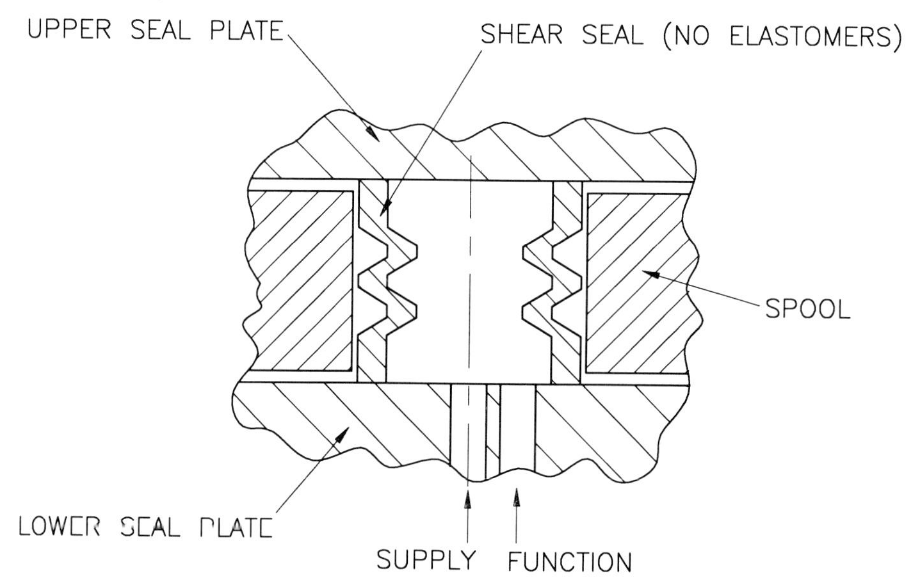

FIG 9 ONE PIECE SHEAR SEAL

14th International Conference on Fluid Sealing, Firenze, Italy,
6-8 April 1994. Organised by BHR Group Limited, Cranfield,
Bedford, MK43 0AJ, UK; Tel: 0234 750422

Seal Failure in Fluid Power Applications

- Causes and Appearances -

Dipl.-Ing. Helmut Weiss
Busak + Shamban GmbH & Co.
D 70565 Stuttgart

Abstract

The aim of every designer is to achieve the highest possible degree of functional reliability in his machines and systems. The interplay between design criteria, material selection, function, operation-related influences as well as the operation and service of the systems shows clearly how many factors have to be taken into consideration in advance.

Despite all his expert knowledge, malfunctions are still a regular part of day-to-day operation. The reasons for this are complex and not always identifiable at first sight.

This article considers damage to hydraulic systems characterised initially by the failure of the seal as the weakest link in the functional chain.

The specialist can draw conclusions as to the primary causes of the seal failure from the symptoms of the damage exhibited by the seals and the surrounding functional surfaces. This analysis of the cause of the failure then leads to concrete improvement measures.

This article is intended to allow problem areas to be recognized in good time.

1. Introduction

In many cases the failure of a hydraulic system is preceded by damage to or diminishing effectiveness of the seals. Although seals generally have to be regarded as wear elements, this does not necessarily mean that they are the primary cause of the failure.

As experience in practice shows, 70 to 80% of malfunctions in hydraulic systems are attributable to the poor condition of the hydraulic fluids. These statements are backed up by the investigations conducted by the Massachusetts Institute of Technology [1] in the early 1970's. They were later confirmed in an extensive study by the British Hydromechanics Research Association. This research work, conducted in conjunction with the National Engineering Laboratory, examined the causes of failure in 100 different hydraulic systems. It showed that solid contaminants were responsible for approx. 50% of the failures [2]. This article thus focuses particular attention on this cause of failure.

In addition, air and water must also be considered as damaging to the system as we will see later.

Determination of the design criteria, material selection, observance of design rules, knowledge of the functions of the components and the operation-related influences and proper installation, start-up and operation allows possible causes of failure to be eliminated at the planning stage.

In practice, malfunctions are often not recognised early enough as they occur in a closed system. In order to be able to determine the actual causes of the malfunctions it is often necessary to draw conclusions from analyses of failed seal elements and the surrounding components in order to discover the original cause of the damage.

Precise knowledge of the causes of the failure are necessary before concrete improvements can be introduced.

2. Summary of the Causes of Damage

As already pointed out, the causes of damage and malfunctions have a very wide range of possibilities. It would go beyond the scope of this article to discuss all of them here.

Table I shows a summary of possible causes of damage and their consequences and indicates where and at what point a failure is predestined or can be avoided by utilising the technical competence of the seal manufacturer. These are primarily the areas

– Design

– Choice of the seals

– Installation and start-up

– Operation and monitoring of the systems

The modern seal, scraper and guide elements are today developed as high-performance machine elements. With proper choice and installation, they ensure trouble-free operation with a long service life expectation.

Influenced by	Cause of failure	Consequence
Design	Excessively large gap	Gap extrusion
	Pressure relief too fast with gases	Explosive decompression
	Incorrect installation dimensions, tolerances	Excessive friction, low preloading, blow-by effect
	Flow velocity too high	Erosive wear, air bubble cavitation
	Contact pressure too high	Adhesive wear
	Surfaces too rough	Wear of seals
	Build-up of back pressure	Overloading of the seal elements
	Grooves too tight	No pressure activation, blow-by effect,
	Grooves too large	Wear by blast jet
Choice of the seals	Thermal overloading	Brittleness of elastomers Melting or scarring of plastics
	Chemical overloading	Swelling, shrinking, hardening
	Seals too soft	Gap extrusion, plastic deformation
	Unsuitable materials or type of seal	Stick-slip phenomena, loud noises
Installation and start-up	O-ring twisted – due to installation, or – due to long stroke	Leakage due to spiral breakage
	No lead-in chamfers	Shearing off of the seal
	System not bled	Dieseling effect, charring of the seal
	Folding over of seal edges	Leakage during start-up
	Unsuitable installation tools	Premature failure
Soiling in the system	Scoring of rods due to dirt particles or surface flaws	Leakage due to wear of the seals Entry of dirt via wiper seals Impaired functions due to guide wear
	Scores in the cylinder barrel	Premature failure of the seals
	Water in the oil	Hydrolysis, e.g. of polyurethane
	Solid particles in the system	Abrasive wear

Table I   Summary of Causes of Damage in Hydraulic Systems

3. Design

The reliable operation of a system can be predetermined to a great extent by careful design, choice of the sealing systems and choice of the seal materials.

From the wide range of possible design-related faults, I would like to concentrate here on the main causes of damage in hydraulic systems. These are:

3.1. Unsuitable Surfaces

A mating surface material not matched to the operating conditions, the application and the seal material, excessive surface roughness or insufficient hardness result in premature wear of the seal and contact surface due to abrasive and adhesive wear.

3.2. Gap Extrusion

Elastromer or thermoplastic-based seal material tend to flow under pressure.

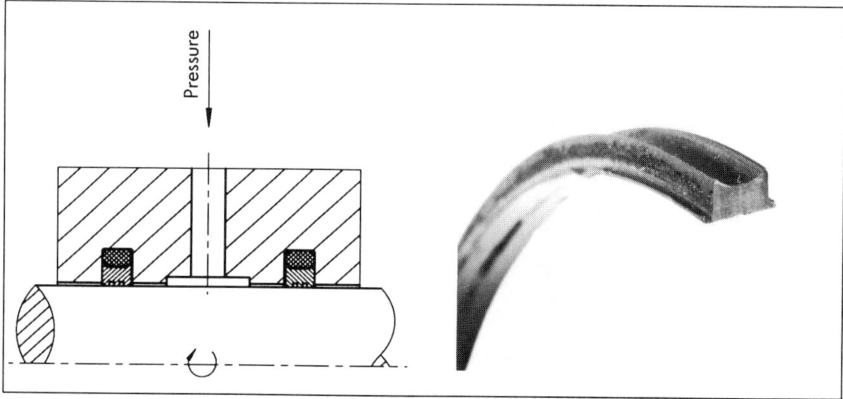

Figure 1   Gap Extrusion at Rotational Seal Element

Excessively large gaps not adapted to the pressures allow the material to extrude into the gap. Figure 1 shows the extrusion lips of a rotational seal element. The material loss leads to decreasing O-Ring preload and thus to leaks. In this seal, used in a rotary joint, several possible sources of faults can be the cause of a failure. Excessively large gap, overwide grooves, pressure build-up due to a grease packing in the secondary chamber or overheating due to dry running.

3.3. Backpressure

Long hydraulic cylinder strokes cause backpressure in narrow gaps. If the medium drawn out is further throttled by a further narrow gap, e.g. the gap of a second seal or of the scraper, pressure is built up by the excessive amount of fluid. The level of this pressure can exceed the system pressure and thus cause the seal e.g. to fold over or even be pressed out of its seat. Figure 2 shows a seal element which has tilted due to impermissibly high build-up of pressure and is worn. The excess pressure has forced the scraper out of its seat.

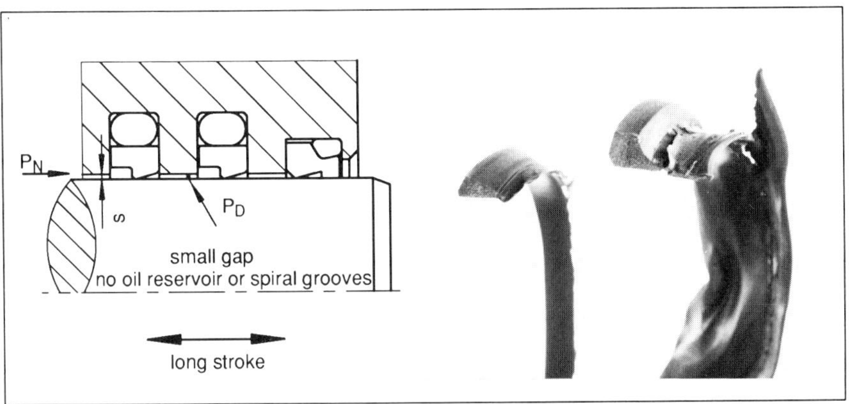

Figure 2   Extrusion caused by Drag Pressure

From the drag pressure formula

$$P_D = P_N + \frac{6 \cdot v \cdot Ls \cdot \eta}{s^2}$$

we can see that the velocity v, the gap length L_s, the viscosity η and in particular the gap size s are responsible for the pressure increase.

Dependent on each actual situation the pressure increase of drag pressure P_D can reach 2 – 5 times of nominal (working) pressure P_N.

The build-up of drag pressure can be prevented by design measures and the choice of suitable seal elements, e.g. with hydrodynamic pumping back effect.

3.4. Installation and Start-up

After installation, it is generally no longer possible to check the correct seating and position of the seal. Installation must therefore be carried out with great care and the design rules applicable to seal installation must be observed. Installation aids should be used, where necessary.

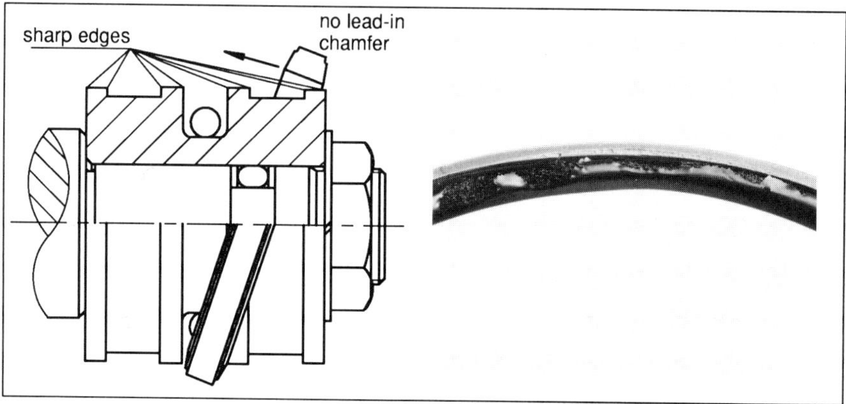

Figure 3   Extrusion and Nibbling-Effect

Figure 4   Spiral-Cut at O-Ring

Furthermore, sheared or damaged seal edges become noticeable only during start-up or after a period of operation in the form of leaks, although the reason for the leaks is not identifiable at this stage.

Figure 3 shows a rubber preloaded piston seal in which the O-Ring has been nibbled away as a result of gap extrusion. The reason for this was the lack of a lead-in chamfer on the piston. The inside diameter of the seal face ring was sheared. The O-Ring then extruded into the gap between ring and groove flank at each change in pressure and became damaged.

The O-Ring shown in Figure 4 was drilled during assembly. As a consequence the spiral cut arised by change of pressure.

4. Gaseous Contaminants

Air and gases can prove to be considerable problem factors in the hydraulic system. At atmospheric pressure, up to 9% air can be dissolved in the oil. At higher pressures, far more air or gas can be dissolved in the oil and is given off again when the pressure is decreased.

Hydraulic systems must be carefully bled before start-up. Any air or gas remaining in the system can lead to serious consequential faults.

4.1. Cavitation

By cavitation we mean the implosion of vapour bubbles in the system. Compressed air not dissolved in the oil which passes the seal gap (blow-by effect) expands on the low pressure side with considerable energy.

The surface roughness intensifies the process which causes gaps and scratches not only on the seal element. It affects also the adjacent metal surface of the groove and cylinder bore.

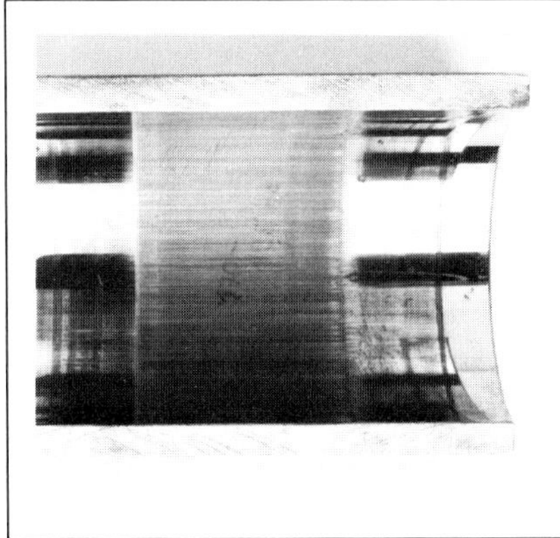

Once the seal surface has been damaged by the pittings and blistered craters, the hydraulic fluid flows at high velocity and with high acceleration through the longitudinal scores and causes the wear patterns illustrated in Figure 5.

At first sight, a surface scored in this way could be analysed as having been subjected to abrasive wear.

Such processes can be triggered off by vibrations, flows, high pressures and vacuums.

Figure 5   Micro-Cavitation and Jet
Erosion on Cylinder Bore

4.2. Dieseling Effect

A further consequence of air bubbles in hydraulic fluids can be the "dieseling" effect. If the pressure in the system increases sharply with a short period of time, the air bubbles are heated to such an extent that spontaneous ignition of the air/gas mixture in the bubbles occurs.

If, for example, an air bubble of 25 mm diameter is compressed from atmospheric pressure to 50 MPa within a few milliseconds, a temperature of 2,500 °C occurs at the core of the bubble.

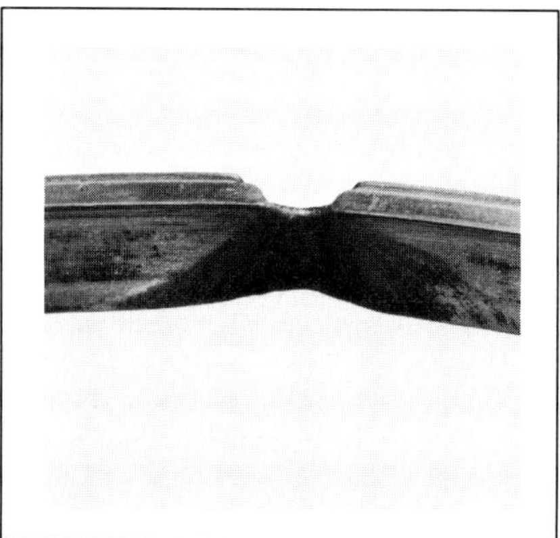

Figure 6   Wear Ring melted by
            Dieseling Effect

If this effect occurs in the vicinity of seals or guides, the guide rings and seals are charred. In addition to the possible direct failure of the elements, crumbling hard particles of the rings or seals can cause problems in the system. For this reason, particular attention should be paid to the use of materials which leave no harmful combustion residues for such applications.

Figure 6 shows a plastic wear ring damaged by the dieseling effect. The modified PTFE material used, however, produces no further abrasive combustion residues which could disturb operation.

5.   Solid Contaminants in the System

When we talk about solid contaminants in hydraulic systems we think first and foremost about the soiling of pumps, valves, cylinders and other components and concern ourselves with their wear behaviour.

The seals themselves, however, are subject to considerable danger from the contaminated media and also require particular consideration. They are, after all, responsible for ensuring that the hydraulic fluids remain in the system under very high pressures and heavy duty conditions.

The harmful effect of contaminants generally comes from the abrasive effect of the solid particles. These solids are the cause of the wear to components and seals. Even where optimum filtering techniques are used, the system operator will always assume a permissible degree of contamination of the medium. This level, however, is frequently reached only after a start-up and running-in period, i.e. the initial degree of contamination is higher.

"Fresh oil" drawn from barrels or oil tanks can have a 20 to 50 times higher level of solid contaminants than is permissible for the system [3].

Even during start-up of the cylinder, a slight scoring of the piston rod is often observed and arouses grave doubts in the mind of the operator. The causes of this are among other things coarse and hard particles such as grinding dust, mould sand, scale, welding beads, etc. which have already entered the system during the manufacture of the individual parts and the assembly of the components.

Figure 7 shows the diagram of a hydraulic system and indicates possible weak points. The potential locations of filters in the system are illustrated. All the systems can be operated either individually or in combination. Even using the very finest filter mesh size, however, it will not be possible to remove all the particles from the system - not least due to the fact that new particles are continuously produced as a result of the wear of mechanical components in the system.

For example, metal or scale particles form on drums or tanks, condensation leads to rust and impurities can also be carried into the system by the atmosphere.

Particular attention is demanded when refilling hydraulic systems in the field under extremely dirty conditions, for example in mobile hydraulics. Sludging of the fluid can occur in the actual container. If the hydraulic oil is not changed at sufficiently short intervals or the oil is overloaded, shear stresses can occur which cause a chain degradation of the oil.

Particular dirt hazards can be directly related to the processes in the sector in which the hydraulic systems are used, e.g. dust, scale or dirt in foundries, rolling mills, mines and in mobile hydraulics. Particles can be brought into the system via venting filters or by the piston rods.

The seal and scraper system on the piston rod, in particular, must be adapted to the particular application. As far as the scraper seal is concerned, a compromise has to be made between the 100% scraping effect - which at the same time means dry running - and the function under optimum sealing conditions.

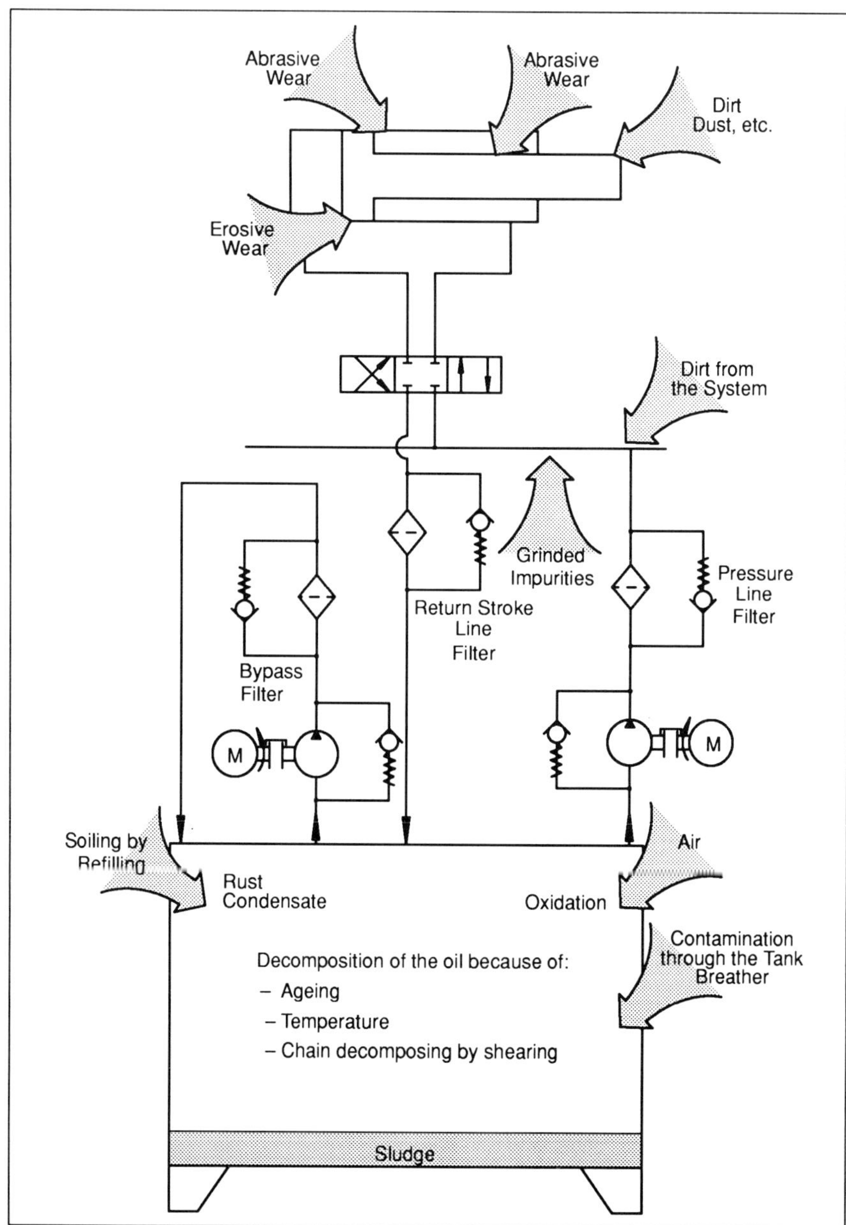

Figure 7   Hydraulic Circuit with Danger Areas for Impurities

6.   Effects of the Contaminants

The consequences of contaminants are the early failure of seals due to wear and the premature replacement or remachining of mating surfaces.

These effects must always be considered in conjunction with the forces acting in the system, such as frictional forces, flow forces or spring forces and the influences of the operating conditions such as pressure, temperature, medium, frequency, cyclic duration factor, etc.

6.1. Wear due to Abrasion

Figure 8   Wear due to Abrasion at a Wear Ring

Abrasive wear occurs between parts which perform a movement relative to one another. Where there is insufficient lubrication between the seal and mating surface, the surface particles which are in contact have a large contact surface area. Less-than-optimum surface structures with roughness peaks due to poor machining, in particular, cause particles to break out of the metallic or chrome-plated surface. They become embedded in the softer seal material and result in abrasive wear of the harder mating surface. This can set a wear mechanism in motion which can no longer be stopped, as shown in Figure 8. Particles entering the gap from the outside accelerate this process further.

6.2. Wear due to Erosion

Erosive wear occurs when particles strike the surfaces of seals, guides and components at high velocity. This is the case, for example, when a fluid contaminated with abrasive particles flows through a narrow gap at high speed. The high flow velocities are generally caused by large pressure differentials and small flow cross-sections.

7.   Influence of Surface Finish

The functional reliability and service life of a seal depend to a very great extent on the quality and surface finish of the mating surface to be sealed. This can also be a significant factor in reciprocating seal failure.

Scores, scratches, pores, concentric or spiral machining marks are not permitted. Higher demands must be made on the surface finish of dynamic mating surfaces than of static mating surfaces.

The characteristics most frequently used to describe the surface microfinish R_a, R_z and $R_{max.}$ are defined in DIN 4762/ISO 4287/1 and DIN 4768. These characteristics alone, however, are not sufficient for assessing the suitability in seal engineering.

In addition, the material contact area M_r (previously percentage contact area tp) in accordance with DIN 4762/ISO 4287/1 should be demanded. The significance of this surface specification is illustrated in Figure 9. It shows clearly that specification of R_a and R_z alone does not describe the profile form accurately enough and is thus not sufficient for assessing the suitability in seal engineering. The material ratio M_r is essential for assessing surfaces, as this parameter is determined by the specific profile form. This in turn is directly dependent on the machining process employed.

The material ratio M_r should be approx. 50 to 70%, determined at a cut depth $c = 0.25 \cdot R_z$, relative to a reference line of $c_{ref.}$ 5%.

Surface profiles	Ra	Rz	Mr
	0.2	1.0	70%
	0.2	1.0	15%

Figure 9   Profile forms of surfaces

A general recommendation for surface finishes of reciprocating seals is given in the table II.

Surface Microfinish	Mating Surface		Groove
	PTFE based Material	Rubber and Polyure-thane	
R_{max}	0,63 – 2,50	1,00 – 4,00	< 10,0
$R_{z\,DIN}$	0,40 – 1,60	0,63 – 2,50	< 6,3
R_a	0,05 – 0,20	0,10 – 0,40	< 1,6

Table II   Surface Finishes

8. Evaluation of Seal Failures

Since cause and effect (pattern of the damage) are often very far apart, every observation made during the dismantling of the system is of the greatest significance.

A separation must be made, in particular, of foreign particles. Even particles embedded in the seal material must be subjected to a material test. The chemical analysis of the particles then allows conclusions to be drawn as to their origin. The detection of silicone particles, for example, would indicate the impermissible use of a grinding cutter in the vicinity of the cylinder assembly.

The analysis of the oil specimen provides further information on the degree of contamination, particle size and material composition. Chemical and physical examinations give indications of possible changes in the seal materials which, in turn, indicate chemical or thermal conditions.

Excessive cleanliness often makes analysis of the cause of failures very difficult or even impossible. The most important pieces of evidence are destroyed by the cleaning rag.

The specialist should - whenever possible - be present during removal of the seals. Only in this way can incorrect installation or the presence of sources of problems be discovered, analysed or reconstructed.

Without exact knowledge of the cause of a failure, finding the appropriate remedy is simply a game of chance.

492

References

[1]   Dr. E. Rabinowicz   "The Loss of Usefullness of Components
                                             in Hydraulic and Lubrication Systems",
                                             Vortrag zum Workshop "Bearings" of the
                                             American Society of Lubrication
                                             Engineers, 1971

[2]   Department of Industry   "Research Programme on Contamination
                                             Control on Fluid Power System
                                             1980–1983"
                                             Volume I "Field Studies", Volume V
                                             "Filter Behaviour" DIT, National
                                             Engineering Laboratory, East Kilbride,
                                             Glasgow, UK, 1983

[3]   H. Essers   Ölpflege mit System
                                             Paper for 9. Fachtagung Hydraulik und
                                             Pneumatik, September 1993, Dresden

14th International Conference on Fluid Sealing, Firenze, Italy,
6-8 April 1994. Organised by BHR Group Limited, Cranfield,
Bedford, MK43 0AJ, UK; Tel: 0234 750422

Hydraulic Rod Seals with laser-structured Back-Surface

Ulrich Frenzel, Heinz K. Müller

Institut für Maschinenelemente , Universität Stuttgart, Germany

Optimum rod seals of hydraulic cylinders require effective wiping on the outstroke and, ideally, a complete inward pumping of the adhering film on the instroke. This is achieved by an asymmetric profile which creates a high maximum contact pressure gradient at the sealing edge on the fluid side and a corresponding low gradient near the outside end of the contact. In practice, however, by rising the fluid pressure inward-pumping diminishes which is due to a continuing increase of the maximum pressure gradient at the outside end of the seal contact. Based on the experience that a second circumferentially closed edge does not prevent the increase of leakage, the authors pursued the idea to support the back-surface with a large number of small asperities forming small interconnected gaps which boost inward pumping and reduce friction. Experimental results on the sealing capacity and friction of various back-surface structures are presented, and the design characteristics of optimum rod seals are discussed.

Nomenclature

d   -   rod diameter

Δh -   difference of film thickness between instroke and outstroke

p   -   pressure

u   -   velocity

w   -   maximum pressure gradient in sliding direction

α, β -   seal edge angle

η -   dynamic viscosity

Index

i   -   instroke

o   -   outstroke

Abbreviations

PTFE -   polytetrafluor ethylene

PU   -   polyurethane

Introduction

Modern hydraulic rod seals are characterized by a high standard of sealing capacity. *Figure 1* shows the functional areas and details of a hydraulic rod seal. Almost every detail of the rod seal has been thoroughly investigated and described in literature. Up to now there is, however, a lack of knowledge on the importance of a particularly efficient detail, the back-surface.

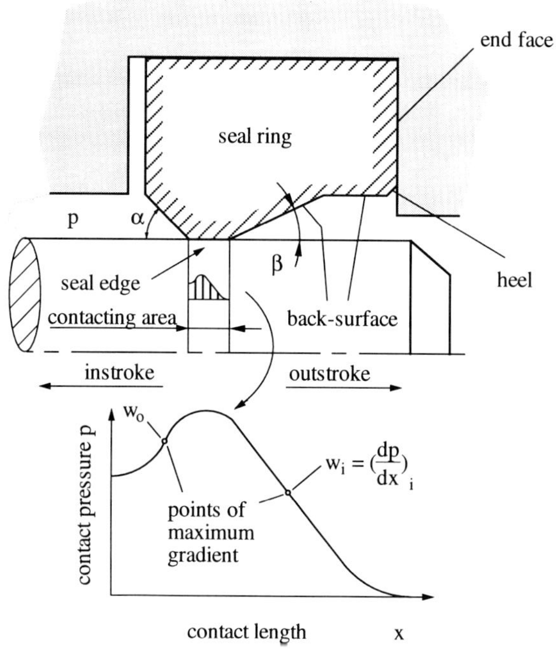

Figure 1. Functional areas of hydraulic rod seals

The elastic property of the seal material in conjunction with an interference of seal to rod diameter creates an initial pressure at the seal contact. By increasing the contact pressure with rising fluid pressure the incompressibility of the elastomer material makes the seal to work "automatically". The initial contact length of the flattened seal edge is some tenth of a millimetre only. Most important is the flank angle α on the high pressure seal edge which determines the maximum contact pressure gradient w_o on the fluid side. According to Blok's inverse solution of Reynolds equation /1/, a high value of w_o ensures optimum wiping on the outstroke which results in a minimum film thicknes h_o

$$h_o = \sqrt{\frac{2}{9} \cdot \frac{\eta \cdot u_o}{w_o}} \qquad (1)$$

To ensure leak-free operation of one single seal a complete inward pumping of the adhering fluid is required on the instroke. This is most likely achieved if the maximum pressure gradient w_i at the back-surface is as small as possible. This is realized by a small flank angle β, so that $\beta \ll \alpha$. Under these conditions a fluid film with the maximum thickness

$$h_i = \sqrt{\frac{2}{9} \cdot \frac{\eta \cdot u_i}{w_i}} \qquad (2)$$

is pumped back into the cylinder. Furthermore, a non-contacting part of the back-surface hydrodynamically boosts the inward pumping on the instroke by rising the fluid pressure in the slender wedge. A chamfered or rounded outside end of the back-surface, called heel, prevents extrusion and shearing-off of the elastomer on the outstroke. On the instroke the heel must boost but not wipe the fluid.

Two kinds of rod seals

Industrial development has led to two different kinds of elastomer rod seals /2/, both using equal elasto-hydrodynamic principles regarding the teamwork of seal-edge and back-surface. The one kind of modern rod seals is the "U-seal". Different existing shapings of the U-seal are shown in *figure 2a-c*. Normally, U-seals are made completely of elastomeric material, preferably of polyurethan. In practice U-seals have different lip length which influences the stiffness of the seal lip and, hence, the intensity of wiping. In practice, a U-seal having an short lip is called a "Compact-seal". The "Edge-seal" /3/, /4/ is the other basic design of rod seals. Originally, Edge-seals were made of stiff PTFE compounds, thereby preserving the asymmetric pressure distribution and high pressure gradient on the fluid side and preventing contact at the heel, over a wide range of pressure and temperature. In the meantime, also polyurethane edge-seals are available.

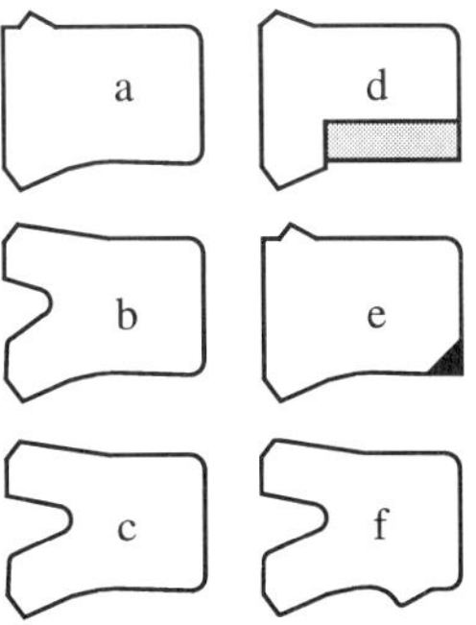

Figure 2. U-seal-shapings

Sealing capacity

The performance of rod seals is characterized by the sealing capacity and the frictional behaviour. While sealing is predominantly influenced by the contact pressure, distribution friction also depends on the area of contact. *Figure 1* shows the initial contact pressure distribution with its higher maximum pressure gradient w_o on the fluid side and a lower maximum pressure gradient w_i on the back-surface.

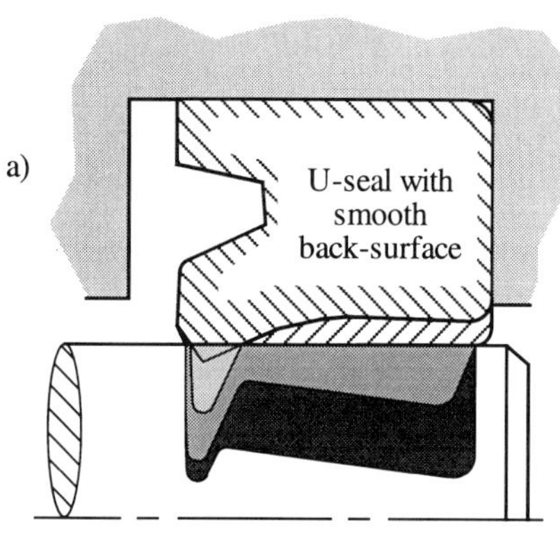

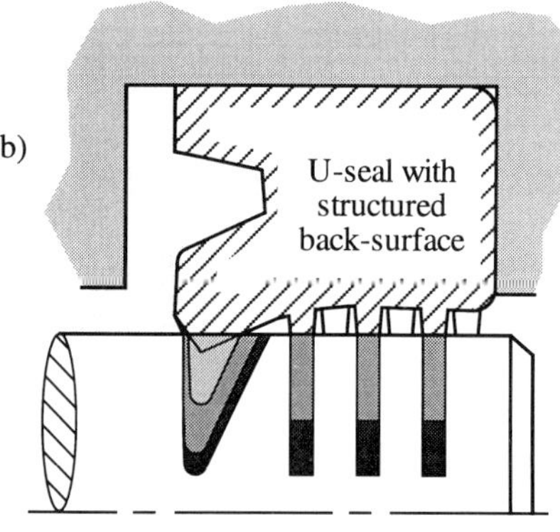

Figure 3. Contact pressure distribution for various values of fluid pressure

In the course of rising fluid pressure p, as shown in figure 3a, the axial length of contact at the back-surface and hence the area of contact increase. Up to a certain value of pressure p the gradients w_o and w_i remain approximately constant. At higher fluid pressure, however, the pressure distribution at the back-surface begins to change its pattern. The location of w_i shifts towards the heel and the gradient rises to a higher value w_i which may well exceed w_o. When, for example, the velocities at outstroke and instroke are equal, the heel on the instroke begins to wipe off part of the the fluid film which had been drawn out before.

Regarding the equation for difference Δh of film thickness between instroke and outstroke

$$\Delta h = h_i - h_o = \sqrt{\frac{2\eta}{9}} \cdot \left(\sqrt{\frac{u_i}{w_i}} - \sqrt{\frac{u_o}{w_o}}\right)$$

which reflects the sealing ability, the strategy for an improvement of the dynamic sealing capacity is now well defined. Because the pressure gradient w_o realized in modern rod seals cannot be increased any more all efforts have to concentrate on improving the inward pumping capacity of the seal.

Friction

Parallel to all efforts intended to improve sealing, it should be kept in mind that friction must remain as low as possible, not at least because of the temperature limit of polyurethane material. The friction of rod seals with standard back-surface (*figures 6 ,7*) is characterized by a different behaviour on instroke and outstroke. On outstroke friction is relatively low and approximately independent of pressure and velocity. On the instroke, however, the available fluid film thickness depends on how many fluid adheres to the rod surface. Therefore, friction depends both on the thickness of the incoming film and on the magnitude of the gradient w_i. Generally, the friction of a conventional rod seal is higher on the instroke than on the outstroke.

At first sight a simple method to lower the friction seems to be to reduce the dimensions of the back-surface (*figure 2d*) and hence of the contact area. However, this would mean to loose the boosting effect, to increase the gradient w_i , and thereby worsen the sealing capacity. A similar disadvantage appears when a "back-ring" is installed, (*figure 2e*), which at a higher pressure acts as a wiper and, therefore, impedes inward pumping. Another conventional approach, as shown in *figure 2f*, is to support the radial force by a single circumferentially closed toroidal bulge on the back-surface. In various experiments the authors confirmed the experience known from practical application that in essence the installation of a "second edge" does not improve the friction behaviour but worsens the sealing capacity. At higher pressures the entire back-surface, including the bulge, contact the rod and friction again is approximately as high as of a single-edged seal. Furthermore it was shown that due to the generation of a high gradient w_i at the bulge a detrimental secondary wiping occurs during the instroke /6/.

Best Sealing at Lowest Friction

Theoretically, there are two conceivable methods to improve inward pumping. One is to take measures which set limits to the maximum of the gradient w_i. The other is to keep partially the back-surface on a very close distance to the rod, thereby creating powerful boosting on the instroke. The authors [7] have developed a concept which allows to put into effect both methods at the same time.

As shown in *figure 3b*, the principle is to place on the back-surface a large number of small supporting asperities which form a sort of "piers" to avoid that the areas of the back-surface between the asperities contact the rod. The distance of the "piers" as well as their height (*figure 4*) are to be well-suited to the stiffness of the seal material and the maximum pressure p to be handled by the seal. Finite-Element-calculation is applied to adapt the stiffness of the asperities to the pressure-dependent radial load considering the temperature-dependent elasticity of the polyurethane material. Practically the manufacturing process to erode the negatives of the asperities into the rubber mold was performed alternatively by a conventional electro-erosive procedure or by an Excimer-Laser. Existing rubber molds can be treated in such a way.

498

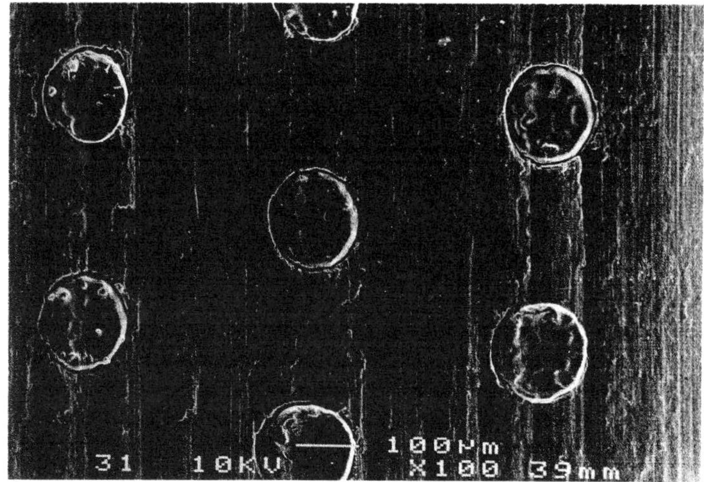

Electronic-microscope image

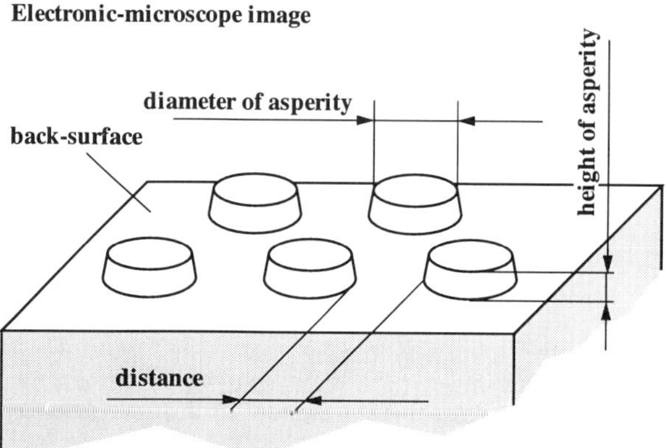

Figure 4. Laser-structured back-surface

At the heel the supporting asperities during the instroke keep an entrance open to the fluid film. The "piers" are formed either by the asperities of a relatively coarse roughness or by numerous well-distanced single asperities. In the vicinity of the seal edge the seal surface is smooth on both sides.

The back-surface and the rod together form a labyrinth-like gap which, by the sealing edge, is closed towards the fluid side. In this gap, on the instroke, the fluid pressure is boosted to a high value. Thereby, the seal edge is relieved and the actual value of the gradient w_i becomes very low. The result is that the fluid film is easily pumped back into the pressure space.

Because the contact area on the back-surface is restricted to lubricated asperities which are globally relieved by the boosting pressure on the instroke the friction is lower than that of a conventional seal. On the outstroke, however, the interspace between the asperities is unpressurized and the asperities support the entire radial load /5/. Contrary to the global lubrication of a conventially smooth back-surface the elasto-hydrodynamic lubrication is restricted locally to the micro-structures of the back-surface. Therefore, the friction of an asperity-supported seal at the outstroke is higher than of the conventional seal. Virtually the friction of an asperity-supported seal is independent of the direction of the rod motion. To achieve minimum friction the asperities are dimensioned and distributed so that at low pressure only a part of the asperities contact the rod; the others step by step get in touch at higher pressure.

Experiments

Sealing and friction of rod seals are governed by hydrodynamical effects. To obtain meaningful results within a short period all test rigs were operated with high viscous oil HLP 68, tempered to 25°C.

Sealing

To evaluate the sealing capacity of various rod seals the difference of film thickness Δh in dependence of the pressure to be sealed and particular ratios of velocities is determined on a particular test rig, which was earlier described by Prokop /4/. A negative value of Δh would mean a leaky seal. A rod seal indicated by a positive difference Δh remains leak-free.

The sealing behaviour of two kinds of back-surfaces is compared in *figure 5*. Both have identical geometries of sealing edges, so that wiping is identical. The difference of the sealing capacity is determined by the inward pumping ability only.

At low pressure (2 MPa) the pumping of the asperity-supported back-surface is appreciably higher compared to the smooth surface. Rising fluid pressure decreases the sealing capacity of both kinds of surfaces, but the smooth surface is reaching its limit already at 10 MPa and this particular velocity ratio. At 10 MPa, the laser-structured surface has approximately the same sealing capacity as the smooth surface has at 2 MPa.

The inward pumping ability of the laser-structured back-surface is further improved at higher pressure by adapting the height or distribution of asperities to the working conditions. There is no possibility to improve the pumping ability of smooth back-surfaces.

Friction

A test rig described by Prokop /4/ was applied to measure the friction of the seals. During friction measurement on the outstroke the seal deposites a fluid film according

500

to the particular sliding velocity needed for the following instroke measurement. A sliding velocity of $u_i = u_o = 150$ millimeters per second was sufficient to prevent stick-slip, and ensure full lubrication.

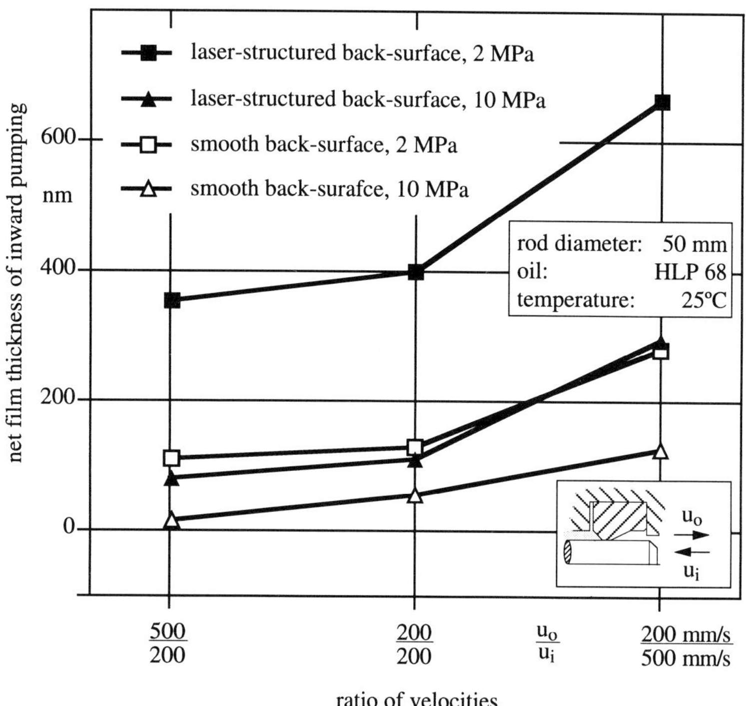

Figure 5. Sealing capacity of structured U-seals

In general the experimental results shown in *figure 6* and *7* confirm theoretical expectations. The smooth back-surface causes different behaviour dependent on the sliding direction, whereby the friction on the outstroke is relatively low. Initial contact of a second edge (*figure 2f*) at pressure p = 0 results in higher friction on both sliding directions. Because all kinds of U-seals have the same pressure distribution at the seal edge contact it is assumed that the different magnitudes of friction depend on the different back-surfaces.

On the instroke the friction of a U-seal with smooth back-surface with rising pressure increases very steep and reaches a high limit of approximately 600 N already at 10 MPa which is due to the increasing back-surface contact (*figure 6*). Although starting with a higher value at low pressure, the friction of a U-seal with a second edge reaches the same level as the U-seal with smooth surface. The delayed contacting of the entire back-surface due to the second edge causes a relatively shallow slope of the friction course up to 15 MPa. The laser-structured U-seal starts with low friction which increases but moderately when pressure rises. The asperities on the surface reduce the upper limit of friction at high pressure by preventing contact of the entire back-surface.

As shown in *figure 7*, on the outstroke the U-seal with smooth back-surface has relatively low friction, because the circumferentially constant film thickness under the back-surface causes a low viscous shear.

Remembering that by normal operation of a hydraulic cylinder the pressure on the outstroke is low, decreasing friction at high pressure on the instroke and increasing friction at low pressure on the outstroke results in an overall reduction of the frictional loss.

PU-Edge-seals, structured on their back-surface, and tested under identical conditions showed a similar friction and sealing behaviour.

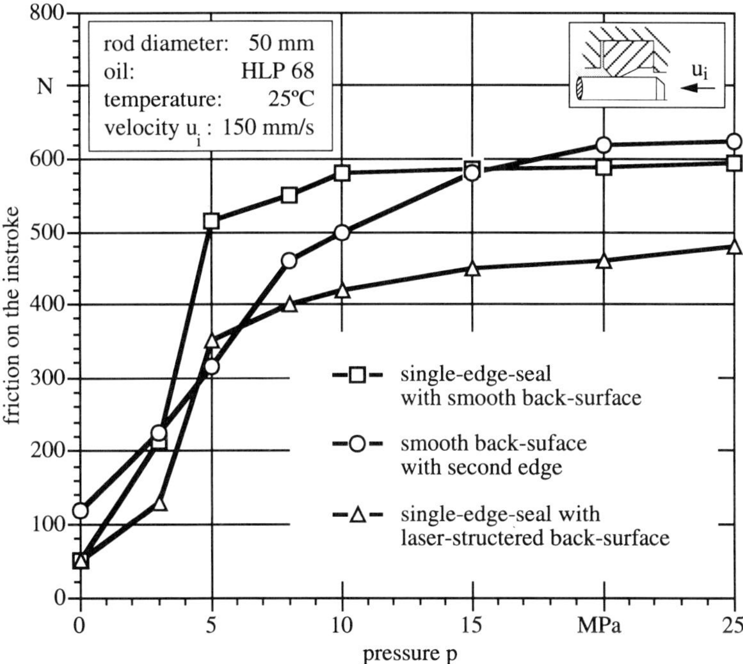

Figure 6. Friction on the instroke of U-seals

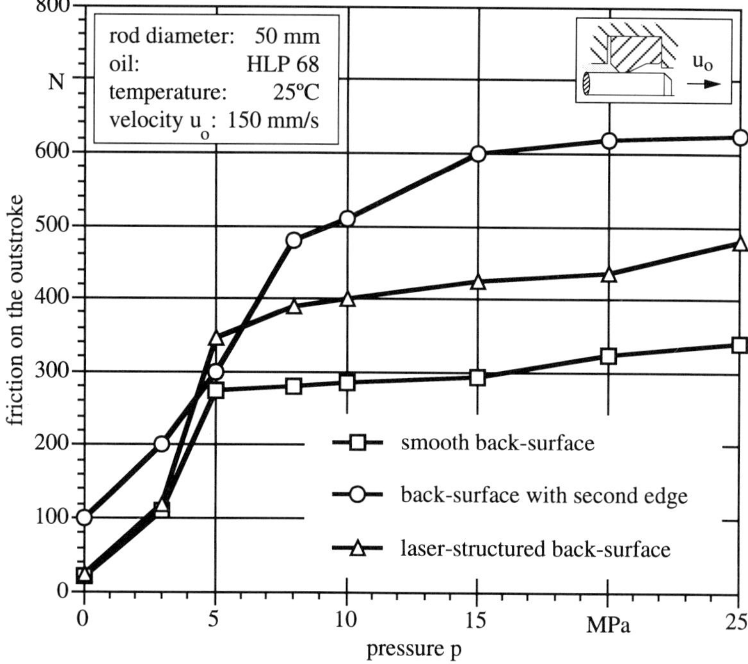

Figure 7. Friction on the outstroke

Design characteristics of optimum elastomer rod seals

According to the authors' results polyurethane rod seals with conventional overall dimensions, having an asymmetrically flanked seal edge close to the fluid side of the seal ring are made to optimum sealing capacity and low friction by the following measures:

* The seal surface is smooth in the vicinity of the seal edge only.

* Numerous asperities are placed on the back-surface of the seal.

* Contact of asperites is made favourable to elasto-hydrodynamic lubrication.

* The height and the distance of the asperities are made such that

    ** when the seal is installed at zero pressure difference a part only of the asperities do contact the rod,

    ** the asperities must not impair the maximum pressure gradient at the fluid side of the seal edge,

** up to the highest pressure and temperature the areas between the asperities must not contact the rod, particularly not at the heel of the seal.

Further theoretical and experimental work on seal optimization along these lines is under way.

Conclusions

The sealing capacity as well as the frictional behaviour of conventional polyurethane rod seals is improved by providing the back-surface with particular patterns of asperities. The negatives of the asperities can be eroded or laser-etched into existing rubber molds. Besides of favouring inward-pumping on normal operating of hydraulic cylinders the asperities also reduce the overall friction loss. The improvement of the seals has been demonstrated experimentally. The variables of the ongoing optimizing process are the shape, the height and the distribution of the asperities as well as the hardness of the polyurethane material.

Acknowledgements

The investigations were supported by *Deutsche Forschungsgemeinschaft (DFG)*, Bonn-Bad Godesberg, Germany.

References

/1/ Blok, H.: Inverse Problems in Hydrodynamic Lubrication and Design Directives for Lubricated Flexible Surfaces. Proc. Symp. on Lubrication and Wear, Houston, Texas, 1963

/2/ Müller, H.K.: Abdichtung bewegter Maschinenteile. Waiblingen, 1990

/3/ DE 32 25 906 (Germ. Pat.), 1984

/4/ Prokop, H.-J.: Zum Abdicht- und Reibungsverhalten von Hydraulik-Stangendichtungen aus Polytetrafluoräthylen, Dissertation, Universität Stuttgart, 1989

/5/ Kanters, A.F.C.; Visscher, M.: Lubrication of Reciprocating Seals; Experiments on the Influence of Surface Roughness on Friction and Leakage. Proc. 15th Leeds-Lyon Symp. on Tibology, Leeds, 1988

/6/ Tao, J.: Untersuchung der pyhsikalischen Vorgänge im Dichtspalt und des Reibverhaltens von Hydraulik-Stangendichtungen, Dissertation, TH Aachen, 1991

/7/ DE 43 00 889.5 (Germ. Pat. Appl.), 1993

14th International Conference on Fluid Sealing, Firenze, Italy,
6-8 April 1994. Organised by BHR Group Limited, Cranfield,
Bedford, MK43 0AJ, UK; Tel: 0234 750422

Effects of Surface Finish on Reciprocating Seal Performance.

R K Flitney and B S Nau
BHR Group Limited, Cranfield, Bedford, U.K.

SUMMARY   *Details are given of an experimental investigation into the significance of a variety of
surface texture paramters in relation to the performance of reciprocating polyurethane lip
seals on rods and bores finished by a range of different processes. Surface parameters
examined include Ra, Rtm. Rpm, Rku, and Rsk. Seal performance measurements included
leakage, friction and wear. In addition , the inter-dependence of the surface parameters for
surfaces finished by the processes studied, is discussed.*

1   INTRODUCTION

In recent years there has been increasing interest in the surface roughness and, more
generally, in counterface texture requirements for dynamic elastomer and plastic seals.
Considerable work has been reported for rotary shaft lip seals, however these contact on a
limited area of shaft and operate at relatively high speed and low pressure.   By contrast,
reciprocating seals have a considerable wiped area of counterface, either piston rod or
cylinder wall, and compared with rotary shaft seals operate at low speed and high
pressure.   Resaerch on lubrication of reciprocating seals (e.g Austin et al. 1977 ; Field et
al. 1973, and 1975) shows that the oil film generated between seal and counterface can be
thinner than surface texture irregularities.   There has been little systematic investigation
into the behav- iour of reciprocating seals on surfaces having different texture
characteristics, we know of only one reference where quantified surface texture has been
related to reciprocating seal performance (Holmberg 1974).   Several factors have
stimulated current interest in surface roughness and texture:

(a) Advances in micro-computers have reduced costs, and increased availability and
capability of stylus instruments to measure surface texture, hence quantitative surface-
texture specification for manufacturing processes is now feasible.

(b) User demand for more consistent and predictable seal performance necessitates that
the seal and equipment manufacturer understand the effects of different surface textures,
in order to provide the required standard of performance over an extended service life.

(c) Due to the large investment in manufacturing equipment required for producing
finished piston rods and cylinder bores, it is important for equipment manufacturers to

be able to identify the most cost effective production processes, balancing seal performance against manufacturing costs.

(d) Surface measurement techniques can be very inconsistent, for instance very different values were obtained when measurements were repeated in different establishments (Flitney 1986).

To address these matters a consortium was set up to undertake a research programme. The consortium comprised: BHR Group, a number of seal and hydraulic equipment manufacturers, a major supplier of surface texture measuring equipment, and the Department of Trade and Industry. The work reported here was mainly carried out at BHR Group, some support and verification testing was carried out by one of the sponsors using a similar test facility.

2   TEST METHODS

2.1 Test Rigs

Two similar reciprocating seal test rigs were used at BHR Group for the majority of the test programme. The test rod reciprocates horizontally through a pressure vessel which has a stationary rod-seal at each end, the system is pressure balanced to permit measurement of seal friction. Alternative housings were fitted for rodseal and cylinder seal ('piston seal') tests. An hydraulic drive provides constant speed through the stroke, the test fluid circuit is independent of the drive circuit.

Figure 1   Double-ended, reciprocating-seal test rig, seal housing at centre left.

For rod seal tests, the housing incorporated bearings to locate the rod with respect to the housing. As interaction between seal and bearing was a concern, the spacing between seal and bearing in relation to rod stroke was chosen so that seal and bearing each slid on a separate area of rod. In a few tests, bearing and seal shared a common area of rod. Figure 1 is a photograph of the rig and Fig. 2 shows cross–sections of the pressure vessel assembly for rod and cylinder configurations.

For cylinder seal tests, two seals were mounted on the piston and test oil circulated between the seals.

The sealed pressure is cycled for alternate stroke directions, seal A has high pressure on the out-stroke and low pressure on the return; seal B is the reverse, low pressure on the out-stroke and high pressure on the return (Figure 2). It was found that seal B is more sensitive to surface texture than seal A, therefore only results for seal B will be included here.

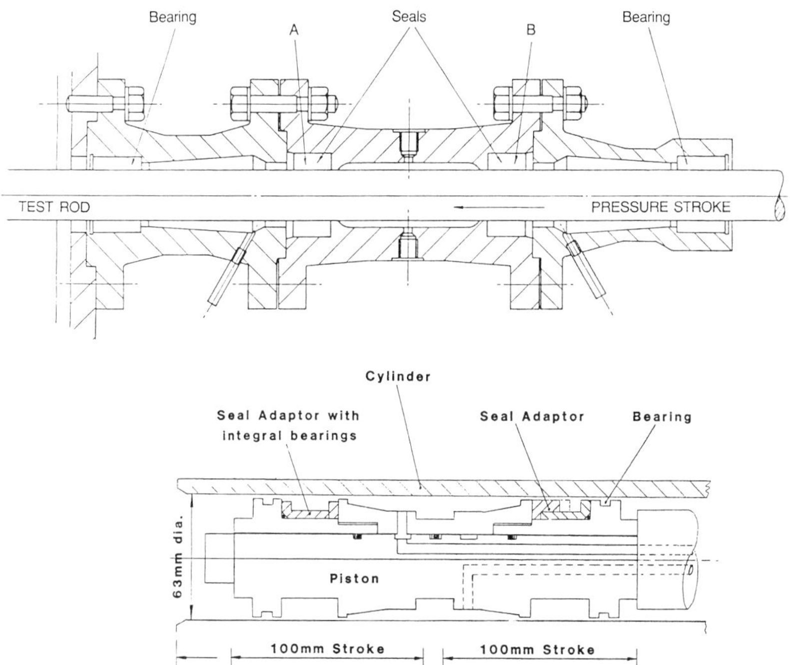

Figure 2   Test housing for rod seals (above) and cylinder seal (below)

2.2 Measurements

Leakage : was collected in measuring cylinders, separately for each seal, wipers were not used. Accuracy was ±0.2 ml.

Friction : a load cell was fitted in line with the drive shaft and its output fed to a chart recorder. Both static (breakout) and dynamic friction were recorded.

Seal -wear : was determined by measuring the seal cross-section optically (±0.02 mm).
Surface texture : measurements were made at Rank Taylor Hobson using "Talysurf"®
equipment having full parameter-readout capability. Profiles were measured axially,
at 90° circumferential intervals,near the end of the rod and at each seal and bearing
wiping zone, before and after test. The following were recorded:

Average Roughness, R_a : is arithmetic mean departure of profile from a mean line.

Kurtosis, R_{ku} : measures the extent to which the profile distribution curve is
broader or narrower than a Normal distribution.

Skew, R_{sk}: measures the asymmetry of the profile : e.g. a sinusoidal profile has
value zero; if peaks are lapped off then skew becomes negative.

Mean Peak Height, R_{pm}: is the mean of Rp for a set of sample lengths; Rp is the
maximum height of the profile above the mean, in a sample length.

Mean Peak-to-Valley Height, R_{tm}: is the mean of all peak-to-valley heights (c.f R_z
is the mean of the five largest).

Slope Parameter, Δq: is the r.m.s slope of the profile.

The bearing-length ratio, t_p, is also of interest but is not a single-valued parameter, its
value depends on the depth at which the bearing length is measured. For surfaces
produced by very different processes it is likely that the relevant depth would differ, for
this reason it was considered that t_p was outside the scope of the study.

Surface texture was monitored during tests using Rank Taylor Hobson "Surtronic"®
equipment, R_a and R_{tm} were measured on seal and bearing tracks, without removing the
rod.

3   SEALS AND COUNTERFACES

3.1 Seals
Four types of polyurethane U-seals (Fig.3) were tested:
type P and type W were a cylinder and a rod seal from
a UK manufacturer, the design was similar except for
orientation; type S and type U were equivalent rod seals
from two other manufacturers. The results for all four
types have been grouped together for the analysis.

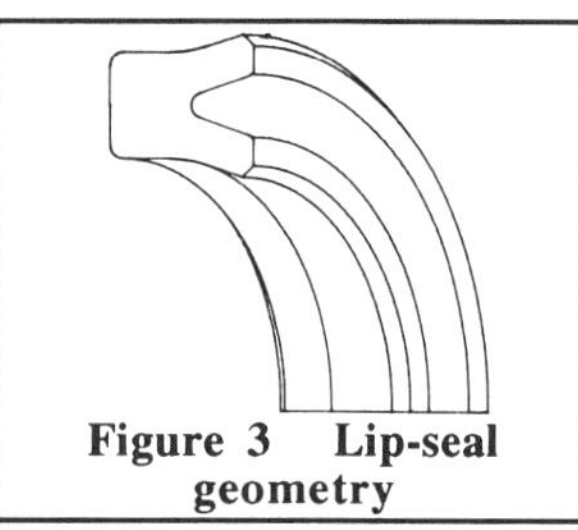

**Figure 3   Lip-seal
geometry**

3.2 Counterface surface treatments
Table 1 lists surface treatments used in the test programme, with identification codes. In
addition to rods from commercial suppliers, which covered the range 0.05 to 0.5 μm R_a,
some rods were specially treated: a specially smooth rod was produced by polishing with
a Meehanite bearing to give an R_a of approximately 0.03 μm; peened rods were produced
with an R_a of 1.0 to 1.4 μm.   The peened surfaces "as received" had a considerably
higher roughness than normal for reciprocating seal applications, about 1.4 μm R_a, and
were used both "as received" and after polishing with 1000 grit paper to reduce R_a.

Difficulty was experienced in the procurement of rods and cylinders to specific surface
finishes, R_a often deviated widely from the nominal. Eventually it was necessary to visit
suppliers to measure and select from material in stock. At one supplier, two tonnes of
chromed and ground rods were all 0.1 μm R_a or less. At another, there were two bundles
of rods, three tonnes in all, in one bundle R_a varied from 0.1 to 0.54 μm whilst in the
other all were 0.12 μm or less. Similar difficulties resulted in only four different cylinder
bore finishes being tested. At the time, contacts in the industry were under the impression
that a drawn-over-mandrel bore finish (DOM - 'smooth bore' or 'mirror bore') was
nominally 0.8 μm R_a, however all samples proved to be 0.2 μm or less. Skived-and-

burnished bores presented similar problems. Honed bores to 0.1 and 0.4 µm R_a were readily obtained.

3.3 Test parameter combinations

The test programme was designed to provide the maximum of performance information with a limited number of test-runs, a summary of the combinations of test parameter combinations is presented in Table 2.

3.4 Test fluid

The test fluid throughout was Mobil DTE, a 32 mm²/s mineral oil of ISO type HL.

Table 1a   Rod finish details.

Ref	Treatment	Ra µm
A	Ground + polished + chrome-plated	0.1
B	Ground + polished + chrome-plated	0.25
C	Ground + chrome-plated	0.4
D	Induction-hardened + ground + chrome-plated	0.1
E	Peened	1.4
F	Ground + polished + chrome-plated	0.4
G	Ground	0.25
H	Burnished	n/a
I	as D, used	0.08 - 0.02
J	as C, used	0.31 - 0.08
K	as E + polished (1000 grade wet & dry paper)	1.0
L	as A + lapped (Meehanite bearing + alumina powder)	0.03

Table 1b   Bore finish details.

A'	Drawn over mandrel ('DOM')	0.2
B'	Honed	0.1
C'	Honed	0.35
D'	Skived and burnished	0.07

4   SURFACE TEXTURE OF RODS AND CYLINDERS

4.1 Classification and definition

Surface texture parameters can be divided into three main groups:–

Amplitude parameters
Spatial parameters
Hybrid parameters

Amplitude parameters include the widely used averaging parameters:– R_a ('cla' or 'AA') and R_q ('rms') together with numerous peak-height measurements of which R_t and R_z (R_{tm}) are used widely in some countries. Other amplitude parameters are two which measure the shape of the profile skew (R_{sk}) and kurtosis (R_{ku}). Spatial parameters provide a measure of either the number of peaks or their spacing. Numerous hybrid parameters are being progressively introduced as statistical analysis of profiles progresses, two well-established hybrid parameters are slope (Δq) and bearing ratio (t_p). The mathematical basis of various texture parameters is presented in Appendix 1.

4.2 Finishing processes

The following manufacturing methods were included:

a) rods

    grinding and polishing (most rods)
    peening,
    additional polishing (worn rods)
    abrasive treatment to increase R_a

b) cylinders

    drawn-over-mandrel ('DOM'),
    honing (two types)
    skived and burnished.

Table 2 gives more detail and reference codes for the treatments, and Figure 4 shows examples of measured surface profiles, shown on the same scale to aid comparison.

Table 2 Test parameter combinations

Seal	-	S ———————		U ———————				W ———————				
Rod	-	A	A	C	C	F	F	D	D	C	C	D	D	F
Pressure bar		160 ————————————————————
Speed m/s		0.4 ————————————————————

Seal	-	W ————————————————————											
Rod	-	F	F	G	G	E	J	L	A	A	B	B	A	A
Pressure bar		160 ———————————————— 160 —————————
Speed m/s		0.4 ———————————————— 0.04 —————————

Seal	-	W ————————————									-	-	-
Rod	-	B	B	C	D	D	C	D	D	C	C	-	-	-
Pressure bar		315 —— 160 ————————————————									-	-	-
Speed m/s		0.04 —— 0.4 ————————— 0.04 —————————									-	-	-

5 INTERDEPENDENCE OF PARAMETERS

Not all surface texture parameters give unique information, depending on surface finishing process various parameters may be closely correlated. The interdependence of surface tex- ture parameters for the above finishes will now be considered. For this purpose it is appropriate to take R_a as the reference for comparison since this is the parameter conventionally measured. Of particular interest are those parameters which are independent of the value of R_a since these will not be controlled by specifying R_a alone. If, in addition, such parameters also affect seal performance then this indicates that the parameter <u>should</u> be controlled.

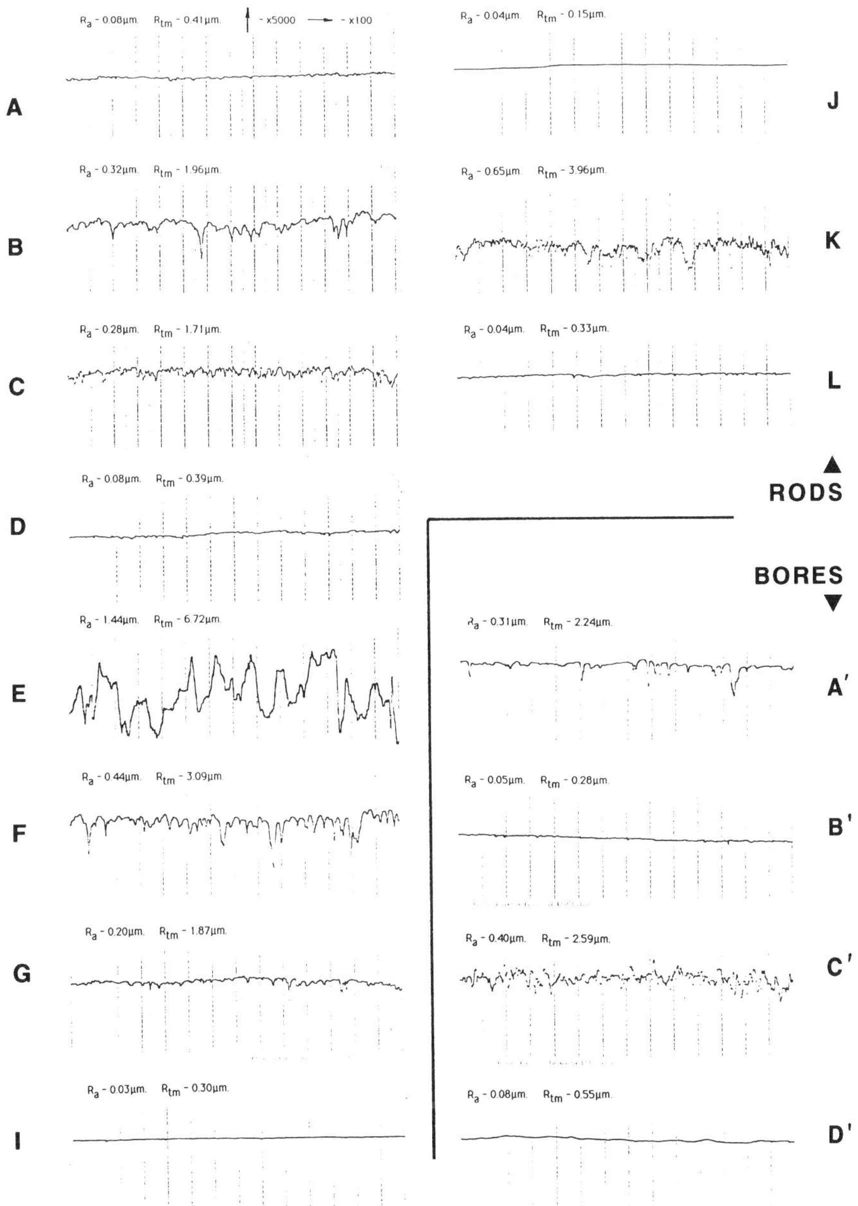

Figure 4   Examples of surface profiles of rods and bores.

512

5.1 Ra and Rtm

In Figure 5a, Ra and Rtm, commonly used parameters, are plotted against each other for all surfaces. The data lie close to a straight line, only DOM cylinders deviate slightly.

5.2 Ra and Rpm

Rpm (analogous to Rtm) is a calculated from the maximum height of peaks above the mean line and might be expected to be a good measure of degree of polishing (reduced peak heights). However, Figure 5b shows that in most cases Rpm correlates very closely with Ra. The largest deviation is for ground-and-polished rods ('FR') of high Ra. The trace for this (Figure 4) has a distinctly flat-topped texture appearance, which might be expected to be indicated bythe value of Rpm , however Rpm is not greatly affected and it would therefore be difficult to make use of this parameter as a measure of polish.

5.3 Ra and Rsk

Figure 5c is a plot of skew (Rsk) against Ra. The lack of correlation indicates that these parameters are indeed independent for the present suite of finishes. Notice that zero skew occurs for 0.1 and for 1.4 μm Ra, although the highest skews mostly occur at low Ra. Ground rod skew ranges from 0 to −1.5, overlapping ground-and-polished rods of 0.5 μm Ra (rods F) which cluster around −1.5. In general, the rod skew tends to cluster within a finish category, the same is true for honed cylinders. The largest skews are for cylinders drawn-over-mandrel or skived-and-burnished (A′ and D′); the skew values were extremely variable, DOM varied from −1 to −5.5, and skived-and-burnished from near zero to −5.0. This wide variation indicates that the test cylinders did not provide a consistent surface for comparative seal performance evaluation, even though Ra and Rpm were well controlled. This may in part be due to the relatively low values of Ra.

5.4 Ra and Δq

Figure 5d shows the slope parameter Δq compared with Ra, again there is a strong correlation. An increase in surface roughness leads to steeper slope-angles. However, it can be seen that there is much less dependence on Ra than for Rpm or Rtm. Of particular interest here is that two rod categories can be distinguished:
  i) Peened rods which had a high Ra, types ER and KR, did not have a high slope.
  ii) Type C rods, which were abraded to increase Ra, had a particularly steep angle,
       significantly higher than other rods in the same Ra range.

5.5 Other literature

Comparison of the above with an authoritative discussion by Nowicki (1985) indicates that the present results are consistent with that author's findings on the correlation between ampl- itude parameters, skew and slope.

6   SEAL PERFORMANCE EFFECTS

Seal performance parameters have been analysed in relation to the surface texture parameters listed above.

Seal performance parameters considered were seal wear, total leakage for the duration of the test run, friction at the start of the test, friction change during the test. In the present paper only the more interesting cases are presented in Figs.6 and 7.

The results for polyurethane seals, three types of rod seal and one type of cylinder seal, show that Ra, or a related peak–height parameter, is the surface texture parameter having most effect on seal performance. However, current recommendations for the value of this,

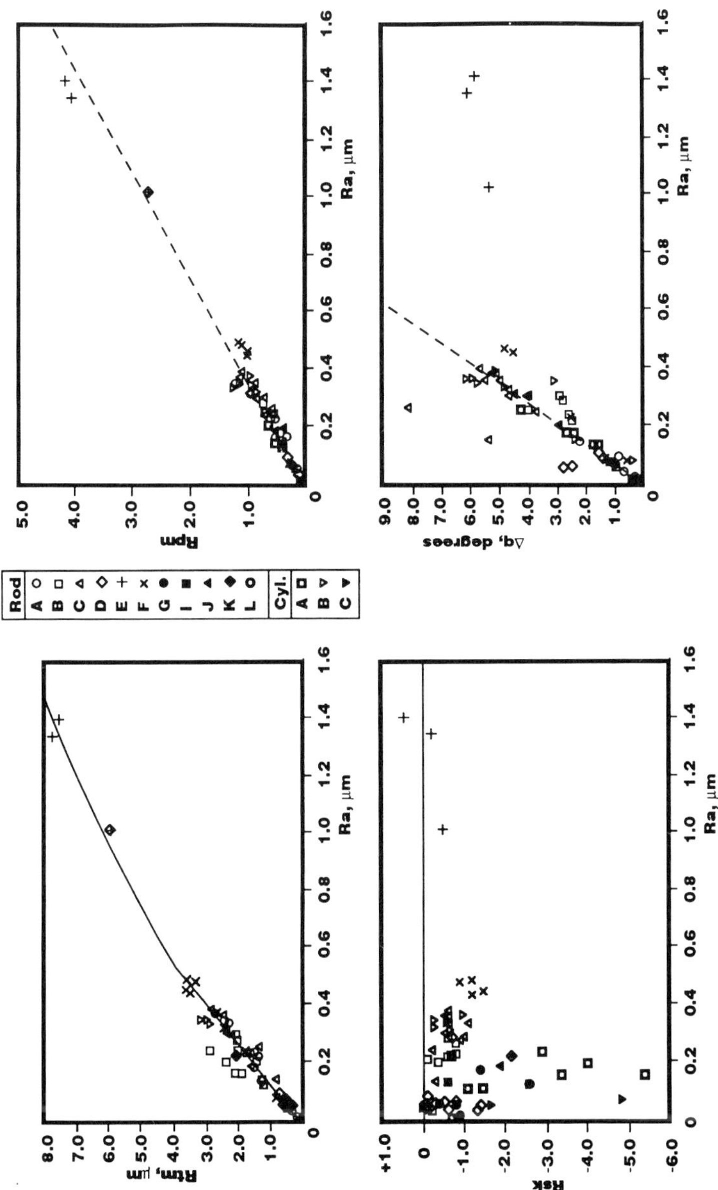

Figure 5   Relationships between surface parameters.

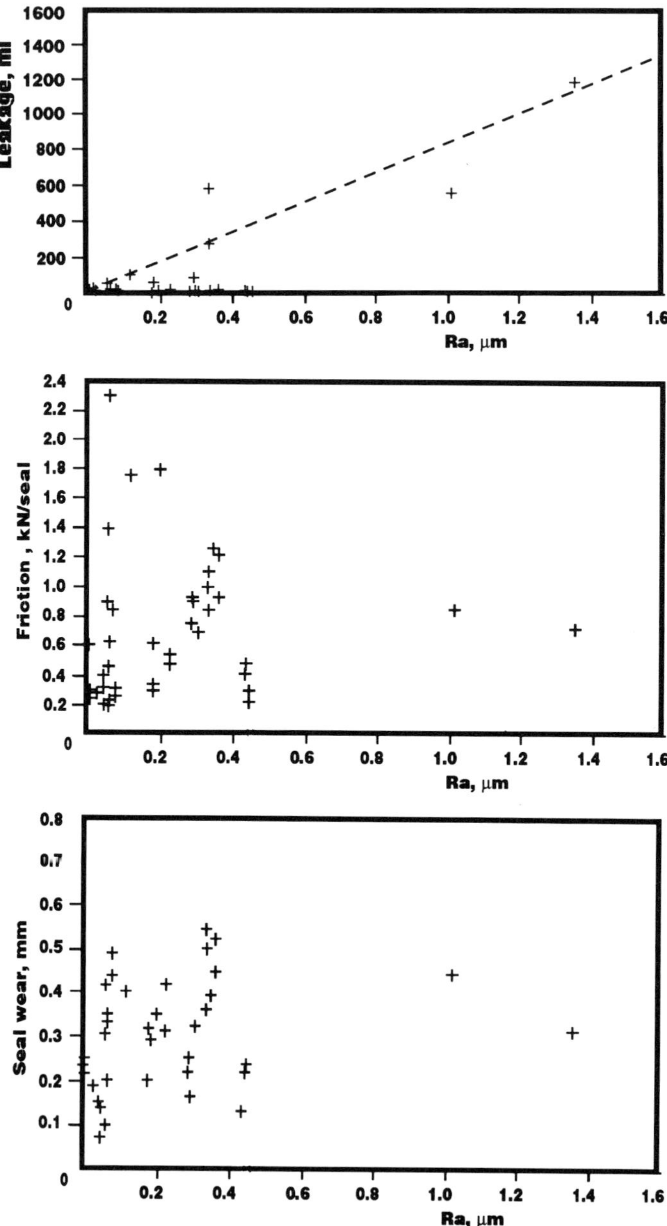

Figure 6   Effects of Ra on seal performance.

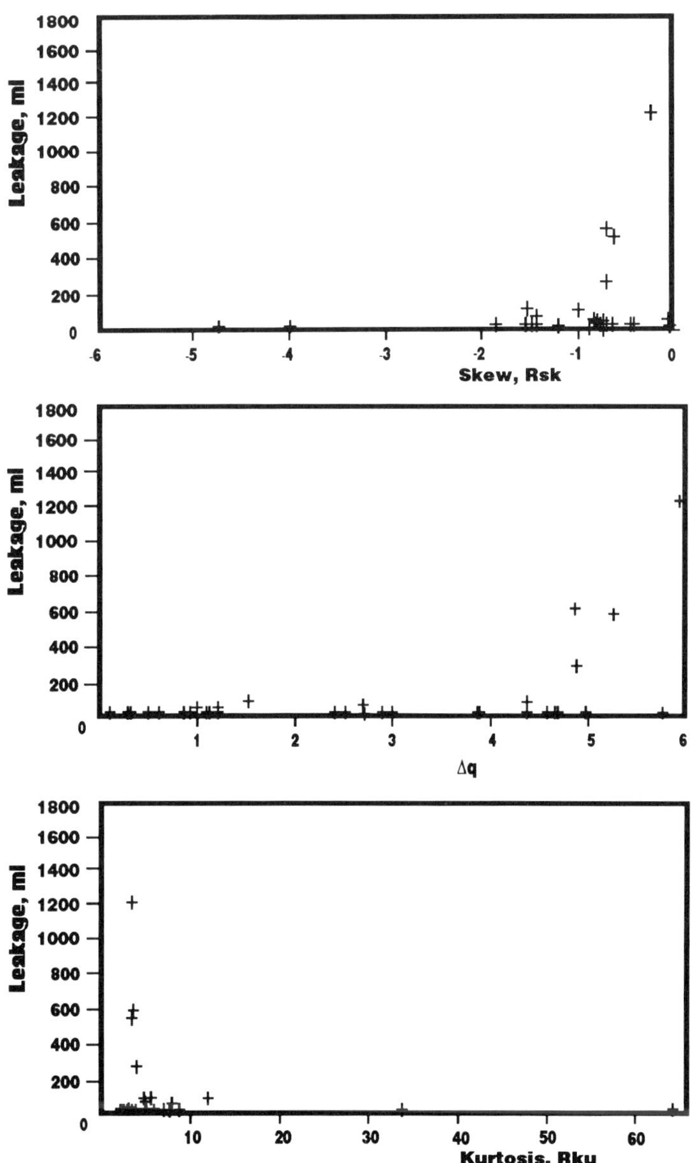

Figure 7   Effects of skew, slope and kurtosis on leakage.

typically 0.1 to 0.4 μm Ra, are too wide to give consistent seal performance without further qualification.

Seal wear and friction exhibit considerable scatter, Figures 6b,c. High skew surfaces did not appear to affect seal wear, leakage or friction.

The leakage results, Figure 6a, show leakage increasing with Ra. This is not totally determined by Ra as there was zero leakage with rod type F, a high Ra but polished. Polishing affects second-order surface texture characteristics. Leakage for rod K, peened then polished, was similar to that of the peened-only rod, type C. Seal wear figures show that the rod type C caused the highest seal wear. Further investigation of wear results shows that high wear also occurred at low values of kurtosis and skew, indicat- ing that an Ra towards the top of the current recommendations but without any polishing i.e. a fairly angular profile, may be detrimental to seal performance.

The smoothest rods, 0.03 μm Ra, polished by the Meehanite bearing, had: seal wear comparable with most other seals tested, zero leakage, amongst the lowest friction recorded and also very low friction change during the test. It appears that although an excessively smooth rod is often considered detrimental to seal performance, with polyurethane seals these rods can provide the most consistent and reliable performance.

Leakage results also show that an increase in slope parameter Δq is associated with increasing leakage, again indicative of angularity adversely affecting performance.

7 CONCLUSIONS

i)   Ra is indeed the dominant factor controlling performance of reciprocating polyurethane lip seals, but is not the only factor. Also, Ra is a reliable indicator only within a relatively narrow range of values, certainly less than found in current specifications. Outside this range other parameters become important and tighter definition of the surface is required.

ii)   Commercial rods typically have Ra in the range 0.05 to 0.56 μm, i.e. wider than the normally quoted range (0.1 to 0.4 μm Ra).

iii)   The required Ra range for seals, as commonly quoted by seal suppliers is too wide. It allows considerable variation in seal performance. Combined with the even wider range of surface finishes on material procured from stockists, an even wider variation in performance reoults.

iv)   Textures above 0.3μm Ra can cause an increase in leakage wit polyurethane seals. There is a more definite tendency for increasing leakage with increasing Ra above this value. The seals could in fact be used satisfactorily to quite high Ra, for example 1.0 μm Ra on the peened and polished rod. However, less satisfactory performance was achieved on other rods in the range 0.3 –0.5 μm Ra. Further test work would be necessary with varying texture at a single high Ra to isolate the relevant parameters.

v)   The parameters required for successful use of these seals at high Ra are more critical than for nitrile seals. It appears prudent to restrict the maximum Ra for polyurethane seals to about 0.25 to 0.3 μm Ra.

vi)   The polyurethane seals appeared to work consistently on the very smooth surface, 0.03 μm Ra.

vii) No differences in performance were noted between the tests with chromed and non–chromed surfaces. To achieve definitive correlation it would be necessary to achieve surfaces with very similar R_a and surface textures in chromed and non–chromed conditions and this is technically extremely difficult.

viii) Within the constraints of current knowledge, to achieve consistent performance, control over the manufacture of the seal counterface is required. Definition of the surface can be achieved by ensuring that three characteristiscs are controlled:

(a) *Surface roughness* : R_a in the range 0.1-0.2 μm.
(Or another roughness parameter such as R_{tm} , with appropriate range).

(b) *Asperity slope* : Δq less than 3.0.

(c) *Skew* : R_{sk} in the range -2.0 to -4.0.
(Or another parameter dependent on bearing-area ; a function of peak heights, such as R_{pm}/R_{tm} , is likely to be less sensitive than skew.)

ACKNOWLEDGEMENTS

The authors wish to acknowledge the support of the Department of Trade and Industry, and the industrial sponsors :–

Automotive Products	Dowty Seals	Freudenberg Simrit
Hallite Seals International	Parker Hannifin	Rank Taylor Hobson
Shamban Europa	Vickers Systems	James Walker

REFERENCES

Austin, R.M., Flitney, R.K, and Nau, B.S., 1977, Research into factors affecting reciprocating rubber seal performance. *BHRA Report* RR 1449.

Field, G.J. and Nau, B.S., 1973, The lubrication of rectangular rubber seals under conditions of reciprocating motion: An instrumented seal study. *BHRA Report* RR 1200.

Field,G.J., Flitney, R.K, and Nau, B.S., 1975, The lubrication of reciprocating rubber 'U' – seals.
BHRA Report RR 1315.

Flitney, R.K, 1986, Reciprocating seal test procedure:Report of round robin tests.
BHRA Report DCR 2666.

Flitney, R.K. and Nau, B.S., 1989, The effect of surface texture on reciprocating seal performance: Literature Review. *BHRA Report* CR 3068.

Holmberg, B., 1974 The influence of surface finish on wear of hydraulic cylinder seals.
Proc. 5th Int. Fluid Sealing Conf., BHRA, Cranfield, April 1974.

Nowicki, B., 1985, Multiparameter representation of surface roughness. *Wear* , **102** , 161-176

THE EFFECTS OF MULTISTAGE CONTACT PRESSURE DISTRIBUTION ON SEALING AND FRICTIONAL CHARACTERISTICS IN RECIPROCATING SEALS

Y. Kanzaki[*], Y. Kawahara[*] and M. Kaneta[**]

[*]Engineering Division, NOK Corporation,
4-3-1, Tsujidoshinmachi, Fujisawa, Kanagawa, Japan

[**]Department of Mechanical Engineering, Kyushu Institute of Technology,
1-1, Sensuicho, Tobata, Kitakyushu, Japan

SUMMARY

Fundamental sealing and frictional characteristics of reciprocating seals with multistage contact pressure distribution are examined using a newly developed testing device composed of a stationary rod with multi-grooves, in which lip seals are installed, and a transparent glass cylinder. The results obtained are explained on the basis of the inverse theory of hydrodynamic lubrication. It is found that although the leakage for multistage seals is mainly controlled by a seal which has the most striking sealing ability and is mounted more on the oil side, the air side seal also affects significantly the amount of leakage. Furthermore, details of the interseal pressure which is built-up between seals are discussed.

NOMENCLATURE

B = total contact width
D = inner diameter of cylinder
f = coefficient of friction defined by $(f_M+f_P)/2$
f_M = coefficient of friction at central portion of motoring stroke
f_P = coefficient of friction at central portion of pumping stroke
F = friction force
G = duty parameter defined by $\pi\eta Sv/(Pr/\pi D)$
P = contact pressure
Pa = average contact pressure
P_{is} = interseal pressure
Pr = total contact load
q = fluid flow rate
Q = dimensionless flow rate defined by $\eta q/PrB$
S = length of stroke
V_{isp} = increasing rate of interseal pressure
x = coordinate in the direction of motion
η = viscosity
v = number of cycles of reciprocation

INTRODUCTION

The role of reciprocating seals is to prevent leakage with a good reliability and a long life. The friction force must be low so that a good working performance as well as reliability and long life is given to the machinery. The sealing and frictional characteristics of reciprocating rubber seals may be explained by the inverse theory of hydrodynamic lubrication [1-3], where the distribution of film pressure is given and the hydrodynamically corresponding profile of the film is required. The film pressure distributions for reciprocating seals, such as U-ring seals under high sealing pressures, are not necessarily simple. Furthermore, a tandem arrangement of seals is often used in the reciprocating machinery, where one seal would not be considered adequate from a safety point of view or as a precaution against foreign dust. It is well known that in such seals a relatively high interseal pressure is built-up between seals [4-6], and it may cause damage to the seals.

The purpose of this investigation is to make clear the sealing and frictional characteristics of reciprocating rubber seals with multistage contact pressure distribution. In order to directly observe the sealing surface, a new testing device was constructed. It is composed of a stationary rod with multi-grooves, in which lip seals with an asymmetric configuration are installed, and of a transparent glass cylinder. The results obtained are explained physically by applying the inverse theory.

EXPERIMENTAL PROCEDURE

Testing Machine and Specimens

Figure 1 shows a schematic diagram of a newly developed testing device for reciprocating seals. It consists of a stationary rod with multi-grooves, in which lip seals are installed, and a reciprocating cylinder of 110 mm inner-diameter. The cylinder was driven sinusoidally by means of an oil hydraulic pump.

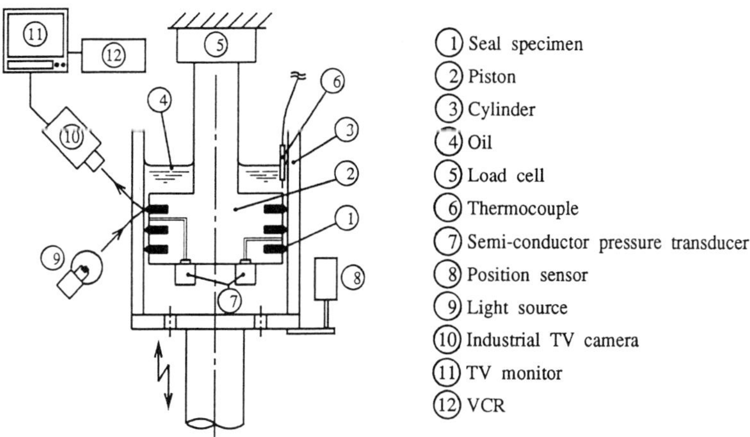

Fig.1   Schematic diagram of testing device

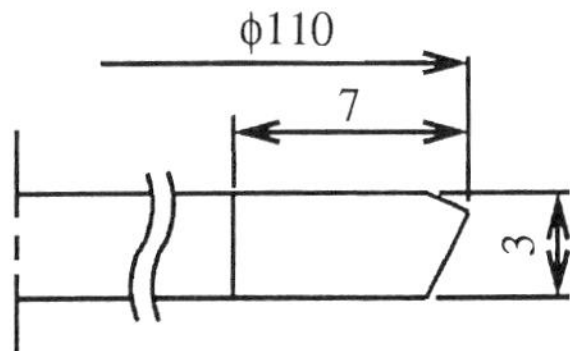

Fig.2 The shape and size
of a lip seal

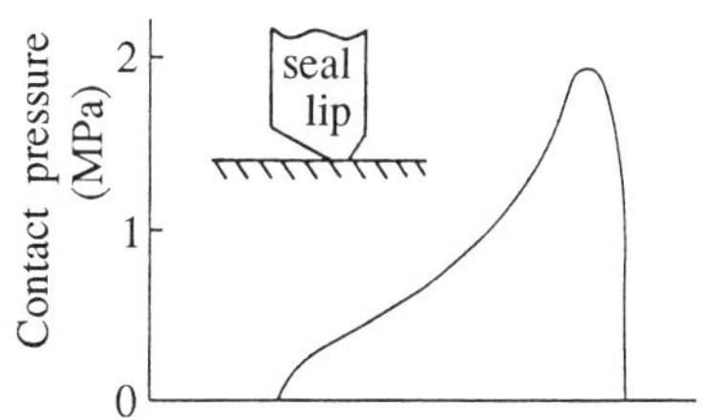

Fig.3 Contact pressure distribution
calculated by the FEM

Table 1 Mechanical properties of
nitrile rubber material

Hardness	: 71	(IRHD)
Young's modulus	: 4.5	(MPa)
Tensile strength	: 20.8	(MPa)
Elongation	: 340	(%)

Table 2 Details of a single
seal specimen

Rubber material	: NBR	
Interference	: 0.3	(mm)
Total contact load	: 110~140(N)	
Contact width	: 0.2~0.3	(mm)

Figure 2 shows the shape and size of a lip seal used. The sealing action takes place on the external diameter of the lip seal. The shape of the lip is asymmetric so that an asymmetric contact pressure distribution can be obtained. Figure 3 shows the contact pressure distribution calculated by the finite-element method. The material was nitrile rubber (NBR). Its mechanical properties are listed in Table 1. The interference in the radial direction was 0.3 mm and the contact width was in the range of 0.2 to 0.3 mm as shown in Table 2.

Each part between the grooves, in which lip seals were installed, has a hole of 1 mm diameter drilled through it in the radial direction. The holes were connected with two partly tapped holes fabricated in the piston, respectively. A semi-conductor pressure transducer was screwed into each tapped hole in order to measure the interseal pressure. The distance between grooves was 6 mm.

The friction force between the sealing surfaces was measured by means of a load cell attached to the end of the stationary rod. Variations of the friction force and the interseal pressure, as well as the displacement of the cylinder, were monitored by a synchroscope and simultaneously recorded by a computer.

The surface of the cylinder, made of a 0.45% carbon steel, was plated by a hard chromium and its roughness was less than 0.3 μm R_{max}. In order to facilitate observation of the behavior of oil on sealing surfaces during the experiment, a transparent glass cylinder was also used instead of the carbon steel cylinder. The image of the sealing surface was obtained by making the incident light reflect almost perfectly from the sealing surface as seen in Fig.1. The image was recorded with a VCR through an industrial TV camera with a shutter speed of one thousandth of a second.

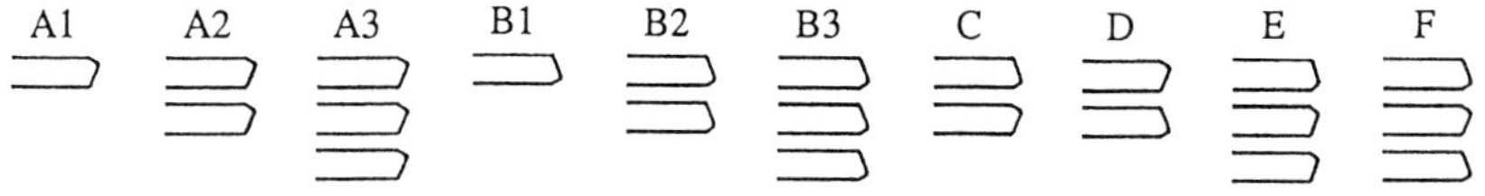

Fig.4 Schematic view of combination types of lip seals

Table 3 Properties of paraffinic mineral oils

Oils	Kinematic viscosity (mm^2/s)		Specific gravity
	313(k)	373(k)	
a	2.0	1.0	0.8750
b	10	2.5	0.8773
c	47	6.9	0.8768
d	320	24	0.8769

Table 4 Test conditions

Length of stroke	: 10~50 (mm)
Number of cycles	: 0.3~3 (Hz)
Oil used	: four kinds of paraffinic mineral oils
Temperature	: room temperature

Test Conditions

In order to investigate the effects of the pressure distribution or seal-surface configuration on the friction and sealing characteristics, ten types of combinations of lip seals as schematized in Fig.4 were chosen on the basis of the inverse theory described later. The upper side of each figure corresponds to the oil side and the lower side corresponds to the air side. In this paper, the oil side seal may be called the first stage seal, and the second and third stage seals follow.

The properties of paraffinic mineral oils used as sealed liquid are listed in Table 3. The viscosity of the oil was calculated from oil temperature measured by a trailing thermocouple at the position shown in Fig.1.

The experiments were carried out at room temperature under sinusoidal reciprocation with a stroke length of 10 to 50 mm, and a number of cycles ranging from 0.3 to 3 Hz as shown in Table 4. The duration of each test was mainly 30 minutes, during which the wave form of the friction force became stable. The amount of oil leaked was determined by wiping the oil stuck to the inner wall of the cylinder thoroughly with a sheet of filter paper. Thus the amount of leakage less than 10^{-13} m^3/s contains a fairly large margin of error.

INVERSE THEORY

In describing the direction of motion of the cylinder, the terms "pumping" and "motoring" strokes are used in this paper, and they are designated as "P" and "M". The pumping stroke means that the cylinder moves towards the air side. The motoring stroke means that the cylinder moves towards the oil side.

According to the inverse theory [1,2], the necessary condition for the existence of stable oil film in reciprocating motion is that the pressure distribution curve has a point of inflexion on the ascending side, where the pressure gradient $|dp/dx|$ takes a maximum value, i.e., $|dp/dx|_{max}$. Even if there is no inflexion point on the stationary contact pressure distribution curve, the inflexion point is likely to be formed automatically in the boosting zone under operation [1], because of the boosting pressure caused by the wedge action of hydrodynamic lubrication. The film thickness at the point of maximum pressure at the centers of both pumping and motoring strokes are given by

$$h_{m\,{}^P_M} = \sqrt{\frac{8(P_r/\pi D)}{9\,|dp/dx|_{max\,{}^P_M}}}\ G^{1/2} \tag{1}$$

where G is the duty parameter defined as

$$G = \pi\eta Sv/(P_r/\pi D) \tag{2}$$

Coefficients of friction at the centre of both on the pumping and motoring strokes are given by

$$f_{{}^P_M} = \sqrt{\frac{9}{8}\left|\frac{dp}{dx}\right|_{max\,{}^P_M}\frac{B}{P_a}}\cdot J_{{}^P_M}\cdot G^{1/2} \tag{3}$$

where J is defined by the following integral:

$$J_{{}^P_M} = \int_0^B \left\{3\left(\frac{h_m}{h}\right)^2_{{}^P_M} - 2\left(\frac{h_m}{h}\right)_{{}^P_M}\right\}dx \tag{4}$$

The net leakage per unit time, q, under sinusoidal reciprocation can be estimated without serious error by using the following equation obtained for the reciprocation with uniform average velocity [3]:

$$Q = \frac{\eta\,q}{P_r B} = \frac{2}{3\,\pi^{3/2}}\left(\sqrt{\frac{P_a/B}{|dp/dx|_{max,P}}} - \sqrt{\frac{P_a/B}{|dp/dx|_{max,M}}}\right)G^{3/2} \tag{5}$$

Consequently, the sealing condition can be expressed as follows:

$$|dp/dx|_{max,P} < |dp/dx|_{max,M}\ :\ \text{leak} \tag{6a}$$

$$|dp/dx|_{max,P} = |dp/dx|_{max,M}\ :\ \text{seal} \tag{6b}$$

$$|dp/dx|_{max,P} > |dp/dx|_{max,M}\ :\ \text{negative leak} \tag{6c}$$

The pressure distribution for type A_1 can be illustrated schematically as shown in Fig.5. Since this satisfies the relationship (6c), the oil in the air side flows into the oil side; the leakage is prevented. On the other hand, in type B_1, the net leakage takes place, since the relationship (6a) is satisfied. As a result, in the case of type C, the oil is accumulated between two lip seals and positive interseal pressure is produced. On the contrary, in the case of type D, interseal pressure decreases or becomes negative.

524

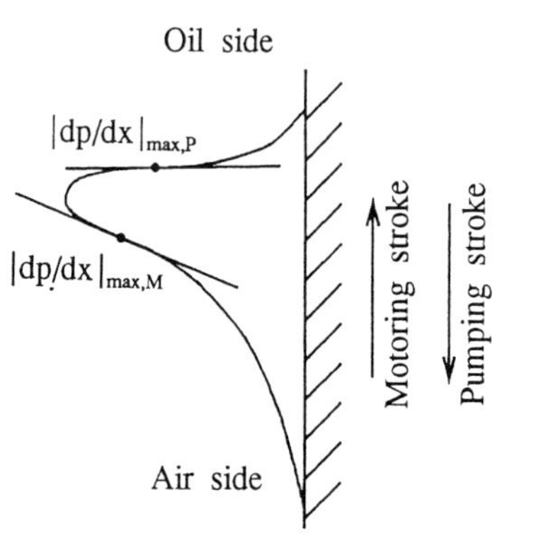

Fig.5 Strokes and pressure distribution in single seal

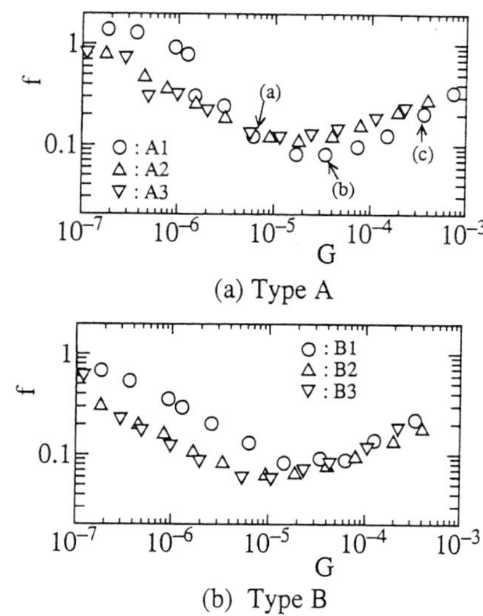

(a) Type A

(b) Type B

Fig.6 The relationship between coefficient of friction f and duty parameter G

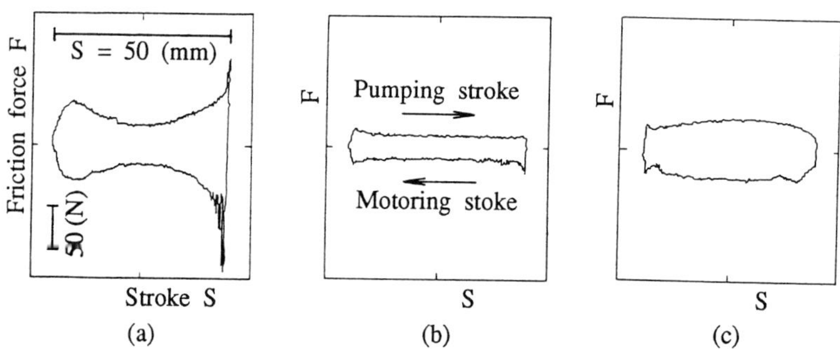

Fig.7 Typical examples of the relationship between friction force and stroke
(Type A)

EXPERIMENTAL RESULTS AND DISCUSSION

Friction Force

Figure 6 plots the coefficient of friction $f=(f_P + f_M)/2$ for types A and B against the duty parameter G. In the region of $G>10^{-5}$, the relation $f \propto G^{1/2}$ corresponding to equation (3) is established irrespective of the seal arrangements. This indicates that the region of $G>10^{-5}$ is dominated by full hydrodynamic lubrication. Variations in friction force with stroke corresponding to the points (a), (b) and (c) in Fig.6(a) are given in Fig.7. Figures 7(a) and 7(c) are typical examples corresponding to the non-hydrodynamic and full hydrodynamic lubrication regimes, respectively.

In type A, the coefficient of friction in the region of $G>10^{-5}$ tends to increase slightly with the number of lip seals installed in the groove. In type B, however, it hardly depends on the number of lip seals. This can be explained on the basis of direct observations of sealing surface. Photographs which were taken at the central portion of the pumping stroke are shown in Fig.8, and, as a typical example, compare type A_2 with type B_2. In type A, the leakage seems to be mostly prevented by the first stage seal or the oil side seal, because there is little oil on the contact surfaces of the second stage seal and between seals as seen in Fig.8. This seems to result in the increase in the coefficient of friction. In type B, however, a large quantity of oil was observed on the surfaces of lip seals and between seals. This suggests that each seal exhibits the same frictional characteristics.

In types C, E and F, a build-up of interseal pressure was observed in the range of $G>10^{-6}$ as described later. When the interseal pressure rose more than 3 MPa, the extrusion of the seals occurred. Hence, the experiments were finished when its value reached about 2 MPa. The coefficient of friction increased with the increase of interseal pressure. This seems to be due to the fact that the contact load Pr increases as the interseal pressure increases, though the authors could not measure the contact load while it was pressurized.

Type D exhibited almost the same coefficient of friction as type A_2. Although the oil was observed on the outside, i.e., air side, of the second stage seal, it was not observed between both lip seals.

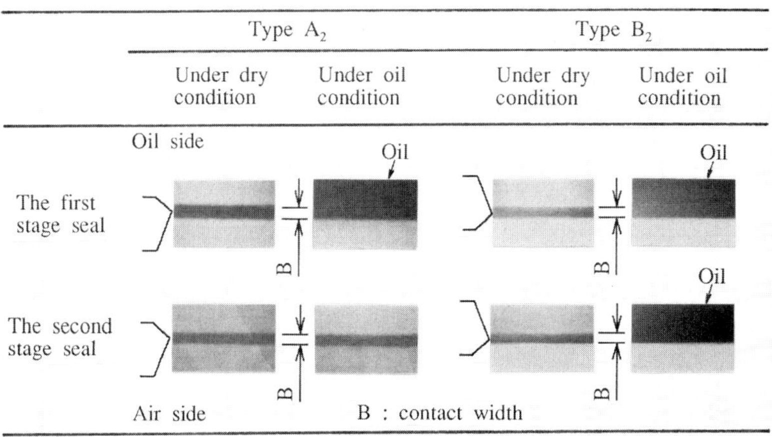

Fig.8 Photographs of seal surface at the central
portion of the pumping stroke (Types A and B)

526

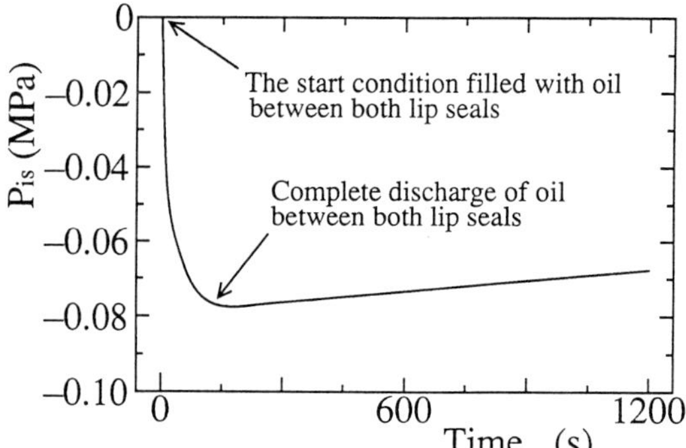

Fig.9 Generation test of negative interseal pressure P_{is}
(Type D)

Interseal Pressure

In types A and B, the occurrence of the interseal pressure was hardly detected. This means that the sealing ability of each seal is almost the same. If there is a difference in the amount of leakage between lip seals, any interseal pressure will be built-up in the tandem arrangement.

In type D, it can be expected by the inverse theory that the interseal pressure becomes negative. Direct observation of the sealing surface also suggests this fact as described in the previous section. However, the actual values of interseal pressure were relatively low and within the range of 0 to -0.01 MPa. In order to make sure of the prediction obtained by the inverse theory, the space between both lip seals was filled with oil in advance. The experimental result is shown in Fig.9, where the interseal pressure is plotted as a function of time. As the oil between both lip seals is discharged, the interseal pressure decreases and takes a minimum value of -0.08 MPa, at which the oil is almost completely discharged, and then increases gradually. This result shows that the prediction by the inverse theory is correct.

The positive interseal pressure P_{is1} produced between the first and second stage seals for types C, E, and F within 30 minutes from the beginning of the test are shown in Fig.10 as a function of the duty parameter G. As has been described, the experiments were stopped when the interseal pressure reached about 2 MPa. The interseal pressure P_{is1} is produced in the range of $G>10^{-6}$, which is less than $G>10^{-5}$ corresponding to the transition point from non-hydrodynamic to hydrodynamic lubrication regimes [7]. These facts suggest that the positive interseal pressure P_{is1} is produced only when sufficient oil is accumulated between seals.

The interseal pressures P_{is1} for types C, E and F are plotted in Fig.11, as a function of time. These results were obtained by changing the oil to be sealed and the frequency of reciprocation for a constant stroke length of S=50 mm. The interseal pressure P_{is1} increases rapidly as time passes, during which it was observed that the oil is accumulated between lip seals. As seen from these figures, the increasing rate of the interseal pressure increases with increasing in the frequency of reciprocation for the same oil or the same viscosity. However, its increasing rate is not necessarily a function of the duty parameter G (for example, compare results marked by the arrows in Fig.11). From equation (5), the net leakage per unit time,

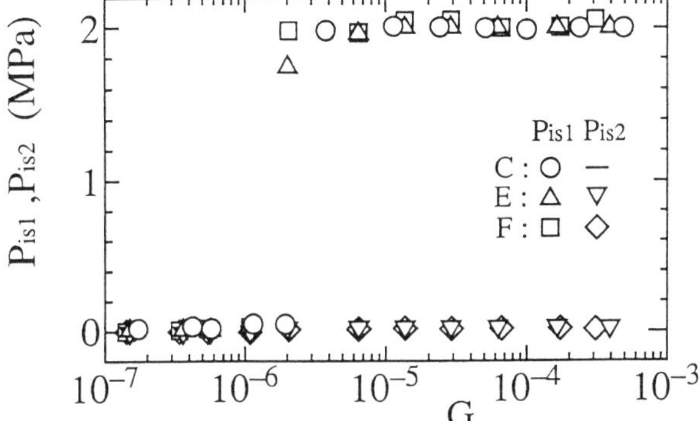

Pis1 : interseal pressure btween the first and second seals
Pis2 : interseal pressure btween the second and third seals

Fig.10 The relationship between interseal pressure
Pis and duty parameter G (Type C,E and F)

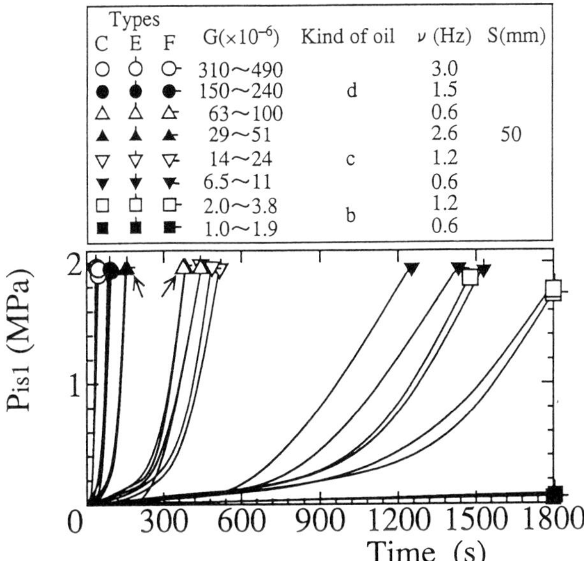

Fig.11 The relationship between the interseal pressure
Pis1 and testing time   (Type C,E and F)

528

q, for the single seal can be expressed as

$$q \propto \eta^{1/2}(Sv)^{3/2} \tag{7}$$

For the sake of convenience, let us define the increasing rate of the interseal pressure, V_{isp}, by dividing the maximum interseal pressure reached under a certain condition by the time required to reach its maximum pressure. Figure 12 shows the relationship between V_{isp}, and $\eta^{1/2}(Sv)^{3/2}$. It is clearly recognized that V_{isp} increases in proportion to the amount of leakage per unit time.

Comparing the results for types C, E and F, it can be seen that the increasing rate of the interseal pressure V_{isp} is not influenced remarkably by the third stage seal. The interseal pressure P_{is2} between the second and third seals was low in magnitude as compared with that between the first and second seals. As the latter pressure P_{is1} increased the former pressure P_{is2} increased and reached about 0.03 MPa.

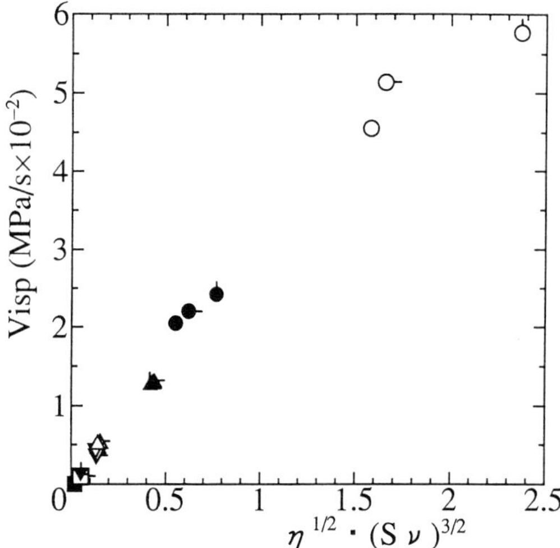

Fig.12 The relationship between increasing rate
of interseal pressure Visp and $\eta^{1/2} \cdot (Sv)^{3/2}$

Leakage

Figure 13 shows the relationship between the dimensionless leakage $Q=\eta q/P_r B$ and the duty parameter G. The results for type F are somewhat scattered. For the other types of seal arrangements, however, the relationship $Q \propto G^{3/2}$, which is induced for the single seal by the inverse theory, is established.

As has been expected from the inverse theory, there is a very large difference in the amount of leakage between types A and B. For type A as well as type B, the amount of leakage is not influenced remarkably by the tandem arrangement. This means that, in type A, oil is sealed by the first stage seal which is installed nearest to the oil. This fact is confirmed by the observation of the sealing surface shown in Fig.8.

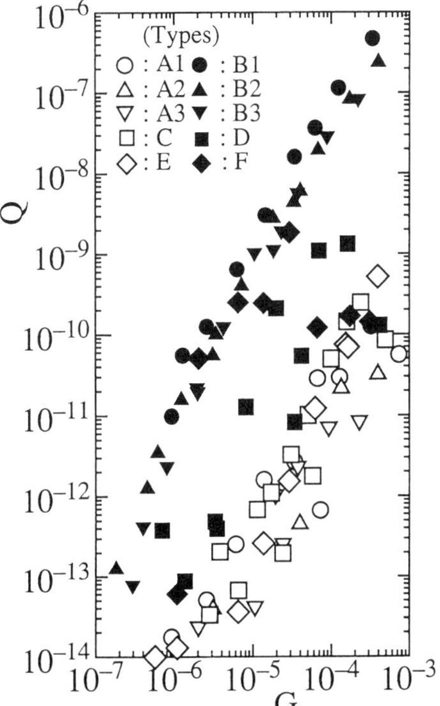

Fig.13 The relationship between dimensionless
leakage Q and duty parameter G (All types)

The amount of leakage obtained is arranged in descending order magnitude as follows:

$$\text{type B} > \text{type F} > \text{type D} > \text{type E} \doteqdot \text{type C} \doteqdot \text{type A} \qquad (8)$$

It seems to be worthwhile to discuss the amount of leakage for types C, D, E and F. Two seals existing on the air side i.e., the second and third stage seals, of type F have the same arrangement of lip seals as type D. The amount of leakage is larger for type F than for type D. From this result, it is considered that the oil film on the pumping stroke is thicker for the second stage seal of type F than for the first stage seal of type D. The pressure distribution for the second stage seal of type F is influenced by the interseal pressure between the first and second stage seals. This consideration may be applied for type C and E. Two seals existing on the oil side, i.e., the first and second stage seals, of types E and F have the same arrangement of lip seals as type C. The point in which types E and F differ from each other is the air side lip seal. In type E, the maximum pressure gradient for the air side seal on the pumping stroke is larger than that on the motoring stroke.

530

On the contrary, the air side seal of type F has the reverse relation. That is, the former, which corresponds to type A_1, prevents the leakage and the latter, which corresponds to type B_1, promotes the leakage. The amount of leakage is larger for type F than for type E. It should also be noted that the amount of leakage of type F is larger than that of type C. That is, although the leaked oil is again drawn back by the action of the air side seal of type C, in the motoring stroke, the air side seal of type F prevents its action. These results show that the leakage of multistage seals is influenced significantly by the air side seal.

It should be furthermore noted that type C has almost the same amount of leakage as type A_2 in which the leakage is controlled by the first stage seal. The amount of leakage for type E is also almost the same as that for type C; the air side seal of type E hardly contributes to the leakage. These results show that the leakage is mainly controlled by a seal which has a remarkable sealing ability and is mounted more on the oil side.

Figure 14 summarizes the results. The appearance of oil observed using the transparent cylinder and the amount of leakage are schematically illustrated. The blacker the circle, the larger the leakage. It can be concluded that the amount of leakage of multistage seals is influenced significantly by the air side seal though the leakage is mainly controlled by a seal which has a remarkable sealing ability and is mounted more on the oil side.

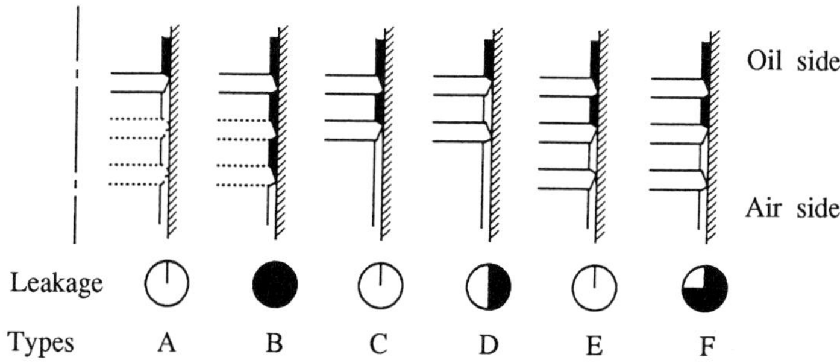

Fig.14 The appearance of oil and the amount of leakage

CONCLUSIONS

Basic friction and sealing characteristics of reciprocating seals with multistage contact pressure distribution have been clarified through the direct observations of sealing surfaces and the inverse theory of hydrodynamic lubrication. The main results obtained are summarized as follows:
(1) The mechanisms of sealing and the occurrence of the interseal pressure can be explained by the inverse theory.
(2) The amount of leakage of multistage seals is influenced significantly by the air side seal though the leakage is mainly controlled by a seal which has most remarkable sealing ability and is mounted more on the oil side.

REFERENCES

1. Blok, H., "Inverse Problems in Hydrodynamic Lubrication and Design Directives for Lubricated Flexible Surfaces", Symp. Lubric. & Wear, Houston, USA (1963).
2. Hirano, F and Kaneta, M., "Theoretical Investigation of Friction and Sealing Characteristics of Flexible Seals for Reciprocating Motion", Proc. 5th Int. Conf. on Fluid Sealing, BHRA, G2 (1971).
3. Hirano, F. and Kaneta, M., "Experimental Investigation of Friction and Sealing Characteristics of Flexible Seals for Reciprocating Motion", Proc. 5th Int. Conf. on Fluid Sealing, BHRA, G3 (1971).
4. Field, G. J. and Nau, B. S., "Interseal Pressure between Reciprocating Rectangular Rubber Seals", Proc. 5th Int. Conf. on Fluid Sealing, BHRA, D2 (1971).
5. Müller, H. K. and Messner, N., "PTFE-Seals for Reciprocating Rods", Proc. 9th Int. Conf. on Fluid Sealing, BHRA, J2 (1981).
6. Kaneta, M., "Sealing Characteristics of Double Reciprocating Seals", J. JSLE, Int. Ed., No.7, pp.141-146 (1986).
7. Kawahara, Y., Ohtake, Y. and Hirabayashi, H., "Oil Film Formation of Oil Seals for Reciprocating Motion", Proc. 9th Int. Conf. on Fluid Sealing, BHRA, C2 (1981).

14th International Conference on Fluid Sealing, Firenze, Italy,
6-8 April 1994. Organised by BHR Group Limited, Cranfield,
Bedford, MK43 0AJ, UK; Tel: 0234 750422

STUDY OF THE BEHAVIOUR OF AN HYDRAULIC ACCUMULATOR SEAL

G.MORIN and S.DELATTRE

PRINCIPIA RECHERCHE DEVELOPPEMENT S.A.
Zone Portuaire de Brégaillon
83507 LA SEYNE/MER CEDEX   FRANCE

ABSTRACT

This paper presents work concerning the detailed study of the behaviour of a particular seal fitted on an hydraulic accumulator. The study was based on material testing, numerical simulation and experiments on the seal at full scale. The association of these three methods was essential to perform a detailed analysis of the seal behaviour. It was made possible by the association of several partners : the accumulator industrial user (DCN INGENIERIE, which depends on the French Ministry of Defence), the seal manufacturer (EFJ MASSOT), a laboratory specialized in polymer materials (LRCCP) and a specialist of numerical simulation (PRINCIPIA).

This work allowed significant progress in the understanding of the seal behaviour. The causes of some troubles encountered under working conditions were pointed out. The importance of the non-linear aspects of the behaviour, including buckling and materials viscoelasticity, was confirmed. At the end of the study, some modifications of the seal design could be determined to improve its reliability.

I - INTRODUCTION

DCN INGENIERIE is an engineering department of the French Ministry of Defence. Its mission is to design and control the building of navy ships. Some ships are equipped with hydraulic accumulators. In these accumulators, a seal separating gas and hydraulic fluid plays a very important role. In 1992, DCN INGENIERIE decided to undertake a large program of study in order to improve this seal performance and reliability. It appeared fundamental to first analyse precisely the physical behaviour of the seal before trying to improve its design. It was decided to perform this analysis by numerical simulation allied with material testing and correlated with experiments on the seal at full scale.

The management of the project and the numerical simulations were carried out by PRINCIPIA, an engineering society specialized in structural and fluid mechanics. The material tests were carried out by LRCCP, a laboratory specialized in rubber and polymer materials testing. The experiments on the seal at full scale were performed by DCN CHERBOURG, which is an industrial department of the French Ministry of Defence in charge of navy shipbuilding. The seals and material samples for testing were supplied by EFJ MASSOT, the company which designs and manufactures the seal.

II - METHOD OF STUDY

II.1 - Numerical simulation

The numerical simulations were performed with the industrial finite element software ABAQUS version 5.2. This allowed to take into account all the particularities of the seal behaviour and especially the non-linearities. Two materials exist in the seal : rubber and PTFE. The rubber behaviour was modelled by an hyperelastic law giving the strain energy potential as a polynomial of the invariants of the strain tensor. Extensive tests have been made in order to determine the best form of this law. A law of second degree, with five coefficients, was retained. The PTFE behaviour was modelled by an elastic law. Hybrid elements, which compute the hydrostatic pressure independantly from the nodes displacements, were used to solve the numerical difficulties caused by the very low compressibility of rubber. Interface elements were used on the contact areas between the seal and the accumulator.

As the geometry of the seal is axisymmetric, axisymmetric detailed models have been used to compute stresses, strains and contact pressures during the different steps of loading of the seal.

These steps are the setting of the seal in the accumulator, the application of the gas pressure and the increase of temperature. They are applied statically. A fourth step is added to study the effects of the material viscoelasticity on the relaxation of the seal in time. The observation of the contact pressure around the seal gives a direct evaluation of the airtightness.

Due to the gas pressure, the seal is under compression. As any structure undergoing compression, it is apt to buckle. If it buckles, the buckling mode will certainly not be axisymmetric. To study the risk of buckling, two tridimensional models were used. A complete model, with simplified geometry and linearized materials behaviour, was used to determine the buckling modes apt to exist under the seal working conditions. Then, a partial model, with symmetry conditions compatible with these modes, was used with non-linear materials behaviour and contact conditions. This model allowed us to determine the lost of airtightness due to buckling.

II.2 - Material testing

Both materials, rubber and PTFE, were extensively tested under a wide range of strains and over a range of temperature corresponding to the seal working conditions. These tests were essential to determine precisely the behaviour law for each material. The viscoelastic behaviour was defined from relaxation tests at several temperatures. The simulations showed that the viscoelasticity plays a great role in the seal airtightness. So, relaxation tests are necessary in this kind of study.

II.3 - Experiments at full scale

Experiments at full scale were performed on the accumulator. The accumulator could not be easily instrumented. The results of these experiments were more qualitative than quantitative. Nevertheless, the measurement of leakage level evolution with time allowed us to confirm the results of the relaxation simulations. On the other hand, buckling simulations gave a new insight on some experimental observations concerning sudden loss of airtightness.

III - PHYSICAL BEHAVIOUR OF THE SEAL

III.1 - Behaviour under standard working conditions

The geometry of the seal is shown on figure 1.

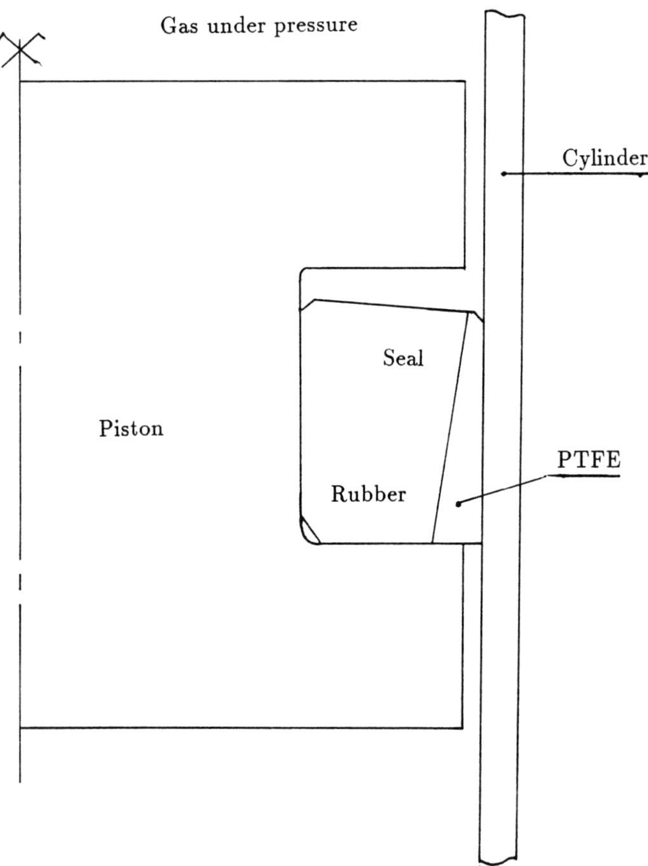

Figure 1 - Geometry of the seal

The deformed shape of the seal, for the different steps of analysis with the axisymmetric model is shown on figure 2.

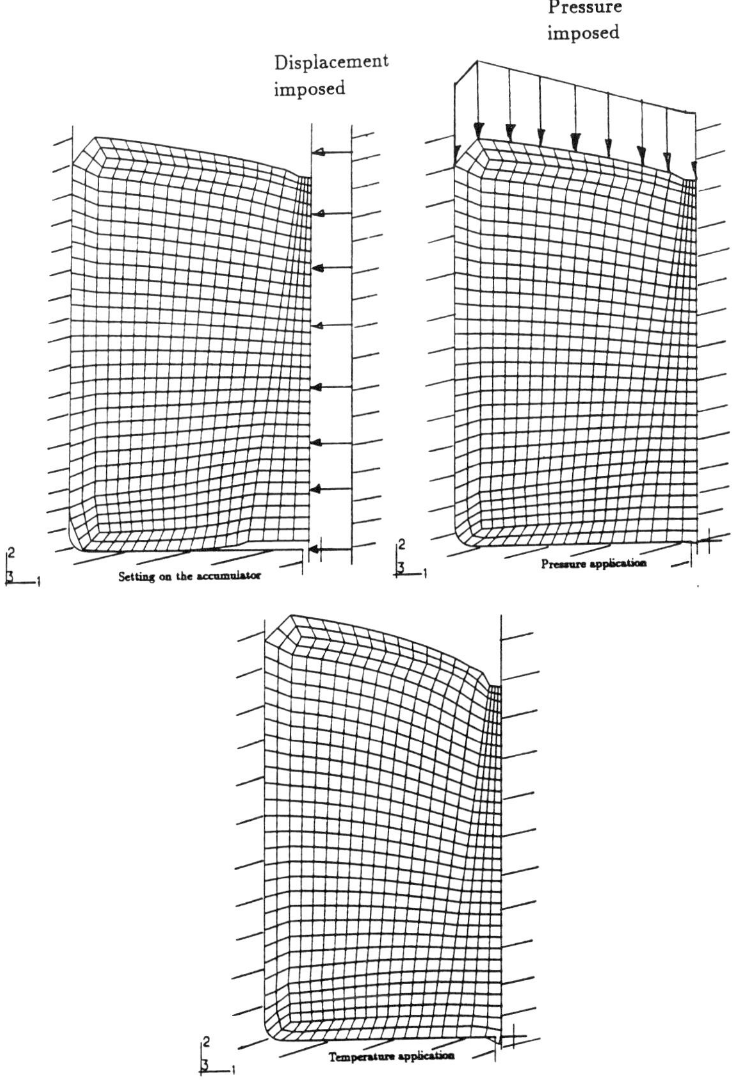

Figure 2 - Deformed shape

538

The figure 3 shows the repartition of contact pressure around the seal under working conditions. The pressure is almost constant, equal to the gas pressure. The airtightness is provided by a local peak pressure, at the top of the outer surface of the seal, which overcomes the gas pressure by several tens of bars. The role of the rubber, under the compression imposed by the gas, is to push the PTFE part on the outer cylinder of the accumulator and to cause this peak of pressure.

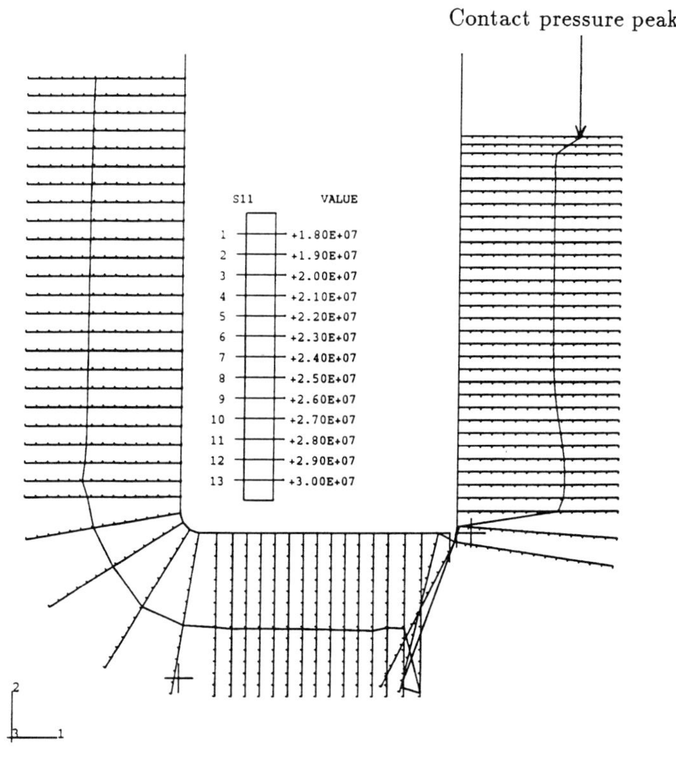

Figure 3 - Contact pressure under working conditions

The relaxation tests on rubber showed an important decrease of stiffness and stress with time when strain is maintained constant. Figure 4 plots the decrease of normalized elasticity modulus versus time.

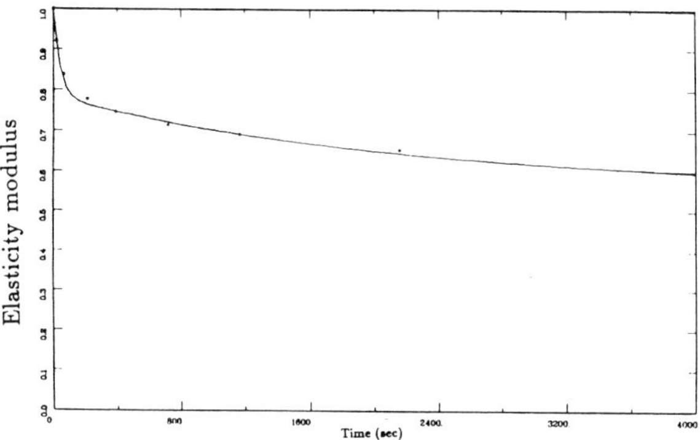

Figure 4 - Relaxation test of rubber

The simulation of relaxation on the seal showed a corresponding decrease of the peak pressure with time. On figure 5, contact pressure on two points of the seal is plotted versus time. The decrease of pressure is very important at the beginning of the simulation so the curve is almost vertical. This explained some decrease of the seal airtightness and was confirmed by some quasi-static experiments performed on the accumulator.

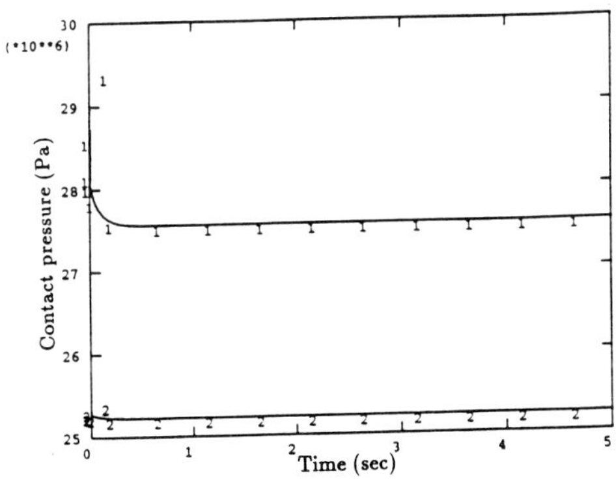

Figure 5 - Contact pressure during relaxation

III.2 - Buckling analysis

The figure 6 shows the buckling modes obtained with the simplified and linearized model under various conditions : seal free, seal mounted on the accumulator, seal under working conditions.

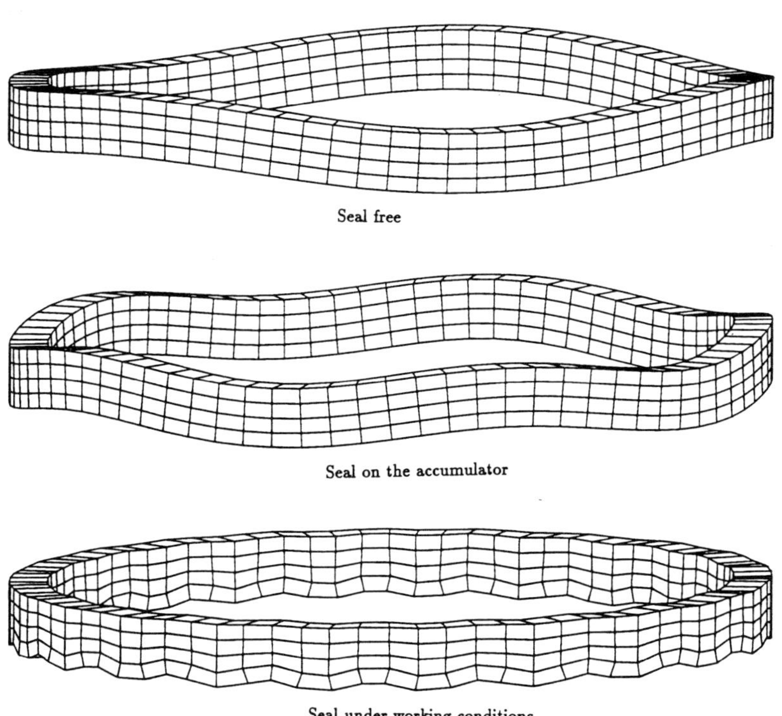

Figure 6 - Buckling modes

The first plot on figure 6 shows the buckling mode of the seal, not mounted on the accumulator, when the temperature rises up to the working temperature (around 100 °C). The circumferential order of this mode is three (i.e. three maxima around the circumference of the seal). When the seal is mounted on the accumulator, with no pressure applied, and temperature rises to the working temperature, the circumferential order of the buckling mode is five (second plot on figure 6). When the working pressure is applied (more than 200 bars), and when the temperature rises to the working temperature, the buckling mode of the seal has a circumferential order around twenty (third plot on figure 6). This proves that, when the seal is under working conditions, only modes of high circumferential order can be excited.

During the experiments at full scale on the accumulator, a sudden loss of airtightness occured several times. The seal was then unmounted and a mode of deformation of circumferential order three was observed. The simulations show that this mode cannot exist under working conditions and is caused by the fast decrease of pressure and temperature during unmounting. It is a consequence of the unmounting of the seal and not the cause of the sudden loss of airtightness. The numerical simulations allowed us to correct the interpretation of the experiments.

The dynamic analysis of buckling, taking account of the exact geometry and the non-linearities, was performed on a model of a sixth of the seal circumference. The figure 7 shows the deformed shape of the seal when the buckling occurs. This deformed shape has two maxima in the radial direction and three maxima in the circumferential direction. The buckling mode is of high order (2 in the radial direction and around 20 in the circumferential direction for the complete seal). It confirms the results obtained with the simplified model.

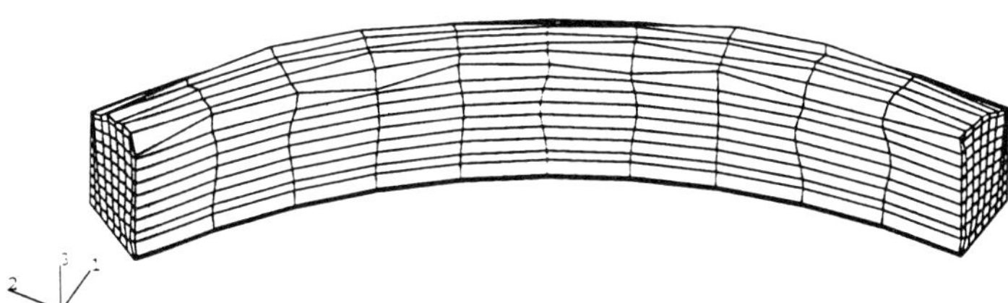

Figure 7 - Deformed shape during buckling

542

Taking into account the non-linearities, especially on the contact conditions, allows us to follow the evolution of the contact pressure during the buckling. The figure 8 shows this evolution, on the top of the external contact area, where the airtightness is ensured, at five points equally distributed on the circumference.

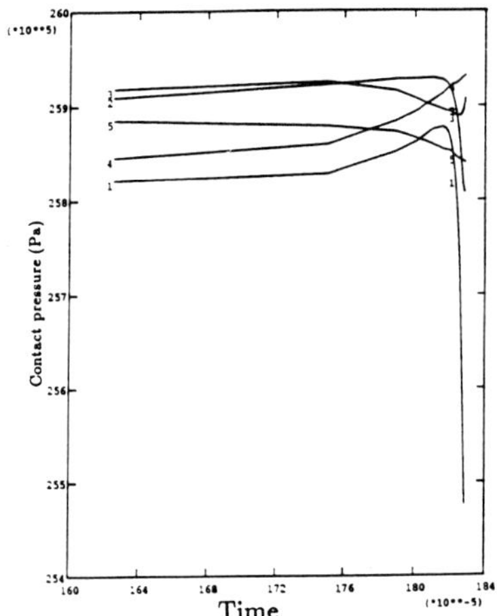

Figure 8 - Contact pressure decrease during buckling

Along the circumference, the contact pressure alternatively increases and decreases, accordingly with the buckling mode. The contact pressure decrease can be sufficient to provoke a loss of airtightness. These simulations allowed us to make decisive progress in the understanding of some loss of airtightness of the seal under extreme working conditions. Furthermore, it has been possible, with the help of these simulations, to design and to test numerically an original modification of the seal to prevent buckling.

IV - CONCLUSION

During this study, a complete method of a seal mechanical behaviour analysis could be developed. This method is based on the union of numerical simulation, material testing and experiments at full scale. It allowed us to make significant progresses in the understanding of the seal behaviour. The causes of some troubles encountered under working conditions were pointed out. Relaxation related with materials viscoelasticity causes a slow decrease of airtightness. Buckling can provoke sudden loss of airtightness under extreme working conditions. During a second step, some modifications of the seal design could be determined in order to improve its reliability.